IoT Multi Sensors

IoT Multi Sensors

Editors

Alexandru Lavric
Liliana Anchidin
Adrian I. Petrariu

MDPI • Basel • Beijing • Wuhan • Barcelona • Belgrade • Manchester • Tokyo • Cluj • Tianjin

Editors

Alexandru Lavric
Computers, Electronics and Automation Department
Stefan cel Mare University of Suceava
Suceava
Romania

Liliana Anchidin
Computers, Electronics and Automation Department
Stefan cel Mare University of Suceava
Suceava
Romania

Adrian I. Petrariu
Computers, Electronics and Automation Department
Stefan cel Mare University of Suceava
Suceava
Romania

Editorial Office
MDPI
St. Alban-Anlage 66
4052 Basel, Switzerland

This is a reprint of articles from the Special Issue published online in the open access journal *Sensors* (ISSN 1424-8220) (available at: www.mdpi.com/journal/sensors/special_issues/IoTMS).

For citation purposes, cite each article independently as indicated on the article page online and as indicated below:

LastName, A.A.; LastName, B.B.; LastName, C.C. Article Title. *Journal Name* **Year**, *Volume Number*, Page Range.

ISBN 978-3-0365-7751-7 (Hbk)
ISBN 978-3-0365-7750-0 (PDF)

Contents

About the Editors

Alexandru Lavric

Alexandru Lavric (Member, IEEE) was born in 1987. He received a bachelor's degree in administrative sciences, an M.Sc. degree in computer and communication networks, and a Ph.D. degree in computer technology from the Stefan cel Mare University of Suceava, Suceava, Romania, in 2008, 2011 and 2013, respectively. He published more than 80 scientific papers. His research interests include WSN, the IoT systems, smart sensors, artificial intelligence and wireless communication technologies.

Liliana Anchidin

Liliana Anchidin - Researcher born in 1987, received a bachelor degree in Telecommunication Systems and Technologies (2012), a master's degree in Integrated Communication Circuits and Systems (2014), and a Ph. D. degree in Electronics and Telecommunications (2018) from the University of Bucharest. Her field of interest includes antenna theory, more specifically near-field characterization techniques. Dr. Anchidin is currently with the Faculty of Electrical Engineering and Computer Science, "Ştefan cel Mare" University of Suceava. Recently, her activity is focused on design, simulations and measurements of antennas for Internet of Things applications, antennas for mobile communications, wireless communication technologies, neural networks, machine-vision artificial intelligence and deep-learning techniques. Dr. Anchidin has published more than 20 scientific papers, concerning antenna theory, analysis, design and measurements, Internet of things systems and machine learning techniques.

Adrian I. Petrariu

Adrian I. Petrariu (Member, IEEE) was born in 1984. He received an M.Sc. degree in electronics and a Ph.D. degree in computer technology from the Stefan cel Mare University of Suceava, Suceava, Romania, in 2009 and 2013, respectively. Since 2016, he has been a Lecturer in the Computers, Electronics, and Automation Department, Stefan cel Mare University of Suceava, where he is also a member of the MANSiD Research Team. His points of interest include RFID systems, antenna design, impedance matching and wireless communication (ZigBee, Bluetooth, LoRa, SigFox, and NB-IoT).

Preface to "IoT Multi Sensors"

In recent years, we have witnessed continuous discussion about the Internet of Things (IoT) concept, which involves connecting the various objects that surround us in everyday life to the Internet. In order to cope with the new challenges and IoT applications, low-power wide-area networks (LPWANs) have been created. The IoT concept is currently the focus of the entire academic community.

The main purpose of the IoT concept, which is closely related to the Smart City topic, is to increase quality of life by contributing to the efficient use of resources and environment protection. IoT technologies are sufficiently enhanced to enable the development of integrated solutions for multi-sensors design.

This reprint book focuses on state-of-the-art technologies, the latest findings, and current challenges in IoT with emphasis on healthcare, transportation, antenna design and disease detection.

The included papers cover numerous topics of interest that include:

- IoT communication protocols;
- LPWAN for IoT (Sigfox, LoRa, etc.);
- Antenna design for IoT applications;
- Large-scale, high-density IoT networks and architectures;
- IoT applications and multi-sensors for transportation and traffic control;
- IoT convergence for Smart Health;
- Machine-learning/deep-learning algorithms for sensing IoT;
- Machine-learning-based healthcare applications and disease detection using IoT architectures;
- Applications and examples of use.

Alexandru Lavric, Liliana Anchidin, and Adrian I. Petrariu
Editors

sensors

MDPI

Article

Massive Data Storage Solution for IoT Devices Using Blockchain Technologies

Alexandru A. Maftei *, Alexandru Lavric *, Adrian I. Petrariu and Valentin Popa

Computers, Electronics and Automation Department, Stefan cel Mare University of Suceava, 720229 Suceava, Romania

* Correspondence: alexandru.maftei@usm.ro (A.A.M.); lavric@usm.ro (A.L.)

Abstract: The Internet of Things (IoT) concept involves connecting devices to the internet and forming a network of objects that can collect information from the environment without human intervention. Although the IoT concept offers some advantages, it also has some issues that are associated with cyber security risks, such as the lack of detection of malicious wireless sensor network (WSN) nodes, lack of fault tolerance, weak authorization, and authentication of nodes, and the insecure management of received data from IoT devices. Considering the cybersecurity issues of IoT devices, there is an urgent need of finding new solutions that can increase the security level of WSNs. One issue that needs attention is the secure management and data storage for IoT devices. Most of the current solutions are based on systems that operate in a centralized manner, ecosystems that are easy to tamper with and provide no records regarding the traceability of the data collected from the sensors. In this paper, we propose an architecture based on blockchain technology for securing and managing data collected from IoT devices. By implementing blockchain technology, we provide a distributed data storage architecture, thus eliminating the need for a centralized network topology using blockchain advantages such as immutability, decentralization, distributivity, enhanced security, transparency, instant traceability, and increased efficiency through automation. From the obtained results, the proposed architecture ensures a high level of performance and can be used as a scalable, massive data storage solution for IoT devices using blockchain technologies. New WSN communication protocols can be easily enrolled in our data storage blockchain architecture without the need for retrofitting, as our system does not depend on any specific communication protocol and can be applied to any IoT application.

Keywords: blockchain; consensus mechanisms; secure data management; data storage; Internet of Things; wireless sensor networks

Citation: Maftei, A.A.; Lavric, A.; Petrariu, A.I.; Popa, V. Massive Data Storage Solution for IoT Devices Using Blockchain Technologies. *Sensors* **2023**, *23*, 1570. https://doi.org/10.3390/s23031570

Academic Editor: Alessandra Rizzardi

Received: 30 November 2022
Revised: 18 January 2023
Accepted: 29 January 2023
Published: 1 February 2023

1. Introduction

The Internet of Things (IoT) concept involves connecting devices via the internet and forming a network of objects that can collect information from the environment without human intervention. According to [1], the number of IoT devices deployed from year to year is constantly increasing and it is estimated to reach 25.44 billion by 2030. There have been many surveys [2,3] conducted regarding the usage of different blockchain applications. Areas such as healthcare [4], agriculture [5], the smart grid [6], the smart city [7], or traffic management [8] are all contributing to the rapid growth of active IoT devices, as it seeks to automate the tasks of collecting data from the environment for further processing and storage.

The use of IoT devices in our daily life brings benefits such as the increase in automated tasks, obtaining low energy consumption and efficient resource management, automated data retrieval, information storage, and remote control of devices installed on the field. The IoT concept is usually implemented through wireless sensor networks (WSNs). Although the IoT concept offers some advantages, it also has some issues that are usually associated with cyber security risks, such as the lack of detection of malicious WSN nodes [9], lack of fault tolerance, weak authorization, and authentication of nodes [10], and the insecure management of retrieved data from IoT devices. Considering the cybersecurity issues for IoT devices, there is an urgent need of finding new solutions that can increase the security level of WSNs.

A WSN refers to a group of spatially dispersed wireless sensors distributed over a large geographical area that collect data from the environment and then send them to a sink node for processing and storage purposes. Using different visualization algorithms, the collected information is presented to users. The architecture of such a wireless sensor network can be seen in Figure 1. The structure includes a centralized sink node (Gateway) that is usually connected to the internet, and wireless sensors nodes, which is a communication mechanism that allows the information to be transmitted to the end users.

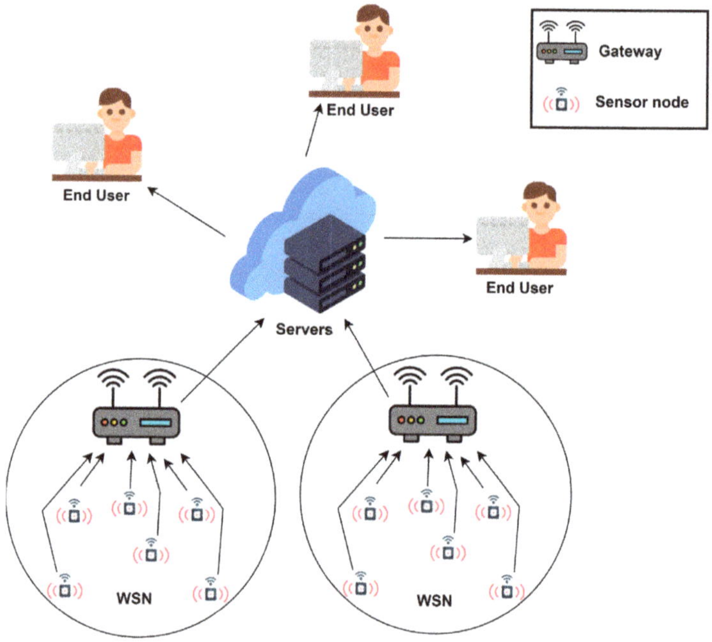

Figure 1. Classic wireless sensor network architecture.

The communication protocol is specific to the application IoT type and usually is represented by Bluetooth [11], Wi-Fi [12], LoRaWAN [13], ZigBee [14], or NB-IoT [15] when we are considering low power consumption efficiency. Using multiple types of wireless communications protocols in the same wireless network can be both an advantage and a disadvantage. One advantage is the ability to use area-specific types of communications, such as Wi-Fi and Bluetooth for shorter-range communications, 10–50 m, or LoRaWAN and NB-IoT for long-range communications, 1–10 km, in urban environments [8]. One disadvantage that can arise from using multiple communication protocols in the same IoT network is interoperability [16]. This can become an issue because each communication protocol has its security policies for securing data, which can lead to congestion and slow down the entire network. Taking into account the large number of wireless communication protocols used in the IoT, it is somewhat difficult to maintain and update but also to detect problems that may occur in an IoT network due to the use of a large number of wireless protocols.

When we talk about WSNs, we consider networks where the number of active IoT devices can be hundreds of thousands or even millions, which will result in a large number of data packets that need to be stored securely. In a classic WSN network, all information is sent to a central data storage point. In terms of cybersecurity, centralized storage of large amounts of data is becoming a secure target for malicious entities. Another aspect of wireless sensor networks is that they have limited resources such as memory, computing power, and limited energy/battery capacity [17]. However, the role of WSN IoT devices is to operate for long periods, with as little human intervention as possible, being mostly standalone devices that are disposable at the end of their life.

The rest of the paper is organized as follows: Section 3 presents the blockchain technology along with its mechanisms that can provide security to data from IoT devices. Section 4 presents the proposed architecture for securing IoT data. Section 5 presents a proof-of-concept practical implementation of the proposed architecture, and the last section presents the conclusions.

The main contributions of this paper are the following:

- We present a solution for the secure management and storage of the data retrieved from IoT devices using blockchain technology.
- The data storage process is performed by using a dual blockchain topology that includes a lightweight blockchain (local blockchain) and a public blockchain. The lightweight blockchain stores temporarily all the IoT information acting as a buffer that stores the node identity ledgers and the hash addresses that point to the data packets located in the public blockchain acting as a register. The public blockchain permanently stores the entire IoT data stream sent through the entire WSN architecture.
- Our novel blockchain architecture uses an IoT authentication process that allows new WSN nodes to be accepted using a voting process that integrates the PoS consensus mechanism specifically used by blockchain architectures.
- Another advantage is related to the scalability of the proposed architecture, which can integrate a very large number of IoT devices without decreasing the performance level of the system. This IoT device can join the network and contribute to its maintenance by implementing the consensus mechanism.
- We propose an architecture where different WSN entities (e.g., gateways, sensors) have blockchain capabilities and functionalities to achieve a massive data storage solution.
- The proposed architecture is scalable and is not locked on a particular IoT communication protocol. New IoT sensors and communication protocols can be easily enrolled in our blockchain architecture without the need for retrofitting.
- Our massive blockchain data storage solution can also be used in a hybrid manner by classic IoT WSN networks with no enhanced capabilities.

2. Related Works

Blockchain technology can be used to implement different security aspects such as access control, authorization, and authentication processes of IoT devices in a wireless sensors network, methods to detect malicious devices or DDOS attacks in a network or achieve secure storage of IoT data. There are a few solutions in the literature that attempt to solve each problem individually. Thus, in this section, we present some solutions that aim to solve the problem of secure storage of data from IoT devices, solutions based on blockchain technology with all its features such as consensus mechanisms, smart contracts, decentralization, pseudo-anonymity (even anonymity in some cases), and immutability.

In [18], Liu et al. propose a Data Integrity as a Service (DIaaS) framework based on blockchain technology for verifying data coming from IoT devices. The authors attempt to solve two problems: the first is to eliminate the requirement of trust in third-party auditors (TPA) and increase the reliability of the data integration service. The second problem is that of verifying data without relying on a single third-party auditor, so protocols are proposed to verify data integrity in a fully decentralized environment. The main disadvantage of this approach is the fact that the integrity of the data packets needs to be verified by a trusted third-party authority.

In [19], Li et al. propose a scheme using blockchain technology for storing and protecting large amounts of data from IoT devices. Their proposed method guarantees data protection by having a large number of blockchain miners handle data from IoT devices, eliminating centralized servers. The proposed architecture uses edge computing to take over the task of processing the data and later sending it to the Distributed Hash Tables (DHT). A final proposal of the authors in their scheme is to use certificate-less cryptography. Certificateless cryptography reduces the redundancy that is brought about by traditional Public Key Infrastructure (PKI) and provides an efficient way of authentication for IoT devices. However, this scheme may have problems if there is a necessity for implementing a system that uses more complicated access control policies.

In [20], Shafagh et al. propose a system based on blockchain technology that provides access control and data management in a distributed manner. Among the contributions presented in the paper, we can mention designing a secure cryptographic method of sharing data with frequent key updates, the ability to revoke access to data, an efficient search method in a compressed chunked data stream, and a location-aware level of data storage. One drawback of the proposed architecture is the consensus mechanism used, namely the Proof of Work mechanism. PoW is a resource-intensive mechanism and suffers from the so-called 51% attack [21] that can occur if the proposed architecture does not have many users and hash power is abundantly available.

In another paper [22], Ren et al. propose an architecture for securely storing data from edge devices. This architecture uses blockchain technology; specifically, the authors implement two blockchain networks, one local and the other global. The local blockchain network has limited storage space and is created by the main edge nodes and stores all data coming from IoT devices. The global blockchain network is built using cloud servers and stores all data coming from local blockchain networks. The cloud servers calculate hash values for the data uploaded to the global blockchain, and to ensure the integrity of the data, a periodic check is performed using the hash values already calculated.

Ren et al. [23] propose the use of blockchain technology for storing data from Wireless Body Area Networks (WBANs). A modification added to the blockchain-based storage system is to implement a sequential aggregate signature scheme (DVSSA). This scheme ensures that data can only be accessed by assigned individuals and that the privacy of users in a WBAN network is protected. DVSAA can also compress the data stored in the blockchain, thus solving the problem of limited storage space.

Our blockchain data storage system differs from the works mentioned above in that it can accept other WSNs and IoT devices that do not have blockchain capabilities but wish to use the proposed ecosystem for secure and traceable data storage. All verification processes of the proposed architecture are performed by the IoT devices that join the network and want to contribute to the better functioning of the network; therefore, the ecosystem can work without interruption and user intervention.

In addition, in our work, we propose the use of the "Proof of Stake" consensus mechanism, a mechanism that eliminates the need to use devices with a high computing power because the nodes that add transactions in blocks and the blocks in the blockchain become validators that stake a part of their reputation rather than needing high computational power. Once again, the system proposed in this paper is a scalable one where a large number of IoT devices can join the network, contribute to its maintenance by implementing the consensus mechanism, or use the network only to send and securely store sensor data. The security of the system is also high because the more nodes there are that put their reputation on the line to validate blocks and transactions, the more secure your network will become. The number of transactions is also high, and this is again due to the PoS consensus mechanism, which does not require a high computing power, and due to this, energy consumption also decreases, thus reducing the carbon footprint.

3. Blockchain Technology for IoT Concept

Blockchain technology [24] involves chronologically saving data in the form of a blockchain called a distributed ledger where each block of data is linked to each other using cryptographic methods, thus making the entire system secure, immutable, and tamper-proof. A distributed ledger can be seen as a database that saves data in chronological order, but the difference is that of the permissions set on the data stored in the ledger. The main purpose of this paper is to propose an architecture of data storage using blockchain technology that can be a secure data management solution for the IoT concept.

Blockchain technology became known in 2008 when Satoshi Nakamoto presented it for the first time [25] as a way to introduce a virtual currency-sharing system through a Peer-to-Peer (P2P) network, thus eliminating centralized tertiary institutions. Over time, blockchain technology has evolved positively and has been increasingly adopted in various fields, with one of them being the Internet of Things. The implementation of the blockchain in different fields is due to the way the data are stored but also to the fact that this technology has several features that offer increased security. Each block in turn has three basic elements:

Hash of the current block. This is the result of applying a hash function to the data and information that is stored inside the block. A hash function is a one-way function that involves transforming a text of variable length into a text of constant size. The result of a hash function is also called a hash value or just a hash. Hash functions are used because they are computationally difficult to reverse. It should be noted that hash values are very sensitive to changes, that is, when a single character or number is changed, the whole hash value will become completely different.

The encapsulated data in each block. The data are passed through the hash function so that they cannot be read by anyone. Each block contains several data transactions that are measured in transactions per second (TPS), and this differs from system to system. In the case of the Bitcoin blockchain, which only saves information on the transfer of BTC virtual currencies, the number of transactions per second is approximately 7 TPS [26]. To add blocks of data, they must first go through a mining process, where users of a network compete to find the result of a math problem, in the case of the Bitcoin network. This mining process is due to the feature called the consensus mechanism. In the IoT field, the PoW mining process or PoS validation process can be performed by IoT devices. This operation of implementing consensus mechanisms can be applied by the IoT nodes of the network, thus providing the opportunity for IoT devices to manage the addition of data packets in blocks and then their addition to the blockchain.

Hash of the previous block. This is an important element for the security of the blockchain. Each block in the blockchain points to the hash of the previous block. If a block's previous hash value is wrong, that block and its successors will become invalid. To modify a block, the hash value of the block must be recalculated together with other blocks, and this is performed using consensus mechanisms that can be difficult from a computational point of view.

Therefore, the way data are stored in the blockchain is a secure solution for saving data from wireless sensor networks. In addition, blockchain technology also has some features that can be beneficial for data storage, such as:

A. Immutability

The blockchain is similar to a database but differs in the way the data are managed, and this is due to the immutability feature. Immutability refers to the fact that once data have been stored in a block, it cannot be modified in any way. Thus, the impossibility of modifying the data in a block introduces benefits such as traceability or audibility.

B. Decentralization and distributivity

In most cases, these two terms, decentralization and distributivity [27], are used interchangeably, although their meanings are different from each other. These two features are very important for the blockchain because any system that wants to implement this technology will automatically benefit from both features, so there is no need to implement other third-party technologies to act as data storage and management.

At present, most data storage and management applications use a server-client architecture similar to IoT classical data storage. In such an architecture, the transfer of data between two or more clients is performed through a server that acts as an intermediary for forwarding information. Thus, if the server suffers a cyber-attack, the entire system may be left without the central data distribution system. To solve this problem, the distributive feature of blockchain technology can be used, which uses a peer-to-peer (P2P) architecture where the transfer of data between two clients is performed in a point-to-point manner. From the point of view of storing data in a P2P network, it is performed in several places to provide redundancy, which means that a data packet is in several places in the IoT network. Thus, in case of a cyber-attack on any node in the network, the data will be able to be reconstructed through the other copies from the network.

The other feature is decentralization. In classic server-client structures, the server acts as a central data distribution system, while in a P2P structure, this task is managed by the entire network. Thus, the decentralization feature appears, where several nodes have decisions about what is happening in the network.

C. Consensus mechanism

A consensus mechanism [28] is a way to gain agreement, trust, and security in a decentralized network based on blockchain technology. The role of consensus mechanisms in a network that uses blockchain as a core component is to ensure that all nodes in the network follow the same rules. If a blockchain has no consensus mechanism, it means that the blockchain has no rules and can become an easy target for malicious entities. Users of a blockchain that works without a set of rules will find it difficult to prove the validity and integrity of the data that are added and shared on the network.

There are currently three consensus mechanisms [29] with the highest percentage of use, being reference mechanisms for the implementation of a blockchain. This is mainly due to the field of application of the technology, namely the financial field. The three consensus mechanisms are Proof of Work (PoW), Proof of Stake (PoS), and Delegated Proof of Stake (DPoS).

Even though blockchain is a technology that has many positive features, it also has some limitations such as large storage capabilities, low scalability, and high energy consumption. When we talk about storage capacity as one of the limitations of blockchain technology, we refer to the fact that the blockchain is constantly growing, i.e., a block of data of a certain size (1 MB in the case of Bitcoin blockchain or 4 MB in the case of Ethereum blockchain) is constantly added to the blockchain, and its size can easily reach the terabyte range. We consider that using state-of-the-art storage solutions, this disadvantage can be overcome.

The low scalability and the energy consumption issues can be solved by the integration of the different consensus mechanism such as PoS that allows support for integrating many IoT devices with low computational recourses.

4. Data Storage for IoT Devices Using Blockchain Technologies

Blockchain technology will allow IoT networks to move away from the classic structure of a WSN network, where nodes communicate with a centralized server, used for data storage, authentication, and sensor authorization, to a more secure and flexible approach. Blockchain offers a new method of decentralized and distributed management of both data and nodes in a P2P network. In our system, WSN nodes have full control over the network, which means that any new WSN that wants to use the network must be accepted by the other WSN enrolled in the consensus mechanism. The architecture of the wireless sensor network based on the blockchain technology proposed can be seen in Figure 2. The proposed architecture integrates a novel modular technique that involves using a lightweight blockchain and public blockchain for WSN data storage.

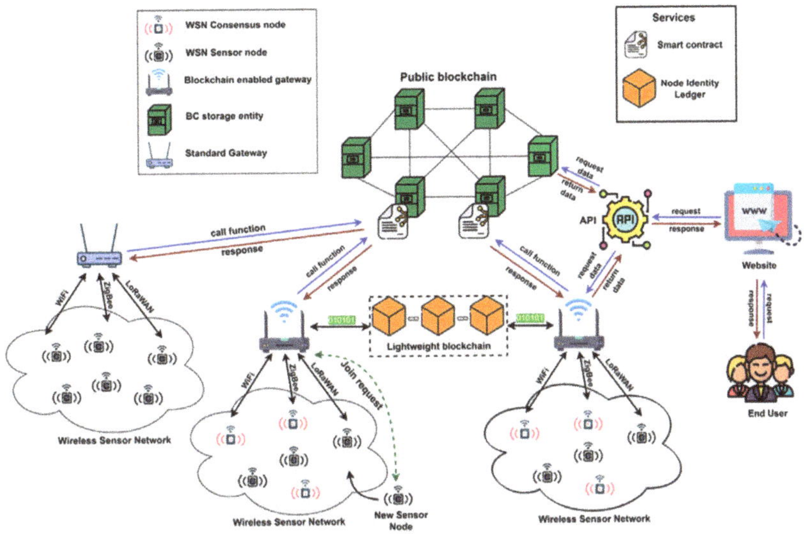

Figure 2. Proposed blockchain WSN architecture for data storage.

The main entities of the developed architecture are:

Blockchain Capabilities: As shown in Figure 2, in the proposed architecture, there are two blockchains, a lightweight blockchain and a public blockchain. The lightweight blockchain is located at the level of the IoT gateways with blockchain capabilities and has the role of storing a light copy of the public blockchain for backup reasons and acts as a register. The public blockchain stores all the data that are sent by the WSN nodes. The lightweight blockchain integrates the node identity ledgers and stores the hash address that points to the data packet itself in the public blockchain. Each time a new block is added to the public blockchain, only the information about the total number of data packets, the validator node ID, and the address of the added block will be saved in the lightweight blockchain. The data block will be saved in the public blockchain.

The public blockchain functions as a database that stores all the data received from the WSN nodes but also stores the information related to the authentication and registration of nodes in the network. In addition, the public blockchain is a P2P system represented by BC storage entities where each one of them has a complete copy of the entire system. The need to have such an entity that has complete copies of the entire blockchain is a method of redundancy in such a system. Usually, these entities have high computing and storage capabilities. Therefore, if a large part of the network nodes become inaccessible and data are lost, the entire system can be rebuilt using a single node that holds a complete copy of the blockchain.

Blockchain-enabled gateway (BCeGW): BCeGWs have the role of communicating directly with the WSN to collect the data sent and redirect them further for storage to the public blockchain. Communication between the BCeGW and consensus WSN nodes is performed when a new node wants to join the wireless sensor network and a validator WSN node must be chosen from the list of available ones. The validator WSN node in a concessions WSN node is mains-powered and has a higher computational power than a regular WSN node. At that point, BCeGWs will send the information requested by the consensus node about the new node that wants to join. In addition, the BCeGWs have a local copy of the blockchain, more precisely a lightweight copy of the public blockchain. Another role of BCeGWs is to implement the network consensus mechanism, which is initiated when a new block is added to the public blockchain. As with consensus nodes, a BCeGW will be chosen through the PoS mechanism to validate and add a block to the public blockchain.

WSN nodes: At the WSN level, there are 2 types of nodes, sensor nodes and consensus nodes. Each type of node has its role in the network.

Sensor nodes: These WSN nodes are registered in the network by the user and have the role of collecting data from the environment. The WSN sensor nodes are devices with low computing power and limited battery level, which transmit information to the BCeGW at user-specified time intervals. The communications of these types of nodes are bidirectional because they will receive specific information from the WSN consensus nodes.

Consensus nodes: These WSN nodes are different from sensor nodes in that they also collect data from the environment but are used to implement the consensus mechanisms. These nodes are usually mains-powered. Although the consensus mechanism used in this architecture does not require high computing power, the use of sensor nodes for the implementation of the PoS mechanism would lead to unwanted power consumption. The communications of these nodes are bidirectional, i.e., they can send but also receive information from the BCeGW.

Consensus model: The consensus model that is used in the proposed IoT blockchain is the core of the entire P2P network. The consensus mechanism used in a WSN determines the performance of the entire system, which includes throughput, security, and delay. Given that IoT devices do not have high computing power, the implementation of the PoW consensus mechanism is not the most appropriate solution when it comes to IoT devices. To take advantage of many IoT devices from WSNs, in our architecture, we propose the use of a customized PoS consensus mechanism. In the case of the PoS mechanism, the consensus nodes in the WSN must stake their reputation points to increase the chance of being chosen as validators for the data packets that are going to be added in blocks. Once a WSN consensus node has been selected as a WSN validator node, it has the task of validating all the data packets in the block, and then it will have to add the new block to the blockchain. After the consensus node adds the new block to the blockchain, it will be rewarded with reputation points. If the validator WSN node accidentally accepts data packets that contain unreal information, it will lose some of the reputation points, thus making the node less trustworthy. Therefore, the PoS consensus mechanism does not require much computing power, due to its mode of operation, and is a good security solution for a P2P system where nodes are stimulated to make decisions that benefit the entire network and its users. PoS is slowly becoming the most used consensus mechanism among public state-of-the-art blockchains. The PoA mechanism has the below disadvantages [30]:

- The identity of the validating nodes in the blockchain network is known and this can cause manipulation and interference by third parties for their own benefit.
- The PoA consensus architecture model has a lower degree of decentralization due to the use of validators nodes. The use of PoA affects the scalability and high throughput of the blockchain architecture.
- The PoA consensus model is prone to "Sybil" [31] and "Cloning" [32] attacks, where attackers can manipulate a large number of validator nodes by forging multiple identities.

Although the PoA consensus mechanism can be used in public blockchains, its applicability is still largely used in private blockchains that require permissions and test nets such as Kovan [33], Goerli [34], and Rinkeby [35].

Smart contract mechanism: In general, a smart contract (SC) [36] is a computer program that is in the blockchain and involves the execution of predefined functions. Once the SC has been implemented in the blockchain, it will receive a unique address (hash value) so that its functions can be called. Smart contracts are also visible to all network participants, but this is not a problem, because they are in compiled bytecode format and cannot be modified.

In our proposed architecture, the SC is implemented at the public blockchain level. The SC functions are built for the WSN ecosystem, for example, WSN sensor nodes registration and the communication between the public blockchain and BCeGW. By using a smart contract, the interaction steps between the BCeGW and the public blockchain are automated and secure. Another important aspect of SC is that the result of the interactions will be a predetermined one, and the possibility of errors is quite small. Because the SC is in the blockchain, there is no way to upgrade or add new features to the source code. If, in the future, there is the need to add new features to the smart contract, this is only possible by changing and relaunching the modified SC in the blockchain. Once the new contract has been launched on the blockchain, all the entities in the system proposed by us will have to use the hash address of the new smart contract to use its new functions.

Tertiary entities: In our architecture, there is also the possibility to add standard WSN gateways, which will be known as tertiary entities, without blockchain capabilities that receive information from wireless sensor networks. In the case of these types of WSNs, the standard WSN gateway will also communicate with the smart contract to send data from the sensor nodes for storage. These gateways will not be involved in enforcing consensus mechanisms when new blocks are added to the public blockchain, but the consensus mechanism will be used by the BCeGW when these standard gateways want to add new data packets to the blockchain. Joining new nodes to a classic WSN is performed without the need to use consensus nodes to implement the PoS consensus mechanism. In addition, the type of communication protocols that can be used depends on the standard WSN gateway, ranging from WiFi to NB-IoT. As our proposed system uses both blockchain-enabled gateways (BCeGWs) and traditional gateways without blockchain capabilities, scalability is not an issue. Regardless of the type of gateway but also of the communication protocol, any IoT provider or WSN can join, benefit, and also contribute to the security and smooth functioning of the whole ecosystem easily.

Implementing the blockchain technology with characteristics such as immutability, timestamping, unanimity, distributivity, and decentralization, and with components such as consensus mechanisms, decentralized networks, and the smart contract is a good solution because it can alleviate the security and scalability concerns for IoT.

The blockchain ledger is distributed. This aspect provides some certainty that the data in the blocks will not be altered or deleted, because there is no single entity in control of the network. Blockchain provides transparency. All data transactions made through the blockchain are in plain sight and can be viewed by anyone. This can be a method of identifying a particular source that has added data to the blockchain or even identifying the source where there are data leaks.

The use of blockchain can be another layer of security for the IoT, a layer that third parties need to overcome in order to alter or steal user data.

Thousands, hundreds of thousands, or even millions of IoT packets can be carried out in an automated, secure, and transparent manner without human user intervention. This can be performed through smart contracts.

WSN authentication scheme

As previously specified, both consensus nodes and BCeGWs will implement the network consensus mechanism. Consensus nodes implement the PoS mechanism when a new node wants to join the network, and the BCeGW when a new block is added to the public blockchain. The authentication scheme and the steps performed by each entity can be seen in Figure 3.

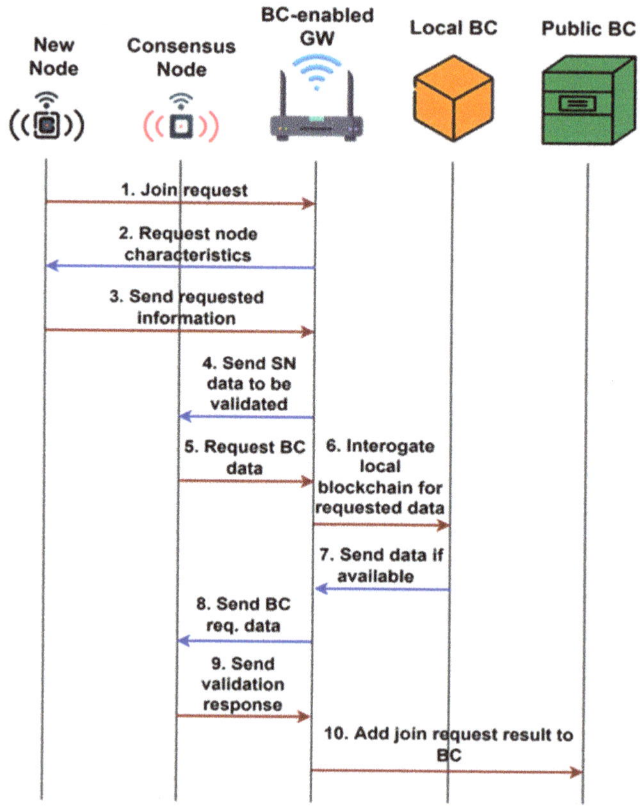

Figure 3. New nodes join the request procedure to a blockchain-enabled WSN.

1. When a new node wants to join a WSN, it will first send a Join Request to the BCeGW and a secure radio communication channel will be established.

2. Once a communication channel has been established, the BCeGW will query the new node to send it a specific data set, such as MAC address, unique ID, and manufacturer ID.

3. The new node will send the entire list of its characteristics requested by the BCeGW.

4. The BCeGW will send the new node's characteristics to be validated by the WSN consensus node, which will be chosen by applying the PoS consensus mechanism. Also here, one WSN consensus node will be chosen as a validator to verify the information from the node that wants to join the network.

5. After a WSN consensus node has been selected as a validator, it will ask the BCeGW to check if the data for the new node are not already in the local blockchain.

6. The BCeGW will interrogate the local blockchain for the data requested by the WSN consensus node.

7. The local blockchain will return the data if any are available; otherwise, nothing will be returned.

8. The BCeGW will send to the WSN consensus node the result queried from the local blockchain.

9. The WSN consensus node will compare the result from the new node with the result that came from the local blockchain and then it will return a response if the new node is or is not accepted to join the WSN.

10. The BCeGW will send the join request result to be stored on the public blockchain.

Users: In the proposed architecture, users can register WSN nodes, BCeGWs, and even BC storage entities that have the role of maintaining a complete copy of the public blockchain. An important thing to note is that the user can only register their BCeGW or full node after using the network for some time and its reputation level is a positive one. To register a BCeGW or a full node, the other users of the network must verify the request, and this is performed automatically through smart contracts that are at the level of the public blockchain. If the application for registration of a BC storage entity or BCeGW is rejected, the user cannot repeat the same application until after a certain period, for example, one week or 6 months. In addition, any registration of a BC storage entity or BCeGW must be paid by the user using reputation points, and if the application is rejected, the points will be lost. When it comes to recording a WSN sensor node, the process to be accepted is one with a higher success rate and no reputation points are needed.

Depending on the number of data packets transmitted by an IoT device using the proposed architecture, the device owner is required to pay a fee for storing the data in the blockchain. This aspect is applied only to the sensor nodes. Another way to pay for the possibility of storing data in the blockchain is by converting reputation points into messages that can be transmitted.

Reward system: To motivate users to enroll validator WSN nodes, a rewarding system, in the form of reputation points, can be used in the network. The rewarding system will be implemented at the WSN level, BCeGW level, and full node level. The nodes in the WSN network will receive a certain number of points depending on the type of node and the number of tasks performed in the blockchain architecture. These points can be converted into messages that give the user the possibility to store more data packets in the blockchain.

API: The query of the blockchain will be performed through an application programming interface (API), which has the role of returning the data packets requested by users. The only type of permission that the API has in the proposed architecture is to query the blockchain (it only makes GET requests), without the ability to add data (it cannot use POST requests). The necessary information that the API needs to return the data requested by the user is the gateway ID where the sensor is located, and the sensor ID. To return data as soon as possible, it is recommended that the user provides both parameters required by the API.

5. Performance Evaluation

For the performance evaluation, we conducted a test setup where we used blockchain technology to securely store data from sensor nodes in WSNs. To perform this test setup, BlockSim [37] was modified and used on a computing system that has an AMD Ryzen 5 1600X processor and 16 GB DDR4 of RAM, over a period of 5 days in total. Thus, as has been specified in Section 4, the proposed architecture does not consider the communication protocol integrated into the IoT network. However, in the case of this paper, to create a test setup as close to reality as possible, we took it into consideration for the emulation of the LoRaWAN communication protocol. According to [38], only 3 gateways are needed to cover an urban area with a radius of about 15 km. A single LoRaWAN gateway is capable of handling up to 100,000 WSN nodes transmitting a data packet of 50 bytes once per hour. The developed architecture considers the analysis of a data storage system whose security is provided by blockchain technology regardless of the communication protocol used. To evaluate the performance level of the proposed architecture, we developed two scenarios that consider both the latency and the throughput of the blockchain architecture. The first metric, latency, is the elapsed time of a data packet that was received by the gateway and when it was added to the blockchain. This performance metric is associated with the processing speed of the proposed architecture. The higher the latency, the more difficult it will be to add data packets into blocks and scale the proposed architecture.

The throughput of the proposed blockchain architecture shows us the obtained performance level when the number of blockchain WSN nodes is increased per BCeGW and represents the total number of processed transactions by the blockchain. As different functionalities of the blockchain are given to BCeGWs, it is important to analyze a load of architecture in different operating conditions to evaluate its scalability. The parameters used for our test setup to evaluate the performance of the proposed architecture included a total number of 5 gateways where the total number of blockchain WSN nodes varied from 500 to approximately 20,000. The size of the block in which the data packets are added was set to a size of 1 MB and the payload had a size of 50 bytes [38].

The performance metric measured in our architecture was the average latency of accepting a data packet in the blockchain, and this included validating and adding it in a block and transaction throughput. As can be seen in Figure 4a, the average latency of accepting a single data packet increases, because the gateway must validate the data packets sent by the blockchain WSN nodes and then add the data packets in blocks. The duration of the validation process of a single data packet increases because the validation process is not performed instantly. Therefore, the data packets will be placed on a waiting list so that they can be validated and then added in blocks, thus increasing the process of validating a single data packet. The validation process for all data packets differs from one data packet to the next. For a more accurate visualization of the performance level, in Figure 4a,b, we can observe the average latency for each performance test.

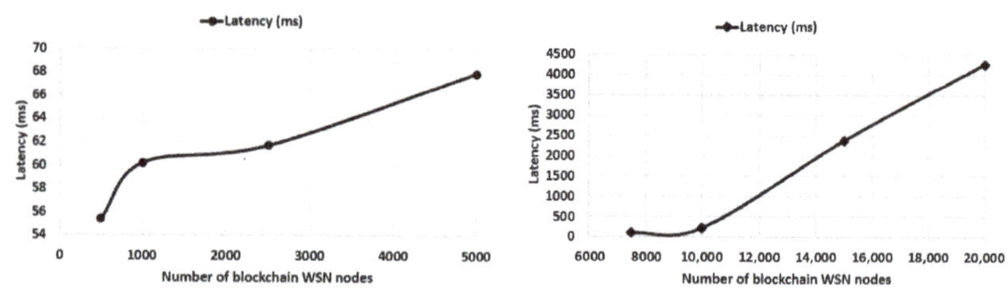

(a) (b)

Figure 4. Average latency metric for the proposed blockchain architecture: (**a**) 500, 1000, 2500, 5000 blockchain WSN nodes; (**b**) 7500, 10,000, 15,000, 20,000 blockchain WSN nodes.

Following the evaluations, for 500 blockchain WSN nodes, we have an average latency of 55.4 ms, and an average latency of 67.9 ms in the evaluation of 5000 blockchain WNS nodes. In Figure 4b, we can observe that the latency in the test where 7500 blockchain WNS nodes are used is 108.5 ms, and 4261 ms in the test with 20,000 blockchain WSN nodes used.

Figure 5a,b show the transaction throughput, which is the number of transactions that a blockchain architecture can process. In our case, transactions are the data packets sent by the blockchain WSN nodes. In our testing, we have approximately 496.31 TPS for a total of 500 blockchain WSN nodes, and 16,006.73 transactions for a total of 20,000 blockchain WSN nodes.

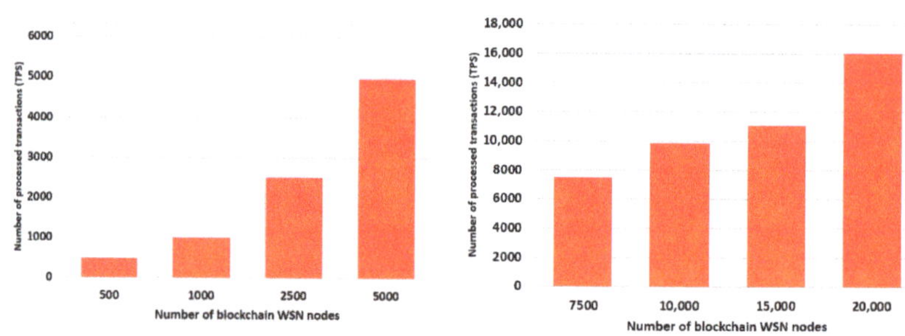

(a) (b)

Figure 5. Throughput: (**a**) 500, 1000, 2500, 5000 blockchain WSN nodes; (**b**) 7500, 10,000, 15,000, 20,000 blockchain WSN nodes.

The final step was to determine the size of the data stored in the blockchain. In this case, parameters such as data packet size, total number of nodes, and frequency of sending data packets were used. The used parameters have characteristics where the data packet size is 50 bytes (mean size of an IoT data packet), the total number of nodes is 20,000, and the frequency of sending data packets is once an hour for each node in the network [38]. If a total of 20,000 nodes each send every hour a 50-byte IoT data packet, in 24 h, we will have a total data volume of 22.8 MB collected from IoT devices. Each IoT packet is stored in the blockchain by means of a transaction.

The data packet transaction contains various information such as sequence ID, input counter, transaction inputs and outputs, output counter, and lock time, which lead to a dimension of about 100 bytes according to [39]. Following this information, the total size of a data packet storage transaction will be about 150 bytes. To summarize, for each 50 bytes of IoT data, we will have 150 bytes stored in blockchain.

Another important aspect that must not be neglected is that the storage capacity of a block in blockchain is not entirely available for storing the IoT data. The block dimensions of 1 MB include information such as block size description, block header, and data transactions counter; thus, approximately 0.95 MB can be used for the data storage transactions [40,41]. Each block can store up to 6640 packet storage transactions.

From the performed evaluation, we consider the developed storage process to be efficient and scalable due to the PoS consensus integrated mechanism. The proposed architecture can integrate hundreds of thousands of IoT devices distributed over a large geographical area.

6. Conclusions

In this paper, we propose an architecture based on blockchain technology that aims to manage and store large amounts of data from IoT devices in wireless sensor networks. By using blockchain technology, the large amounts of data that are sent by the IoT devices will be managed and stored securely, and this is due to the characteristics such as immutability, decentralization, distributivity, and consensus mechanism. The data packets are stored in a distributed manner, thus eliminating the classical centralized storage entity.

The entire architecture is governed by a P2P network, which must ensure proper operation and keep the system functioning. We also consider the fact that IoT devices have limited resources such as memory, computing power, and limited battery capacity, so we propose using the Proof of Stake consensus mechanism that does not require a high level of resources but offers the same security as PoW or DPoS. We also use a smart contract (SC) technique to ensure that the outcome of any information transfer is predefined, thus eliminating the chance of malicious communications.

The main contribution of the proposed architecture is that its scalability is also not being locked on a particular wireless communication protocol. New wireless networks can be easily enrolled in our blockchain architecture without the need for retrofitting. Data storage is performed by using a lightweight blockchain (local blockchain) and a public blockchain. The node joining request to a network is performed using blockchain technologies. We use the Proof of Stake consensus mechanism for the WSN authentication scheme. We propose an architecture where different entities (e.g., gateways) that manage the IoT devices have blockchain capabilities.

In addition, for the architecture proposed in this paper, two characteristics are analyzed, latency and throughput. According to the performance evaluation, the proposed architecture offers low latency, with an average of 55.4 ms for a total of 500 blockchain WSN nodes, and an average of 4.2 s for a total of 20,000 blockchain WSN nodes. In the case of throughput, if we increase the number of blockchain WSN nodes in the network, the architecture scales and can integrate a high number of blockchain WSN nodes. The proposed blockchain architecture uses an IoT authentication process that allows new WSN nodes to be accepted using a voting process that integrates the PoS consensus mechanism. Another advantage is related to the scalability of the proposed architecture, which can integrate a very large number of IoT devices without decreasing the performance level of the system. This IoT devices can join the network and contribute to its maintenance by implementing the consensus mechanism. Our massive blockchain data storage solution can also be used in a hybrid manner by classic IoT WSN networks with no enhanced blockchain capabilities.

Table 1 presents a performance evaluation of the proposed architecture considering other solutions presented in the scientific literature. From the obtained results, the proposed architecture ensures a high level of performance and can be used as a massive data storage solution for IoT devices using blockchain technologies.

Table 1. Performance evaluation of the proposed architecture.

Parameters	Liu et al. [18]	Li et al. [19]	Shafagh et al. [20]	Ren et al. [22]	Ren et al. [23]	Our Approach
Consensus mechanism	PoW	-	PoW	-	PoW	PoS
TPS	12–15	-	12–15	-	12–15	100+
Block Time	-	-	-	-	-	~15 s
Scalable	No	-	No	-	No	Yes
Security	High	High	High	High	High	High
Computational Power	High	-	High	-	High	Low
Storage	High	-	High	-	High	High

As future work, we plan to further test the proposed architecture in real operating conditions in order to evaluate its scalability and efficiency. For better and more rigorous testing, we intend to use application-specific techniques using blockchain technology. Some of these techniques can be found in [41,42].

Author Contributions: Conceptualization, A.L. and A.A.M.; methodology, A.L., A.A.M. and V.P.; software, A.A.M. validation, A.I.P. and A.L.; formal analysis, A.L.; investigation, A.L., A.I.P. and A.A.M.; resources, A.I.P., A.L. and A.A.M.; data curation, A.I.P. and A.L.; writing—original draft preparation, A.L., A.A.M. and A.I.P., writing—review and editing, A.L.; visualization, A.I.P.; project administration, A.L. and A.A.M.; funding acquisition, A.L. and V.P. All authors have read and agreed to the published version of the manuscript.

Funding: This work was supported in part by the project Integrated Center for Research, Development, and Innovation in Advanced Materials, Nanotechnologies, and Distributed Systems for Fabrication and Control (MANSiD), Sectoral Operational Program for Increase of the Economic Competitiveness co-funded from the European Regional Development Fund, under Contract 671/9 April 2015 and in part by project "Interdisciplinary Cloud and Big Data Center at Ștefan cel Mare University of Suceava", POC/398/1/1, 343/390019, co-founded by the European Union.

Institutional Review Board Statement: Not applicable.

Informed Consent Statement: Not applicable.

Data Availability Statement: This study did not report any data.

Conflicts of Interest: There is no conflict of interest associated with this article.

References

1. IoT Connected Devices Worldwide 2019–2030. Available online: https://www.statista.com/statistics/1183457/iot-connected-devices-worldwide/ (accessed on 7 March 2022).
2. Jaoude, J.A.; Saade, R.G. Blockchain applications—Usage in different domains. *IEEE Access* **2019**, *7*, 45360–45381. [CrossRef]
3. Krichen, M.; Ammi, M.; Mihoub, A.; Almutiq, M. Blockchain for Modern Applications: A Survey. *Sensors* **2022**, *22*, 5274. [CrossRef] [PubMed]
4. Lavric, A.; Petrariu, A.I.; Mutescu, P.M.; Coca, E.; Popa, V. Internet of Things Concept in the Context of the COVID-19 Pandemic: A Multi-Sensor Application Design. *Sensors* **2022**, *22*, 503. [CrossRef] [PubMed]

5. Muangprathub, J.; Boonnam, N.; Kajornkasirat, S.; Lekbangpong, N.; Wanichsombat, A.; Nillaor, P. IoT and agriculture data analysis for smart farm. *Comput. Electron. Agric.* **2019**, *156*, 467–474. [CrossRef]
6. Petrariu, A.I.; Coca, E.; Lavric, A. Large-Scale Internet of Things Multi-Sensor Measurement Node for Smart Grid Enhancement. *Sensors* **2021**, *21*, 8093. [CrossRef]
7. Bhushan, B.; Khamparia, A.; Sagayam, K.M.; Sharma, S.K.; Ahad, M.A.; Debnath, N.C. Blockchain for smart cities: A review of architectures, integration trends and future research directions. *Sustain. Cities Soc.* **2020**, *61*, 102360. [CrossRef]
8. Al-Shammari, B.K.J.; Al-Aboody, N.; Al-Raweshidy, H.S. IoT Traffic Management and Integration in the QoS Supported Network. *IEEE Internet Things J.* **2018**, *5*, 352–370. [CrossRef]
9. She, W.; Liu, Q.; Tian, Z.; Chen, J.; Wang, B.; Liu, W. Blockchain trust model for malicious node detection in wireless sensor networks. *IEEE Access* **2019**, *7*, 38947–38956. [CrossRef]
10. Janes, B.; Crawford, H.; Oconnor, T.J. Never ending story: Authentication and access control design flaws in shared IoT devices. In Proceedings of the 2020 IEEE Symposium on Security and Privacy Workshops, SPW 2020, San Francisco, CA, USA, 21 May 2020; pp. 104–109. [CrossRef]
11. Nair, K.; Kulkarni, J.; Warde, M.; Dave, Z.; Rawalgaonkar, V.; Gore, G.; Joshi, J. Optimizing power consumption in iot based wireless sensor networks using Bluetooth Low Energy. In Proceedings of the 2015 International Conference on Green Computing and Internet of Things, ICGCIoT 2015, Greater Noida, India, 8–10 October 2015; pp. 589–593. [CrossRef]
12. Samijayani, O.N.; Darwis, R.; Rahmatia, S.; Mujadin, A.; Astharini, D. Hybrid ZigBee and WiFi Wireless Sensor Networks for Hydroponic Monitoring. In Proceedings of the 2nd International Conference on Electrical, Communication and Computer Engineering, ICECCE 2020, Istanbul, Turkey, 12–13 June 2020. [CrossRef]
13. Singh, R.K.; Aernouts, M.; de Meyer, M.; Weyn, M.; Berkvens, R. Leveraging LoRaWAN Technology for Precision Agriculture in Greenhouses. *Sensors* **2020**, *20*, 1827. [CrossRef]
14. Wang, H.; Dong, L.; Wei, W.; Zhao, W.S.; Xu, K.; Wang, G. The WSN Monitoring System for Large Outdoor Advertising Boards Based on ZigBee and MEMS Sensor. *IEEE Sens. J.* **2018**, *18*, 1314–1323. [CrossRef]
15. Dangana, M.; Ansari, S.; Abbasi, Q.H.; Hussain, S.; Imran, M.A. Suitability of NB-IoT for Indoor Industrial Environment: A Survey and Insights. *Sensors* **2021**, *21*, 5284. [CrossRef]
16. Burg, A.; Chattopadhyay, A.; Lam, K.Y. Wireless communication and security issues for cyber- physical systems and the internet-of-things. *Proc. IEEE* **2018**, *106*, 38–60. [CrossRef]
17. Augustin, A.; Yi, J.; Clausen, T.; Townsley, W.M. A Study of LoRa: Long Range & Low Power Networks for the Internet of Things. *Sensors* **2016**, *16*, 1466. [CrossRef]
18. Liu, B.; Yu, X.L.; Chen, S.; Xu, X.; Zhu, L. Blockchain Based Data Integrity Service Framework for IoT Data. In Proceedings of the 2017 IEEE 24th International Conference on Web Services, ICWS 2017, Honolulu, HI, USA, 25–30 June 2017; pp. 468–475. [CrossRef]
19. Li, R.; Song, T.; Mei, B.; Li, H.; Cheng, X.; Sun, L. Blockchain for Large-Scale Internet of Things Data Storage and Protection. *IEEE Trans. Serv. Comput.* **2019**, *12*, 762–771. [CrossRef]
20. Shafagh, H.; Burkhalter, L.; Hithnawi, A.; Duquennoy, S. Towards blockchain-based auditable storage and sharing of iot data. In Proceedings of the CCSW 2017—2017 Cloud Computing Security Workshop, Co-Located with CCS 2017, Dallas, TX, USA, 3 November 2017; pp. 45–50. [CrossRef]
21. Yang, X.; Chen, Y.; Chen, X. Effective scheme against 51% attack on proof-of-work blockchain with history weighted information. In Proceedings of the 2019 2nd IEEE International Conference on Blockchain, Blockchain 2019, Atlanta, GA, USA, 14–17 July 2019; pp. 261–265. [CrossRef]
22. Ren, Y.; Leng, Y.; Cheng, Y.; Wang, J. Secure data storage based on blockchain and coding in edge computing. *Math. Biosci. Eng.* **2019**, *16*, 1874–1892. [CrossRef]
23. Ren, Y.; Leng, Y.; Zhu, F.; Wang, J.; Kim, H.J. Data Storage Mechanism Based on Blockchain with Privacy Protection in Wireless Body Area Network. *Sensors* **2019**, *19*, 2395. [CrossRef]
24. Yaga, D.; Mell, P.; Roby, N.; Scarfone, K. *Blockchain Technology Overview*; National Institute of Standards and Technology: Gaithersburg, MD, USA, 2019. [CrossRef]
25. Nakamoto, S. Bitcoin: A Peer-to-Peer Electronic Cash System. Available online: www.bitcoin.org (accessed on 9 March 2022).
26. Li, C.; Li, P.; Zhou, D.; Xu, W.; Long, F.; Yao, A.C.-C. Scaling Nakamoto Consensus to Thousands of Transactions per Second. *arXiv* **2018**, arXiv:1805.03870.
27. Vergne, J. Decentralized vs. Distributed Organization: Blockchain, Machine Learning and the Future of the Digital Platform. *SAGE* **2020**, *1*, 263178772097705. [CrossRef]
28. Zhang, P.; Schmidt, D.C.; White, J.; Dubey, A. Consensus mechanisms and information security technologies. *Adv. Comput.* **2019**, *115*, 181–209. [CrossRef]
29. Hazari, S.S.; Mahmoud, Q.H. Comparative evaluation of consensus mechanisms in cryptocurrencies. *Internet Technol. Lett.* **2019**, *2*, e100. [CrossRef]
30. Joshi, S. Feasibility of Proof of Authority as A Consensus Protocol Model. *arXiv* **2021**, arXiv:2109.02480.
31. Douceur, J.R. The sybil attack. In *Lecture Notes in Computer Science (Including Subseries Lecture Notes in Artificial Intelligence and Lecture Notes in Bioinformatics)*; Springer: Berlin/Heidelberg, Germany, 2002; Volume 2429, pp. 251–260. [CrossRef]

32. Hu, Y.; Tian, G.; Jiang, A.; Liu, S.; Wei, J.; Wang, J.; Tan, S. A Practical Heartbeat-based Defense Scheme against Cloning Attacks in PoA Blockchain. *Comput. Stand. Interfaces* **2023**, *83*, 103656. [CrossRef]
33. Kovan Testnet. Available online: https://kovan-testnet.github.io/website/ (accessed on 11 January 2023).
34. Goerli Testnet. Available online: https://goerli.net/ (accessed on 11 January 2023).
35. Rinkeby: Network Dashboard. Available online: https://www.rinkeby.io/#stats (accessed on 11 January 2023).
36. Zheng, Z.; Xie, S.; Dai, H.N.; Chen, W.; Chen, X.; Weng, J.; Imran, M. An overview on smart contracts: Challenges, advances and platforms. *Future Gener. Comput. Syst.* **2020**, *105*, 475–491. [CrossRef]
37. Alharby, M.; van Moorsel, A. BlockSim: An Extensible Simulation Tool for Blockchain Systems. *Front. Blockchain* **2020**, *3*, 28. [CrossRef]
38. Yousuf, A.M.; Rochester, E.M.; Ousat, B.; Ghaderi, M. Throughput, Coverage and Scalability of LoRa LPWAN for Internet of Things. In Proceedings of the 2018 IEEE/ACM 26th International Symposium on Quality of Service, IWQoS 2018, Banff, AB, Canada, 4–6 June 2018. [CrossRef]
39. Zheng, Z.; Xie, S.; Dai, H.; Chen, X.; Wang, H. An Overview of Blockchain Technology: Architecture, Consensus, and Future Trends. In Proceedings of the 2017 IEEE 6th International Congress on Big Data, BigData Congress 2017, Honolulu, HI, USA, 25–30 June 2017; pp. 557–564. [CrossRef]
40. The Blockchain—Mastering Bitcoin. Available online: https://www.oreilly.com/library/view/mastering-bitcoin/9781491902639/ch07.html (accessed on 17 January 2023).
41. Gao, J.; Liu, H.; Li, Y.; Liu, C.; Yang, Z.; Li, Q.; Guan, Z.; Chen, Z. Towards automated testing of blockchain-based decentralized applications. In Proceedings of the 2019 IEEE/ACM 27th International Conference on Program Comprehension (ICPC), Montreal, QC, Canada, 25–26 May 2019; pp. 294–299. [CrossRef]
42. Lahami, M.; Maalej, A.J.; Krichen, M.; Hammami, M.A. A Comprehensive Review of Testing Blockchain Oriented Software. In Proceedings of the International Conference on Evaluation of Novel Approaches to Software Engineering, ENASE, Online, 25–26 April 2022; pp. 355–362. [CrossRef]

sensors

Article

The Design and Development of a Microstrip Antenna for Internet of Things Applications

Liliana Anchidin *,†, Alexandru Lavric *,†, Partemie-Marian Mutescu, Adrian I. Petrariu and Valentin Popa

Computers, Electronics and Automation Department, Stefan cel Mare University of Suceava, 720229 Suceava, Romania
* Correspondence: liliana.achitei@usm.ro (L.A.); lavric@usm.ro (A.L.)
† These authors contributed equally to this work.

Abstract: The Internet of Things (IoT) has become a part of modern life where it is used for data acquisition and long-range wireless communications. Regardless of the IoT application profile, every wireless communication transmission is enabled by highly efficient antennas. The role of the antenna is thus very important and must not be neglected. Considering the high demand of IoT applications, there is a constant need to improve antenna technologies, including new antenna designs, in order to increase the performance level of WSNs (Wireless Sensor Networks) and enhance their efficiency by enabling a long range and a low error-rate communication link. This paper proposes a new antenna design that is able to increase the performance level of IoT applications by means of an original design. The antenna was designed, simulated, tested, and evaluated in a real operating scenario. From the obtained results, it ensured a high level of performance and can be used in IoT applications specific to the 868 MHz frequency band.By inserting two notches along x axis, we find an optimal structure of the microstrip patch antenna with a reflection coefficient of −34.3 dB and a bandwidth of 20 MHz. After testing the designed novel antenna in real IoT operating conditions, we concluded that the proposed antenna can increase the performance level of IoT wireless communications.

Keywords: IoT devices; wireless sensor networks; IoT applications; small patch antenna; SigFox

Citation: Anchidin, L.; Lavric, A.; Mutescu, P.-M.; Petrariu, A.I.; Popa, V. The Design and Development of a Microstrip Antenna for Internet of Things Applications. *Sensors* **2023**, *23*, 1062. https://doi.org/10.3390/s23031062

Academic Editor: Antonio Guerrieri

Received: 23 December 2022
Revised: 12 January 2023
Accepted: 14 January 2023
Published: 17 January 2023

1. Introduction

The Internet of Things (IoT) has become a part of modern life where it is used for data acquisition and long-range wireless communications. The IoT concept is usually implemented by means of sensors that are able to transmit information in a wireless network configuration. Applications based on wireless sensor networks (WSN) technologies range from healthcare systems, the military, smart city, and smart homes to agricultural applications [1–7] focused on increasing efficiency and improving resource allocation.

This IoT penetration is sustained by various wireless communication protocols that enable data transmission such as: Bluetooth (BLE) [8], LoRa (long range) modulation [9], SigFox [10] or 5G [11] technologies.

Regardless of the IoT application profile, every wireless communication transmission is enabled by highly efficient antennas. The role of the antenna is very important and must not be neglected. The antenna must be easy to manufacture, have small dimensions, be lightweight, have a low manufacturing cost, and be compatible with different integrated-circuit configurations. A microstrip patch antenna (MPA) can achieve these constraints and be easily designed for different configurations. Different types of microstrip patch antennas can be designed in various geometric forms such as rectangular, square, or circular [12]. Considering the high demand of IoT applications, there is a constant need to improve antenna technologies, including new antenna designs, in order to increase the performance level of the WSNs and increase communication efficiency.

The main challenges of the IoT sensors are related to the following issues:

- IoT sensors are usually battery-powered and the low-energy consumption is an important constraint [13];
- The communication environment is characterized by urban non-LoS (Line of Sight) propagation conditions and the antenna must enable a low error-rate communication link;
- The communication range must be as large as possible in order to ensure connectivity over large geographical areas such as the surface of a city;
- Wireless communication protocols must coexist since many of the IoT technologies operate in unlicensed communication frequency bands [14].

IoT applications require a low-consumption profile and more efficient antennas that can improve the performance level of the communication link, as well as provide a long-range information transfer. Many of the IoT communication protocols operate in the ISM (Industrial, Scientific, and Medical) and SRD (Short-Range Devices) non-licensed frequency bands, where the 868 MHz band is the one most commonly used by LoRa and SigFox IoT communication protocols. The designed IoT antenna must therefore ensure a high level of performance in the 868 MHz frequency band.

In this paper, we make the following contributions:

- the design, development and test of a highly efficient antenna that is suitable for IoT applications;
- the antenna is produced using an FR-4 substrate, is low-cost, and relatively easy to manufacture;
- the novel design can be directly integrated into IoT devices operating in the 868 MHz frequency band;
- the performance level of the antenna was evaluated using laboratory equipment and a real operating scenario, being compared with two reference IoT antennas;
- from the obtained results we concluded that the proposed antenna ensures a high level of performance and can increase the efficiency of SigFox communications.

The paper is organized as follows: first, a brief introduction related to the state-of-the-art, followed by Section 2, where the main antenna design for IoT challenges is presented in detail. The main goal was to obtain the highest level of performance for IoT communications. In Section 3, the performance level of the designed IoT antenna is presented and discussed in detail. The antenna was evaluated in real operating conditions and different performance metrics were recorded and analyzed.

The final section of the paper consists of the conclusions and the overall performance discussion of the developed antenna. From the obtained results presented in this paper, we conclude that the proposed antenna design ensures a high-level performance in the 868 MHz frequency band and can be easily integrated into many IoT applications.

2. IoT Technologies and Antenna Design

In recent years, the research interest in wearable antennas inserted on flexible materials [15–17] has increased. Biomedical sensor networks for health care applications have drawn great attention mainly due to the recent pandemic context; therefore, wireless IoT communication systems with large range communication capabilities are in high demand. When evaluating the performance level of IoT technologies, another important aspect that should not be neglected is related to the communication protocols integrated at the application level.

In this paper, we focus our attention on Sigfox communication protocol that is a Low-Power Wide-Area Network type used for IoT applications in the free-licenses SRD 868 MHz frequency band. To evaluate the quality of a system, three major criteria are considered, i.e., the communication range, the communication speed and the power consumption needed to provide the first two. Sigfox addresses these problems with some advantages: low power, long range capability and high robustness to interferences. Another aspect is related to the dimensions of the antenna that must be reduced and provide support for small-form wireless sensors integration. An electrically small antenna is defined as an antenna area equal or less than 0.1λ [18].

Due to the low dimensions of WSNs modules, microstrip antennas are usually preferred in many IoT applications. A microstrip antenna is characterized by its small size and a narrow frequency-band behavior antenna [19]. Low radiation efficiency usually resumes to a decreased gain. An improvement on bandwidth of the antenna structure can be achieved by increasing the width using a high dielectric substrate, introducing specific slots [20] or notches on the patch, or by increasing the number of propagation modes of the antenna.

As shown by Chu [19], small dimension antennas usually have a narrow frequency bandwidth because of a high Q-Factor parameter. In the literature, several works have used matching circuits to increase the frequency bandwidth [16].

Figure 1 presents the design methodology used for the developed IoT antenna. The first step was to analyze the IoT concept antenna requirements, empirically design the antenna, and evaluate its performance level through simulations by performing optimizations, modifications and parameter inspections. The next step was to manufacture the selected configuration and test it using laboratory equipment to perform calibrations. The final step included the integration of the designed antenna in a real IoT application and evaluating the level of performance.

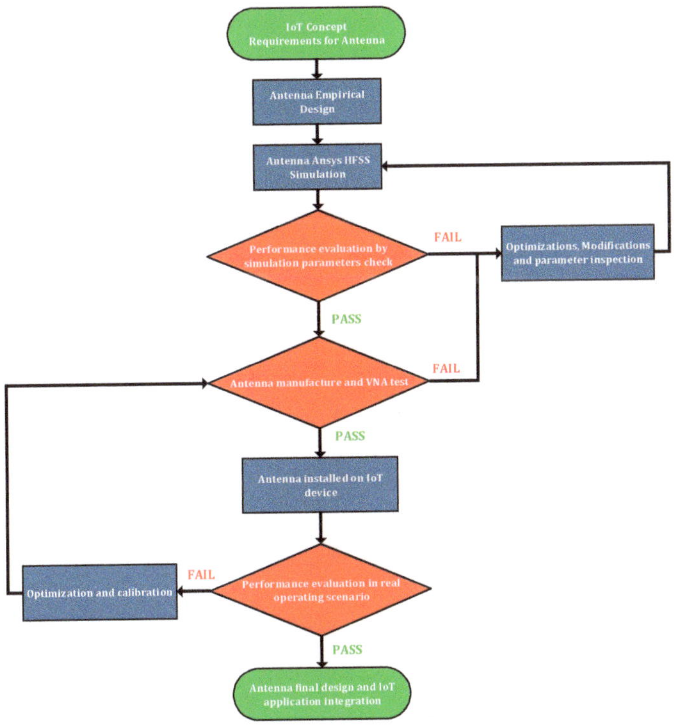

Figure 1. Design methodology.

In order to increase the radiated power and to decrease the input impedance, we increased the substrate height; consequently, the radiation resistance was approximately 300 Ω [21]. Another technique to decrease the input impedance is to increase the width dimension of the antenna structure.

There are some methods to increase the performance level such as: short-circuits, increasing electrical length [22], using magnetic or high permittivity substrates, and using superstrates or a combination of them in order to design electrically small antennas that are able to operate at high frequencies.

Another concern about the antenna is the excitation method of the patch, i.e., transmission line, probe feed and aperture coupling. When a transmission line is used, the energy is converted by Joule effect into heat and an aperture coupling increases the difficulty of the practical implementation. The feeding point must be placed at a distance at least equal to one third of the distance between the two slots; however, the location point should be adjusted where the antenna input impedance is equal to the characteristic impedance of the connecting coaxial cable in order to comply with the impedance's match. In this paper, we focused on the design of a microstrip patch antenna for long-range IoT communications protocols that are suitable for urban non-LoS (Line of Sight) up to 2 km [23].

3. The Design, Development, and Implementation of the IoT Antenna

In the literature there are many microstrip antenna design techniques [24–26] that use the cavity model and transmission line model because of their low complexity. Considering that the ratio height (h) over width (W) is very small for our antenna design requirements, the current densities moving around the principal conductor can be achieved at the edges of the patch antenna. Furthermore, since the substrate thickness is very small compared to the dielectric wavelength ($h << \lambda_g$), the electric field can be considered normally distributed on the patch surface. In that case, we considered that the microstrip antenna could be modeled as a resonant cavity where the top and bottom are electrical walls, and the sides are magnetic walls [26,27].

In this section, we present the design and development of the proposed antenna considering different IoT application requirements specific to the 868 MHz frequency band. The proposed microstrip patch antenna (Figure 2a) was manufactured on an Fr-4 substrate with a relative permittivity of 5.1 and a loss tangent parameter of 0.02. The substrate height was 1.5 mm and the thickness of the antenna copper was about 0.035 mm. In order to achieve the resonant frequency for a patch antenna of 50 mm × 50 mm, two notches were inserted in the structure along the x axis with the role of increasing the electrical length of the patch. The antenna's size was 0.13λ × 0.13λ, with the substrate and the ground plane dimension of 0.14λ × 0.14λ. Antenna size was reduced by 1 mm in respect to the ground in order to increase the active power output.

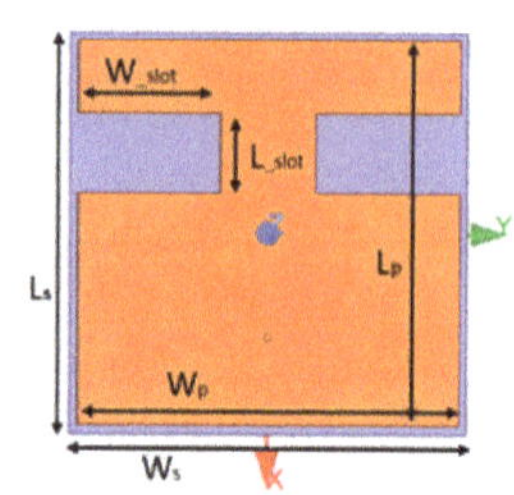

L_S	50 mm
W_S	50 mm
L_P	48 mm
W_P	48 mm
$L_{\_slot}$	11 mm
$W_{\_slot}$	18.5 mm
h	1.5 mm

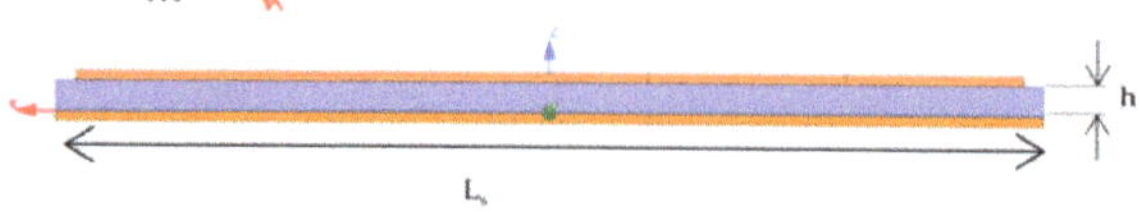

(a)

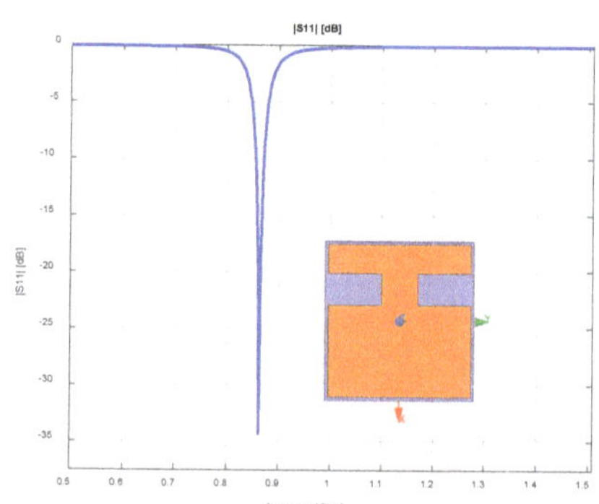

(b)

Figure 2. Antenna dimensions: (**a**) Front view and side view with dimensions; (**b**) Reflection coefficient.

The entire structure of the antenna was simulated using the Ansys HFSS [28] environment to verify the influence of the notches on the electrical parameters of the antenna. Figure 2b represents the reflection coefficient of the proposed antenna with the value at

868 MHz of −34.3 dB. This section presents in detail the design choices that led to the antenna with the above-presented dimensions. The design steps of the IoT antenna included the determination of the optimal influence of the notches' width and length on the antenna parameters, the evaluation of the notches' position, the feeding point location in respect with the reference center and the overall dimensions of the antenna.

We verified the influence of the notches' width; the evaluated values were 16, 17, 18.5, 20 and 21 mm. At 16 and 17 mm, the resonant frequencies achieved 0.98 GHz and 0.94 GHz, respectively, and for 20 and 21 mm, the resonant frequency decreased to 0.79 GHz and 0.74 GHz, respectively.

Figure 3 shows the variation of the reflection coefficient (a) and input impedance (b), respectively, when the notches' width dimension varied.

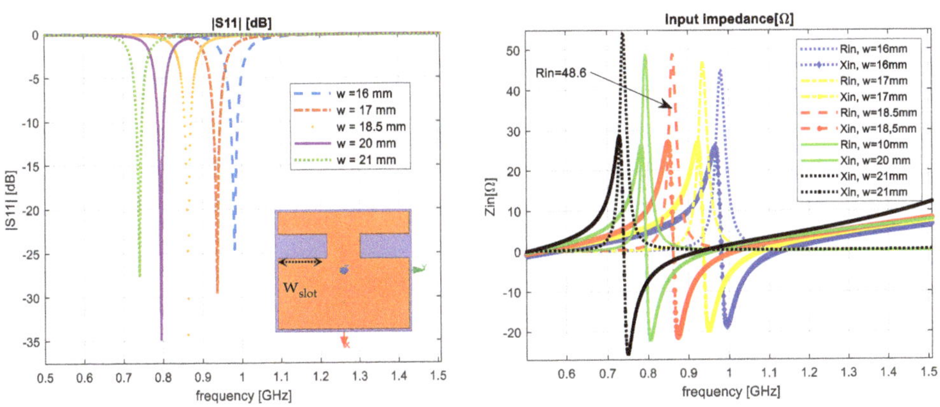

(a) (b)

Figure 3. Varying the notches' width: (**a**) reflection coefficient; (**b**) input impedance.

The reflection coefficient for the notches' length size of 7, 9, 11, 13 and 15 mm is presented in Figure 4a. For 7 and 9 mm, we obtained a resonant frequency of 0.94 GHz and 0.9 GHz, and for 13 and 15 mm we achieved 0.84 GHz and 0.82 GHz. In this particular case, the real part of the input impedance achieved 42.69 Ω, 44.13 Ω, 46.89 Ω, 48.6 Ω and 54.96 Ω as shown in Figure 4b.

From the obtained results, we observed that when increasing and decreasing the size of the notches, we varied the electrical length of the patch antenna. We achieved an optimal antenna dimension for the size of 18.5 mm × 11 mm with a reflection coefficient of −34.3 dB.

In the second evaluation scenario, we simulated the influence of the notches' position on the antenna's principal parameters. The position was moved in respect with the axial direction of propagation. From the obtained results, the notches' size dimensions were set to 18.5 mm × 11 mm. When the two notches move further from one another, the resonant frequency increases. The variation was 5, 10, 12.5, 15 and 20 mm, respectively, from the center of the reference. Figure 5a shows the reflection coefficient variation from 0.77 GHz to 1.01 GHz, and Figure 5b shows the input impedance.

For a small size antenna, the input impedance is around 300 Ω. In order to decrease this value, we had to find the optimal voltage-to-current ratio. We varied the feed-point position to 9, 11, 13, 15 and 17 mm, respectively, from the center of the antenna, with the aim to set the match between the input impedance and the cable connection input. The obtained results are presented in Figure 6.

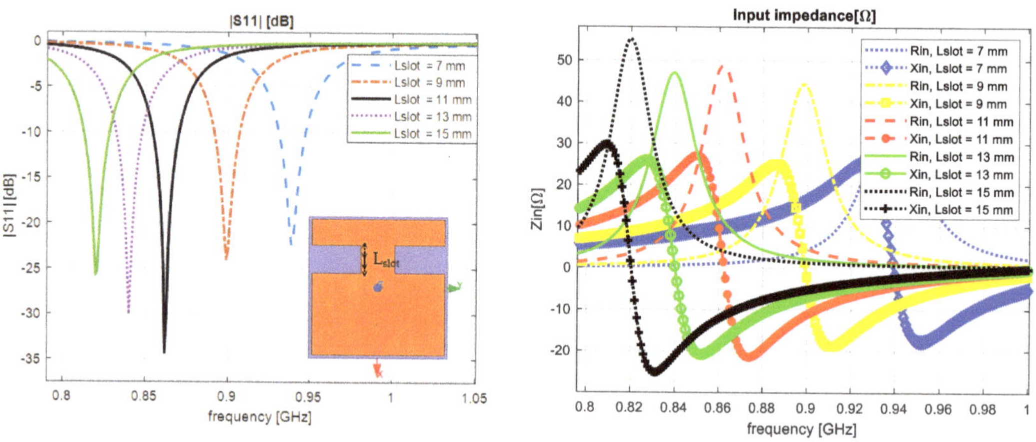

(a) (b)

Figure 4. Varying the notches' length: (**a**) reflection coefficient; (**b**) input impedance.

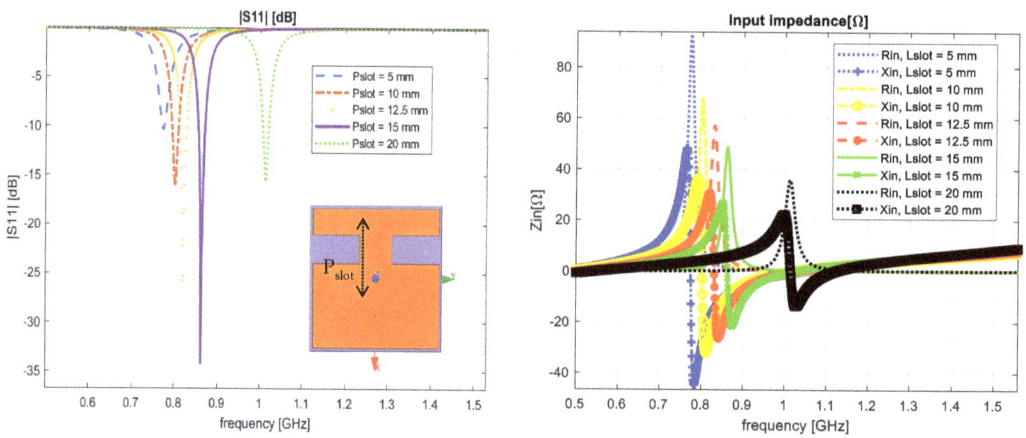

(a) (b)

Figure 5. Patch antenna with two notches. The slot position is moved along the x axis. (**a**) Reflection coefficient; (**b**) input impedance.

From the obtained results, we achieved an optimum value of the input impedance parameter when the feed-point was set at 13 mm from the center of the developed IoT antenna.

The next step was to evaluate the antenna with a notch inserted in the front part of the antenna. The notch position was modified along the Y axis with different positions (−8, −4, 0, 4, 8 mm with respect to the center of the references), in order to evaluate the highest level of performance. In Figure 7a, the reflection coefficient parameter achieved a value of −11.18 dB, and the upper input impedance parameter a value of 27 Ω, respectively, as seen in Figure 7b.

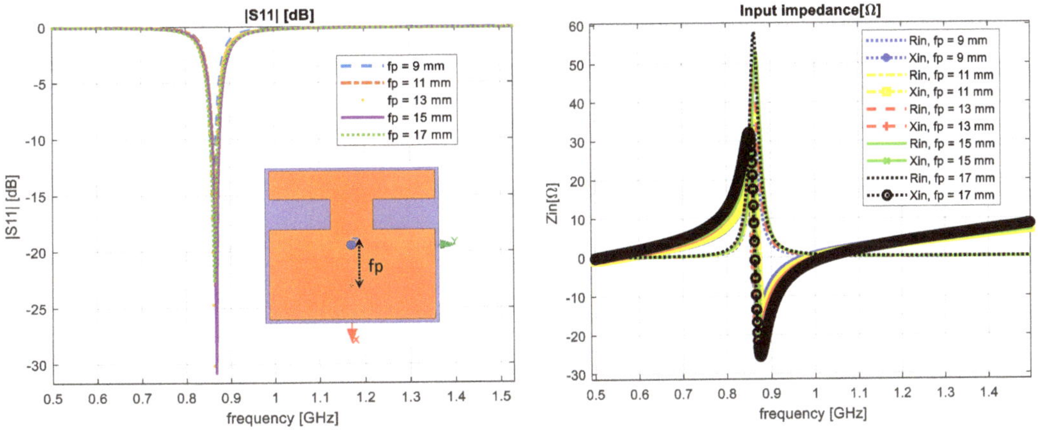

(a) (b)

Figure 6. Feed-point position variation: (**a**) reflection coefficient; (**b**) input impedance.

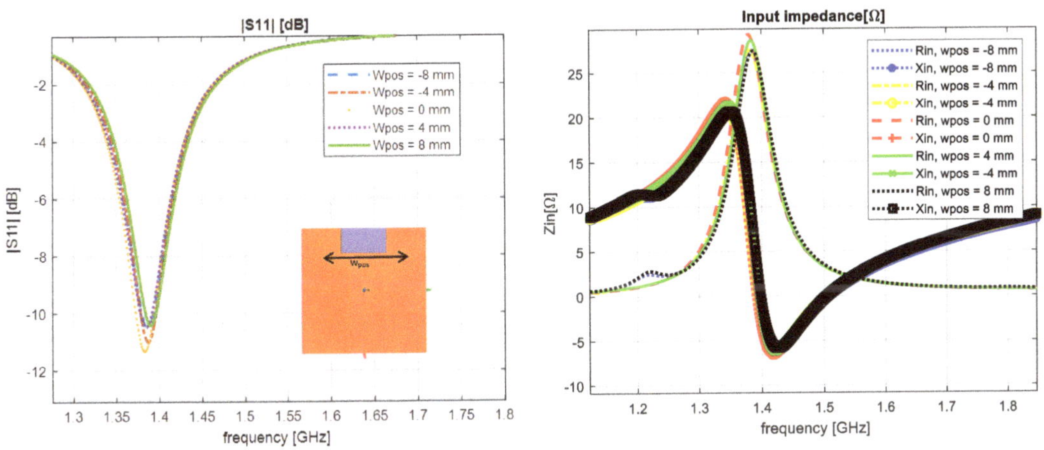

(a) (b)

Figure 7. Varying the notches' position: (**a**) reflection coefficient; (**b**) input impedance.

The next design step was to insert an interior slot of rectangular form that would be able to increase the antenna electrical length, as presented in Figure 8a. We simulated the structure with different dimensions and the slot was moved along the x axis of the patch with a distance of 10, 15, 20 and 25 mm.

Figure 8b shows a reflection coefficient parameter of $|S_{11}| \cong -24$ dB and an input impedance of 55 Ω at a central frequency of 0.9 GHz.

A microstrip patch antenna with 50 mm × 50 mm × 0.035 mm size was evaluated, as presented in Figure 9. The structure was realized on an FR4 substrate with 50 mm × 50 mm × 1.5 mm. The patch and the ground were connected to a probe-feed by an inner and an outer conductor, respectively. From the obtained results, we observed that the ground-plane connection ensured a high level of performance and was thus integrated in the antenna design.

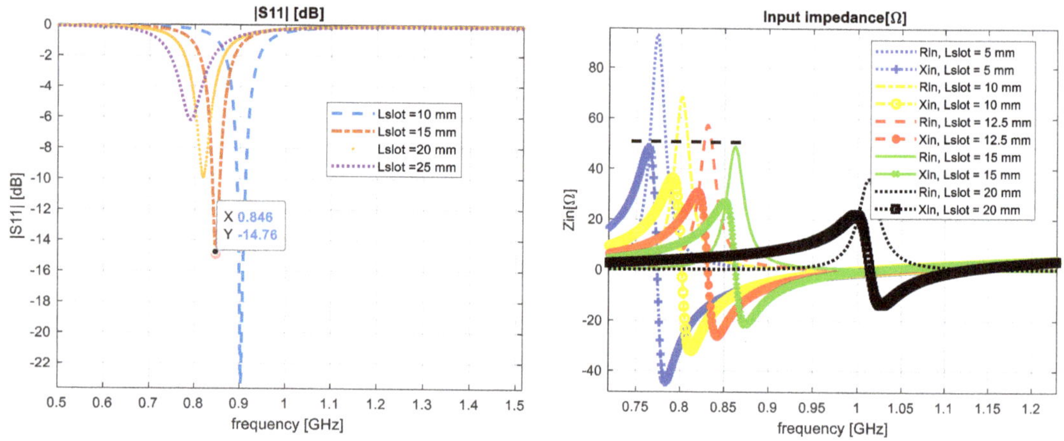

(a) (b)

Figure 8. A centered slot integrated in the antenna design: (**a**) reflection coefficient; (**b**) input impedance.

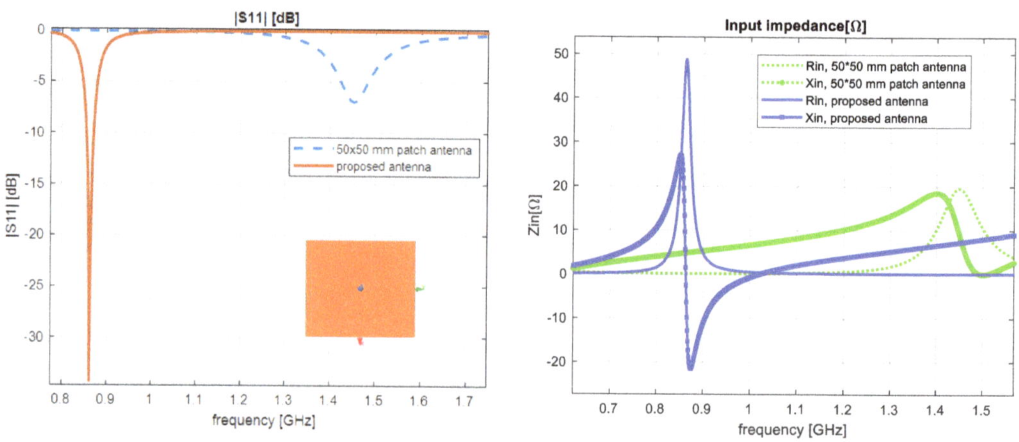

(a) (b)

Figure 9. Rectangular patch antenna: (**a**) reflection coefficient; (**b**) input impedance.

Figure 10a shows the radiation pattern of the designed antenna; the gain value was around −1.49 dB, while the directivity is presented in Figure 10b and achieved a maximum value of −1.16 dB. The radiation efficiency parameter of the designed IoT antenna achieved a value of 51%.

One final step was to analyze the influence of the circuit layer over the antenna parameters. The transceiver placed on the underground plane did not affect the resonant frequency because the size of the ground plane was a smaller dimension compared to the antenna size. We can observe in Figure 11 that, by placing the circuit under the plane, the reflection coefficient decreased to −21.79 dB. The real part of the input impedance decreased to 42.41 Ω and we achieved a VSWR parameter of approximately 1.18.

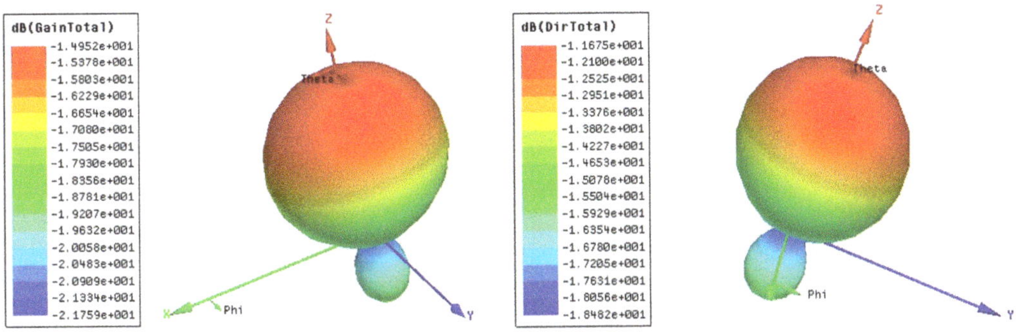

(**a**) (**b**)

Figure 10. The radiation pattern of the developed IoT antenna: (**a**) gain and (**b**) directivity.

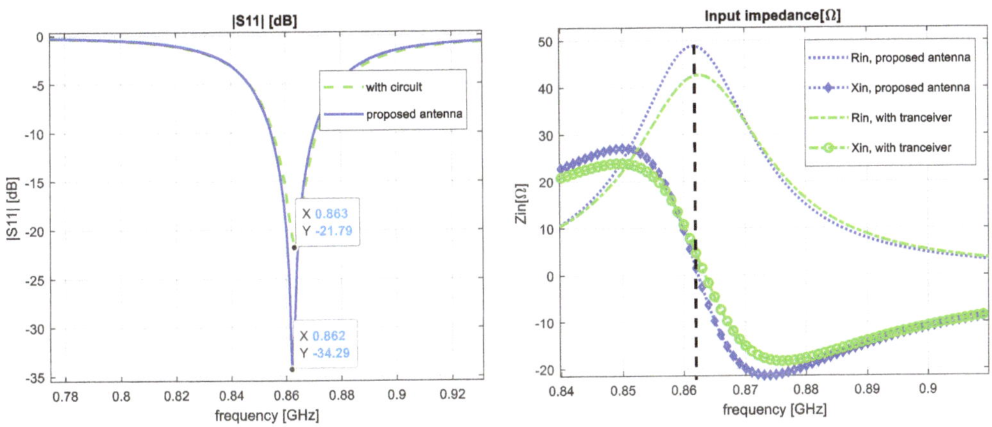

(**a**) (**b**)

Figure 11. Transceiver influence: (**a**) reflection coefficient; (**b**) input impedance.

4. Performance Evaluation of the Proposed IoT Antenna

In this section, we present the performance evaluation of the designed IoT patch antenna. The antenna was tested in a real operating environment and its performance was compared with two reference antennas: one omnidirectional and one microstrip, both designed for the 868 MHz frequency band. Figure 12 presents the developed IoT antenna that was further tested in real operating conditions.

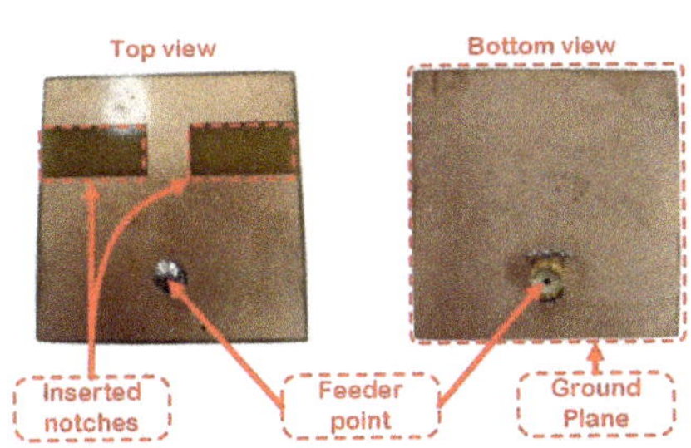

Figure 12. The developed IoT antenna.

As a reference for evaluation purposes, we used an omnidirectional antenna rated for the 868 MHz frequency band included in the ON Semiconductors Sigfox development KIT with a gain of 1 dBi, and a patch antenna respecting the inverted F design [29] with a gain of 0 dB, also rated for the 868 MHz frequency band.

The first step was to perform a vector network analyzer (VNA) analysis of the antennas, in which we measured the S11 parameter in terms of reflection coefficient and VSWR (Voltage Standing Wave Ratio), for frequencies ranging from 500 MHz to 1.5 GHz. Figure 13 presents in detail the test setup used.

(**a**) (**b**) (**c**)

Figure 13. The proposed test setup: (**a**) proposed novel IoT antenna; (**b**) reference omnidirectional antenna; (**c**) reference microstrip antenna.

From the obtained results, in the 868 MHz frequency band, the omnidirectional reference antenna achieved an average reflection coefficient of −2.08 dB and a VSWR value of 8.78; the microstrip reference antenna achieved an average reflection coefficient of −1.81 dB and a VSWR value of 9.6; while the proposed novel IoT antenna achieved an average reflection coefficient of −12.8 dB and a VSWR value of approximately 1.6.

In Figure 14, we observe the reflection coefficient of the reference antennas and the novel designed IoT antenna, as well as the VSWR parameter. The omnidirectional and microstrip reference antennas each present one minimum value around the 704 MHz and 645 MHz frequency mark with the values of −10.13 dB and −13.74 dB, respectively, while our proposed antenna presents one minimum value around the 865 MHz frequency mark with a value of −14.08 dB; all the antennas were rated for the 868 MHz frequency band.

In Figure 14b, we see that the reference antennas have a wideband frequency behavior, with a VSWR value below 10 in the frequency range of 570 MHz to 900 MHz with an absolute minimum VSWR value of 1.9 around the 715 MHz frequency mark for the omnidirectional antenna, and a minimum VSWR value of 1.51 around the 645 MHz frequency mark for the reference patch antenna, while the proposed antenna has a narrowband behavior, with an absolute minimum VSWR value of 1.49 around the 865 MHz frequency mark. These characteristics prove without a doubt that our designed IoT antenna performed better for IoT specific applications.

The next step was to further evaluate the IoT-designed antenna in a real operating scenario for Sigfox communications, as shown in Figure 15. Sigfox [30] is an ultra-narrow band LPWAN (Low Power Wide Area Network) communication protocol, operating in the 868 MHz frequency band on multiple 100 Hz communication channels.

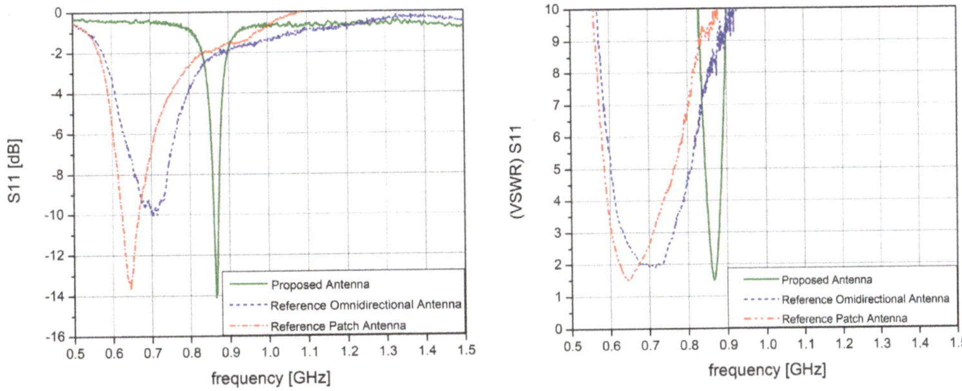

(**a**) (**b**)

Figure 14. Test results: (**a**) reflection coefficient; (**b**) VSWR.

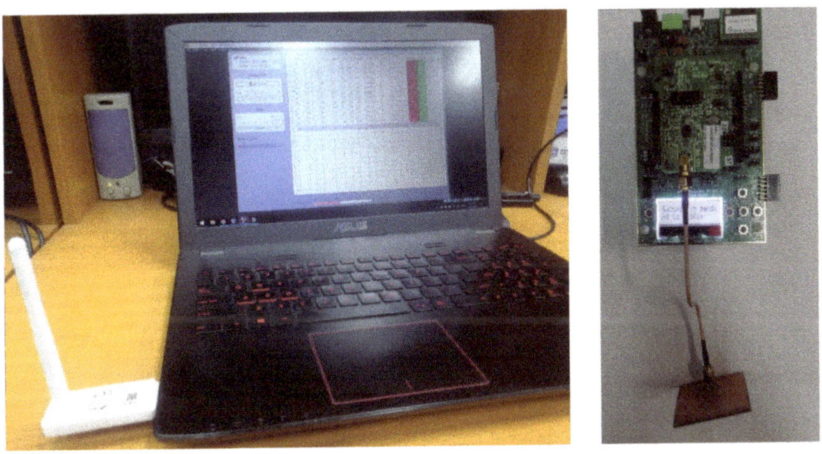

(**a**) (**b**)

Figure 15. Real operating scenario test setup: (**a**) Sigfox SDR certification Kit with the SigFox Radio Signal Analyzer App; (**b**) ON Semiconductors Sigfox Development Kit with the proposed antenna attached.

The main advantage of Sigfox communication protocol is its redundancy mechanism in which a single message is sent three times, at three different timestamps, on three different communication channels, ensuring diversity both in time and frequency.

The RSSI (Received Signal Strength Indicator) is a parameter that indicates the quality of a radio link between the transmitter and the receiver and is a measure of the performance level. Therefore, the higher the RSSI values, the stronger the power of the radio signal. Although the SigFox receiver selects the packet with the highest RSSI at the receiving end, the arrival of all three data packets points to a high-quality link that corresponds to the best-case scenario.

For the real operating scenario test setup, we used the Sigfox SDR certification kit [31] and the ON Semiconductors Sigfox transceiver, as presented in Figure 16. The tests included three different scenarios, two where the SigFox ON transceiver used the reference antennas and one scenario that included the use of the proposed IoT antenna mounted on the SigFox transceiver.

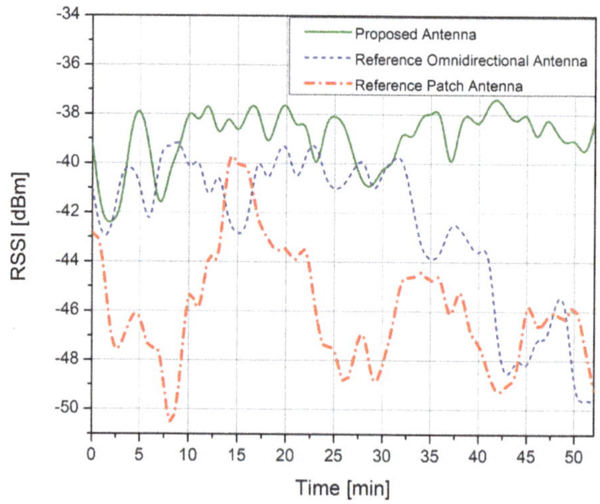

Figure 16. RSSI parameter variation obtained in the real operating scenario.

The evaluation was performed in a non-line of sight (non-LoS) propagation conditions indoor scenario that included a distance of approximately 30 m between the receiver and the transmitter. We used the Sigfox SDR kit to monitor the received data packets while at the same time recording the RSSI parameter of each individual packet. From the obtained results, the reference omnidirectional antenna achieved an average RSSI of −44.56 dBm, the reference patch antenna achieved an average RSSI of −46.56 dBm, and the proposed novel IoT antenna achieved an average RSSI of −38.56 dBm. Figure 15 presents the RSSI parameter variation for the reference antennas and the novel designed IoT antenna. From the obtained results, we see that the proposed IoT antenna had the highest performance level.

5. Conclusions

In this paper, we present the design, development, and performance evaluation of a microstrip patch antenna with small sizes for Internet of Thing applications. The antenna was integrated in a SigFox transceiver, designed to operate at 868 MHz. For this frequency, antennas have large sizes, i.e., L = 82.15 mm and W = 105.1 mm; in this case, we introduced two notches on the axial direction of propagation with the aim to increase the electrical length of the patch. By inserting two notches along the x axis, we found the microstrip patch antenna to be an optimal structure with a reflection coefficient of −34.3 dB and a bandwidth of 20 MHz.

The antenna was manufactured on an FR-4 substrate with electrical permittivity, εr = 5.1 and loss tangent, $\tan\sigma$ = 0.02. The antenna was placed on the upper side of the configuration, and the ground plane on the bottom side of the configuration. The final antenna had a square form of 48 mm × 48 mm with a thickness of 0.035 mm and a substrate high of 1.5 mm with two notches inserted on both sides. The feeding point location varied in order to determine the optimum point for matching the input impedance and co-axial cable impedance. After the antenna was designed and simulated, it was physically produced and tested in a real IoT operating condition.

From the obtained results, we concluded that the proposed antenna can increase the performance level of SigFox wireless communications.

The antenna was tested in a real operating environment and its performance was compared with two reference antennas: one omnidirectional and one microstrip antenna, both designed for the 868 MHz frequency band.

This paper thus proposes a new antenna design that is able to increase the performance level of IoT applications by means of an original design. The antenna was designed, simulated, tested and evaluated in a real operating scenario and from the obtained results, it ensures a high level of performance that can be used in IoT applications specific to the 868 MHz frequency band regardless of the wireless communication protocol used.

Author Contributions: Conceptualization, A.L. and L.A.; methodology A.L. and L.A.; software, A.L. validation, A.L. and L.A.; formal analysis, A.L.; investigation, A.L. and L.A.; resources, A.L., L.A. and V.P.; data curation, A.L., A.I.P. and P.-M.M.; writing—original draft preparation, A.L., L.A., A.I.P. and P.-M.M. writing—review and editing, A.L. and L.A; visualization, A.L. and L.A.; project administration, A.L. and L.A.; funding acquisition, A.L., L.A. and V.P. All authors have read and agreed to the published version of the manuscript.

Funding: This paper has been financially supported within the project entitled "DECIDE-Development through entrepreneurial education and innovative doctoral and postdoctoral research", project code POCU/380/6/13/125031, project co-financed from the European Social Fund through the 2014–2020 Operational Program Human Capital.

Institutional Review Board Statement: Not applicable.

Informed Consent Statement: Not applicable.

Data Availability Statement: Not applicable.

Conflicts of Interest: The authors declare no conflict of interest.

References

1. Lavric, A.; Petrariu, A.I.; Mutescu, P.M.; Coca, E.; Popa, V. Internet of Things Concept in the Context of the COVID-19 Pandemic: A Multi-Sensor Application Design. *Sensors* **2022**, *22*, 503. [CrossRef] [PubMed]
2. Qian, Z.; Lin, Y.; Jing, W.; Ma, Z.; Liu, H.; Yin, R.; Li, Z.; Bi, Z.; Zhang, W. Development of a Real-Time Wearable Fall Detection System in the Context of Internet of Things. *IEEE Internet Things J.* **2022**, *9*, 21999–22007. [CrossRef]
3. Mahamuni, C.V.; Jalauddin, Z.M. Intrusion Monitoring in Military Surveillance Applications using Wireless Sensor Networks (WSNs) with Deep Learning for Multiple Object Detection and Tracking. In Proceedings of the 2021 International Conference on Control, Automation, Power and Signal Processing (CAPS), Jabalpur, India, 10–12 December 2021. [CrossRef]
4. Ali, A.; Jadoon, Y.K.; Changazi, S.A.; Qasim, M. Military Operations: Wireless Sensor Networks based Applications to Reinforce Future Battlefield Command System. In Proceedings of the 2020 IEEE 23rd International Multitopic Conference (INMIC), Bahawalpur, Pakistan, 5–7 November 2020. [CrossRef]
5. Khalifeh, A.; Darabkh, K.; Khasawneh, A.; Alqaisieh, I.; Salameh, M.; AlAbdala, A.; Alrubaye, S.; Alassaf, A.; Al-HajAli, S.; Al-Wardat, R.; et al. Wireless Sensor Networks for Smart Cities: Network Design, Implementation and Performance Evaluation. *Electronics* **2021**, *10*, 218. [CrossRef]
6. Madhu, S.; Padunnavalappil, S.; Saajlal, P.P.; Vasudevan, V.A.; Mathew, J. Powering Up an IoT-Enabled Smart Home. *Int. J. Softw. Sci. Comput. Intell. IJSSCI* **2022**, *14*, 1–21. [CrossRef]
7. García, L.; Parra, L.; Jimenez, J.; Parra, M.; Lloret, J.; Mauri, P.; Lorenz, P. Deployment Strategies of Soil Monitoring WSN for Precision Agriculture Irrigation Scheduling in Rural Areas. *Sensors* **2021**, *21*, 1693. [CrossRef] [PubMed]
8. Collotta, M.; Pau, G.; Talty, T.; Tonguz, O.K. Bluetooth 5: A Concrete Step Forward toward the IoT. *IEEE Commun. Mag.* **2018**, *56*, 125–131. [CrossRef]
9. Raychowdhury, A.; Pramanik, A. Survey on LoRa Technology: Solution for Internet of Things. *Adv. Intell. Syst. Comput.* **2020**, *1148*, 259–271. [CrossRef]
10. Lavric, A.; Petrariu, A.I.; Popa, V. Long Range SigFox Communication Protocol Scalability Analysis under Large-Scale, High-Density Conditions. *IEEE Access* **2019**, *7*, 35816–35825. [CrossRef]
11. Vaezi, M.; Azari, A.; Khosravirad, S.R.; Shirvanimoghaddam, M.; Azari, M.M.; Chasaki, D.; Popovski, P. Cellular, Wide-Area, and Non-Terrestrial IoT: A Survey on 5G Advances and the Road Toward 6G. *IEEE Commun. Surv. Tutor.* **2022**, *24*, 1117–1174. [CrossRef]
12. Trinh, L.; Nguyen, T.Q.K.; Phan, D.D.; Tran, V.Q.; Bui, V.X.; Truong, N.V.; Ferrero, F. Miniature antenna for IoT devices using LoRa technology. *Int. Conf. Adv. Technol. Commun.* **2017**, *2017*, 170–173. [CrossRef]
13. Mutescu, P.M.; Petrariu, A.I.; Lavric, A. Wireless Communications for IoT: Energy Efficiency Survey. In Proceedings of the 2021 12th International Symposium on Advanced Topics in Electrical Engineering (ATEE), Bucharest, Romania, 25–27 March 2021. [CrossRef]
14. Fadeyi, J.; Markus, E.D.; Abu-Mahfouz, A.M. Technology Coexistence in LPWANs-A Comparative Analysis for Spectrum Optimization. *IEEE Int. Symp. Ind. Electron.* **2019**, *2019*, 2244–2249. [CrossRef]
15. Chandran, A.R.; Conway, G.A.; Scanlon, W.G. Compact low-profile patch antenna for medical body area networks at 868 MHz. In Proceedings of the 2008 IEEE International Symposium on Antennas and Propagation and USNC/URSI National Radio Science Meeting, APSURSI, Taipei, Taiwan, 2–5 November 2008; pp. 9–12. [CrossRef]

16. Atanasova, G.; Atanasov, N. Small antennas for wearable sensor networks: Impact of the electromagnetic properties of the textiles on antenna performance. *Sensors* **2020**, *20*, 5157. [CrossRef] [PubMed]
17. Wang, X.; Xing, L.; Wang, H. A wearable textile antenna for LoRa applications. In Proceedings of the 2021 IEEE 4th International Conference on Electronic Information and Communication Technology, ICEICT, Qingdao, China, 7–9 August 2021. [CrossRef]
18. Breed, G. Basic Principles of Electrically Small Antennas. *High Freq. Electron.* **2007**, *6*, 50–53.
19. Chu, L.J. Physical limitations of omni-directional antennas. *J. Appl. Phys.* **1948**, *19*, 1163–1175. [CrossRef]
20. Ossa-Molina, O.; López-Giraldo, F. A simple model to compute the characteristic parameters of a slotted rectangular microstrip patch antenna. *Electronics* **2022**, *11*, 129. [CrossRef]
21. Moná, D.F.; Sakomura, E.S.; Nascimento, D.C. Circularly polarised rectangular microstrip antenna design with arbitrary input impedance. *IET Microw. Antennas Propag.* **2018**, *12*, 1532–1540. [CrossRef]
22. Nguyen, T.Q.K.; Lizzi, L.; Ferrero, F. Dual-Matching for Single Resonance Miniaturized Antenna for IoT applications. In Proceedings of the 2018 IEEE Antennas and Propagation Society International Symposium and USNC/URSI National Radio Science Meeting, APSURSI 2018-Proceedings, Boston, MA, USA, 8–13 July 2018. [CrossRef]
23. Centenaro, M.; Vangelista, L.; Zanella, A.; Zorzi, M. Long-range communications in unlicensed bands: The rising stars in the IoT and smart city scenarios. *IEEE Wirel. Commun.* **2016**, *23*, 60–67. [CrossRef]
24. Balanis, C.A. *Antenna Theory Analysis and Design*; Wiley & Sons: Hoboken, NJ, USA, 2015; pp. 1–27.
25. Fang, D.G. *Antenna Theory and Microstrip Antennas*; CRC Press: Boca Raton, FL, USA, 2015.
26. Garg, R.; Bhartia, P.; Bahl, I.J.; Ittipiboon, A. *Microstrip Antenna Design Handbook*; Artech House: Norwood, MA, USA, 2001.
27. Hoefer, W. Equivalent Series Inductivity of a Narrow Transverse Slit in Microstrip. *IEEE Trans. Microw. Theory Tech.* **1977**, *25*, 822–824. [CrossRef]
28. Ansys HFSS | 3D High Frequency Simulation Software. Available online: https://www.ansys.com/products/electronics/ansys-hfss (accessed on 22 December 2022).
29. Kervel, F. Design Note DN023 868 MHz, 915 MHz and 955 MHz Inverted F Antenna. Tex. Instrum. 2011. Available online: https://www.ti.com/lit/an/swra228c/swra228c.pdf?ts=1673811355472&ref_url=https%253A%252F%252Fwww.google.com%252F (accessed on 22 December 2022).
30. Lavric, A.; Petrariu, A.I.; Popa, V. SigFox Communication Protocol: The New Era of IoT? In Proceedings of the 2019 International Conference on Sensing and Instrumentation in IoT Era, Lisbon, Portugal, 29–30 August 2019. [CrossRef]
31. SDR Dongle | Sigfox Build. Available online: https://build.sigfox.com/sdr-dongle (accessed on 22 December 2022).

sensors

MDPI

Article

A Secure Long-Range Transceiver for Monitoring and Storing IoT Data in the Cloud: Design and Performance Study

Nurul I. Sarkar [1,*], Asish Thomas Kavitha [2] and Md Jahan Ali [2]

1 Department of Computer Science and Software Engineering, Auckland University of Technology, Auckland 1010, New Zealand
2 School of Engineering, Computer and Mathematical Sciences, Auckland University of Technology, Auckland 1010, New Zealand
* Correspondence: nurul.sarkar@aut.ac.nz; Tel.: +64-211-758390

Abstract: Due to the high demand for Internet of Things (IoT) and real-time data monitoring and control applications in recent years, the long-range (LoRa) communication protocols leverage technology to provide inter-cluster communications in an effective manner. A secure LoRa system is required to monitor and store IoT data in the cloud. This paper aims to report on the design, analysis, and performance evaluation of a low-cost LoRa transceiver interface unit (433 MHz band) for the real-time monitoring and storing of IoT sensor data in the cloud. We designed and analyzed a low-cost LoRa transceiver interface unit consisting of a LoRa communication module and Wi-Fi module in the laboratory. The system was built (prototype) using radially available hardware devices from the local electronics shops at about USD 150. The transmitter can securely exchange IoT sensor data to the receiver node at about 10 km using a LoRa Wi-Fi module. The receiver node accumulates the sensor data and stores it in the cloud for processing. The performance of the proposed LoRa transceiver was evaluated by field experiments in which two transmitter nodes were deployed on the rooftop of Auckland University of Technology's Tower building on city campus (New Zealand), and the receiver node was deployed in Liston Park, which was located 10 km away from the University Tower building. The manual incident field tests examined the accuracy of the sensor data, and the system achieved a data accuracy of about 99%. The reaction time of the transmitter nodes was determined by the data accumulation of sensor nodes within 2–20 s. Results show that the system is robust and can be used to effectively link city and suburban park communities.

Keywords: LoRa; transceiver; modules; IoT; LPWAN

Citation: Sarkar, N.I.; Kavitha, A.T.; Ali, M.J. A Secure Long-Range Transceiver for Monitoring and Storing IoT Data in the Cloud: Design and Performance Study. *Sensors* **2022**, *22*, 8380. https://doi.org/10.3390/s22218380

Academic Editors: Alexandru Lavric, Liliana Anchidin and Adrian I. Petrariu

Received: 11 September 2022
Accepted: 27 October 2022
Published: 1 November 2022

1. Introduction

LoRa (long-range) communication protocols are the heart of Internet of Things (IoT) applications and connect cloud devices for the real-time monitoring and storage of data for efficiency and productivity purposes. The growing number of IoT sensors makes efficient transmission difficult because of the high-cost infrastructure [1–3]. Emerging LoRa technology can be more attractive for long-distance data transmission than the existing wide area network (WAN) communication systems. Most current technologies consume much energy and consequently decrease the battery lifetime of IoT devices. LoRa technology offers simplicity in securing and monitoring IoT sensor data [4]. In addition, this technology can be used for the efficient transmission of IoT sensor data and to enable the system to operate in both outdoor and indoor scenarios [5,6]. Cloud technology delivers a robust infrastructure for controlling information hubs that are tailored to deal with a significant volume of data. Furthermore, a cloud infrastructure can provide storage and surveillance for a large volume of sensor data and can be accessed from anywhere in the world. Existing technologies such as ZigBee [7], NB-IoT [8], and Wi-Fi [6] are not used in the LoRa transceiver system because of data losses, low coverage, and inefficient transmission.

A cost-effective LoRa transceiver system, therefore, is crucial in solving the problems of information transmission in current systems, and such a system assists in the storage and transmission of IoT device information through the execution of a LoRa connector [3,8–10]. LoRa systems offer a low-power WAN communication that propose several services for implementation in IoT applications [11]. This LoRa technology can transmit data over a long distance, which is enough for most modern IoT applications [12].

The main objective of this paper is to report on the design (prototype), analysis, and performance testing of a low-cost LoRa transceiver system (433 MHz band) for long-distance communication. Our LoRa transceiver can provide long-range communication without using traditional local and wide area networking technologies. We tested the performance of our proposed LoRa transceiver system in the laboratory as well as using field experiments to quantify the performance gain. Furthermore, we obtained a cloud interface for secure storage, live monitoring, and access to the IoT sensor data from anytime and anywhere across the globe. For the performance testing of the proposed LoRa transceiver, we conducted field experiments at the rooftop and in the laboratory of the Auckland University of Technology Tower building. We also report herein on the data accuracy performance and time efficiency of the proposed LoRa transceiver over 10 km.

Various scientific methods address the issues of data transfer from measurement nodes to destinations using Wi-Fi or ZigBee technologies [13–16]. A gigahertz band was used in the transceiver system to transfer IoT sensor data over 2.5 km, and the Iridium Satellite Constellation was used in another transceiver for successful communications [8]. We adopted the real-time field measurement method while conducting this research work. The research questions/challenges and research contributions are discussed next.

1.1. Research Challenges

In this study, we addressed the following three research questions/challenges.

- Research Question 1: What LoRa infrastructure can be developed to store and retrieve IoT sensor data in the cloud?

To address Research Question 1, we developed a cloud database to transfer IoT sensor data into the database using the Wi-Fi module that we developed and are reporting on in this paper. This Wi-Fi module is integrated into the LoRa system to provide an internet connection to access the cloud services. However, IoT sensor data are stored in the appropriate tables and columns for each sensor node in the database. The system performance is verified by uploading IoT data in the cloud for various climatic and distance conditions.

- Research Question 2: What can be done to monitor real-time IoT sensor data in the system?

To address Research Question 2, we developed a LoRa transceiver system by integrating an LCD unit into the LoRa master receiver node. This allows us to monitor and control real-time IoT data more efficiently. The output data collected from various sensors are verified using an LCD monitor; the data accessibility from the cloud and data synchronization at regular intervals are also verified.

- Research Question 3: What are the main features that affect data loss between the receiver and transmitter end nodes?

To address Research Question 3, herein we identify and discuss the key factors influencing data losses between the receiver and transmitter end nodes, including energy loss, less signal coverage, and low data rates. The proposed LoRa transceiver system provides a secure long-distance communication with low power consumption. For instance, the proposed system consumes about 472 mA of current per year; that is excellent for sustainable communication without any data losses. The signal coverage of the LoRa device is up to 10 km, which is capable of linking city and suburban communities. The data loss in LoRa is less due to its long signal coverage and maximum data rate of 50 kbps.

1.2. Research Contribution

The main contributions of this paper are summarized as follows.

- We designed (prototyped) a secure LoRa transceiver system in the laboratory for linking city and park communities at a distance of about 10 km. To this end, we designed, analyzed, and evaluated a LoRa transceiver system.
- We designed and configured a Wi-Fi module to be used in the system for sending and retrieving IoT data to and from the cloud. We evaluated and validated the system performance through various field experiments, including real-time IoT data monitoring and storage in the cloud.
- A secure private cloud database was developed for storing and retrieving of IoT sensor data. The system performance was validated through real-time IoT data captured through various sensors used in the study.

1.3. Structure of the Article

The rest of this paper is organized as follows. The related works on LoRa transceivers are presented in Section 2. The LoRa transceiver design and analysis are presented in Section 3. The research methodology is discussed in Section 4. The system evaluation and test results are presented in Section 5; the practical implications are also discussed in this section. Finally, the paper is concluded in Section 6. Table 1 lists the abbreviations used in this paper.

Table 1. List of abbreviations used in this paper.

Abbreviation	Definition	Abbreviation	Definition
IoT	Internet of Things	3G	Third Generation Cellular network
IP	Internet Protocol	GPS	Global Positioning System
LCD	Liquid Crystal Display	WSN	Wireless Sensor Network
LED	Light-Emitting Diode	PV	Photovoltaic
LoRa	Long-Range	DHT	Digital Temperature Humidity
LPWAN	Low-Power WAN	PLC	Power Line Communication
LoRaWAN	Long-Range WAN	IDE	Integrated Development Environment
LTE	Long-Term Evolution	GUI	Graphical User Interface
M2M	Machine-to-Machine	AQI	Air Quality Index
MAC	Media Access Control	IAQ	Indoor Air Quality
MIC	Message Integrity Code	LPG	Liquefied Petroleum Gas
NBIoT	Narrow-Band IoT	EnMoS	Environmental Monitoring System

2. Related Work

The advances in IoT aim to provide real-time monitoring and intelligent services. The innovative range communication systems such as 4G, 5G, and GPS technologies have been widely used to control and regulate the real-time environments [17–20]. An outstanding classification of IoT arrangements includes short battery-powered network components such as end nodes, which are equipped with actuators and sensors that wirelessly interact. A common practice implies IoT settings, including the end nodes which collect social knowledge of the surroundings and transmit the data into a gateway, where the data is processed for the end-users [21–26]. The IoT technology ensures that the meaning significantly increases over the production activities within real-time monitoring, including its authority if connected by the wireless intelligent method instead of being generally identified as part of the wireless sensor networks (WSNs) [27].

In WSNs, information can be received by a single node, such as a humidity sensor node. Every sensor node inside the system can gather data about its surroundings to collect data [28]. The limited processing provides a capacity, so the nodes must play an economic determination role to obtain the confidence of its limitations from the overall. Hence, all sensor nodes within a network can possess an independent and sensible system, delivering resolutions via an onboard microcontroller [29]. LoRa signifies the entirety of alternate interaction modules that implement settings to machine-to-machine interfaces that are still external from the presence of networks such as 3G/4G [7,30]. A battery-based optimal communication environment has been similarly suggested as an optimal environment. The communication environment needs the least bat and most minor usage to convey a broadcast through the possibility of holding transmission-provided security [31–35].

More recent attention has focused on providing efficient LoRa communication system by deploying LoRa gateways to transfer the information to the remote devices. However, each sensor needs a single-hop LoRa gateway to ensure simultaneous communication because of the various obstacles between LoRa gateways and IoT sensors. A recent study developed wireless mesh networking for the continuous monitoring of IoT sensors, and their application was successful [4].

Most of the previous studies have focused on LoRa performance in scenarios such as single-hop, multi-hop, or hybrid combinations of wireless communication to establish a simultaneous and robust communication system for the devices.

While numerous research articles have been published in networking literature in recent years, very few researchers have addressed the issues of security, energy efficiency, and low-cost design and implementation of LoRa communication systems. For instance, studies [1,5,27] focused on cloud-based communications without elaborating on security and energy efficiency aspects.

The main objective of our study was to design and analyze a low-cost LoRa transceiver communication system that could effectively provide secure cloud-based services with good energy efficiency. We have conducted a thorough literature review from various credible outlets, including MDPI's *sensors* and *electronics* journals, *Computer Communications* (Elsevier), *IEEE Access*, *IEEE Transactions on Instrumentations and Measurements*, and *IEEE Transactions on Vehicular Technology*. The summary of the related work on long-range transceivers is presented in Table 2. The main research contribution and year of publication are listed in Column 3 and 2, respectively. For each main contribution, we examined the aspects of security (low, moderate, and high), energy efficiency, cloud-based implementation, and system implementation cost. The security level, energy efficiency, cloud-based setup, and low-cost are listed in Column 4–7, respectively.

Table 2. Summary of the related work on long-range transceivers.

Reference	Year	Main Contribution	Security Level	Energy Efficient?	Cloud-Based?	Low-Cost?
[4]	2018	Developed a mesh network framework for monitoring IoT applications.	Moderate	Yes	Yes	No
[28]	2018	Designed a LoRa transceiver system with improved characteristics.	Low	Yes	No	No
[36]	2018	Focused on an automatic key generation for long-range wide area communication.	High	Yes	No	No
[31]	2018	Validated LoRa experimentation using commercial devices and software-defined radios.	Low	No	No	No
[37]	2018	Developed a narrowband interference suppression technique to improve BER.	Moderate	Yes	No	No
[22]	2018	Developed an energy-efficient network architecture for IoT.	Low	No	Yes	Yes

Table 2. *Cont.*

Reference	Year	Main Contribution	Security Level	Energy Efficient?	Cloud-Based?	Low-Cost?
[38]	2018	Developed a scheme for overcoming the bandwidth limitation of LoRa.	High	Yes	No	No
[39]	2017	Developed a multi-hop network based on low-power wide area technology.	Low	No	No	No
[12]	2017	Proposed an architecture of IoT LoRa wireless radio-based information display system.	Moderate	Yes	No	Yes
[1]	2022	LoRa communication system for the development and implementation of a smart multi-sensor system for monitoring air quality remotely.	Low	Yes	Yes	No
[5]	2022	LoRaWAN technique to provide a low-power livestock localization and monitoring system	Moderate	No	Yes	Yes
Our work		A low-cost secure system for monitoring and storing of IoT data in the cloud that can be used to link city and suburban park communities effectively.	High	Yes	Yes	Yes

3. Methods: System Design and Analysis

The block diagram of the proposed LoRa transceiver system is shown in Figure 1. We designed and built (prototype) the interface unit, which includes the LoRa (Ra-02) communication module and a Wi-Fi Module. The hardware of the LoRa transceiver mainly consists of the ATMega328P Microcontroller, Nokia 5110 graphic LCD screen (Arduino Compatible), and Arduino Nano board. The sensors used in this study are the DS18B20 fire sensor, DHT11 temperature-humidity sensor, MQ-2 gas sensor, LM393 vibration sensor, and capacitive soil moisture sensor. The data from the IoT sensor nodes are transferred into the cloud using the Wi-Fi module in the transceiver system. The data are on an LCD display positioned beside the server and the gateway.

Figure 2 shows the LoRa transceiver unit, consisting of the master receiver node (Ra-02), an LCD display, and a Wi-Fi module. The Ra-02 LoRa is shown in the top left corner. The system was designed and built using two LoRa-enabled transmitter slave nodes and a master receiver node. The two transmitters (nodes) are connected to various IoT sensors that are suitable for indoor and outdoor conditions. The Ra-02 LoRa module is used for communications between transceiver nodes and the outside world. For instance, the Ra-02 receives IoT sensor data remotely. The data are pushed into the cloud using the Wi-Fi module. In addition, the receiver node is connected to an LCD screen for monitoring the IoT data.

The proposed system was built around a LoRa transceiver unit containing the ATMega328P Microcontroller, LCD screen, and an Arduino Nano board (Nokia, Japan). The sensors used in the device are the DS18B20 fire sensor, DHT11 temperature-humidity sensor, MQ-2 gas sensor, LM393 vibration sensor, and capacitive soil moisture sensor (Texas Instruments, USA). The data collected from the IoT sensor nodes are transferred into the cloud using a Wi-Fi module built on the system.

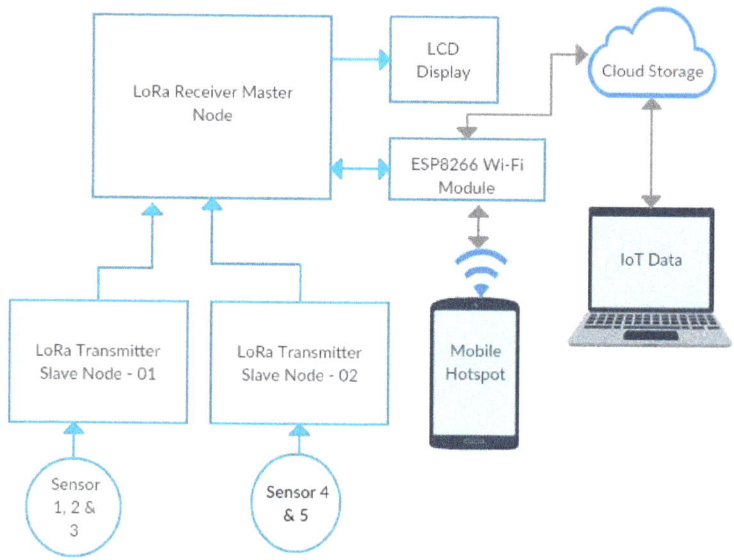

Figure 1. Block diagram of the proposed LoRa transceiver system.

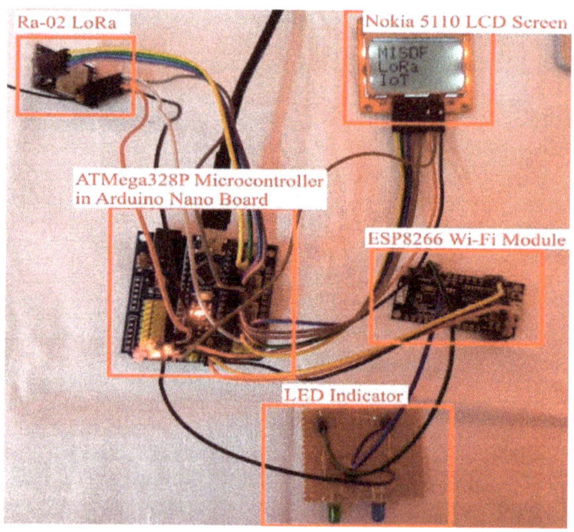

Figure 2. LoRa transceiver (prototype) with LCD screen and Wi-Fi module.

3.1. LoRa Parameters

The main LoRa parameters include the spreading factor, bandwidth, and code rate, which can be arranged by the data rate specifications, sensitivity, and communication range [23]. The LoRa signal strength ranges from −4 to 20 dBm, and the transmitted carrier frequency varies from 137 to 1020 MHz. In the proposed LoRa system, we use 433 MHz bands for transmissions. The LoRa manages the bandwidth of 125, 250, or 500 kHz. The high bandwidth allows for a higher data rate through digital signal processing. The high coding rate extends the presence in the air and increases the messages. The spreading factor (SF) allows several bits to be encoded at the respective symbol, ranging from 6 or 12. The increased SF raises the limit, increased SNR, and higher power consumption.

3.2. LoRa Master Receiver Node

We designed the master receiver node (Ra-02 LoRa) on a nano board controlled by the ATMega328P Microcontroller. An LCD screen is attached to the system for displaying the live sensor data. Moreover, the structure of the LoRa master receiver node is accomplished by including a Wi-Fi module to the board. This Wi-Fi module is a low-power unified Wi-Fi solution with a speed of up to 8 Mbps. It enables the receiver node to interface with the Internet and exchange the sensor data into the cloud.

The master receiver node is one of the main components of this LoRa transceiver system functioning as a beneficiary for receiving all sensor information from the transmitter nodes. The incoming sensor data are received in this master node with the assistance of the Ra-02 LoRa module. The Ra-02 module is an exceptionally viable device for receiving data from long distances without any Internet access. The data received in this node are managed by the microcontroller, and it displays the sensor data in the LCD screen appended to the board. The AVR architecture of the microcontroller has memory spaces, including data and program memory, to store and retrieve adequate information. The Wi-Fi module allows the master node to link to the outside world through the Internet; the Internet access assists the node in transferring the sensor data to cloud storage. The master node updates sensor data in the cloud storage within seconds by utilizing the accelerated Wi-Fi module.

The master receiver node and transmitter nodes are coded in the programming language C++. The coding methodology begins with the master receiver node. Two sets of instructions have been given to this node, one for the receiver and the other for the Wi-Fi module. After connecting the master node to the system, the board type, processor, and port must select the Arduino 1.8.9 IDE for transferring the code to board. The master node is coded as a receiver that collects the sensor information from all of the transmitter nodes. The Arduino software can be downloaded at https://www.arduino.cc/en/Main/Software (accessed on 30 October 2022). The software is installed in a workstation, and each node is connected to that system for uploading the instructions by means of a USB cable [29]. We verified the programming code before loading it to the master receiver node. After successfully programming the master node, only the power supply is required to run this node.

3.3. LoRa Transmitter Node 1

Figure 3 shows the LoRa transmitter slave node–01. The design and operation of the transmitter nodes of the LoRa Transceiver system are entirely different from the master receiver node. The primary function of the transmitter nodes is to collect the sensor data and transmit the information to the master node. The LoRa transmitter node–01 is structured by interfacing the hardware components of the microcontroller, Ra-02 LoRa Module, MQ-2 gas sensor, DS18B20 fire sensor, and LM393 vibration sensor on an Arduino Nano board. The association between the microcontroller and the Ra-02 Lora module is similar to the master node, yet the working of the device is distinctive. In this node, the Ra-02 LoRa module acts as a transmitter that only sends the sensor information to the master node.

There are three sensors linked with this node, and each sensor performs its own operation. The performance of the LoRa transceiver system must be analyzed both indoor and outdoor conditions; therefore, three sensors have been deployed for internal monitoring. This node is intended for detecting fire, gas, and vibrations inside the building. The fire sensor utilized for this node is the DS18B20 digital thermometer, which gives 9-bit–12-bit Celsius temperature estimations and performs well with any sort of microcontroller. It can withstand a temperature range between −55 and +125 °C and also has a user-programmable alarm system. The MQ-2 gas sensor is the most competent sensor for distinguishing any sort of gas spillage. The quick response time and high sensitivity of this equipment is appropriate for detecting alcohol, smoke, and LPG. An onboard potentiometer in the LM393 vibration sensor can be adjustable by the user-defined threshold level of the device to identify the vibrations. These extraordinarily intelligent sensors attached to the transmitter node–01 provide accurate sensor data, and the information will be rapidly processed and instantly transmitted to the master node with the backing of Ra-02 LoRa module.

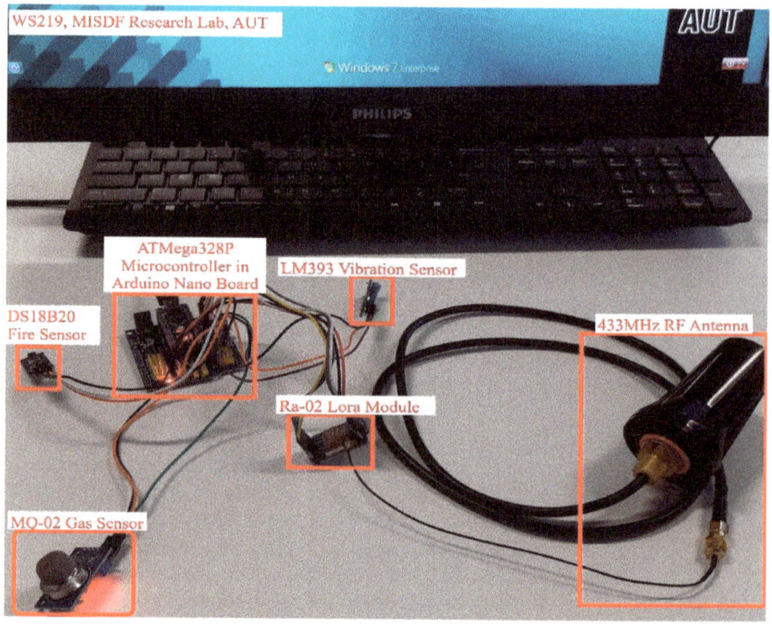

Figure 3. Illustrating the LoRa transmitter node–01 (prototype).

3.4. *LoRa Transmitter Node 2*

Figure 4 shows the deployment of the LoRa transmitter node–02 at the rooftop of the University building. An additional transmitter node–02 was designed to assess the productivity of the transceiver system in open-air conditions. The structure of the transmitter node–02 is practically the same as transmitter node–01, and the progressions made for this node is the utilization of different sensors. In transmitter node–02, the sensors that are associated with the microcontroller are the DHT11 temperature-humidity sensor and capacitive soil moisture sensor. The considerable achievement of the temperature-humidity sensor is that it can refresh the reading every 2 s. This low-cost, efficient sensor conveys a digital signal on the data pin. The capacitive soil moisture sensor provides information on the volumetric water content in the soil and is compatible with any type of microcontroller.

The sensors fixed in this node are reasonable for investigating the performance of the LoRa transceiver system in outdoor conditions. The functioning of this node is similar to that of the transmitter node–01. The sensors affixed to this node collect the digital data of the outdoor conditions such as moisture, temperature, and humidity. At that point, it exchanges the information to the master receiver node at a long distance by means of the Ra-02 LoRa module.

The performance of the LoRa transmitter node–02 has been tested both in the lab and in outdoor environments. The maximum range accomplished by this node was 10 km, and the sensor data was updated within 1 s. The DHT11 temperature-humidity sensor and capacitive soil moisture sensor are fixed in this node to evaluate the performance of the LoRa transceiver system. The 433 MHz water-resistant RF antenna relates to this node to transmit signals in rainy conditions. We tested the performance of this node using sensor data collected.

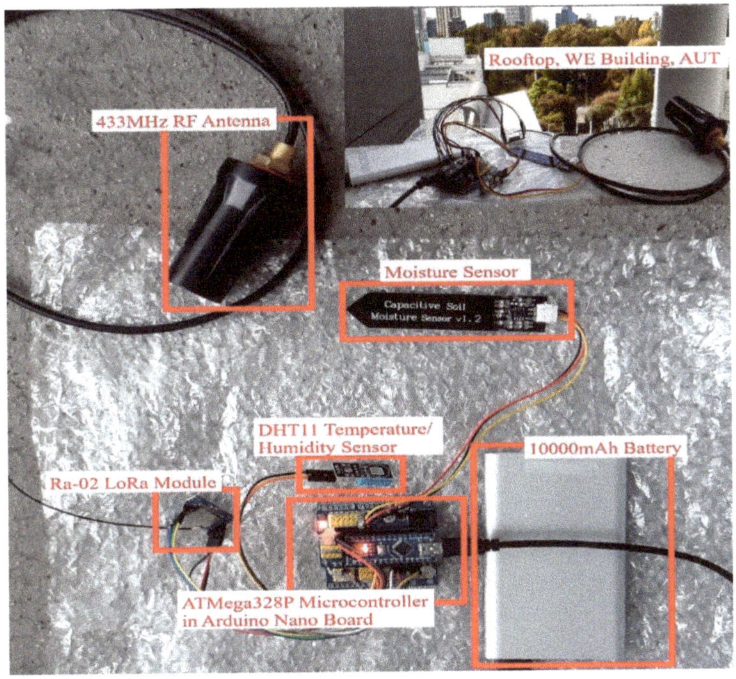

Figure 4. Deployment of LoRa transmitter node–02 at the rooftop of the AUT building.

3.5. Programming the Transmitter Node

The transmitter node–01 is programmed to act as the transmitter that transmits the sensor data to the master node. The significant sections and functions that are utilized in the transmitter node–01 coding include all the required libraries at first. At that point, the sensors connected in the node are characterized by a pinout setup, and the frequency of the LoRa transceiver system of 433 MHz is also defined in the code. The transmitter node–02 is programmed similar to that of node-01. The focus is to characterize the sensors (e.g., Temperature, Humidity, Moisture) in the code.

3.6. Wi-Fi Module Configuration

The ESP8266-based Wi-Fi module facilitates an Internet connection to the master receiver node for transmitting the sensor data into the cloud for storage. We develop and configure a low-cost Wi-Fi module to be used with a microcontroller with an integrated TCP/IP protocol stack. The multifunctional Wi-Fi module can act as an access point (AP) to form a Wi-Fi hotspot. The system is programmed using C++ and configured for optimum performance.

4. Methodology

The master receiver node is one of the main components of the proposed LoRa transceiver system. It operates on a frequency of 433 MHz. The input power supply to this node is 7–12 V, and the power consumption is 9 mA. In the experiments, two transmitter nodes were used to evaluate the performance of the LoRa transceiver at various locations. We first evaluated the performance of LoRa transmitter slave node–01 in the laboratory at Auckland University of Technology (AUT). We also evaluated the performance of the LoRa master receiver node in the laboratory. Next, we then tested the performance of LoRa transmitter node–02 in the outdoor environment to distinguish temperature, humidity, and moisture.

Finally, the LoRa transceiver system was tested for distance coverage by placing the master node on the rooftop of AUT's Tower building. This site was selected as a primary location suitable for testing the node against wind, rain, temperature, and moisture. The secondary location was Liston Park, which is about 10 km away from the primary location (AUT). A power bank (10,000 mAh) was used to operate the LoRa system during the field experiments.

5. Results and Discussion

The results from the field trials show that the proposed LoRa system can be used for a network coverage of about 10km. The spreading factor holds some notable influence on the network coverage and data transmission rate. The system serves the purpose of achieving low-power consumption over long-distance coverage. The system also displays accuracy through LCD screen alterations from specific demands that can be tracked in the future. It is an IoT and LoRa wireless module that promotes consecutive secure monitoring applications. We verified and analyzed the output data collected from the various sensors using an LCD monitor. The data accessibility from the cloud and data synchronization at regular intervals were also verified.

5.1. Real-Time IoT Data Monitoring and Storage

The results from the field trial measurements were obtained in two ways. First, we monitored real-time IoT sensor data by connecting the LoRa master receiver node to an LCD screen. Second, we accessed the IoT sensor data in the cloud. For system performance testing, a database was created in the cloud, and IoT sensor data were transferred into the database using the ESP8266 Wi-Fi module. The verification process involved storing IoT sensor data (received from the transmitter nodes) in the cloud database. This sensor data was stored in the appropriate tables and columns for each sensor node, ensuring that the cloud storage was functioning. We also tested the system performance in various climatic and distance conditions. The IoT sensor data was monitored by an LCD unit at the receiver node. Finally, we successfully uploaded IoT sensor data into the cloud.

5.2. Data Access in the Cloud

The IoT sensor data is automatically updated in the cloud that verifies the storage. The test results proved that the sensor data can easily be accessible from cloud storage using an Internet-enabled device such as a computer, laptop, or smartphone. The LoRa node1 displays the information from the transmitter node–01 that consists of fire, vibration, and gas sensor data. Another link for LoRa node2 shows the information of the transmitter node–02, which contains the sensor information of temperature, humidity, and moisture. An Internet connection is required to access the sensor data of both nodes from anywhere in the world. The monitoring and controlling of the LoRa-enabled sensors from any remote location is a dynamic achievement of this experimental work.

5.3. Efficiency and Data Accuracy

Figure 5 shows the laboratory test results for the DS18B20 Fire Sensor. More than 1000 data sets were collected from every sensor node to evaluate the time efficiency and data accuracy of the transceiver system. The fire sensor in transmitter node–01 delivers the temperature (measured in Celsius). If the temperature inside the building abnormally rises, a warning sound is activated in the sensor to notify the incident. The collected data indicate that this sensor provides the ordinary room temperature of 20 to 25 °C. For testing the accuracy and response of the sensor, a manual flame was placed near the sensor, and the temperature level was recorded at about 55 °C.

Figure 6 shows the data accumulated from the vibration sensor (LM393). The LM393 vibration sensor in transmitter node1 is utilized to recognize the vibrations over a threshold point. For evaluating the performance and accuracy of the vibration sensor, 1025 stored sensor data sets were extracted from the cloud. The sensor counted 976 times as 'No Vibration', and the vibrations happened 42 times at a specific time. An error reading was also recorded seven times. Hence, the accuracy rate of this sensor in transmitter node1 was determined as 99.31%, and the error rate was 0.68%.

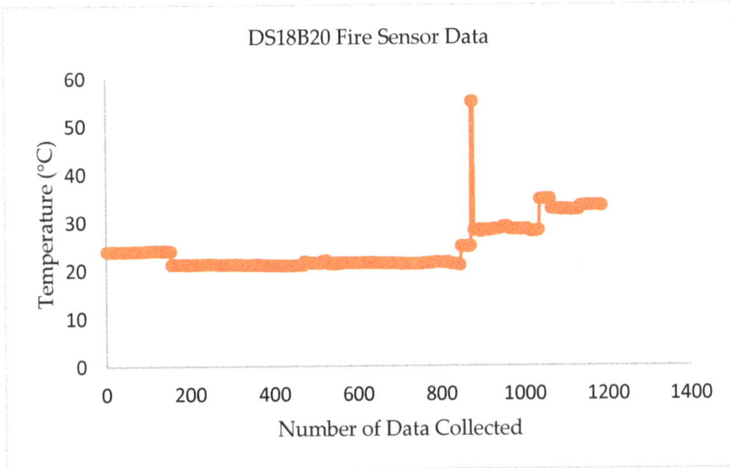

Figure 5. Test results for DS18B20 fire sensor.

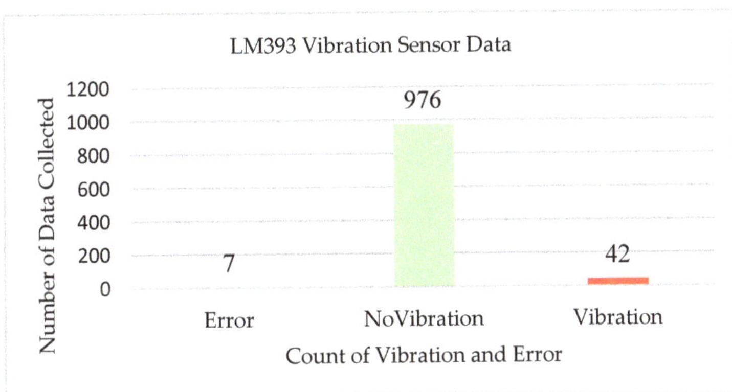

Figure 6. Test results for LM393 vibration sensor.

Figure 7 shows the test results for the gas sensor. To analyze the system efficiency, we recorded 1025 sensor data sets using the MQ-2 gas sensor. We manually inserted smoke and gases into the system and performed various tests for system accuracy. We observed that for 1025 sensor data sets, the system identified 19 instances of abnormal gases and 11 cases of no gases. The system accuracy was found to be 99% with an error rate of 1%.

Figure 8 shows the test results for the DHT11 temperature sensor. We deployed temperature sensors in Liston Park (Auckland), which is about 10 km away from Auckland City. The transmitter node was installed at the rooftop of a seven-story building. The first 500 data points distinguished the temperature range from 20 to 25 °C at the Liston Park, and the remaining data at a temperature ranging from 25 to 33 °C at the rooftop of the AUT building. Hence, the temperature sensor in the LoRa transceiver system could successfully transmit IoT sensor data accurately up to a distance of 10 km.

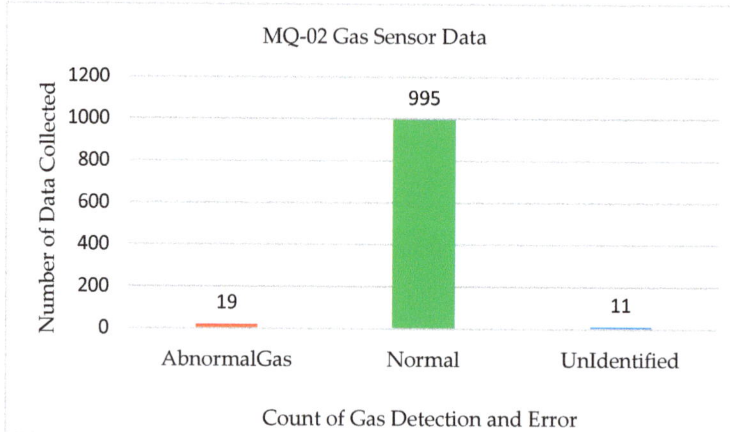

Figure 7. Test results for the MQ-2 gas sensor.

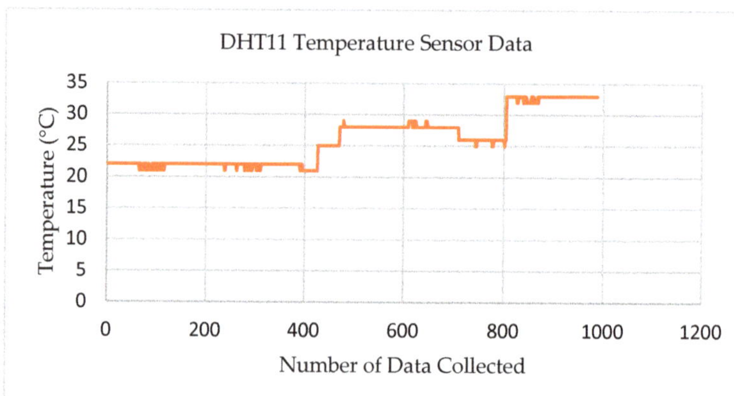

Figure 8. Test results for the DHT11 temperature sensor.

Figure 9 shows the test results for the DHT11 humidity sensor. The humidity sensor is also part of the temperature sensor in transmitter node–02, but this sensor delivers the atmospheric humidity level. The sensor measures the relative humidity (RH) in percentage by calculating the amount of water content present in the atmosphere. The data were collected by placing the transmitter node–02 in various locations. The sensor reading showed that the dissimilarity in the values according to the location change.

At first, the data were gathered from the sensor by locating the node in Liston Park, Ellerslie, Auckland. The humidity level was recorded as 70–80% because of the environmental condition in the park. Later, the node was placed at the rooftop of the AUT building, where the humidity level was 40–60%. The sensor accurately transferred sensor data from both locations.

Figure 10 shows the data collected from the soil moisture sensor. The capacitive soil moisture sensor is another sensor included in the transmitter node–02. This sensor measures the volumetric water content in the soil.

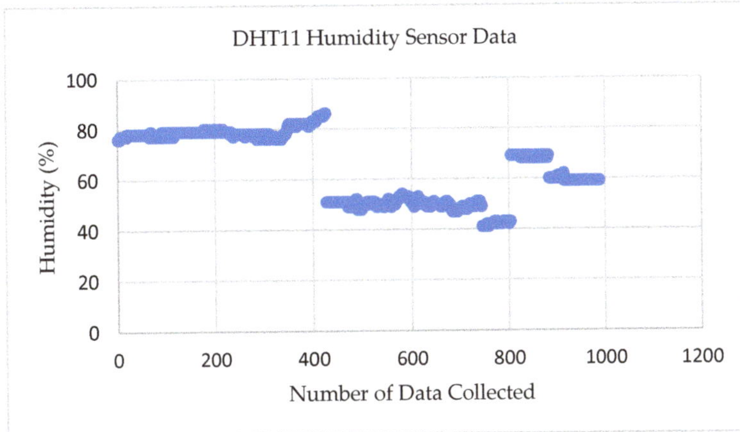

Figure 9. Test results for the DHT11 humidity Sensor.

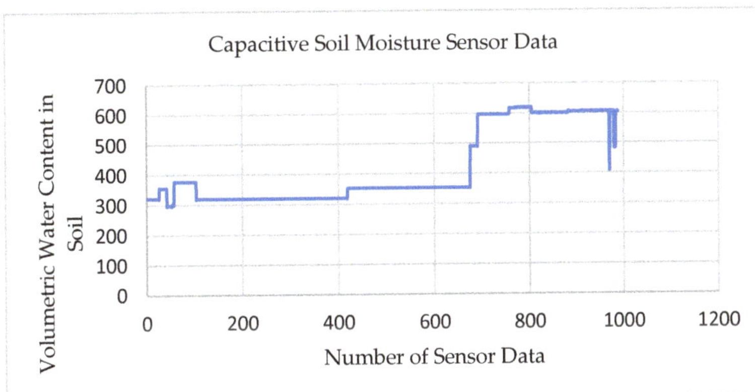

Figure 10. Test results for the soil moisture sensor.

The moisture level in the soil is categorized into three levels to quickly distinguish the condition of the soil. A moisture level of 260–350 indicates high water content in the soil, and 350–430 indicates the normal wet condition. However, moisture level 430–520 shows that the soil condition is dry. The initial 600 sensor data was accumulated from the Liston Park, Auckland, and the moisture level in the soil varied from 300 to 400. Then, the transmitter node–02 was placed in the rooftop of the AUT building and obtained a moisture level of 500, indicating no moisture at all. We observed a variation in the results, which indicated that the transmitter nodes were accurately functioning at a distance of up to 10 km in diverse meteorological conditions. Therefore, the sensors attached in transmitter node–02 can be used to effectively transmit data to the master node over 10 km.

5.4. Transmitter Efficiency Test Results

The time efficiency of the transmitter node was scrutinized by giving manual inputs to the sensors. The response time of the sensor was observed to determine the efficiency of the transmitter node. Table 3 shows the time efficiency of the accumulated sensor data variations with date and time. This field experiment was conducted in the laboratory at AUT.

Table 3. Time efficiency of the vibration sensor in transmitter node–01.

Fire	Vibration	Gas	Date	Time
28.44	No Vibration	Normal	11 April 2019	1:08:59 a.m.
28.37	Vibration	Normal	11 April 2019	1:08:58 a.m.
28.37	Vibration	Normal	11 April 2019	1:08:50 a.m.
28.37	Vibration	Normal	11 April 2019	1:08:49 a.m.
28.37	Vibration	Normal	11 April 2019	1:08:48 a.m.
28.37	Vibration	Normal	11 April 2019	1:08:43 a.m.
28.37	No Vibration	Normal	11 April 2019	1:08:17 a.m.
28.44	No Vibration	Normal	11 April 2019	1:06:27 a.m.
28.5	Vibration	Normal	11 April 2019	1:06:15 a.m.
28.5	Vibration	Normal	11 April 2019	1:06:14 a.m.
28.5	Vibration	Normal	11 April 2019	1:06:13 a.m.
28.5	Vibration	Normal	11 April 2019	1:06:11 a.m.
28.5	Vibration	Normal	11 April 2019	1:06:10 a.m.

We observed the efficiency of the vibration sensor at recognizing the vibrations and idle state of the sensor and transmitting the data to the master node within seconds. The results indicated a variation in the sensor data being updated to the master node in the range of 1–20 s.

Table 4 shows the time efficiency of the node to report variations in the gas sensor. The MQ-2 gas sensor data were accumulated over several days at various locations to determine the efficiency of the node. The sensor data accurately updated in the cloud within 10–20 s. Both transmitter nodes were examined in dissimilar conditions for evaluating the efficiency, accuracy, and speed of the data transfer. The results show that the IoT sensor data were successfully transferred from the transmitter to the receiver (10 km) within a few seconds. In summary, our proposed LoRa transreciver system provides accurate sensor data in linking city and suburban communities. The cloud deployment in the transceiver system allows us to access the sensor data anytime and anywhere.

Table 4. Time efficiency of the gas sensor in transmitter node-01.

Fire	Vibration	Gas	Date	Time
21.12	NoVibration	Normal	27 April 2019	11:24:14 a.m.
21.06	NoVibration	AbnormalGas	27 April 2019	11:23:58 a.m.
21.06	NoVibration	AbnormalGas	27 April 2019	11:23:33 a.m.
21.06	NoVibration	AbnormalGas	27 April 2019	11:23:22 a.m.
24.94	NoVibration	Normal	23 April 2019	4:00:53 a.m.
28.37	NoVibration	Normal	11 April 2019	12:56:42 p.m.
28.37	NoVibration	AbnormalGas	11 April 2019	12:56:32 p.m.
28.31	NoVibration	Normal	11 April 2019	12:56:17 p.m.

5.5. Practical Implications

The results presented in Section 5 provide some insights into the practical implementation aspect of the LoRa transceiver System. This research provides a solution to the problems that smart homes and cities have in connecting to 250 sensor nodes through a single gateway. This research also provides a clear perception of designing a secure low-cost LoRa transceiver system. The field experiments prove that this system can be used to link city and suburban communities covering about 10 km at no cost (no need to go through service providers). The spreading factor holds some notable influence on the network coverage and data transmission rate. The system serves the purpose of achieving a low-power consumption over a long-distance coverage. The system also displays real-time IoT data through an LCD screen. In this research, we developed, configured, and tested the LoRa transceiver, Wi-Fi module, and LCD display unit in the laboratory, and found the setup to be robust. This research can be taken into the next step for commercialization as well as production.

6. Concluding Remarks

A LoRa transceiver system with two transmitter nodes has been designed and built in the laboratory at a cost of about USD 150, which can be used for the monitoring and storing of IoT sensor data in the cloud. We also provided a solution for sending and retrieving IoT data to and from the cloud by designing and configuring a Wi-Fi module. The system performance was tested (both indoor and outdoor conditions) using field data and was found to be robust. The results obtained have shown that the transmitter nodes perform well for up to 10 km, and the receiving data accuracy is found to be 99%. The reaction time of the transmitter nodes is determined by the sensor data accumulation as within 2–20 s. The system is noticeably time-efficient and provides accurate sensor data effectively.

The future expansion of our LoRa transceiver system is also possible through the implementation of an image sensor in the transmitter node to transmit photographs of the incident or even provide a livestream of the climate conditions. Designing a robust software module to facilitate the detection and retransmission of dropped packets is also suggested as a future work.

Author Contributions: Conceptualization, A.T.K. and N.I.S.; methodology, A.T.K. and N.I.S.; software, N.I.S.; validation, N.I.S. and A.T.K.; formal analysis, N.I.S. and A.T.K.; investigation, A.T.K. and N.I.S.; resources, N.I.S.; data curation, N.I.S. and A.T.K.; writing—original draft preparation, A.T.K.; writing—review and editing, N.I.S. and M.J.A.; visualization, N.I.S. and A.T.K.; supervision, N.I.S.; project administration, N.I.S. All authors have read and agreed to the published version of the manuscript.

Funding: This research received no external funding.

Institutional Review Board Statement: Not applicable.

Informed Consent Statement: Not applicable.

Data Availability Statement: Not applicable.

Conflicts of Interest: The authors declare no conflict of interest.

References

1. Camarillo-Escobedo, R.; Flores, J.L.; Marin-Montoya, P.; García-Torales, G.; Camarillo-Escobedo, J.M. Smart Multi-Sensor System for Remote Air Quality Monitoring Using Unmanned Aerial Vehicle and LoRaWAN. *Sensors* **2022**, *22*, 1706. [CrossRef]
2. Lentz, J.; Hill, S.; Schott, B.; Bal, M.; Abrishambaf, R. Industrial Monitoring and Troubleshooting based on LoRa Communication Technology. In Proceedings of the IECON 2018—44th Annual Conference of the IEEE Industrial Electronics Society, Washington, DC, USA, 21–23 October 2018; pp. 3852–3857. [CrossRef]
3. Devalal, S.; Karthikeyan, A. LoRa Technology—An Overview. In Proceedings of the 2018 Second International Conference on Electronics, Communication and Aerospace Technology (ICECA), Coimbatore, India, 29–31 March 2018; pp. 284–290. [CrossRef]
4. Lee, H.-C.; Ke, K.-H. Monitoring of Large-Area IoT Sensors Using a LoRa Wireless Mesh Network System: Design and Evaluation. *IEEE Trans. Instrum. Meas.* **2018**, *67*, 2177–2187. [CrossRef]
5. Ojo, M.O.; Viola, I.; Baratta, M.; Giordano, S. Practical experiences of a smart livestock location monitoring system leveraging gnss, lorawan and cloud services. *Sensors* **2022**, *22*, 273. [CrossRef]
6. Fujdiak, R.; Mikhaylov, K.; Pospisil, J.; Povalac, A. Insights into the Issue of Deploying a Private LoRaWAN. *Sensors* **2022**, *22*, 2042. [CrossRef]
7. Saavedra, E.; Mascaraque, L.; Calderon, G.; Campo, G. A Universal Testbed for IoT Wireless Technologies: Abstracting Latency, Error Rate and Stability from the IoT Protocol and Hardware Platform. *Sensors* **2022**, *22*, 4159. [CrossRef]
8. Sinha, R.S.; Wei, Y.; Hwang, S.-H. A survey on LPWA technology: LoRa and NB-IoT. *ICT Express* **2017**, *3*, 14–21. [CrossRef]
9. Sanchez-Iborra, R.; Liaño, I.G.; Simoes, C.; Couñago, E.; Skarmeta, A.F. Tracking and Monitoring System Based on LoRa Technology for Lightweight Boats. *Electronics* **2018**, *8*, 15. [CrossRef]
10. Leonardi, L.; Battaglia, F.; Patti, G.; Bello, L.L. Industrial LoRa: A Novel Medium Access Strategy for LoRa in Industry 4.0 Applications. In Proceedings of the IECON 2018—44th Annual Conference of the IEEE Industrial Electronics Society, Washington, DC, USA, 21–23 October 2018; pp. 4141–4146. [CrossRef]
11. Ayele, E.D.; Das, K.; Meratnia, N.; Havinga, P.J.M. Leveraging BLE and LoRa in IoT Network for Wildlife Monitoring System (WMS). In Proceedings of the 2018 IEEE 4th World Forum on Internet of Things (WF-IoT), Singapore, 5–8 February 2018; pp. 342–348. [CrossRef]
12. Reda, H.T.; Daely, P.T.; Kharel, J.; Shin, S.Y. On the Application of IoT: Meteorological Information Display System Based on LoRa Wireless Communication. *IETE Technol. Rev.* **2018**, *35*, 256–265. [CrossRef]

13. Rahman, M.A.; Asyhari, A.T.; Kurniawan, I.F.; Ali, M.J.; Rahman, M.M.; Karim, M. A scalable hybrid MAC strategy for traffic-differentiated IoT-enabled intra-vehicular networks. *Comput. Commun.* **2020**, *157*, 320–328. [CrossRef]
14. Rahman, M.A.M.A.M.A.; Kabir, M.N.M.N.; Azad, S.; Ali, J. On mitigating hop-to-hop congestion problem in IoT enabled Intra-Vehicular communication. In Proceedings of the 2015 4th International Conference on Software Engineering and Computer Systems, ICSECS 2015: Virtuous Software Solutions for Big Data, Kuantan, Malaysia, 19–21 August 2015; pp. 213–217. [CrossRef]
15. Rahman, M.A.M.A.; Ali, J.; Kabir, M.N.M.N.; Azad, S. A performance investigation on IoT enabled intra-vehicular wireless sensor networks. *Int. J. Automot. Mech. Eng.* **2017**, *14*, 3970–3984. [CrossRef]
16. Ali, A.I.; Partal, S.Z. Development and performance analysis of a ZigBee and LoRa-based smart building sensor network. *Front. Energy Res.* **2022**, *10*, e933743. [CrossRef]
17. Hejazi, H.; Rajab, H.; Cinkler, T.; Lengyel, L. Survey of platforms for massive IoT. In Proceedings of the 2018 IEEE International Conference on Future IoT Technologies (Future IoT), Eger, Hungary, 18–19 January 2018; pp. 1–8. [CrossRef]
18. Khutsoane, O.; Isong, B.; Abu-Mahfouz, A.M. IoT devices and applications based on LoRa/LoRaWAN. In Proceedings of the IECON 2017—43rd Annual Conference of the IEEE Industrial Electronics Society, Beijing, China, 29 October–1 November 2017; pp. 6107–6112. [CrossRef]
19. Leonardi, L.; Bello, L.L.; Patti, G. LoRa support for long-range real-time inter-cluster communications over Bluetooth Low Energy industrial networks. *Comput. Commun.* **2022**, *192*, 57–65. [CrossRef]
20. Shafiq, M.; Gu, Z.; Cheikhrouhou, O.; Alhakami, W.; Hamam, H. The Rise of 'internet of Things': Review and Open Research Issues Related to Detection and Prevention of IoT-Based Security Attacks. *Wirel. Commun. Mob. Comput.* **2022**, *2022*, 8669348. [CrossRef]
21. Wang, S.-Y.; Chen, Y.-R.; Chen, T.-Y.; Chang, C.-H.; Cheng, Y.-H.; Hsu, C.-C.; Lin, Y.-B. Performance of LoRa-Based IoT Applications on Campus. In Proceedings of the 2017 IEEE 86th Vehicular Technology Conference (VTC-Fall), Toronto, ON, Canada, 24–27 September 2017; pp. 1–6. [CrossRef]
22. Piyare, R.; Murphy, A.L.; Magno, M.; Benini, L. On-Demand LoRa: Asynchronous TDMA for Energy Efficient and Low Latency Communication in IoT. *Sensors* **2018**, *18*, 3718. [CrossRef]
23. Bardyn, J.-P.; Melly, T.; Seller, O.; Sornin, N. IoT: The Era of LPWAN is starting now. In Proceedings of the ESSCIRC Conference 2016: 42nd European Solid-State Circuits Conference, Lausanne, Switzerland, 12–15 September 2016; pp. 25–30. [CrossRef]
24. Wu, F.; Redouté, J.-M.; Yuce, M.R. WE-Safe: A Self-Powered Wearable IoT Sensor Network for Safety Applications Based on LoRa. *Wearable Implant. Devices Syst. IEEE Access* **2018**, *6*, 40846–40853. [CrossRef]
25. Kassanuk, T.; Mustafa, M.; Phasinam, K.; Santosh, T. An Internet of Things and Cloud Based Smart Irrigation System. *Ann. Rom. Soc. Cell Biol.* **2021**, *25*, 1583–6258. Available online: http://annalsofrscb.ro (accessed on 26 October 2022).
26. Mustafa, M.; Abbas, A.; Bsoul, Q.; Shabbir, A. Smart Irrigation System Based on the Internet of Things and the Cloud. *Int. J. Mod. Trends Sci. Technol.* **2021**, *7*, 19–24. [CrossRef]
27. Bouguera, T.; Diouris, J.-F.; Chaillout, J.-J.; Jaouadi, R.; Andrieux, G. Energy Consumption Model for Sensor Nodes Based on LoRa and LoRaWAN. *Sensors* **2018**, *18*, 2104. [CrossRef]
28. Jovalekic, N.; Drndarevic, V.; Darby, I.; Zennaro, M.; Pietrosemoli, E.; Ricciato, F. LoRa Transceiver With Improved Characteristics. *IEEE Wirel. Commun. Lett.* **2018**, *7*, 1058–1061. [CrossRef]
29. Seneviratne, P. *Beginning LoRa Radio Networks with Arduino: Build Long Range, Low Power Wireless IoT Networks*; Apress: Berkeley, CA, USA; New York, NY, USA, 2019. [CrossRef]
30. Sarkar, N.I.; Ho, P.H.; Gul, S.; Zabir, S.M.S. TCP-LoRaD: A Loss Recovery and Differentiation Algorithm for Improving TCP Performance over MANETs in Noisy Channels. *Electronics* **2022**, *11*, 1479. [CrossRef]
31. Croce, D.; Gucciardo, M.; Mangione, S.; Santaromita, G. Impact of LoRa Imperfect Orthogonality: Analysis of Link-Level Performance. *IEEE Commun. Lett.* **2018**, *22*, 796–799. [CrossRef]
32. S⊘ndrol, T.; Jalaian, B.; Suri, N. Investigating LoRa for the Internet of Battlefield Things: A Cyber Perspective. In Proceedings of the MILCOM 2018—2018 IEEE Military Communications Conference (MILCOM), Los Angeles, CA, USA, 29–31 October 2018; pp. 749–756. [CrossRef]
33. Mustafa, M.; Alshare, M.; Bhargava, D.; Neware, R.; Singh, B.; Ngulube, P. Perceived Security Risk Based on Moderating Factors for Blockchain Technology Applications in Cloud Storage to Achieve Secure Healthcare Systems. *Comput. Math. Methods Med.* **2022**, *2022*, 6112815. [CrossRef]
34. Kollu, P.K.; Saxena, M.; Phasinam, K.; Kassanuk, T.; Jawarneh, M. Blockchain Techniques for Secure Storage of Data in Cloud Environment. *Turk. J. Comput. Math. Educ.* **2021**, *12*, 1515–1522.
35. Kumar, N.; Madhuri, J.; Channegowda, M. Review on security and privacy concerns in internet of things. In Proceedings of the 2017 International Conference on IoT and Application (ICIOT), Nagapattinam, India, 19–20 May 2017. [CrossRef]
36. Zhang, J.; Marshall, A.; Hanzo, L. Channel-Envelope Differencing Eliminates Secret Key Correlation: LoRa-Based Key Generation in Low Power Wide Area Networks. *IEEE Trans. Veh. Technol.* **2018**, *67*, 12462–12466. [CrossRef]
37. Elshabrawy, T.; Robert, J. The Impact of ISM Interference on LoRa BER Performance. In Proceedings of the 2018 IEEE Global Conference on Internet of Things (GCIoT), Alexandria, Egypt, 5–7 December 2018; pp. 1–5. [CrossRef]
38. Jebril, A.H.; Sali, A.; Ismail, A.; Rasid, M.F.A. Overcoming Limitations of LoRa Physical Layer in Image Transmission. *Sensors* **2018**, *18*, 3257. [CrossRef] [PubMed]
39. Liao, C.-H.; Zhu, G.; Kuwabara, D.; Suzuki, M.; Morikawa, H. Multi-Hop LoRa Networks Enabled by Concurrent Transmission. *IEEE Access* **2017**, *5*, 21430–21446. [CrossRef]

sensors

MDPI

Article

Leveraging IoT-Aware Technologies and AI Techniques for Real-Time Critical Healthcare Applications

Angela-Tafadzwa Shumba [1,2], Teodoro Montanaro [1], Ilaria Sergi [1], Luca Fachechi [2], Massimo De Vittorio [1,2] and Luigi Patrono [1,*]

1 Department of Engineering for Innovation, University of Salento, 73100 Lecce, Italy
2 Istituto Italiano di Tecnologia, Center for Biomolecular Nanotechnologies, Arnesano, 73010 Lecce, Italy
* Correspondence: luigi.patrono@unisalento.it

Abstract: Personalised healthcare has seen significant improvements due to the introduction of health monitoring technologies that allow wearable devices to unintrusively monitor physiological parameters such as heart health, blood pressure, sleep patterns, and blood glucose levels, among others. Additionally, utilising advanced sensing technologies based on flexible and innovative biocompatible materials in wearable devices allows high accuracy and precision measurement of biological signals. Furthermore, applying real-time Machine Learning algorithms to highly accurate physiological parameters allows precise identification of unusual patterns in the data to provide health event predictions and warnings for timely intervention. However, in the predominantly adopted architectures, health event predictions based on Machine Learning are typically obtained by leveraging Cloud infrastructures characterised by shortcomings such as delayed response times and privacy issues. Fortunately, recent works highlight that a new paradigm based on Edge Computing technologies and on-device Artificial Intelligence significantly improve the latency and privacy issues. Applying this new paradigm to personalised healthcare architectures can significantly improve their efficiency and efficacy. Therefore, this paper reviews existing IoT healthcare architectures that utilise wearable devices and subsequently presents a scalable and modular system architecture to leverage emerging technologies to solve identified shortcomings. The defined architecture includes ultrathin, skin-compatible, flexible, high precision piezoelectric sensors, low-cost communication technologies, on-device intelligence, Edge Intelligence, and Edge Computing technologies. To provide development guidelines and define a consistent reference architecture for improved scalable wearable IoT-based critical healthcare architectures, this manuscript outlines the essential functional and non-functional requirements based on deductions from existing architectures and emerging technology trends. The presented system architecture can be applied to many scenarios, including ambient assisted living, where continuous surveillance and issuance of timely warnings can afford independence to the elderly and chronically ill. We conclude that the distribution and modularity of architecture layers, local AI-based elaboration, and data packaging consistency are the more essential functional requirements for critical healthcare application use cases. We also identify fast response time, utility, comfort, and low cost as the essential non-functional requirements for the defined system architecture.

Keywords: internet of things; edge intelligence; healthcare and wellness; piezoelectric sensors; multi-sensor; anomaly detection

Citation: Shumba, A.-T.; Montanaro, T.; Sergi, I.; Fachechi, L.; De Vittorio, M.; Patrono, L. Leveraging IoT-Aware Technologies and AI Techniques for Real-Time Critical Healthcare Applications. *Sensors* **2022**, *22*, 7675. https://doi.org/10.3390/s22197675

Academic Editors: Alexandru Lavric, Liliana Anchidin and Adrian I. Petrariu

Received: 2 August 2022
Accepted: 4 October 2022
Published: 10 October 2022

1. Introduction

The Internet of Things (IoT) paradigm has rapidly gained popularity over the years resulting in billions of connected devices applicable to everyday scenarios in various industries [1]. Researchers and industry players alike have developed applications that leverage IoT-enabling technologies to develop intelligent environments such as smart cities [2–4], smart factories [5–7], and smart homes [8–10]. Consequently, the healthcare and wellness domain has also seen an increase in the use of wearable devices due to a growing demand for personalised healthcare, advances in the development of miniaturised flexible sensing technologies, and the proliferation of IoT technologies in modern society [11,12].

On the other hand, because of the technological advances applied to healthcare to increase the human lifespan, the number of elderly citizens in many developed countries around the world is increasing. The increased elderly population, as a result, puts a significant burden on existing healthcare systems and infrastructures since elderly citizens require constant care, assistance, and monitoring because of the numerous chronic illnesses and conditions related to ageing. In addition to the increasing number of elderly citizens, a general increase in the number of people suffering from chronic illnesses coupled with the worldwide shortage of healthcare workers also contributes to the burden on healthcare infrastructures [13–16]. Several governments have dedicated significant resources to developing innovative technological solutions to provide efficient, affordable, and non-invasive services to improve overall citizen quality of life. Therefore, several architectures based on IoT-enabling technologies and wearable devices have been designed and developed to improve the overall healthcare services offered to citizens and alleviate the burden on existing healthcare infrastructures [17,18].

Some of the developed architectures are based on wearable devices that involve the use of various combinations of physiological and environmental data collected by wearable sensors to diagnose illnesses and provide warnings and intervention solutions in some cases. In most cases, applying Machine Learning (ML) and Artificial Intelligence (AI) algorithms capable of inferring meaningful patterns from the collected data provides diagnoses and intervention solutions. However, in many existing frameworks, computationally expensive and Cloud-reliant methods and algorithms are employed to infer patterns and meaningful information from the collected data [19–21]. Therefore, for these Cloud-based frameworks to function, frequent Cloud access and data transmission from wearable devices to centrally located Cloud data centres are required, raising privacy and latency concerns. Challenges related to achieving secured transmission using Wide Area Network (WAN) communication technologies such as WiFi or 4G wireless technologies usually used to obtain Cloud access largely contribute to privacy concerns [22,23].

In contrast, the large distances between the data sources and Cloud data centres mainly contribute to latency concerns. Slow response times in Cloud-based architectures also arise because warning or intervention solutions originate from the same Cloud data centres located far away from the user [19,24,25]. This property, as a result, confines the application of these frameworks to application use cases where real-time or timely interventions are not functional requirements, thus limiting the range of possible healthcare services offered by the Cloud-based IoT frameworks [18,21,26–28]. In response to these concerns and limitations, the Edge and Fog Computing paradigms facilitated the realisation of IoT-enabled healthcare application frameworks offering better response time and privacy preservation [27,29].

AI techniques such as ML, Deep Learning (DL), Federated Learning (FL), or Continual Learning (CL) algorithms are also applied to Edge/Fog Computing infrastructures to allow for intelligent data processing at the network edge. Intelligent data processing at the network edge further reduces the application response time and improves the privacy offered to users [24,26,28,30]. Improved response time is particularly essential in time-critical applications, such as healthcare architectures, where quickly obtaining usable information from sensor data is extremely important, and delayed intervention may be fatal.

In addition to the computational considerations mentioned above, utilising advanced, high accuracy, and precision sensing technologies in the right application-specific combinations also improves IoT healthcare frameworks' overall capabilities, accuracy, and robustness. To that end, cutting-edge research has been conducted in recent years to facilitate the creation of sensors capable of monitoring pertinent physiological signals with high accuracy and precision while utilising minimal power. Sensors made from bio-compatible materials easily attached to the skin and designed with comfortable form factors that allow them to cause limited to no intrusion to the user's day-to-day activities are crucial in wearable-based healthcare frameworks. Additional information about such sensing technologies can be found in [11,31–33]. Adding multiple advanced sensors to one framework can provide valuable correlating data to make more meaningful and complex predictions. As a result, wearable-based IoT-aware healthcare frameworks based on multiple advanced sensors are more versatile, robust, and trustworthy. A reference for some healthcare application-specific sensor combinations is available in a survey published by Sabry et al. in [34]. Consequently, adopting miniature, flexible, skin-compatible sensor technology introduces the possibility of unobtrusively monitoring physiological parameters that are usually imperceptible using the typical, commercially available wearable devices [35]. Ultimately, this allows the definition, design, and development of IoT-based healthcare frameworks that facilitate the long-term surveillance of critical parameters in prevention, diagnosis, and rehabilitation.

Several frameworks based on Edge/Fog Computing paradigms have been successfully applied in the healthcare domain through telemedicine, e-health, and mobile health applications. Other application domains have, however, seen the addition of on-device intelligence improve response times and privacy preservation in IoT frameworks. This result means that in addition to AI algorithms applied to the framework's Edge/Fog/Cloud nodes, sensing devices equipped with AI capabilities allow some on-device data analysis and, consequently, increase the amount of sensor data that can be exploited to perform the analysis, ultimately improving framework efficiency. Adding on-device intelligence can also limit data transmission between framework components, thus improving privacy preservation and device power efficiency. Typically, IoT frameworks rely on Bluetooth Low Energy (BLE), ZigBee, or other limited-range communication technologies to transmit raw sensor data to Edge/Fog nodes for processing. Therefore, introducing on-device data processing using ML and AI, regardless of complexity, could result in significant response time and power efficiency improvements by reducing the amount of data transmitted over wireless networks. However, this type of local on-device data processing based on AI is still in its infancy in healthcare domain applications. Therefore, it is necessary to define implementation guidelines, tools, and technical requirements for its adoption and integration with existing reference architectures [25].

Due to advances in the growing field of on-device AI, advances in sensing technologies, and the success of Edge/Fog/Cloud-based IoT frameworks, we propose that combining these technologies to develop healthcare domain frameworks can significantly contribute to the definition of reliable personalised healthcare architectures. In this paper, we, therefore, provide a detailed review of existing IoT-based architectures that utilise wearable devices for various healthcare applications. We also describe the requirements and reference architecture for a multi-layer IoT-aware system based on an advanced multi-sensor network leveraging Edge Computing technologies and on-device intelligence for critical or time-sensitive healthcare domain applications. The proposed architecture leverages advanced sensing technologies that allow the measurement of minute and accurate biosignals, low-cost communication technologies to facilitate the development of affordable wearable devices, Edge Computing in conjunction with Edge and on-device Intelligence technologies to facilitate secure real-time applications, and Cloud technologies to facilitate complex data analysis.

The main contributions of this paper are as follows:

- We provide a detailed analysis of the evolution of IoT-based architectural configurations applied to the healthcare domain.
- We define requirements for next-generation intelligent IoT-enabled personalised healthcare architectures. We define functional and non-functional requirements based on observations made from the existing literature and trends related to existing and emerging technologies while also considering the nature of the application scenarios.
- We define a detailed reference architecture configuration that combines high-precision sensing technologies, on-device intelligence, Edge Intelligence, and Cloud Intelligence.

We also introduce technologies applied in the various components of the defined architecture to aid the successful implementation of the defined architecture.
- We present potential use cases and scenarios for which the proposed architecture can be adopted to guide researchers and interested parties.

The rest of the paper is organised as follows: Section 2 provides the review of existing IoT-based healthcare architecture configurations, Section 3 describes the requirements for the proposed IoT system architecture, Section 4 defines the proposed architecture, and Section 5 proposes potential application use cases that can benefit from adopting the proposed architecture. Finally, Section 6 gives conclusions and recommendations for future work.

2. State-of-the-Art

Many authors have presented various architectural configurations for healthcare domain IoT-aware systems based on wearable devices. These configurations can be classified into two main categories based on the approach used to acquire, store and, most importantly, process the collected sensor data. The discussion presented here is, therefore, divided into two main parts describing the two main architectural configuration approaches adopted in the literature. The first, a centralised architecture approach, involves the transmission of raw data directly from the sensing devices to the Cloud without the use of an intermediate layer, while the second, a decentralised architecture approach, involves the use of one or more intermediate layers that perform elaboration of data, provide temporary storage, or application-specific decisions between the sensors and the Cloud. Solutions within the first group usually leverage the Cloud to process data using either domain-specific non-AI algorithms or AI algorithms to obtain intervention decisions, diagnosis conclusions, or recommendations. They typically conform to the structure illustrated in Figure 1. In the first part of this section, we report the works that leverage the Cloud for storage, processing, and decision making and do not offload any of the computational tasks or offer user services through intermediate layers. While in the second part, we describe the works that involve the use of multi-layer architectures to support one or more of the following:

(a) data collection from wearable sensors,
(b) elaboration of data in Edge nodes,
(c) the temporary storage of data in Edge/Fog nodes,
(d) forwarding of information through gateways in the form of routers, switches, mobile phones, or specialised embedded systems, etc.,
(e) use of Artificial Intelligence for data analysis on Edge/Fog nodes,
(f) the exploitation of intelligence resources provided by a Cloud server.

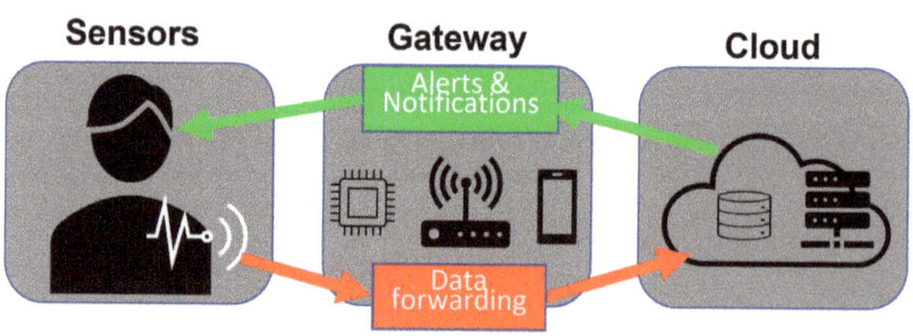

Figure 1. Typical Cloud-based architecture.

The first architecture configuration falling under the first group was presented by Ahamed et al. in [36], who defined a generic architecture that combines IoT-aware wearable devices with Machine Learning and Cloud Computing techniques for the prediction of heart disease. In this architecture, the wearable devices transmit data directly to a Cloud platform containing data processing, storage, and visualisation facilities that can be accessed by the patient or medical practitioners from anywhere. Further, falling under this group is the work by Addante et al. [37] in which a system containing a forearm-worn wearable device that makes use of a combination of accelerometers and gyroscopes to monitor movement and EMG sensors to obtain muscle mass information for the diagnosis of Sarcopenia, an ageing-related muscular disorder, was defined. They also used BLE to transfer data between the measuring device and a mobile device hosting an application to interact with the measured data and function as a gateway to the Cloud database. Another framework where all forms of data processing are performed using Cloud infrastructure for the diagnosis and monitoring of chronic diseases with a focus on diabetes was developed by Abdali-Mohammadi et al. [38]. In their work, a combination of wearable and implantable sensors were used to collect patient physiological parameters, which were then directly transmitted to the Cloud using 3G/4G communication networks. The developed system also included the possibility of identifying emergencies and notifying nearby hospitals, allowing emergency care provision. Further, also falling within this group is the work defined in [39], which described a framework for monitoring, predicting, and diagnosing heart disease using a combination of IoT sensors and Cloud-implemented ML classification algorithms trained using data from existing repositories. Kumar and Gandhi [40] also defined a healthcare monitoring architecture that utilised data from IoT wearable devices. The collected data were directly stored and processed using Cloud-based techniques, namely Apache HBase [41] for data storage and Apache Mahout [42] for the prediction model. Similar to the works described above, several other works, such as the ones presented in [43–45], describe IoT-aware healthcare architectures in which raw data collected from IoT devices are directly forwarded to the Cloud through various gateways using diverse wireless network technologies. As seen from the discussion above, various technology alternatives were adopted to fit specific requirements or application scenarios and improve the efficiency and reliability of Cloud-based IoT infrastructures. However, Cloud-based architecture configurations are characterised by privacy and latency issues mainly because of the centralised Cloud server location and the network infrastructure used for communication and data transfer. Therefore, based on their demand for speed, accuracy, and reliability, time-sensitive real-time solutions cannot be achieved using this approach. Furthermore, since continuous sensing devices produce large amounts of data, architectures solely reliant on Cloud Computing resources to process and analyse all the data put a strain on the network and the sensing devices, which usually have limited power available. The second group of works discussed in this section attempts to address these shortcomings and facilitate robust solutions with improved service delivery. The distributed architectures discussed in this section still use the previously discussed sensor-gateway-cloud 3-layer template; however, the gateway is realised through Edge or Fog Computing paradigms. The intermediate layer(s) perform varying levels of data elaboration, analysis, and application service delivery offering varied scales of improved efficiency, scalability, reliability, latency, privacy, and security. Some other works described in this section also implement AI algorithms at the framework edge, i.e., on nodes located in close proximity to the sensors, further improving the abovementioned parameters.

Gia et al. [46] presented an IoT for healthcare architecture for fall detection and monitoring heart rate variability that exploits Fog Computing technologies to improve the previously mentioned latency and security concerns and the lifetime of the sensing devices. The defined architecture utilises wearable electrocardiogram (ECG), motion, and body temperature sensors with environmental room temperature and humidity sensors to collect data. The sensing devices forward data through a low-cost RF interface to gateways that offer Fog services, including short-term data storage, data filtering, data processing, and generating near-real-time push notifications to inform the user and authorised health professionals of any concerning events or abnormalities. In this application, historical data can be accessed through the Cloud. However, the short-term storage available on the Fog node is also accessible through a local network, thus providing service reliability in the event of a network interruption. Similarly, Hajvali et al. [47] presented a generic software architecture for real-time IoT healthcare systems that contains a partitioned two-level Fog Computing layer. The two layers distribute the services available at the edge to reduce the number of tasks performed by one device, which results in a faster response time. The architecture also includes a smartphone that acts as a gateway between the sensor devices and the Fog layer and, in addition, hosts an application with a GUI that facilitates the accessing of alerts from the Fog node and local data management and visualisation. The architecture also includes a Cloud component responsible for further data processing and storage. The Cloud component also provides users and interested parties access to the database and user interfaces. The authors of this work focused their attention on user mobility; therefore, they emphasise the description of a software set-up that allows a user continuous access to a Fog node even if their physical location changes.

However, as established in previous sections, in addition to simply adding specific data processing algorithms to the Fog/Edge layers, AI algorithms can also contribute to the achievement of fast response times and improvement of the accuracy of the decisions made by IoT-aware health monitoring systems. As a result, several architectures that include AI algorithms applied to the Fog layer have been developed. For instance, the authors of [48] developed an architecture based on RF communication technology that implements temperature, ECG, blood pressure, and blood oxygen measurement with the aid of Fog and Cloud Computing technologies for remote monitoring of pregnant women. In this scenario, the Fog node, in the form of a Raspberry Pi, is responsible for user authentication, feature extraction, classification of collected data using a Bayesian Belief Network (BBN), and issuing alerts to health practitioners if a critical event is detected. On the other hand, the authors of [49] developed an integrated environment that incorporates Deep Learning (DL) algorithms in Fog nodes for an application for coronary disease monitoring and diagnosis. This application has two types of Fog nodes, namely, broker and worker nodes, to distribute the computational tasks. Ribiero et al. [50] also describe an architecture that leverages AI algorithms and advanced mathematical models in both the Fog and Cloud nodes. The proposed architecture accurately performs localised fall detection and classification; however, it does not leverage wearable sensor technology to measure parameters related to fall events.

In a nutshell, IoT-aware health monitoring system architectures fall into two main groups: centralised and decentralised architectures. Centralised architectures were the first to be adopted for periodical monitoring and continuous monitoring frameworks where collected sensor data are forwarded directly to the Cloud for processing. On the other hand, decentralised architectures have distributed data processing and service delivery capabilities in the Edge, Fog, and Cloud nodes per application requirements. In decentralised architectures, the prevailing trend has, until recently, seen AI and ML algorithms applied to Cloud and Edge/Fog nodes to perform data processing and analysis. However, recently, an additional sub-class of decentralised architectures where AI algorithms are implemented directly on wearable devices has emerged. An example of this type of architecture is described by Arikumar et al. [29], who proposed a Person Movement Identification (PMI) framework with AI algorithms for automatic feature extraction on the wearable device and classification in Edge and Cloud nodes. The architecture they defined implements a distributed continuous learning approach that enables on-device processing of data collected from multiple sensors and accounts for differences in user-related features. This emerging architecture allows the realisation of fast, accurate, scalable, and reliable health monitoring architectures suitable for personalised applications. However, as illustrated in Table 1, none of the articles available in the literature provides user services after on-device data processing.

Table 1. Comparative analysis: Existing architectures.

Source	Cloud	User Service	Edge/Fog	User Service	AI	On-Device	User Service	Application
[36]	✓	Data access (GUI)	x	-	Cloud	-	-	Cardiovascular disease
[37]	✓	Data access (Web/Mobile App)	x	-	x	-	-	Sarcopenia
[38]	✓	Patient alerts	x	-	Cloud	-	-	Diabetes diagnosis
[39]	✓	Test Report	x	-	Cloud	-	-	Heart disease prediction
[40]	✓	Medical alerts (doctors) Data access (Web App)	x	-	Cloud	-	-	Heart disease prediction
[43]	✓	Data access (Web App)	x	-	Cloud	-	-	Heart disease prediction
[44]	✓	Data access (GUI) Alerts	-	-	Cloud	-	-	Multiple disease prediction
[45]	✓	-	x	-	Cloud	-	-	Diabetes prediction
[51]	✓	Data access (Web App)	-	-	x	✓PPG HR estimation	-	Elderly citizen health monitoring
[46]	✓	Data access (Web App)	✓	Push notifications Local host GUI	x	-	-	Human fall detection Heart rate variability
[47]	✓	Data access (Web App)	✓	Alerts Local Host GUI	x	-	-	Disease monitoring and prediction
[48]	✓	Authenticated data access	✓	Alerts	Cloud	-	-	Pregnancy e-health
[49]	✓	Data access	✓	Data access	Cloud & Edge	-	-	Heart disease monitoring
[50]	✓	-	✓	Alerts	Cloud & Edge			Human fall classification
[29]	✓	-	✓	-	Cloud & Edge & Device	✓Feature Extraction	-	PMI

Based on the observations made from the conducted research, obtaining some useful information and generating alerts or alarms after the elaboration on-device would significantly improve the response time and, consequently, the efficiency and service delivery offered by the IoT framework. Additionally, to aid the framework design and guide future research in this field, we define requirements and outline the various components of a reference architecture that incorporates on-device intelligence and early service delivery in IoT-based healthcare frameworks.

3. System Requirements

This section, based on the analysed literature and inspired by the solutions proposed in the papers discussed in the previous section, outlines the requirements we defined for a comprehensive and versatile multi-level architecture. The presented architecture can be adopted by researchers interested in utilising an approach that combines IoT, AI, Edge/Fog Computing and multiple advanced sensing technologies to their healthcare domain application solutions. In addition, the presented architecture also allows the inclusion of Cloud Computing technologies since they are still an integral part of IoT architectures, as demonstrated by the trends in the literature.

The first consideration that guided our architecture design is the most commonly used structure in the analysed works. As discussed in previous sections, most of the architectures defined in the literature conform to a 3-layer structure, like the one illustrated in Figure 1, whereby data processing and storage of sensor data are centralised in the Cloud. In some cases, however, the defined architecture includes Edge Computing or Fog Computing layers to overcome the shortcomings of the centralised architecture. It has also been established that the response time, capabilities, efficacy, and fidelity of real-time IoT-aware architectures can be further improved by introducing Artificial Intelligence to the sensing devices. The quality of service offered by the architecture can also be enhanced by providing some insights related to the application scenario after the initial on-device elaboration. Therefore, based on these observations, an improved solution could be a modular system that distributes duties among different modules and utilises sensing devices with AI capabilities. In this case, end-user sensing devices can be used to collect the data from each sensor, perform local elaboration (e.g., filtration, simple analysis, or anomaly detection), and then forward results to a superior Edge/Fog node where further elaboration and analysis can be performed. In this way, user privacy can be preserved by avoiding the direct transmission of raw data collected from sensor devices. In addition, the powerful Edge/Fog node can also contribute to the preservation of privacy by further elaborating the initial results and only sharing inferred results to the last element of the architecture in the Cloud. It is also essential for the system to have the ability to add different, new sensors without affecting the other layers. Based on these considerations, the following list summarises the functional and non-functional requirements we have collated to guide the research community working within this context in the design of a comprehensive architecture.

3.1. Functional Requirements

The functional requirements listed below are essential to achieve the desired functionality.

FR1: Distribution of the duties among different modules/layers.

FR2: Modularity—Independence of each layer to guarantee the possibility of changing each layer with, for instance, a new version of the implementation software at runtime without affecting other layers.

FR3: A first local layer to perform some elaboration of the collected raw sensor data and package the results in a standard packet format to be transmitted to the next layer.

FR4: A second local layer that is able to receive data from the previous layer in the same standard packet format used by the first layer.

FR5: The same second layer should support multiple packet formats.

FR6: The same second layer should locally store the data in a buffer to be robust to connection problems.

FR7: Bidirectional communication between the first and second layers.

FR8: The same second layer should be able to forward the data, in a standard format, after some further elaboration to a Cloud server.

FR9: Cloud server should be able to receive data in a standard format.

3.2. Non-Functional Requirements

The requirements listed in the current sub-section do not affect the technical functionality. However, they are related to the architecture's performance, accuracy, acceptability, utility, and adaptability to application scenarios.

NFR1: Response time and accuracy—considering the nature of the application scenario, slow response time or inaccurate outputs could have fatal consequences; therefore, the system must guarantee accurate output and near-real-time response times.

NFR2: Usability—the system should avail user-friendly interfaces to facilitate easy exploitation of exposed services.

NFR3: Comfort—considering the nature of the scenario, i.e., continuous healthcare monitoring, the system should guarantee the use of highly comfortable monitoring devices not only to ensure user acceptance but to ensure they are willingly accepted.

NFR4: Performance and Interaction—the system should guarantee reactivity to all user requests and interactions.

NFR5: Reliability, Availability, and Maintainability—the system should be able to work and expose services without failures.

NFR6: Scalability—considering the proposed system's mobile nature, it would support the possibility of increasing the number of users and, therefore, sensors in more locations.

NFR7: Low-Cost—considering the application scenario, users and healthcare officials would prefer low-cost architectures to provide affordable healthcare infrastructures.

NFR8: Low-Power—considering the presence of battery-powered mobile healthcare monitoring devices, one of the most important non-functional requirements regards the provision of energy savings.

NFR9: Security—the system should guarantee that all the manipulated data and all system components are protected against malicious attacks or unauthorised access.

4. System Architecture

Based on the analysis reported in the literature review and the consequent observations and requirements defined previously, this section presents the architecture for a real-time IoT-aware healthcare system that incorporates advanced multi-sensing technologies, Edge-based and on-device AI components. The proposed architecture contains three main layers, namely, (i) Intelligent Data Acquisition Layer (iDAL), (ii) Edge Computing Layer, and (iii) Data Visualisation Layer, as illustrated in Figure 2. One of the critical features of the proposed architecture is the modularity and distribution of duties among components to facilitate easy upgrade and fulfil functional requirements, FR1 and FR2.

This section highlights the interactions between the different layers and modules of the proposed architecture and introduces their function and possible composition.

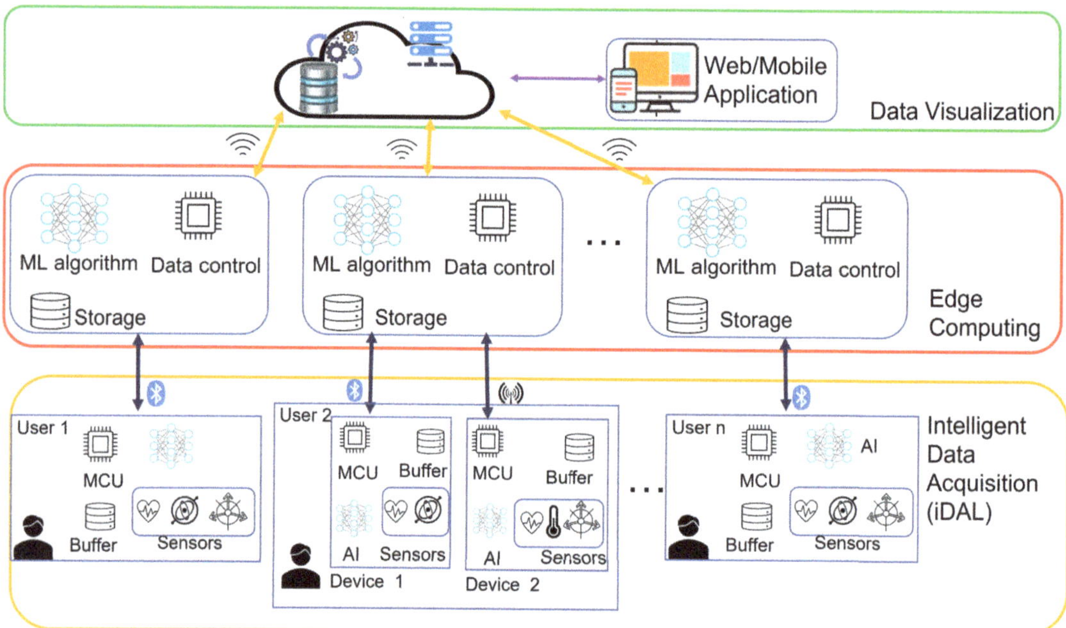

Figure 2. Proposed system architecture.

4.1. *Intelligent Data Acquisition Layer (iDAL)*

The first layer is the Intelligent Data Acquisition layer (Figure 3), with three major components: (a) advanced sensors, (b) a computational and storage unit, and (c) an Artificial Intelligence module.

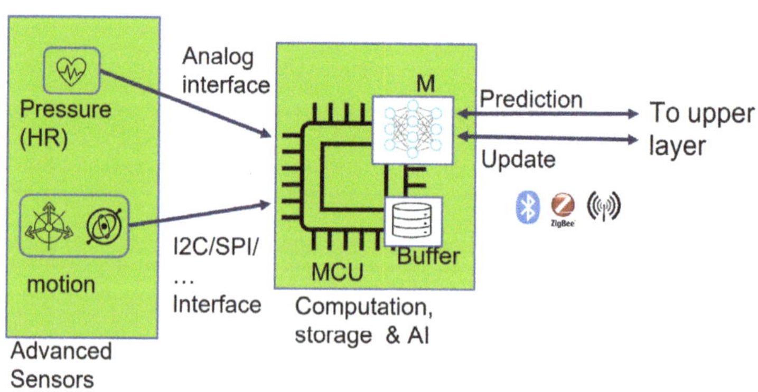

Figure 3. Intelligent Data Acquisition Layer.

4.1.1. Advanced Sensors

Various sensing technologies can be used to collect data relevant to the specific health-related parameters of interest applicable to the use case to which the architecture will be applied. For example, for a cardiac-related AAL application, a combination of skin-compatible piezoelectric or piezoresistive sensors defined in [31–33,52,53] in conjunction with motion, temperature, or positioning sensors, which add contextual information to the physiological measurements, can be used in this layer. The usage of flexible sensors can fulfil NFR3 and facilitate the development of comfortable and reliable wearable devices suitable for long-term surveillance of critical parameters in prevention, diagnosis, and rehabilitation use cases in the healthcare domain. Adopting advanced miniature flexible skin-compatible sensor technologies also introduces the possibility of accurately monitoring tiny physiological parameters that are usually imperceptible using the typical commercial, wearable devices [35], thus increasing the chances of fulfilling NFR1. Incorporating multiple sensors also allows the system to gather valuable correlating data that can be used to make more complex and meaningful deductions, thus making wearable-based IoT-aware healthcare frameworks more robust and trustworthy.

4.1.2. Computation and Data Processing

The second part of the iDAL is the computational unit responsible for controlling the sampling and acquisition of sensor data. To be considered for this function are Commercial-off-The-Shelf (COTS) low-power and low-cost microcontroller units that can support the interfaces and data transfer protocols implemented by the selected sensors, such as analogue inputs, SPI, I2C, etc. Field-Programmable Gate Arrays (FPGAs) or Application-Specific Integrated Circuits (ASICs) can also be considered to achieve higher speed, flexibility, and exclusive control over the functionality, size, or device form factor; however, this would significantly increase the overall cost of development, development time, and, consequently, production.

4.1.3. Artificial Intelligence Module

In addition to the primary signal processing techniques implemented to perform initial signal conditioning and processing, the attached microcontroller unit (MCU) is equipped with specific AI algorithms. The AI algorithms can perform data analysis to facilitate local decision-making by performing functions such as anomaly detection, high-level feature extraction, classification of the measured data, etc., to achieve real-time response and forwarding of processed data to upper layers and, thus, privacy preservation.

Some considerations also need to be made when selecting the Edge Intelligence technologies and, consequently, the level of data processing performed directly on the end devices through Artificial Intelligence. Some of the factors that govern the choice of AI algorithm used, the extent of data analysis, and the eventual output produced by the algorithm include device-related factors such as (a) the power consumption and available computational and storage capacity, (b) algorithm-based factors, such as complexity, computational, and storage requirements or results accuracy, and (c) application specific and operational factors, such as privacy concerns, latency requirements, etc. Several types of AI algorithms can be considered for this stage. For instance, the data-driven techniques for anomaly detection algorithms described in [28,54] could be applied. In cases where multiple sensors are all attached to the same computing/wearable device, instead of transmitting all the raw data from multiple sensors, the ML algorithms can be used to automatically extract the pertinent features from the combined sensor data, thus significantly prolonging the lifetime of the constrained battery-powered devices. Reducing the amount of data to be transmitted, in turn, reduces the required transmission time and, ultimately, the power consumption since communication is usually responsible for most of the device power cost of an IoT system. Pre-trained models that can be deployed to perform on-device inference may be used for simple anomaly detection, whereby the results can then be used to issue warnings if any unusual behaviour is observed or propose a course of action based on the algorithm predictions. In such cases, lightweight algorithms are worth considering based on the resource budget available in the MCU. Alternatively, a predetermined fraction of a large partitioned Neural Network can be implemented to perform partial on-device data elaboration. In this way, only intermediate results are forwarded to higher architecture levels for further elaboration, thus ensuring the preservation of privacy by avoiding the transmission of raw data as required by NFR3. As mentioned earlier, the amount of data to be transferred over a network is proportional to the overall device power consumption and the required bandwidth. Therefore, forwarding intermediate results reduces the bandwidth and power consumption cost, thereby enabling the fulfilment of NFR8. Based on these considerations, further investigation into the implementation and selection of the extent of on-device intelligence can be aided by consulting the architectures described in [30,55]. Sabry et al. [34] also highlight potential issues related to on-device intelligence in healthcare applications and provide some considerations to aid algorithm selection.

Various software tools, libraries, and frameworks have also been developed to facilitate the deployment of AI algorithms on constrained devices. For instance, TensorFlow lite for MCUs (TFLM) [56], an open-source library, can be used to support the deployment of Neural Networks (NNs) on a wide range of MCUs and Digital Signal Processors (DSPs). It has been tested on several Cortex-M series MCUs and can be used to deploy static algorithms trained using TensorFlow [57] to perform on-device inference. STMicroelectronics [58] also developed X-Cube-AI; a software tool that allows the generation and optimisation of AI algorithms developed using the typical ML and AI frameworks such as TensorFlow [57], Keras [59], or PyTorch [60] for deployment on the STM32 family MCUs. In addition, Edge Impulse [61] is a Cloud-based tool that allows the development of both NN and non-NN models for various embedded platforms such as MCUs or mobile phones. This tool allows the collection of sensor data directly from supported devices to train ML models, thus enabling fast prototyping of on-device ML architectures. Another available tool is NanoEdge AI Studio [62], which supports both learning and inference inside the MCU. This tool allows the automatic selection of ML libraries best fitting the provided data, making this tool suitable for developers with little or no AI or ML experience also. This tool contains libraries for the development of anomaly detection algorithms, one-class classification, multiple-class classification, or regression algorithms. The available tools can be selected based on available hardware and the developer's expertise. Other available tools that can be leveraged for the deployment of the selected ML algorithm can be found in [34,63].

Finally, after the elaboration and analysis of data performed by the AI module in the iDAL layer is completed, the results, inferences, or generated alarms are packaged and forwarded to the upper layers for further processing or management. Figure 3 illustrates a graphical summary of an example implementation of this layer consisting of two sensor modules.

4.2. Edge Computing Layer (ECL)

The second layer defined to fulfil FR4 is the Edge Computing Layer, which is primarily responsible for receiving data from the iDAL and providing a gateway to the upper layer. The same layer is also responsible for handling and managing communication with devices that may be equipped with different communication protocols and performing further data analysis. This layer is capable of bidirectional communication with both the lower layer and the upper layer to:

(a) receive data from the iDAL via standard low-power communication protocols such as BLE
(b) send updates to the iDAL
(c) forward data to the upper layer
(d) receive updates or notifications from the upper layer

Based on the AI architecture selected in the previous levels, an AI algorithm can be deployed in this layer in its entirety or as a fraction of a partitioned model. The choice of this model is based on the resources available for computation, storage, or power. The ECL should support various communication protocols since it could be responsible for serving multiple iDAL nodes with diverse protocol requirements. This layer can also be used to contact authorised caregivers, relatives, or health personnel in the event of detected distress or undesired events.

4.3. Data Visualisation Layer

Finally, the third layer is the Data Visualisation layer, which interacts with the storage facilities and facilitates the provision of user interaction services. Authorised users can view available data such as warnings generated by local devices and customised views of historical events through a Web Dashboard exposed by this layer. Healthcare professionals can also use the dashboard to provide recommendations that users can receive through the available communication channels. This layer is also responsible for advanced data analysis facilitated by the Cloud infrastructure. In this case, a more traditional AI algorithm can be considered to facilitate the analysis of historical data. This layer also exposes REST APIs to allow interaction with lower layers and contains a database to store the received data and network configurations.

5. Discussion

As already discussed, in the present paper, we defined an architecture extracted from and inspired by the analysed literature to provide a general and modular architecture that researchers or practitioners in the healthcare domain can take into consideration and exploit in their future works and solutions. Therefore, this section presents examples of the possible scenarios that the designed architecture can serve through interesting use cases inspired by real situations and case studies presented in the literature.

The first scenario to which our architecture can be applied is the continuous monitoring of citizens with chronic heart conditions. Considering the grave implications of heart malfunction to the human body, patients with chronic heart conditions, especially the elderly population, cannot maintain a normal autonomous lifestyle. As a result, they require constant surveillance and must frequently visit hospitals for check-ups. To improve their autonomy and, subsequently, quality of life, the proposed architecture can be applied to perform continuous surveillance on parameters related to heart health. In this application scenario, flexible piezoelectric sensors placed on different body parts (chest, ankle, or wrist) can accurately monitor cardiac function because of their capability to detect minute signal changes. In addition to their accuracy, each sensor, if placed correctly, can also be used to simultaneously monitor multiple parameters, such as heart sounds, blood pressure, and heart rate. Such a framework eliminates periodic blood pressure (BP) checks using the typical cumbersome cuffs by providing continuous BP insights. The frequency of hospital visits to perform periodic heart health check-ups can also be significantly reduced. In addition, an anomaly detection model can be deployed on the device to facilitate the fusion of the sensor data and extraction of parameters. The extracted parameters are transferred to the Edge Computing Layer (ECL), which performs further classification on the detected anomalies. The resulting ECL classifications can then be used to provide recommendations to the user and selected concerned parties. The implementation of Edge and on-device intelligence allows the system to provide real-time notifications, alarms, and recommendations, thus affording users more independence and autonomy.

Another application scenario that could benefit from this architecture is diagnosing, monitoring, and managing patients with neurodegenerative diseases. An Edge AI algorithm can be applied to perform feature extraction and anonymise data collected from a predetermined combination of motion and motor function sensors. The collected information can be used to provide real-time activity recommendations to assist patient rehabilitation or facilitate the provision of real-time feedback while patients are performing their recommended rehabilitation exercises. Caregivers can also remotely monitor patient progress and provide feedback, when necessary, through the web services exposed by the system.

Furthermore, on a larger scale, the architecture can be implemented in nursing homes or communities, especially those populated by the elderly, towards the realisation of self-sustainable smart cities. In this scenario, citizens are equipped with iDAL nodes to measure pertinent physiological parameters. ECL nodes can then be placed in strategic locations within the community to ensure complete coverage. In this case, implementing the Edge Intelligence of the individual iDAL nodes assures the user that their data remain private.

Finally, the presented architecture provides a solution that can be applied to improve the majority of the works discussed in the literature analysis and aid the design of new IoT-based healthcare frameworks for novel application use cases. As mentioned in earlier sections, adding on-device data processing, no matter how limited, can significantly improve various performance parameters of IoT infrastructures for healthcare applications. Therefore, using the results from the on-device analysis could significantly improve the efficacy and efficiency of the defined architectures by signalling any anomalies or concerns as early as possible. For instance, the fall detection architecture defined in [29] that uses federated learning and on-device feature extraction could allow the wearable device to produce an alarm to prevent falls. Additionally, adding Edge and on-device AI and including alarms generated by the pregnancy monitoring system defined in [48] would render the system applicable to critical pregnancies, where the mothers require close monitoring. Any detected abnormalities would be immediately signalled even when the user is far away from a health facility.

In the remainder of this section, we describe a solution we adopted to implement the early notification and anomaly detection functionality we propose for the iDAL layer and how it can interact with the upper layers.

5.1. *iDAL On-Device Intelligence Implementation*

The bottom-most layer of the proposed architecture hosts intelligent sensors capable of data processing and, if required by the application, providing a user service such as notifications and warnings. The state-of-the-art analysis we performed revealed the need for implementing data processing methods to increase the speed at which the system provides user feedback to improve the overall service delivery offered by IoT healthcare frameworks. One of the possible methods of providing user feedback, as has been discussed in earlier sections, is the implementation of anomaly detection algorithms on the sensing device to (a) provide onboard data processing, (b) reduce the amount of the data transmitted to the upper layer, and ultimately, (c) provide quick preliminary user feedback. Therefore, the first experimental work contributing towards implementing the architecture proposed in this work involves implementing and evaluating a prototype of the iDAL layer from our architecture with specific emphasis on the AI section.

This section describes the implementation of an on-device anomaly detection algorithm developed to fulfil FR3 of the proposed architecture. The requirement calls for *"a local layer to perform some elaboration of the collected raw sensor data and package the results in a standard packet format to be transmitted to the next layer"*.

5.1.1. AI Algorithm

Using an ECG monitoring scenario, we designed and deployed the simple pre-trained anomaly detection algorithm with the structure illustrated in Figure 4 on an MCU. The deployed algorithm is composed of an encoder to compress an input sequence into a smaller dimension and a decoder that attempts to reconstruct the input sequence from the compressed data. For this work, we modified the algorithm we evaluated in [64] (Figure 5) by defining and deploying the encoder and decoder as separate models, thus allowing us to access the encoder output in addition to the anomaly prediction during inference. In this way, the encoder output can be transmitted to an upper layer device containing a copy of the decoder to perform the reconstruction, resulting in a reduction in the amount of data to be transmitted by the sensor device over the implemented wireless communication channel. In addition, the inference results can also be made available to the user with a significantly reduced latency. In the testing section, we verify the possibility of implementing the proposed on-device intelligence method to achieve the expected reduced latency and amount of transmitted data.

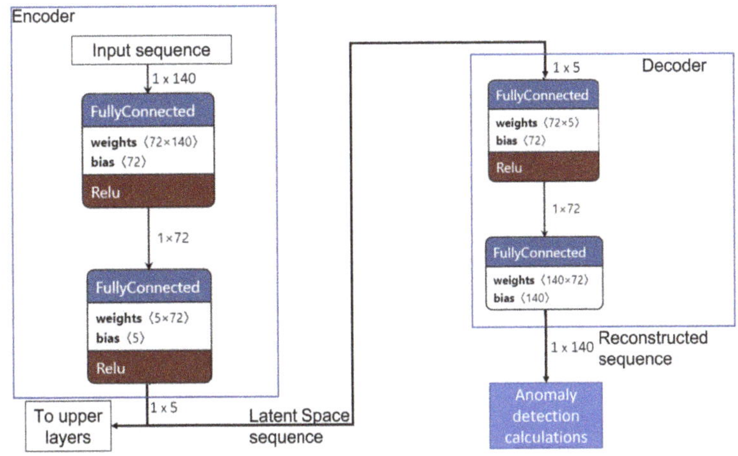

Figure 4. Autoencoder with separated Encoder and Decoder.

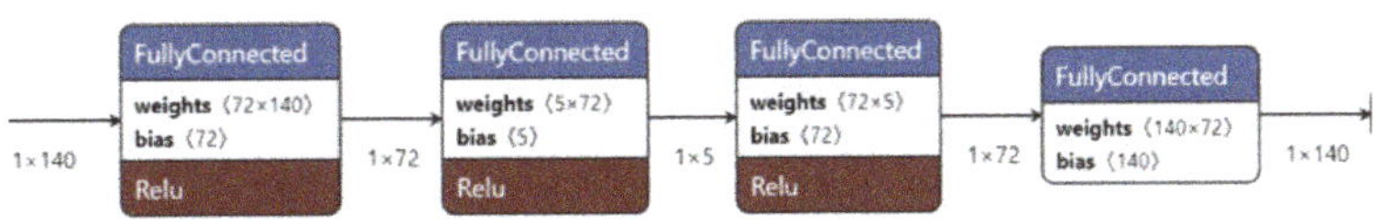

Figure 5. Simple Autoencoder structure.

5.1.2. Test Set-Up

As mentioned earlier, we adopted an ECG monitoring use case; therefore, the first step in the experimental procedure was training an autoencoder in TensorFlow [57] using the publicly available ECG5000 dataset [65]. To perform our tests, we adapted the dataset, which contains 5000 heartbeat samples obtained from monitoring a patient with severe congestive heart failure over 20 h. The original dataset samples are annotated with labels 1–5, where 1 represents a normal heartbeat, and the other labels, 2–5, represent different classes of abnormal rhythms. In the first phase, we focused on the normal samples to determine the autoencoder sample reconstruction ability. Therefore, from the 2919 normal samples, i.e., the samples annotated with label 1, we used 70% to train the model and reserved 30% for validation and testing. In exact numbers, the splitting produced 2043 randomly selected samples for training, and of the remaining 876 samples, the first 438 were used for validation, with the last 438 reserved for testing. The rest of the dataset samples from classes 2–5 were combined and annotated with label 0 to represent abnormal heartbeats to test the algorithm's anomaly detection. We used the Google Colaboratory [66], an online notebook platform that allows browser execution of AI and ML algorithms, to design, validate, and test the algorithm. After training, validation, and testing in Google Colab, we converted the TensorFlow model into C byte arrays that we loaded onto a bare metal MCU to perform inference using the TensorFlow Lite for microcontrollers [56] and for an interpreter [64]. To verify the functionality, i.e., the reconstruction capability of the separated autoencoder, we developed the set-up illustrated in Figure 6. In the defined set-up:

1. The Raspberry Pi represents a sensor device that collects physiological signals in the test scenario.
2. The first nRF52 device represents the on-device intelligence computing module connected to the sensor module on which the complete autoencoder components, i.e., both the encoder and the decoder, are deployed.

3. The second MCU represents the Edge computing layer hosting the separate decoder model.

The inference results from both MCUs are transferred back to the Raspberry PI and compared to the expected results. In this test scenario, we used an SPI bus to transfer data between the on-device intelligence computer and the Edge Computing device to allow fast prototyping and high-speed data transmission. However, BLE, ZigBee, or similar wireless technologies can be used to link between the Edge computing device and the sensor module during the implementation phase.

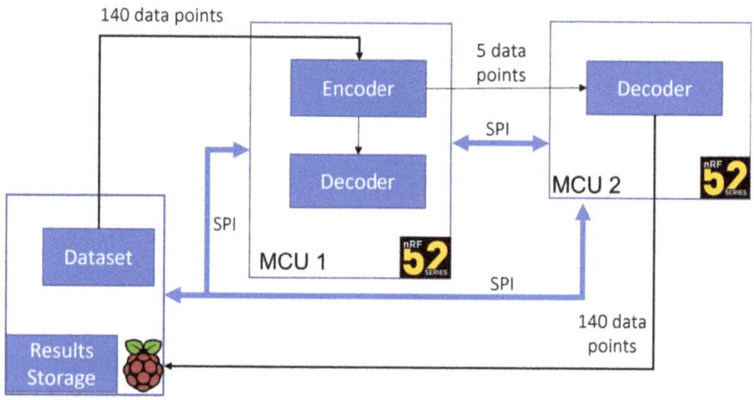

Figure 6. Test set-up.

5.1.3. Results Discussion

One of the tests we performed was comparing the original dataset and the reconstruction obtained from the stand-alone decoder deployed on the Edge computing device. Figure 7 illustrates the reconstruction error between an ECG sequence from the original dataset and the reconstruction from the deployed decoder. We calculated an average reconstruction accuracy of 99.947% using the mean squared error of the difference between the original input sequences and all the reconstructed sequences from 438 test samples.

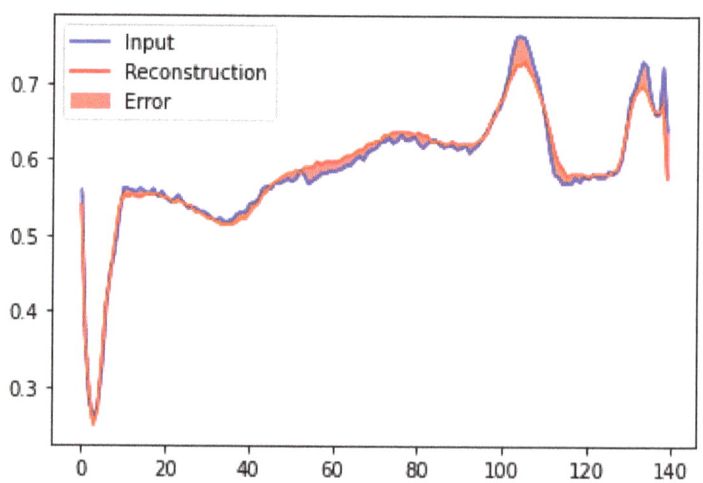

Figure 7. Original input vs. reconstructed decoder output.

The second set of tests we performed was to determine the time required from the moment data are available for processing to the moment usable data for user consumption are made available. In this scenario, we define usable information as the prediction classifying a heartbeat as anomalous or normal. Our tests revealed a latency reduction of 3 ms obtained by considering two test scenarios:

i. when available data samples are transferred over BLE to the Edge computing device when no data processing is performed on-device, and
ii. when the on-device anomaly detection algorithm is implemented on the sensor module using the configuration defined in Figure 3.

To obtain the time required to transmit data between the sensor module and the Edge computing device, we performed throughput tests using the prototype of a custom BLE-enabled sensing device pictured in Figure 8 connected to biocompatible piezoelectric sensors. The throughput test revealed a maximum reliable data rate of 16 kb/s for data transmission with no packet loss. Considering the ECG anomaly detection use case and the data formatting and segmentation parameters used to compile the ECG5000 dataset we used; the anomaly detection algorithm requires 140 data points as input. Therefore, based on this constraint, we obtained the following results from the two scenarios.

- Scenario 1: Based on the throughput tests, and, therefore, a 16 kb/s data rate, transmitting 140 samples from the sensor over BLE would require a minimum of 3.5 ms given a 2-byte digital representation with no additional data overhead. In addition to the transmission time, the total time required to obtain a usable result also includes the processing time required by the Edge computing device to perform inference and anomaly detection. Our tests revealed a minimum average processing time of approximately 0.48 ms to obtain an anomaly prediction. We measured the required inference time by counting the number of CPU cycles used by the MCU from the moment all the data samples are ready for elaboration to the moment the anomaly prediction result is available. Therefore, the total time required to obtain the first usable result and provide feedback to the user is 3.548 ms.
- Scenario 2: In this scenario, we consider the anomaly detection algorithm on the sensor module configured as illustrated in Figure 3. Transmitting the same data considered in the first scenario via SPI to perform on-device anomaly detection requires 0.028 ms with a 1 MHz SPI data rate. With the 0.48 additional milliseconds required for inference, the total time required to obtain a result that can provide valuable information to the user in this scenario is 0.508 ms.

According to the test results, implementing the on-device intelligence method described in this section successfully reduced latency (in this case, in reference to the time required to produce a usable inference result) by 3 ms. As mentioned before, the anomaly detection algorithm requires an input with 140 data points. However, implementing the autoencoder configuration described in this section allows the sensor module to send only five data points obtained from the encoder output to the Edge computing layer for further processing and record keeping. This result signifies a reduction in data that must be transferred over BLE by a factor of 28, from the initial 140 data points per sample to 5 data points per sample. Further data analysis can be performed using classification or other specific mathematical algorithms based on the reconstructed data obtained from the copy of the decoder deployed on the Edge Computing device. A comparison of the minimum timing and data requirements extracted from the two scenarios described above is summarised in Table 2.

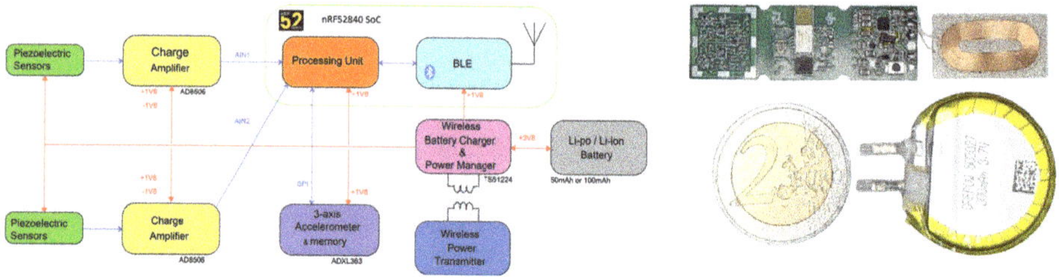

(a) (b)

Figure 8. Custom BLE−enabled sensing module. (**a**) Block diagram, (**b**) prototype device to scale.

Table 2. Tabulated summary of results discussion.

	No On-Device Intelligence	With On-Device Intelligence
Data transmission	3.500 ms	0.028 ms
Inference	0.480 ms	0.480 ms
Total required	3.548 ms	0.508 ms
Number of data points to upper layers	140	5

6. Conclusions

The need for personalised and home-based healthcare architectures has been increasing over the years, driven by factors, including the growing percentage of elderly citizens, the global shortage of healthcare workers, and the need for overall improved healthcare services. In a bid to provide a solution, various researchers provide several technological infrastructures, systems, and frameworks based on technological innovations. However, most of the frameworks utilising IoT technologies developed to date are predominantly based on Cloud infrastructures characterised by problematic issues, such as privacy and latency, which are undesirable for interactive and critical healthcare applications.

In order to contribute to the resolution of these issues, this work presented a modular IoT-aware system architecture that can be applied to numerous application scenarios in the healthcare domain.

The presented architecture encourages the amalgamation of advanced sensing technologies, low-power and low-cost IoT enabling technologies, and emergent AI techniques to develop modular, reliable, and scalable critical healthcare infrastructures. In addition, the modular nature of the architecture permits its suitability for a wide range of use cases since it can be configured based on application requirements, as demonstrated in the discussion. We also discussed and demonstrated the benefits of implementing on-device intelligence from a latency and communication efficiency point of view. The added functionality of providing user alarms or notifications immediately after on-device AI-based data processing has the potential to revolutionise IoT-based healthcare infrastructures.

Further developments can improve the reliability and performance of the system, such as including blockchain technologies to increase scalability while enhancing security to maintain the desired preservation of privacy.

Author Contributions: A.-T.S., T.M., L.P. and M.D.V. formulated the idea and identified the research requirements and objectives. A.-T.S., T.M., I.S. and L.F. designed the software architecture and conceived the system components. A.-T.S., T.M., I.S. and L.P. conducted the study of the literature and identified all the main contributions of each component. L.P., L.F. and M.D.V. supervised the activities. All authors prepared the manuscript. Finally, they all critically edited the manuscript and approved the final draft. All authors have read and agreed to the published version of the manuscript.

Funding: This research was funded by Istituto Italiano di Tecnologia.

Institutional Review Board Statement: Not applicable.

Informed Consent Statement: Not applicable.

Data Availability Statement: Not applicable.

Acknowledgments: This study was partially realised in contribution to the "SMART CARE" project. "SMART CARE" is a research project funded under the Italian "Piano Operativo Salute (FSC 2014–2020)" programme (Traiettoria 2 "E-Health, diagnostica avanzata, medical devices e mini invasività", Linea di azione 2.1 "Creazione di una rete nazionale per le malattie ad alto impatto")—Public Notice: 4 July 2022.

Conflicts of Interest: The authors declare no conflict of interest.

Abbreviations

The following abbreviations are used in this manuscript:

IoT	Internet of Things
BP	Blood Pressure
FL	Federated Learning
DL	Deep Learning
ML	Machine Learning
AI	Artificial Intelligence
iDAL	Intelligent Data Acquisition Layer
ECL	Edge Computing Layer
API	Application Programming Interface
EMG	Electromyography
BLE	Bluetooth Low Energy
MCU	Microcontroller Unit
FPGA	Field Programmable Gate Array
ASIC	Application Specific Integrated Circuit
COTS	Commercial Off The Shelf
SPI	Serial Peripheral Interface
NFC	Near Field Communication
TFLM	TensorFlow Lite for Microcontrollers
DSP	Digital Signal Processor
PMI	Personal Movement Identification
ADC	Analogue to Digital Converter

References

1. State of IoT 2021: Number of Connected IoT Devices Growing 9% to 12.3 Billion Globally, Cellular IoT Now Surpassing 2 Billion. Available online: https://iot-analytics.com/number-connected-iot-devices/ (accessed on 1 March 2022).
2. Fadda, E.; Perboli, G.; Vallesio, V.; Mana, D. Sustainable mobility and user preferences by crowdsourcing data: The Open Agora project. In Proceedings of the 2018 IEEE 14th International Conference on Automation Science and Engineering (CASE), Munich, Germany, 20–24 August 2018; pp. 1243–1248. [CrossRef]
3. Mulero, R.; Aitor, A.; Gorka, A.; Abril-Jiménez, P.; Waldmeyer, M.; Castrillo, M.; Patrono, L.; Rametta, P.; Sergi, I. An IoT-aware approach for elderly-friendly cities. *IEEE Access* **2018**, *6*, 7941–7957. [CrossRef]
4. Sánchez-Corcuera, R.; Nuñez-Marcos, A.; Sesma-Solance, J.; Bilbao-Jayo, A.; Mulero, R.; Zulaika, U.; Azkune, G.; Almeida, A. Smart cities survey: Technologies, application domains and challenges for the cities of the future. *Int. J. Distrib. Sens. Netw.* **2019**, *15*. [CrossRef]
5. Landi, L.; Mödden, H.; Pera, F.; Uhlmann, E.; Meister, F. Probabilities in safety of machinery—Risk reduction through fixed and moveable guards by standardized impact tests, part 1: Applications and consideration of random effects. In Proceedings of the Safety and Reliability—Theory and Applications—Proceedings of the 27th European Safety and Reliability Conference (ESREL 2017), Portoroz, Slovenia, 18–22 June 2017; pp. 2155–2164. [CrossRef]
6. Pancardo, P.; Acosta, F.D.; Hernández-Nolasco, J.A.; Wister, M.A.; López-de Ipiña, D. Real-time personalized monitoring to estimate occupational heat stress in ambient assisted working. *Sensors* **2015**, *15*, 16956–16980. [CrossRef] [PubMed]
7. Fadda, E.; Perboli, G.; Rosano, M.; Mascolo, J.E.; Masera, D. A Decision Support System for Supporting Strategic Production Allocation in the Automotive Industry. *Sustainability* **2022**, *14*, 2408. [CrossRef]

8. Bilbao-Jayo, A.; Almeida, A.; Sergi, I.; Montanaro, T.; Fasano, L.; Emaldi, M.; Patrono, L. Behavior Modeling for a Beacon-Based Indoor Location System. *Sensors* **2021**, *21*, 4839. [CrossRef]
9. Radogna, A.V.; Capone, S.; Di Lauro, G.A.; Fiore, N.; Longo, V.; Giampetruzzi, L.; Francioso, L.; Casino, F.; Siciliano, P.; Sabina, S.; et al. A smart breath analyzer for monitoring home mechanical ventilated patients. In *Lecture Notes in Electrical Engineering*; Springer: Cham, Switzerland, 2019; Volume 539, pp. 465–471. [CrossRef]
10. Bonino, D.; Corno, F.; De Russis, L. Poweront: An ontology-based approach for power consumption estimation in smart homes. In*Lecture Notes of the Institute for Computer Sciences, Social-Informatics and Telecommunications Engineering, LNICST*; Springer: Cham, Switzerland, 2015; Volume 150, pp. 3–8. [CrossRef]
11. Wang, Y.; Yang, B.; Hua, Z.; Zhang, J.; Guo, P.; Hao, D.; Gao, Y.; Huang, J. Recent advancements in flexible and wearable sensors for biomedical and healthcare applications. *J. Phys. D Appl. Phys.* **2021**, *55*, 134001. [CrossRef]
12. De Fazio, R.; De Vittorio, M.; Visconti, P. Innovative IoT Solutions and Wearable Sensing Systems for Monitoring Human Biophysical Parameters: A Review. *Electronics* **2021**, *10*, 1660. [CrossRef]
13. Michel, J.P.; Ecarnot, F. The shortage of skilled workers in Europe: Its impact on geriatric medicine. *Eur. Geriatr. Med.* **2020**, *11*, 345–347. [CrossRef]
14. Džakula, A.; Relić, D. Health workforce shortage—Doing the right things or doing things right? *Croat. Med. J.* **2022**, *63*, 107. [CrossRef]
15. Maresova, P.; Prochazka, M.; Barakovic, S.; Husić, J.B.; Kuca, K. A Shortage in the Number of Nurses—A Case Study from a Selected Region in the Czech Republic and International Context. *Healthcare* **2020**, *8*, 152. [CrossRef]
16. Almeida, A.; Mulero, R.; Rametta, P.; Urošević, V.; Andrić, M.; Patrono, L. A critical analysis of an IoT—Aware AAL system for elderly monitoring. *Future Gener. Comput. Syst.* **2019**, *97*, 598–619. [CrossRef]
17. Dian, F.J.; Vahidnia, R.; Rahmati, A. Wearables and the Internet of Things (IoT), Applications, Opportunities, and Challenges: A Survey. *IEEE Access* **2020**, *8*, 69200–69211. [CrossRef]
18. Metcalf, D.; Milliard, S.T.; Gomez, M.; Schwartz, M. Wearables and the Internet of Things for Health: Wearable, Interconnected Devices Promise More Efficient and Comprehensive Health Care. *IEEE Pulse* **2016**, *7*, 35–39. [CrossRef] [PubMed]
19. Singh, A.; Chatterjee, K. Securing smart healthcare system with edge computing. *Comput. Secur.* **2021**, *108*, 102353. [CrossRef]
20. Dutta, D.L.; Bharali, S. TinyML Meets IoT: A Comprehensive Survey. *Internet Things* **2021**, *16*, 100461. [CrossRef]
21. Deebak, B.D.; Memon, F.H.; Cheng, X.; Dev, K.; Hu, J.; Khowaja, S.A.; Qureshi, N.M.F.; Choi, K.H. Seamless privacy-preservation and authentication framework for IoT-enabled smart eHealth systems. *Sustain. Cities Soc.* **2022**, *80*, 103661. [CrossRef]
22. Muhammad, G.; Rahman, S.M.M.; Alelaiwi, A.; Alamri, A. Smart Health Solution Integrating IoT and Cloud: A Case Study of Voice Pathology Monitoring. *IEEE Commun. Mag.* **2017**, *55*, 69–73. [CrossRef]
23. Henze, M.; Hermerschmidt, L.; Kerpen, D.; Häußling, R.; Rumpe, B.; Wehrle, K. A comprehensive approach to privacy in the cloud-based Internet of Things. *Future Gener. Comput. Syst.* **2016**, *56*, 701–718. [CrossRef]
24. Wang, X.; Magno, M.; Cavigelli, L.; Benini, L. FANN-on-MCU: An Open-Source Toolkit for Energy-Efficient Neural Network Inference at the Edge of the Internet of Things. *IEEE Internet Things J.* **2020**, *7*, 4403–4417. [CrossRef]
25. Mahmud, R.; Koch, F.L.; Buyya, R. Cloud-fog interoperability in IoT-enabled healthcare solutions. In Proceedings of the 19th International Conference on Distributed Computing and Networking, Varanasi, India, 4–7 January 2018. [CrossRef]
26. Merenda, M.; Porcaro, C.; Iero, D. Edge Machine Learning for AI-Enabled IoT Devices: A Review. *Sensors* **2020**, *20*, 2533. [CrossRef]
27. Surati, S.; Patel, S.; Surati, K. Background and Research Challenges for FC for Healthcare 4.0. In *Fog Computing for Healthcare 4.0 Environments: Technical, Societal, and Future Implications*; Springer International Publishing: Cham, Switzerland, 2021; pp. 37–53. [CrossRef]
28. Erhan, L.; Ndubuaku, M.; Mauro, M.D.; Song, W.; Chen, M.; Fortino, G.; Bagdasar, O.; Liotta, A. Smart anomaly detection in sensor systems: A multi-perspective review. *Inf. Fusion* **2021**, *67*, 64–79. [CrossRef]
29. Arikumar, K.S.; Prathiba, S.B.; Alazab, M.; Gadekallu, T.R.; Pandya, S.; Khan, J.M.; Moorthy, R.S. FL-PMI: Federated Learning-Based Person Movement Identification through Wearable Devices in Smart Healthcare Systems. *Sensors* **2022**, *22*, 1377. [CrossRef] [PubMed]
30. Wang, X.; Han, Y.; Leung, V.C.; Niyato, D.; Yan, X.; Chen, X. Convergence of Edge Computing and Deep Learning: A Comprehensive Survey. *IEEE Commun. Surv. Tutor.* **2020**, *22*, 869–904. [CrossRef]
31. Ha, T.; Tran, J.; Liu, S.; Jang, H.; Jeong, H.; Mitbander, R.; Huh, H.; Qiu, Y.; Duong, J.; Wang, R.L.; et al. A Chest-Laminated Ultrathin and Stretchable E-Tattoo for the Measurement of Electrocardiogram, Seismocardiogram, and Cardiac Time Intervals. *Adv. Sci.* **2019**, *6*, 1900290. [CrossRef]
32. Sun, R.; Carreira, S.C.; Chen, Y.; Xiang, C.; Xu, L.; Zhang, B.; Chen, M.; Farrow, I.; Scarpa, F.; Rossiter, J. Stretchable Piezoelectric Sensing Systems for Self-Powered and Wireless Health Monitoring. *Adv. Mater. Technol.* **2019**, *4*, 1900100. [CrossRef]
33. Chen, B.; Zhang, L.; Li, H.; Lai, X.; Zeng, X. Skin-inspired flexible and high-performance MXene@polydimethylsiloxane piezoresistive pressure sensor for human motion detection. *J. Colloid Interface Sci.* **2022**, *617*, 478–488. [CrossRef]
34. Sabry, F.; Eltaras, T.; Labda, W.; Alzoubi, K.; Malluhi, Q. Machine Learning for Healthcare Wearable Devices: The Big Picture. *J. Healthc. Eng.* **2022**, *2022*, 4653923. [CrossRef]
35. Zeng, X.; Deng, H.T.; Wen, D.L.; Li, Y.Y.; Xu, L.; Zhang, X.S. Wearable Multi-Functional Sensing Technology for Healthcare Smart Detection. *Micromachines* **2022**, *13*, 254. [CrossRef]

36. Ahamed, J.; Koli, A.M.; Ahmad, K.; Jamal, M.A.; Gupta, B.B. CDPS-IoT: Cardiovascular Disease Prediction System Based on IoT using Machine Learning. *Int. J. Interact. Multimedia Artif. Intell.* **2021**, *7*, 78–86. [CrossRef]
37. Addante, F.; Gaetani, F.; Patrono, L.; Sancarlo, D.; Sergi, I.; Vergari, G. An Innovative AAL System Based on IoT Technologies for Patients with Sarcopenia. *Sensors* **2019**, *19*, 4951. [CrossRef]
38. Abdali-Mohammadi, F.; Meqdad, M.N.; Kadry, S. Development of an IoT-based and cloud-based disease prediction and diagnosis system for healthcare using machine learning algorithms. *IAES Int. J. Artif. Intell.* **2020**, *9*, 766–771. [CrossRef]
39. Ganesan, M.; Sivakumar, N. IoT based heart disease prediction and diagnosis model for healthcare using machine learning models. In Proceedings of the 2019 IEEE International Conference on System, Computation, Automation and Networking (ICSCAN), Pondicherry, India, 29–30 March 2019. [CrossRef]
40. Kumar, P.M.; Gandhi, U.D. A novel three-tier Internet of Things architecture with machine learning algorithm for early detection of heart diseases. *Comput. Electr. Eng.* **2018**, *65*, 222–235. [CrossRef]
41. Apache Hbase. Available online: https://hbase.apache.org/ (accessed on 27 July 2022).
42. Mahout. Available online: https://mahout.apache.org// (accessed on 27 July 2022).
43. Khan, M.A. An IoT Framework for Heart Disease Prediction Based on MDCNN Classifier. *IEEE Access* **2020**, *8*, 34717–34727. [CrossRef]
44. Verma, P.; Sood, S.K. Cloud-centric IoT based disease diagnosis healthcare framework. *J. Parallel Distr. Comput.* **2018**, *116*, 27–38. [CrossRef]
45. Kumar, P.M.; Lokesh, S.; Varatharajan, R.; Babu, G.C.; Parthasarathy, P. Cloud and IoT based disease prediction and diagnosis system for healthcare using Fuzzy neural classifier. *Future Gener. Comput. Syst.* **2018**, *86*, 527–534. [CrossRef]
46. Gia, T.N.; Jiang, M. *Exploiting Fog Computing in Health Monitoring*; John Wiley & Sons, Inc.: Hoboken, NJ, USA, 2019.
47. Hajvali, M.; Adabi, S.; Rezaee, A.; Hosseinzadeh, M. Software architecture for IoT-based health-care systems with cloud/fog service model. *Clust. Comput.* **2022**, *25*, 91–118. [CrossRef]
48. Beri, R.; Dubey, M.K.; Gehlot, A.; Singh, R.; Abd-Elnaby, M.; Singh, A. A novel fog-computing-assisted architecture of E-healthcare system for pregnant women. *J. Supercomput.* **2022**, *78*, 7591–7615. [CrossRef]
49. Verma, P.; Tiwari, R.; Hong, W.C.; Upadhyay, S.; Yeh, Y.H. FETCH: A Deep Learning-Based Fog Computing and IoT Integrated Environment for Healthcare Monitoring and Diagnosis. *IEEE Access* **2022**, *10*, 12548–12563. [CrossRef]
50. Ribeiro, O.; Gomes, L.; Vale, Z. IoT-Based Human Fall Detection System. *Electronics* **2022**, *11*, 592. [CrossRef]
51. Pinheiro, G.P.; Miranda, R.K.; Praciano, B.J.; Santos, G.A.; Mendonça, F.L.; Javidi, E.; da Costa, J.P.J.; de Sousa, R.T. Multi-Sensor Wearable Health Device Framework for Real-Time Monitoring of Elderly Patients Using a Mobile Application and High-Resolution Parameter Estimation. *Front. Hum. Neurosci.* **2022**, *15*, 836. [CrossRef]
52. Natta, L.; Mastronardi, V.M.; Guido, F.; Algieri, L.; Puce, S.; Pisano, F.; Rizzi, F.; Pulli, R.; Qualtieri, A.; Vittorio, M.D. Soft and flexible piezoelectric smart patch for vascular graft monitoring based on Aluminum Nitride thin film. *Sci. Rep.* **2019**, *9*, 8392. [CrossRef] [PubMed]
53. Natta, L.; Guido, F.; Algieri, L.; Mastronardi, V.M.; Rizzi, F.; Scarpa, E.; Qualtieri, A.; Todaro, M.T.; Sallustio, V.; Vittorio, M.D. Conformable AlN Piezoelectric Sensors as a Non-invasive Approach for Swallowing Disorder Assessment. *ACS Sens.* **2021**, *6*, 1761–1769. [CrossRef]
54. Zhang, Y.; Chen, Y.; Wang, J.; Pan, Z. Unsupervised Deep Anomaly Detection for Multi-Sensor Time-Series Signals. *IEEE Trans. Knowl. Data Eng.* **2021**. [CrossRef]
55. Zhou, Z.; Chen, X.; Li, E.; Zeng, L.; Luo, K.; Zhang, J. Edge Intelligence: Paving the Last Mile of Artificial Intelligence with Edge Computing. *Proc. IEEE* **2019**, *107*, 1738–1762. [CrossRef]
56. TensorFlow Lite for Microcontrollers. Available online: https://www.tensorflow.org/lite/microcontrollers (accessed on 10 February 2022).
57. Tensorflow Library. Available online: https://www.tensorflow.org/ (accessed on 15 December 2021).
58. STMicroelectronics. Available online: https://www.st.com/content/st_com/en.html (accessed on 1 March 2022).
59. Keras. Available online: https://keras.io/ (accessed on 1 March 2022).
60. PyTorch. Available online: https://pytorch.org/ (accessed on 15 January 2022).
61. Edge Impulse. Available online: https://www.edgeimpulse.com/ (accessed on 1 March 2022).
62. NanoEdge AI Studio. Available online: https://www.st.com/en/development-tools/nanoedgeaistudio.html (accessed on 1 March 2022).
63. Ray, P.P. A review on TinyML: State-of-the-art and prospects. *J. King Saud Univ. Comput. Inf. Sci.* **2022**, *34*, 1595–1623. [CrossRef]
64. Shumba, A.T.; Montanaro, T.; Sergi, I.; Fachechi, L.; Vittorio, M.D.; Patrono, L. Embedded Machine Learning: Towards a Low-Cost Intelligent IoT edge. In Proceedings of the 2022 7th International Conference on Smart and Sustainable Technologies (SpliTech), Split/Bol, Croatia, 5–8 July 2022; pp. 1–6. [CrossRef]
65. Dataset: ECG5000. Available online: http://www.timeseriesclassification.com/description.php?Dataset=ECG5000 (accessed on 20 January 2022).
66. Google Colaboratory. Available online: https://colab.research.google.com/ (accessed on 20 January 2022).

sensors

MDPI

Article

A Historical Twist on Long-Range Wireless: Building a 103 km Multi-Hop Network Replicating Claude Chappe's Telegraph

Mina Rady [1,2,3,*], Jonathan Muñoz [4], Razanne Abu-Aisheh [2,5], Mališa Vučinić [2], José Astorga Tobar [2], Alfonso Cortes [2], Quentin Lampin [1], Dominique Barthel [1] and Thomas Watteyne [2,4]

1 Orange Labs, 38240 Meylan, France
2 Inria, 75012 Paris, France
3 CITI, National Institute of Applied Sciences (INSA) Lyon, Inria, University of Lyon, 69621 Villeurbanne, France
4 Wattson Elements/Falco, 75012 Paris, France
5 Nokia Bell Labs, 91620 Nozay, France
* Correspondence: mina.rady@insa-lyon.fr

Abstract: In 1794, French Engineer Claude Chappe coordinated the deployment of a network of dozens of optical semaphores. These formed "strings" that were hundreds of kilometers long, allowing for nationwide telegraphy. The Chappe telegraph inspired future developments of long-range telecommunications using electrical telegraphs and, later, digital telecommunication. Long-range wireless networks are used today for the Internet of Things (IoT), including industrial, agricultural, and urban applications. The long-range radio technology used today offers approximately 10 km of range. Long-range IoT solutions use "star" topology: all devices need to be within range of a gateway device. This limits the area covered by one such network to roughly a disk of a 10 km radius. In this article, we demonstrate a 103 km low-power wireless multi-hop network by combining long-range IoT radio technology with Claude Chappe's vision. We placed 11 battery-powered devices at the former locations of the Chappe telegraph towers, hanging under helium balloons. We ran a proprietary protocol stack on these devices so they formed a 10-hop multi-hop network: devices forwarded the frames from the "previous" device in the chain. This is, to our knowledge, the longest low power multi-hop wireless network built to date, demonstrating the potential of combining long-range radio technology with multi-hop technology.

Keywords: LPWANs; industrial Internet of Things; mesh networks; wide area networks

Citation: Rady, M.; Muñoz, J.; Abu-Aisheh, R.; Vučinić, M.; Astorga Tobar, J.; Cortes, A.; Lampin, Q.; Barthel, D.; Watteyne, T. A Historical Twist on Long-Range Wireless: Building a 103 km Multi-Hop Network Replicating Claude Chappe's Telegraph. *Sensors* **2022**, *22*, 7586. https://doi.org/10.3390/s22197586

Academic Editors: Alexandru Lavric and Francesca Cuomo

Received: 25 July 2022
Accepted: 29 September 2022
Published: 6 October 2022

1. Introduction

Wireless connectivity has been increasingly deployed in diverse industrial and urban applications. It has enabled automated remotes and frequent monitoring of machines and smart meters. This increased adoption has challenged wireless communications to offer higher capacities and longer ranges at lower costs.

In Industry 4.0, wireless is a key component for enabling remote interaction with machinery. Sensors can report temperature, pressure, and vibration [1]. Early performance degradation signs can be detected and the machine can be maintained before anything breaks, with the promise of avoiding unplanned downtime. Predictive maintenance adds *prognosis* to *diagnosis* [2]. An industrial plant has hundreds of sensors or actuators connected to low power networks [3] deployed in a challenging environment. For example, an oil refinery can have more than one million devices spread over an area of several square kilometers, in an environment full of metallic structures [3]. This challenges the range of the wireless network. In these setups, wiring the sensors together is often not an option because of the installation complexity and operational hazards [3,4].

Long-range wireless communication is used to remotely read domestic utility meters (water, electricity, gas). The European Commission mandates their remote readings to save natural resources [5]. The monitoring infrastructure is necessarily wireless, as running a dedicated wired network between houses is too expensive [6]. Similar to industrial use cases, the distance between utility meters is a challenge for wireless, particularly as meters can be installed underground or behind concrete barriers.

These applications have the same challenge: "how can we extend the area these networks are deployed in without installing significant infrastructure?". These networks typically use long-range radios in a star topology, or short-range mesh topologies. Both are limited by the fact that gateways need to be installed for the nodes to report data. Gateways are expensive to install as they need mains power and internet connectivity (Ethernet, cellular, etc.). The goal of both long-range radio technology and multi-hop mesh networking technology is to reduce the number of gateways installed.

Several radio technologies have been developed for low-power long-range wireless networking. The IEEE has standardized a set of 31 PHYs in the IEEE802.15.4g (2012) amendment [7]. They are based on three families of modulations: frequency shift keying (FSK), offset-quadrature phase shift keying (O-QPSK), and orthogonal frequency division multiplexing (OFDM). Long-Range (LoRa) is a proprietary PHY that has become pervasive in low-power long-range networks. It is based on chirp spread spectrum modulation and it is deployed in a star topology under the LoRa wide area network (LoRaWAN) protocol stack. Sigfox is a technology that was available recently for long-range low-power wireless networking. Similar to LoRa, it used a proprietary PHY and protocol stack and was based on star topologies. Cellular networks, standardized by the 3rd generation partnership project (3GPP), introduced narrow-band IoT (NB-IoT) for low-power wireless applications. It is provided as an integrated service in cellular base stations by "slicing" the network into machine-type communications and human-type communications. Recently, satellite communication solutions have been introduced for long-range low-power connectivity to help reach areas where no terrestrial coverage is available. These families of networks are limited by the installation costs of their infrastructure, whether gateways, cellular-based stations, or satellites.

This article recreates the Claude Chappe Telegraph by applying its concept to low-power wireless. Using 11 battery-powered IoT devices, we created a 103 km low-power wireless network. The contributions of this article are three-fold:

- We provide a historical overview of the development of the Chappe Telegraph.
- We describe how we translate that concept to today's low-power wireless technology.
- We demonstrate a wireless network reaching 103 km, and discuss the challenges and the opportunities for long-range multi-hop mesh networks.

The demonstration presented in this article is the longest low-power wireless multi-hop network, to the best of our knowledge.

The remainder of this article is organized as follows. Section 2 presents a historical overview of the Chappe telegraph. Section 3 surveys current IoT technologies. Section 4 introduces the experimental setup we used to replicate the Chappe telegraph. Section 5 describes the experiment of building a 103 km low-power wireless network. Finally, Section 6 discusses the opportunities and challenges for long-range mesh networking.

2. Claude Chappe's 1794 Telegraph

Telegraphy is a word from ancient Greek that means "remote writing". Both the Roman and Persian empires had systems of remote signaling of simple information that date as far back as the second Century B.C. [8]. The intensity of conflicts in Europe that erupted at the end of the 18th century created an urgent need for faster communications. It was at that time that Claude Chappe, a French engineer, developed his telegraph.

Starting in 1792, France entered into conflicts on several fronts with Austria, Prussia, Russia, and Britain. As troops were deployed in different zones, fast relaying of news and commands was necessary for tactical reactions. Claude Chappe indicated: "it is imperative to establish a rapid communication network, with which we can orchestrate the movements, simultaneously, of a million men dispersed on an immense space as if they were at the same place" [translated] [9].

In 1790, Claude Chappe created the first telegraph based on time-synchronized clocks to transmit information from one tower to the next. Afterward, he introduced a new design that did not require time synchronization and relied purely on optical observations. It relied on a wooden structure installed at the top of a tower, made up of one long beam and two shorter beams, see Figure 1. Operators inside each tower rotated the beams using pulleys, in steps of 45 degrees. This allowed for 4 positions of the long beam and 8 positions of each of the short beams, resulting in 256 symbols [8]. Information was encoded in pairs of symbols, resulting in a dictionary of 256 × 256 possible words. Messages were transmitted in a "multi-hop" manner: operators used binoculars to look at the position of the upstream tower and replicated it on theirs.

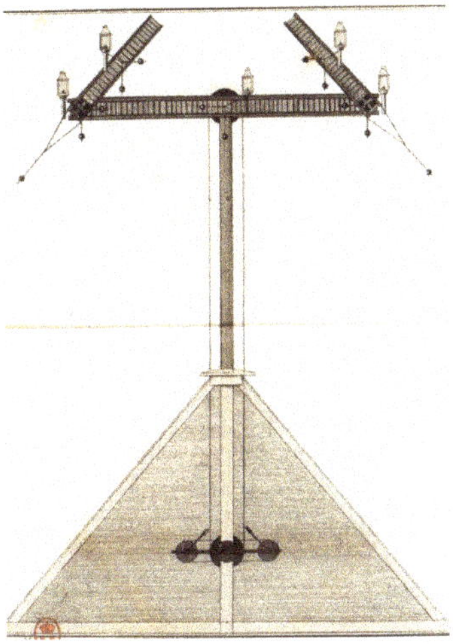

Figure 1. The Chappe telegraph adopted by the French state as depicted in Ignace Chappe's book. Three arms were used to convey signals and each could rotate at steps of 45°. (Source: [9]).

In September 1793, the General Assembly approved the establishment of a line of 15 towers from Paris to Lille. On 16 July 1794, the Paris-Lille line was completed and tested with 18 towers, covering a distance of 190 km with hops between 10 and 15 km. The transmission rate was estimated at one symbol per minute, i.e., one word every two minutes [8].

Chappe introduced further improvements to his telegraph: the tower was redesigned to withstand higher winds, a portable version of the telegraph was introduced for in-battle communication, and lanterns were installed on the beams to allow for operation at night [9]. The deployment of the towers expanded rapidly under Napoleon Bonaparte as he realized their tactical advantage [8]. By 1846, the Chappe telegraph covered most of France (Figure 2).

The telegraph expanded to other countries [9]. It was experimented with in England, Turkey, Sweden, and Russia. It crossed the Mediterranean during the rule of Mohamed Aly Pasha, governor of Egypt, who wanted to establish a communication line between Cairo and Alexandria. He imported models from France and had his architect choose how to place the machines. The messages traveled between Alexandria and Cairo in 40 min, with a line of 17 stations covering nearly 180 km.

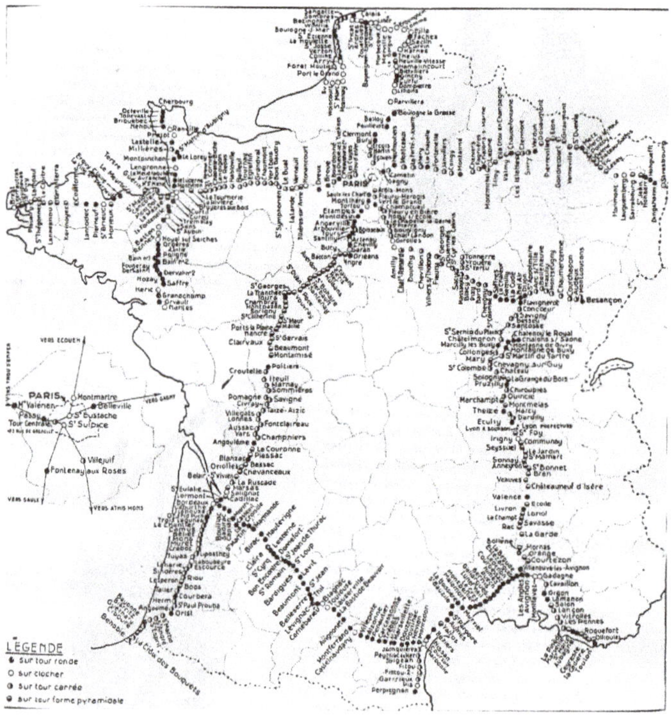

Figure 2. The Chappe Telegraph network deployed between 1794 and 1846. Each dot represents a tower. (Source: Cité des Télécoms).

Samuel Morse experimented with using electricity to communicate, and, in 1837, he introduced a single-wire technique [8] that was far more economical than the Chappe telegraph. In 1849, the electrical telegraph was tested over large distances and fully patented, marking the beginning of the end for the optical telegraph. It was slowly adopted in Europe and France; by 1881, all optical telegraphs had been replaced by the electrical telegraph [8].

3. Survey of Current IoT Technologies

Different IoT technologies offer different trade-off points between the communication range, throughput, and power consumption. Figure 3 gives examples of these technologies, together with an indication of the communication range. This section provides an overview of single-hop networks and multi-hop networks. We discuss the contributions of this article in light of the state-of-the-art.

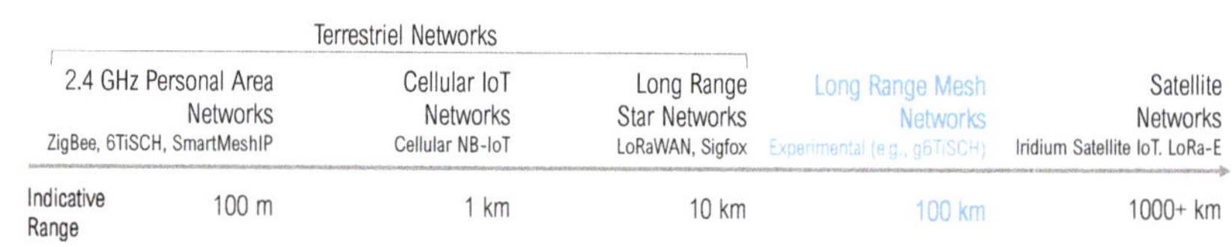

Figure 3. Existing low-power wireless technologies and their indicative ranges.

3.1. Single-Hop IoT Networks

This section discusses the four technologies of single-hop wireless networks for IoT: NB-IoT, cellular networks, LoRaWAN, Sigfox, and satellite networks.

Long-range connectivity is offered by cellular network operators using NB-IoT—an enhancement of long-term evolution (LTE). NB-IoT is standardized in Release 13 of the 3GPP [10]. It improves the communication range by combining sub-GHz frequencies (in the 900 MHz band) with a narrow channel bandwidth of 180 kHz [10]. It uses OFDM on top of O-QPSK modulation [11,12]. The communication range of NB-IoT is $\approx$1 km [13].

Another example is LoRaWAN, which uses a proprietary LoRa PHY [14]. LoRa uses chirp spread spectrum (CSS) FSK modulation, which makes it robust against multi-path fading, sub-GHz frequencies, and narrow-band channel bandwidths of 125 kHz (compared to 2 MHz for O-QPSK 2.4 GHz). Depending on regional regulations, the LoRa modulation may occupy channel bandwidths as narrow as 7.8 kHz. LoRa offers a range of bit rates, based on a spreading factor index (SF) of its CSS sweeps. The fastest bit rate of 5.4 kbps is achieved with SF-7 and the slowest bit rate of 293 bps using SF-12 (assuming 125 kHz bandwidth and 4/5 coding rate) [15]. Range tests show that LoRa PHY offers a 70% packet delivery ratio up to 10 km distance [16] in an urban setting without line-of-sight guarantees.

Sigfox was another LPWAN technology [17]. It featured sub-GHz operations, an ultra-narrowband channel bandwidth of 100 Hz, and a 100 bps bitrate [12]. Sigfox was an operated network: base stations were installed throughout the service area. However, due to the unstable market economy during the recent pandemic, Sigfox declared bankruptcy in early 2022 [18].

LPWAN technologies require base stations to provide connectivity to devices deployed around them in star topologies, typically within 10 km. Installing/maintaining such base stations is costly and may not be an option in remote areas or in complex critical infrastructures that do not allow invasive installations. Moreover, even as LPWANs show high link robustness outdoors or indoors [19,20], network coverage blind spots are inevitable (i.e., places where end devices are completely out of network range), such as in deep indoor devices (e.g., utility meters). An example of such blind spots of a LoRaWAN network was demonstrated in previous work [21]. In this case, it is often not economically viable to install a new base station only to service a few uncovered nodes. Users can still operate their own gateways for some technologies, such as LoRaWAN, but they need to provide their own network server and their own way of retrieving the received data. Therefore, only a minority of end users have the skills and resources to do that.

Low Earth orbiting (LEO) satellites orbit at an altitude between 160 and 2000 km, allowing them to offer network coverage to hundreds of square kilometers [22]. This allows them to be useful for remote sensing and monitoring systems spanning hundreds of kilometers. For example, they are suitable for IoT for over-the-horizon Maritime surveillance [23], environmental monitoring [24], emergency management, and smart grid monitoring [25]. LEO satellites follow two kinds of patterns: a Walker Star formation and a Walker Delta formation. In Walker Star, satellites orbit the earth in a 90° inclination, passing over both poles. They offer global coverage but fly over a given location infrequently, typically four times a day, with a total of 20 min per day of availability for a device. In Walker Delta, a satellite follows an inclined orbit close to the equator. This is interesting for populated areas around the equator but offers less global coverage. Direct connectivity to a satellite is available only once it is over the region of the device. There is a trade-off between network availability and constellation density (therefore, constellation and network cost).

Iridium is one technology for satellite IoT [26]. It relies on LEO satellites orbiting in a Walker Star formation [22] offering global coverage, with an orbital period of 100.13 min. The Iridium constellation contains 77 satellites. The technology relies on a combination of techniques for its PHY layers: an unmodulated tone, a unique word transmission in binary phase shift keying, and data payload transmission in O-QPSK. Reliability is achieved at the medium access control (MAC) layer by use of: (1) FDMA with 240 frequency channels and a 48 kHz channel bandwidth on the 1616 kHz baseband; and (2) TDMA with a 90 ms frame duration. Iridium devices can have as much as 10% of the lifetime of LoRaWAN or Sigfox devices when a message is sent every 10 min because, in part, of the long transmission procedure that can take between 6 and 20 s [26]. Therefore, their deployment is feasible only with solar power or with an electrical outlet. Similar satellite coverage solutions include the LoRaWAN-based Lacuna Space [27] and the proprietary Swarm Space network by SpaceX [28].

LoRa-E is an LPWAN technology designed for dense deployments and satellite connectivity [29]. It is an addition to the LoRa suite of modulations that uses long-range frequency-hopping spread spectrum (LR-FHSS) and provides a >155 dB link budget necessary for LEO satellites. In Europe, it uses a channel bandwidth of 137 or 336 KHz (compared to the typical 125 kHz for LoRa) and it divides the channel into sub-channels of 488 Hz of bandwidth (280 or 688 channels in the EU). Long-range and robustness improvements are achieved by a combination of transmitting duplicate headers, and fast hopping over a subset of the sub-channels, thereby allowing simultaneous transmissions in the same channel. Simulations show that LoRa-E increases the amount of network goodput by nearly two orders of magnitude compared to LoRa; therefore, it is a good candidate in dense environments [29].

Today, satellite connectivity is the only option when base stations cannot be installed.

3.2. Multi-Hop IoT Networks

In this section, we provide an overview of the two kinds of multi-hop networks: short-range and long-range. Some IoT personal area networks (PANs) operate at 2.4 GHz. One example is the IEEE O-QPSK 2.4 GHz PHY of the IEEE 802.15.4 standard. It uses a 2 MHz channel bandwidth and is the PHY of different protocol stacks [30]. The communication range of this PHY is in the order of tens of meters depending on the deployment environment. Networks using this PHY typically use multi-hop mesh topologies to extend their coverage. This includes Internet engineering task force (IETF) 6TiSCH standard protocol stack (IPv6 over the TSCH mode of IEEE 802.15.4e) [31], ZigBee [32], and SmartMesh IP [33].

Smart meter regulations in Europe mandate member states to provide smart meter capabilities as alternatives to old meters "unless [...] this is not cost-effective in relation to the estimated potential savings in the long term" [5]. Since cost-effectiveness is key to IoT networks, high-maintenance gateways and satellites may not be cost-effective, even if they are energy efficient [34]. Some applications have strict constraints on message frequency. For instance, peach orchards require regular reporting of humidity to predict frost, otherwise peach farmers can lose all of their crops [35]. This was witnessed as Argentina lost as low as 85% of its peach crops in 2013. Therefore, when such use cases are out of coverage of 10-km scale terrestrial LPWANs, it would be unpractical to use satellite devices, which offer only an average of 20-min of availability every day [26].

Long-range low-power wireless mesh networking can be an option in such cases. They allow a flexible coverage extension using low-cost battery-powered devices. The IEEE already adopted a family of 31 physical layers for long/short-range IoT connectivity [7]. Their performances were evaluated in an exhaustive range of test campaigns in indoor, urban, agricultural, and remote area scenarios [36]. They offered the bitrate as fast as 800 kbps using OFDM 868 MHz and a range as far as ≈14 km when using FSK 868 MHz.

Recent research has provided experimental evaluations of both single-hop and multi-hop IoT networks. Cattani et al. [37] showed the performances of single-hop LoRa links in three scenarios: indoors, outdoors, and underground. They ran the setup on a university campus and they showed the link quality in the received signal strength indicator (RSSI) and packet reception ratio (PRR). They reported the performance in ranges up to 135 m. They showed that the fastest bitrate PHY had a mean PRR that was only 10% less than the slowest bitrate PHY. Hardie and Donald [20] ran a performance evaluation campaign of single-hop LoRa links in the 433 MHz band in an underground communication scenario and they reported the link quality in RSSI and signal-to-noise ratio (SNR). They showed a maximum range of the LoRa link of 200 m in an underground to above-ground link. Similarly, Cecilio et al. [38] ran experimental evaluations of single-hop LoRa links for flood detection scenarios by placing the transmitters and the gateway at the sides of a lake. They reported link qualities in terms of RSSI, PRR, and SNR for three settings: low tide, high tide, as well as in rural settings. They showed a maximum range of the LoRa link up to 1300 m during the high tide when the transmitter was 2 m above ground but this range was reduced to 650 m when the transmitter was 1 m or less above ground. Valecce et al. [19] tested NB-IoT connectivity in an agricultural setup. They showed that NB-IoT maintained connectivity at 83% PRR to an underground cellar and up to 90% PRR in an open terrace. However, they did not report the maximum range to the base station since this was not provided by that NB-IoT setup. Liando et al. [39] performed the longest single-hop tests of LoRa links, to our knowledge. The authors ran LoRa links in three scenarios: outdoor line-of-sight, outdoor non-line-of-sight, and indoors. They showed (experimentally) that packet reception was possible at a maximum of 10 km using LoRa SF-7. Using regression analysis, they estimated that LoRa could reach up to 17 km using SF-12 (in the outdoor line-of-sight scenario). Finally, Parri et al. [40] carried out performance evaluations of LoRa links at 8.3 km of distance in a marine environment. The authors found that using LoRa SF-7 offered the best trade-off between link robustness and power consumption compared to higher SFs, an observation confirmed also in [37].

Further research provided experimental evaluations for multi-hop IoT networks. Tran et al. proposed a protocol stack for a multi-hop LoRa network based on time-slotted MAC and demonstrated the reliability of 95% in a two-hop network [41]. The authors evaluated it in a simulation in an area of 800×800 m^2 and experimentally in a 400×400 m^2 area, using SF-7 LoRa PHY. Mai et al. proposed a time-slotted MAC for the multi-hop LoRa network with a focus on minimizing latency using parallel transmissions [42]. The authors ran an experimental evaluation using the SF-7 LoRa PHY in a 400×400 m^2 area. Similarly, Basili et al. [43], provided an architecture for a multi-hop linear LoRa network that acted as an extension for a LoRaWAN gateway. They ran the experimental setup and they demonstrated the feasibility of the system in a four-hop network. In previous research, we used the IETF standard 6TiSCH protocol stack with the long-range FSK 868 MHz PHY [44]. We then proposed a generalized 6TiSCH protocol stack (g6TiSCH) to integrate any combination of long-range and short-range PHYs in the same network [45]. The g6TiSCH architecture was evaluated on an indoor testbed of 36 motes in a 100 m^2 area office setup. We showed how a network could improve its reliability by combining long-range and short-range radios in a mesh topology, using heterogeneous 6TiSCH slotframes [46].

The power consumption of the technologies discussed in this section (Figure 3) increases depending on several factors, such as: the current draw of the used radio chips, transmit power, MAC layer configuration, and link quality. Generally, it ranges from tens of milliamps for a 2.4 GHz mesh network or a LoRa transceiver [15,33,44], to hundreds of milliamps for satellite IoT transceiver [47].

The articles cited in this section demonstrate the growing interest in combining long-range PHYs with multi-hop connectivity. There are two common limitations to these articles. First, they did not test the maximum line-of-sight connectivity between the nodes as the maximum range achieved experimentally was for LoRa PHY at 10km [39]. Second, the outdoor long-range setup experiments with proprietary PHYs (i.e., LoRa or NB-IoT), did not consider the potential benefits of standard IEEE 802.15.4 PHYs. In this article, we went one step further by building a 103-km low-power wireless multi-hop network to demonstrate the potential of long-range low-power wireless full-fledged mesh networking. We used the standard IEEE 802.15.4 FSK 868 MHz PHY and we demonstrated that the range limit in line-of-sight could reach up to 12.7 km in the single-hop distance.

4. Experimental Setup

This section presents the planning of the experimental setup in two parts: configuring the network specifications (Section 4.1), and planning the geographical deployment (Section 4.2). A major challenge was to ensure reliability and speedy installation. The full demonstration had to be done in a single day. We only had 6 h of daylight and needed to deploy 11 balloons along a 103 km route. Reliability and speed were our main objectives.

4.1. Network Configuration

We used two sets of hardware, the OpenMote B Figure 4 and the proprietary Falco hardware. We use the generic term "mote" to refer to either. The OpenMote B [48,49] features the Texas Instruments CC2538 System-on-Chip [50] and the Atmel AT86RF215 radio chip [51]. The latter implemented all IEEE 802.15.4g PHYs; we used the IEEE standard FSK Option 1 modulation with forward error correction at the 868 MHz band at 50 kbps. This PHY layer was tested in previous research and showed the highest robustness in the family of the IEEE 802.15.4g PHYs [36].

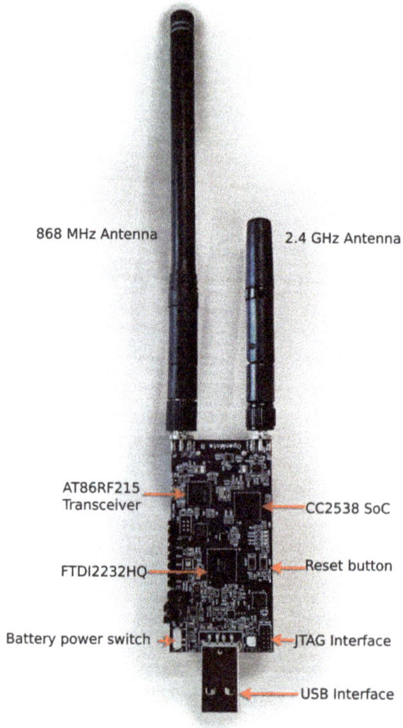

Figure 4. The OpenMote Bused in parts of the experiment.

We used a specially-crafted multi-hop communication protocol, shown in Figure 5. The root node was set to re-transmit a frame every second. At the MAC layer, each relay listened for packets from the previous hop. The relay transmitted the received packet three times, every 20 ms.

A packet included the source and destination fields. Each device statically allocated a unique identifier, from 0 to 10. Device 0 was the root. Each relay incremented the source and destination fields in the relayed packet, resulting in multi-hop routing. This was used to prevent loops and backward-relaying because a node dropped any packet that was not destined for it. For debugging, relay nodes appended the RSSI of the received packet to the relayed packet.

The packet format (Figure 6) consisted of three main parts: the packet header, payload, and footer containing the captured RSSIs along the path. The header contained the source and destination fields, a network ID used to filter out any packets received from outside the network, and a sequence number used to filter out duplicate packets. At the footer, 12 B were used to store RSSI packets of each of the 12 hops (10 hops between balloons and two extra hops to the terminal computers on the ground at both ends) in the network, as well as a 2 B CRC check. A packet had a maximum of 127 B.

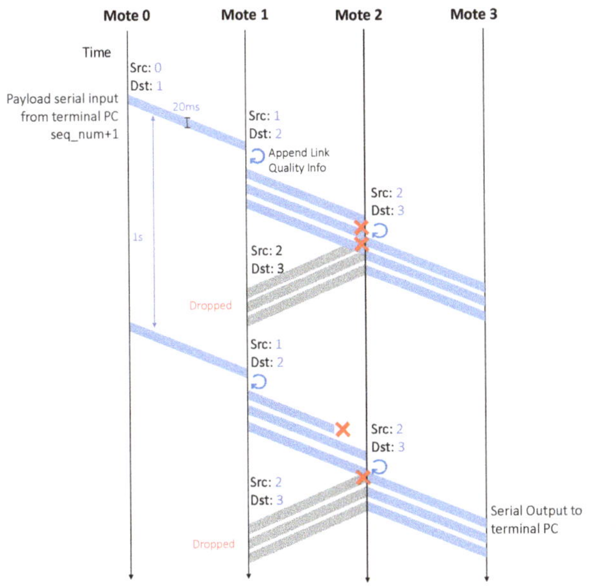

Figure 5. The communication protocol relays transmitted each received packet three times to increase reliability.

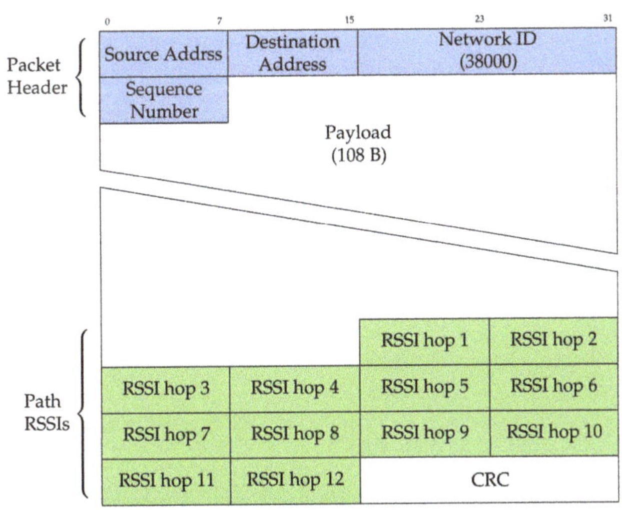

Figure 6. Format of the packet format used. Source and destination addresses are used for hop-by-hop routing.

4.2. Network Planning

We selected a set of locations in the southwest of Paris from the historical map of Claude Chappe towers (Figure 2). The locations are shown in Figure 7, corresponding to the following villages (from north to south): Torfou, Etampes, Angerville, Arbouville, Toury, Artenay, Chevilly, Bucy, Baccon, Cravant, Séris.

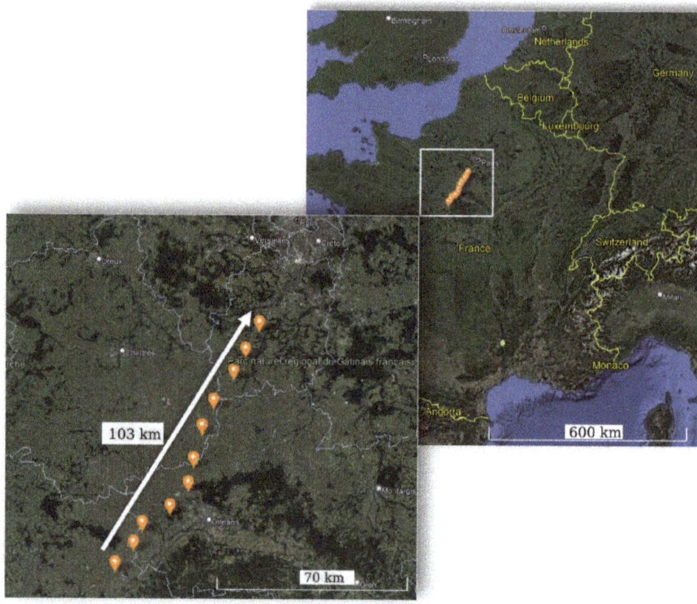

Figure 7. Location of the experiment in the southwest of Paris.

Two factors affected the selection of the exact coordinates for positioning the motes. The first challenge was the variation in the terrain altitude. Figure 8 shows the terrain altitude of the covered distance and the selected locations of the motes. For example, at the first hop, the altitude varied between 75 and 170 m at a distance of 10.5 km. This was enough of a difference to have it impact the communication range; the terrain profile is an important factor to take into account when selecting locations.

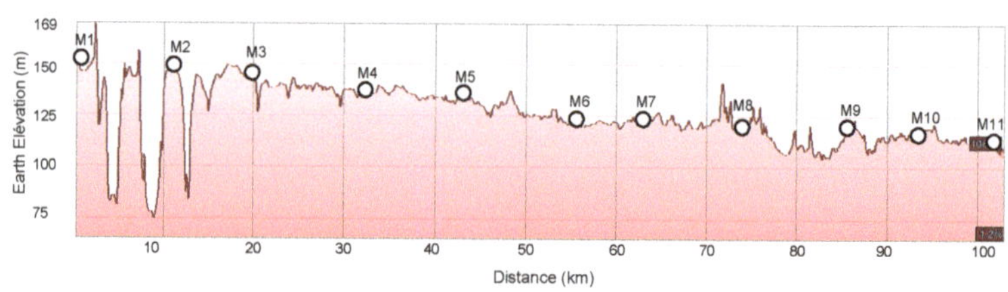

Figure 8. Terrain elevation is an important factor when selecting locations.

The second challenge was to ensure the distances between the motes were within the communication range of the motes. We needed to find locations close enough to one another while avoiding densely populated areas (to avoid inconveniencing residents). Table 1 outlines the distance of each hop along the path, resulting in a total of a 103 km network using 11 motes.

Table 1. Length of each hop in the network.

	Hop	Distance	Fresnel Clearance
1	Torfou–Etampes	10.48 km	30.1 m
2	Etampes–Angerville	8.97 km	27.9 m
3	Angerville–Arbouville	12.67 km	33.1 m
4	Arbouville–Toury	10.25 km	29.7 m
5	Toury–Artenay	11.86 km	32.0 m
6	Artenay–Chevilly	8.74 km	27.5 m
7	Chevilly–Bucy	10.69 km	30.4 m
8	Bucy–Baccon	11.57 km	31.6 m
9	Baccon–Cravant	7.93 km	26.2 m
10	Cravant–Séris	9.97 km	29.3 m
	total distance	**103.13 km**	

After selecting the locations, we needed to place the motes at appropriate heights that allowed for line-of-sight visibility. We were particularly aware of the Fresnel zone clearance: the line-of-sight between the transmitter and receiver needed to be clear of any obstructions by a certain distance called the Fresnel radius. The longer the range, the larger the radius that needed to be clear of obstructions [52]. The Fresnel zone is an elliptical-shaped zone between the transmitter and the receiver, as seen in Figure 9. Similar to the research in [38,40], we calculated the required Fresnel radius clearance. For a distance of d km between the transmitter and the receiver, and using frequency f GHz for communication, the maximum radius of the ellipse r was at the center of the ellipse and it could be calculated as in Equation (1). Using 868 MHz frequency and the distances of each hop, the minimum required Fresnel clearance is shown in Table 1 for each hop. Therefore, a theoretical minimum of 33 m of clearance above ground was needed. Further clearance was also required to account for intermediate bumps in terrain elevation that could reach up to 15 m as seen in the hop between mote 7 and mote 8 in Figure 8.

$$r = 8.656\sqrt{d/f} \tag{1}$$

With distances between motes up to 13 km (Table 1), and as we used the 868 MHz frequency band, a clearance of 33 m above ground or obstructions was needed, according to the Fresnel zone formula [52].

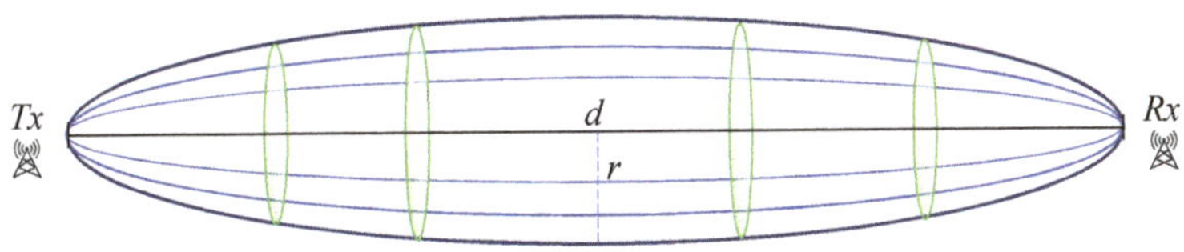

Figure 9. Illustration of the Fresnel zone between the transmitter and receiver.

5. A 103-km Wireless Network

To ensure a good communication range, we used helium balloons to lift the motes in the air. These needed to be large enough to keep the mote high enough even in the event of wind. We tested latex balloons with a diameter between 80 and 100 cm, and chloroprene balloons with a 120 cm diameter. We conducted the experiment in clear weather, at 14 °C, with wind speeds between 12 and 15 km/h, and wind gusts up to 18 km/h. We used helium with 97% purity. The payload consisted of a mote (either OpenMote B or Falco) powered by a pair of AA batteries, weighing a total of ≈100 g. We selected the chloroprene balloon, which we inflated to 100 cm diameter; even with an 80 m rope altitude, they stayed at 50 m or higher in the wind conditions of the day. Figure 10 shows one of the balloons used.

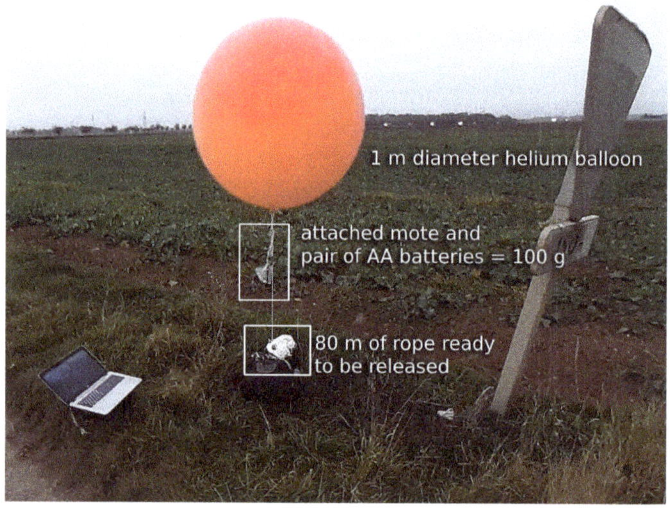

Figure 10. An OpenMote Bwas attached to a helium balloon.

At each location, we inflated the balloon, attached the mote under it, and released up to 80 m of the rope as shown in Figure 11. This allowed each balloon to be within the line of sight of the next, higher than the intermediate trees or the occasional house while respecting the Fresnel zone clearance. The experiment was conducted in rural areas where no trees or buildings were more than 20 m high. Having the balloon float at 50 m left sufficient space for the Fresnel zone at each hop. Therefore, we released enough rope so the balloon would float at 50 m or higher. Depending on the location, we attached the rope to a tree branch, a sign, or anything sturdy.

Figure 11. One out of 11 balloons carrying a mote.

We used two cars for the experiment, driving about 10 min from one another. The first car was in charge of inflating the balloon at the location, the second of attaching and testing the mote. A person stayed at the initial location (Torfou) with a mote connected to a laptop; he could type in a message that would be sent over the chain of motes. At each new location, we used another laptop attached to a mote to verify the message reached that far. We had no technical difficulty throughout the experiment; communication continuously worked at each location. We also had no balloon- or battery-related failures. The deployment of all 11 balloons took 6 h.

Figure 12 is a picture of the laptop screen at the last location (the village of Séris), 103 km away from the transmitting computer. It shows the messages sent by the person in Torfou, as well as the RSSI at each of the hops.

Figure 12. Captured packets 103 km away from the transmitting computer.

Figure 13 shows the recorded RSSI at each hop when using the OpenMote B. The sensitivity of the AT86RF215 when using the FSK PHY is −116 dBm; it is represented as the red bar in Figure 13. We needed the RSSI at each hop to be above that sensitivity. Figure 13 shows the RSSIs of 90 packets captured when we completed the 6 hours of deployment as box plots (median in the red bar and the box limits representing 25% and 75% of the data). We can see that hops 1, 5, and 8 were the "weakest" (the RSSI closest to sensitivity) while hops 2, 7, and 10 were the "strongest" (the motes could have been separated more).

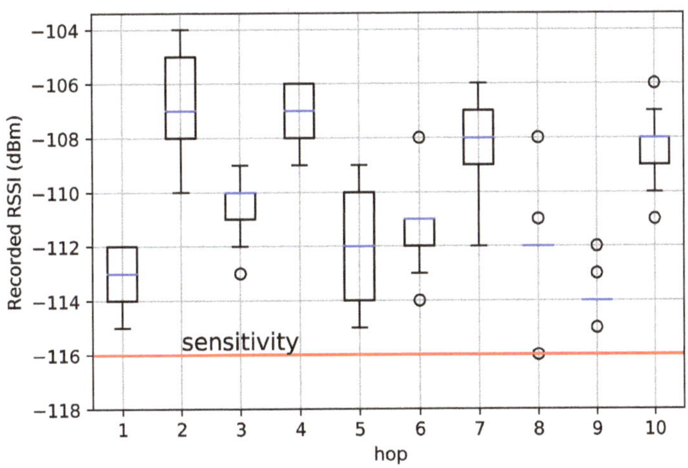

Figure 13. The RSSI at the receiving mote of each hop, when using the OpenMote B. The red bar shows the sensitivity of that radio in the configuration we used: we needed the RSSI of each hop to be above that.

The captured RSSIs showed promising link budgets of the standard FSK 868 MHz link at low-power consumption. The longest hop in the network was hop 3 with a 12.7 km distance. The packets captured at that hop showed median RSSI that was 6 dBm above the receiver sensitivity. It is worth noting the low-power consumption characteristics of the AT86RF215 chip used in this experiment: with 2.5V supply voltage, the current draw was 67 mA in transmission and 28 mA in receive mode [51]. When a similar test was conducted using LoRa SF-7 in line-of-sight, no packets were delivered beyond the 10 km range using the SX1276 radio chip [39]. Even though FSK 868 MHz is nearly five times higher in the data rate than LoRa SF-7, it still maintained packet reception up to 12.7 km with 6 dBm above the receiver threshold. This can be attributed to differences in experimental setups or transmit power configurations. This encourages further experimental evaluations, specifically of FSK 868 MHz compared to LoRa SF-7, especially as LoRa SF-7 is observed to offer the best trade-off between power consumption and link robustness in LoRa configurations [37,40].

6. Conclusions

Wireless networking has become an essential technology that serves several critical activities, such as smart industry, precision agriculture, and utility metering. In this article, we presented a historical overview of the development of long-range telecommunications by Claude Chappe. We discussed how long-range telecommunications through the Chappe telegraph paved the way for modern telecommunications.

As an homage to the Chappe telegraph, we replicated its vision, replacing its visual component with long-range radio communication. We demonstrated a 103 km network that we deployed in the southwest of Paris on a portion of a historical Chappe telegraph line.

The article does not argue for the specific PHY layer we used, nor proposes a specific technical architecture, as we focused on the simplicity and reliability of the network for the sake of the experiment. It serves as a strong highlight of the incredible potential of pushing the boundaries of long-range IoT technologies. Today, solutions such as LoRaWAN are, in fact, incredibly simple: motes transmit their data directly to a nearby base station. These technical choices are rather underwhelming, as they do not take advantage of many innovations in multi-hop mesh networking.

Moreover, while the protocol implemented is simplistic, it shows how a 10 km IoT technology can be used in a 100 km deployment. This is, we believe, where the road lies ahead.

Imagine you are a network operator in charge of the wireless connectivity that is meant to automatically read electricity meters. While the long-range IoT network you installed works well for most of your meters, approximately 1% of them are installed in "deep indoor" environments, such as in underground parking garages. Moreover, in your medium-sized city, this is over 1000 m scattered around. You could of course install many more gateways. Besides the obvious costs of their installation, it is the lengthy trial-and-error, the repeated repositioning, and the unhappy customers that are the real costs of this approach. The more sensible approach, we argue, is to augment the long-range technology with multi-hop capabilities: have motes "help out" the ones that are out of range of the gateway by relaying their data.

While we do not intend to provide a detailed performance evaluation of the network, with an exhaustive range-testing campaign, we find these initial results encouraging to explore the full potential of the standard IEEE 802.15.4g PHYs as opposed to proprietary technologies. Since we had successful packet receptions at distances up to 13 km with 6 dBm above the receiver threshold, we wonder if a standard PHY, such as FSK 868 MHz, can offer a reasonable trade-off between range and chip costs, due to the fact that it is a non-proprietary PHY. This may provide network architects with more diverse options for their networks with IEEE standard PHYs in addition to the proprietary LPWAN technologies.

Author Contributions: Methodology, M.R., J.M., R.A.-A., M.V., J.A.T., A.C., Q.L., D.B. and T.W.; Software, M.R., J.M., R.A.-A., M.V., J.A.T., A.C., Q.L., D.B. and T.W.; Writing—original draft, M.R., J.M., R.A.-A., M.V., J.A.T., A.C., Q.L., D.B. and T.W. All authors have read and agreed to the published version of the manuscript.

Funding: This work was partially funded by French Agence Nationale de la Recherche under contract CIFRE 2019/0605. This work was conducted while the first author was affiliated with Orange Labs and Inria Paris.

Institutional Review Board Statement: Not applicable.

Informed Consent Statement: Not applicable.

Data Availability Statement: Not applicable.

Conflicts of Interest: The authors declare no conflict of interest.

References

1. Civerchia, F.; Bocchino, S.; Salvadori, C.; Rossi, E.; Maggiani, L.; Petracca, M. Industrial Internet of Things Monitoring Solution for Advanced Predictive Maintenance Applications. *J. Ind. Inf. Integr.* **2017**, *7*, 4–12. [CrossRef]
2. Soualhi, M.; El Koujok, M.; Nguyen, K.T.P.; Medjaher, K.; Ragab, A.; Ghezzaz, H.; Amazouz, M.; Ouali, M.S. Adaptive Prognostics in a Controlled Energy Conversion Process Based on Long- and Short-term Predictors. *Appl. Energy* **2021**, *283*, 116049. [CrossRef]
3. Pister, K.; Phinney, T.; Thubert, P.; Dwars, S. Industrial Routing Requirements in Low-Power and Lossy Networks. 2009. Available online: https://www.rfc-editor.org/rfc/rfc5673.html (accessed on 19 February 2022).
4. Izadi, M.; Abd Rahman, M.S.; Ab-Kadir, M.Z.A.; Gomes, C.; Jasni, J.; Hajikhani, M. The influence of Lightning Induced Voltage on the Distribution Power Line Polymer Insulators. *PLoS ONE* **2017**, *12*, e0172118. [CrossRef]
5. The European Council. Directive 2006/32/EC of the European Parliament and of the Council on Energy End-Use Efficiency and Energy Services and Repealing Council Directive 93/76/EEC. *OJ* **2006**, *114*, 64–85.
6. The European Commission. M/441 Standardisation Mandate to CEN, CENELEC And ETSI in the Field of Measuring Instruments for the Development of an Open Architecture for Utility Meters Involving Communication Protocols Enabling Interoperability. 2009. Available online: https://ec.europa.eu/growth/tools-databases/mandates/index.cfm?fuseaction=search.detail&id=421 (accessed on 19 February 2022).
7. IEEE. IEEE Standard for Local and Metropolitan Area Networks—Part 15.4: Low-Rate Wireless Personal Area Networks (LR-WPANs) Amendment 3: Physical Layer (PHY) Specifications for Low-Data-Rate, Wireless, Smart Metering Utility Networks. 2012. Available online: https://ieeexplore.ieee.org/document/6190698 (accessed on 19 February 2022).
8. Selleri, S. Claude Chappe and the First Telecommunication Network (without Electricity). *IEEE URSI Radio Sci. Bull.* **2017**, *2017*, 96–101. [CrossRef]
9. Chappe, I.U.J. *Histoire de la Télégraphie*; V-25091; Bibliothèque Nationale de France, Département Réserve des Livres Rares: Paris, France, 1840.
10. Ratasuk, R.; Mangalvedhe, N.; Zhang, Y.; Robert, M.; Koskinen, J.P. Overview of narrowband IoT in LTE Rel-13. In Proceedings of the 2016 IEEE Conference on Standards for Communications and Networking (CSCN), Berlin, Germany, 31 October–2 November 2016.
11. Foubert, B.; Mitton, N. Long-Range Wireless Radio Technologies: A Survey. *Future Internet* **2020**, *12*, 13. [CrossRef]
12. Mekki, K.; Bajic, E.; Chaxel, F.; Meyer, F. Overview of Cellular LPWAN Technologies for IoT Deployment: Sigfox, LoRaWAN, and NB-IoT. In Proceedings of the 2018 IEEE International Conference on Pervasive Computing and Communications Workshops (PerCom Workshops), Athens, Greece, 19–23 March 2018.
13. Manzoor, B.; Al-Hourani, A.; Al Homssi, B.; Magowe, K.; Kandeepan, S.; Chavez, K.G. Evaluating Coverage Performance of NB-IoT in the ISM-band. In Proceedings of the 2020 27th International Conference on Telecommunications (ICT), Bali, Indonesia, 5–7 October 2020.
14. LoRa Alliance. What Is LoRaWAN® Specification. Available online: https://lora-alliance.org/about-lorawan/ (accessed on 23 September 2022).
15. Semtech Corporation. Datasheet: SX1276/77/78/79—137 MHz to 1020 MHz Low Power Long Range Transceiver. 2020. Available online: https://www.semtech.com/products/wireless-rf/lora-connect/sx1276 (accessed on 18 February 2022).
16. Petajajarvi, J.; Mikhaylov, K.; Roivainen, A.; Hanninen, T.; Pettissalo, M. On the Coverage of LPWANs: Range Evaluation and Channel Attenuation Model for LoRa Technology. In Proceedings of the 2015 14th International Conference on ITS Telecommunications (ITST), Copenhagen, Denmark, 2–4 December 2015.
17. Sigfox Technology. Available online: https://www.sigfox.com/en/what-sigfox/technology (accessed on 8 April 2022).
18. Lunden, I.; TechCrunch. Sigfox, the French IoT Startup That Had Raised More than $300M, Files for Bankruptcy Protection as It Seeks a Buyer. 2022. Available online: https://techcrunch.com/2022/01/27/sigfox-the-french-iot-startup-that-had-raised-more-than-300m-files-for-bankruptcy-protection-as-it-seeks-a-buyer/ (accessed on 20 July 2022).
19. Valecce, G.; Petruzzi, P.; Strazzella, S.; Grieco, L.A. NB-IoT for Smart Agriculture: Experiments from the Field. In Proceedings of the 2020 7th International Conference on Control, Decision and Information Technologies (CoDIT), Prague, Czech Republic, 29 June–2 July 2020.
20. Hardie, M.; Hoyle, D. Underground Wireless Data Transmission Using 433-MHz LoRa for Agriculture. *Sensors* **2019**, *19*, 4232. [CrossRef]
21. Rady, M. Agile Multi-PHY Wireless Networking. Ph.D. Thesis, Sorbonne University, Paris, France, 2022.

22. Fraire, J.A.; Céspedes, S.; Accettura, N. Direct-To-Satellite IoT—A Survey of the State of the Art and Future Research Perspectives. In *Ad-Hoc, Mobile, and Wireless Networks*; Springer: Berlin/Heidelberg, Germany, 2019; pp. 241–258.
23. Petrovic, R.; Simic, D.; Cica, Z.; Drajic, D.; Nerandzic, M.; Nikolic, D. IoT OTH Maritime Surveillance Service over Satellite Network in Equatorial Environment: Analysis, Design and Deployment. *Electronics* **2021**, *10*, 2070. [CrossRef]
24. Moron-Lopez, J.; Cristina Rodriguez-Sanchez, M.; Carreno, F.; Vaquero, J.; Pompa-Pernia, A.G.; Mateos-Fernandez, M.; Aguilar, J.A.P. Implementation of Smart Buoys and Satellite-Based Systems for the Remote Monitoring of Harmful Algae Bloom in Inland Waters. *IEEE Sens. J.* **2021**, *21*, 6990–6997. [CrossRef]
25. De Sanctis, M.; Cianca, E.; Araniti, G.; Bisio, I.; Prasad, R. Satellite Communications Supporting Internet of Remote Things. *IEEE Internet Things J.* **2016**, *3*, 113–123. [CrossRef]
26. Gomez, C.; Darroudi, S.M.; Naranjo, H.; Paradells, J. On the Energy Performance of Iridium Satellite IoT Technology. *Sensors* **2021**, *21*, 7235. [CrossRef]
27. Lacuna. Available online: https://lacuna.space/about/ (accessed on 29 August 2022).
28. Swarm. Available online: https://swarm.space/industries/ (accessed on 29 August 2022).
29. Boquet, G.; Tuset-Peiró, P.; Adelantado, F.; Watteyne, T.; Vilajosana, X. LoRa-E: Overview and Performance Analysis. 2020. Available online: https://hal.inria.fr/hal-03115551/document (accessed on 20 July 2022).
30. IEEE. IEEE Standard for Local and Metropolitan Area Networks—Part 15.4: Low-Rate Wireless Personal Area Networks (LR-WPANs). 2003. Available online: https://ieeexplore.ieee.org/document/1237559 (accessed on 19 February 2022).
31. Vilajosana, X.; Pister, K.; Watteyne, T. Minimal IPv6 over the TSCH Mode of IEEE 802.15.4e (6TiSCH) Configuration. 2017. Available online: https://datatracker.ietf.org/wg/6tisch/documents/ (accessed on 19 February 2022).
32. What Is Zigbee? Available online: https://zigbeealliance.org/solution/zigbee/ (accessed on 8 April 2022)
33. Watteyne, T.; Doherty, L.; Simon, J.; Pister, K. Technical Overview of SmartMesh IP. In Proceedings of the International Workshop on Extending Seamlessly to the Internet of Things (esIoT), Taichung, Taiwan, 3–5 July 2013.
34. Rady, M.; Georges, J.P.; Lepage, F. Can Energy Optimization Lead to Economic and Environmental Waste in LPWAN Architectures? *ETRI J.* **2020**, *43*, 173–183. [CrossRef]
35. Watteyne, T.; Laura Diedrichs, A.; Brun-Laguna, K.; Emilio Chaar, J.; Dujovne, D.; Taffernaberry, C.J.; Mercado, G. PEACH: Predicting Frost Events in Peach Orchards Using IoT Technology. *EAI Endorsed Trans. Internet Things* **2016**, *2*, e2. [CrossRef]
36. Muñoz, J.; Chang, T.; Vilajosana, X.; Watteyne, T. Evaluation of IEEE802.15.4g for Environmental Observations. *Sensors* **2018**, *18*, 3468. [CrossRef] [PubMed]
37. Cattani, M.; Boano, C.; Römer, K. An Experimental Evaluation of the Reliability of LoRa Long-Range Low-Power Wireless Communication. *J. Sens. Actuator Netw.* **2017**, *6*, 7. [CrossRef]
38. Cecílio, J.; Ferreira, P.M.; Casimiro, A. Evaluation of LoRa Technology in Flooding Prevention Scenarios. *Sensors* **2020**, *20*, 4034.
39. Liando, J.C.; Gamage, A.; Tengourtius, A.W.; Li, M. Known and Unknown Facts of LoRa. *ACM Trans. Sens. Netw.* **2019**, *15*, 1–35. [CrossRef]
40. Parri, L.; Parrino, S.; Peruzzi, G.; Pozzebon, A. Low Power Wide Area Networks (LPWAN) at Sea: Performance Analysis of Offshore Data Transmission by Means of LoRaWAN Connectivity for Marine Monitoring Applications. *Sensors* **2019**, *19*, 3239. [CrossRef]
41. Tran, H.P.; Jung, W.S.; Yoo, D.S.; Oh, H. Design and Implementation of a Multi-Hop Real-Time LoRa Protocol for Dynamic LoRa Networks. *Sensors* **2022**, *22*, 3518. [CrossRef]
42. Mai, D.L.; Kim, M.K. Multi-Hop LoRa Network Protocol with Minimized Latency. *Energies* **2020**, *13*, 1368. [CrossRef]
43. Basili, F.; Parrino, S.; Peruzzi, G.; Pozzebon, A. IoT Multi-Hop Facilities via LoRa Modulation and LoRaWAN Protocol within Thin Linear Networks. In Proceedings of the 2021 IEEE Sensors Applications Symposium (SAS), Sundsvall, Sweden, 23–25 August 2021.
44. Rady, M.; Lampin, Q.; Barthel, D.; Watteyne, T. No Free Lunch—Characterizing the Performance of 6TiSCH When Using Different Physical Layers. *Sensors* **2020**, *20*, 4989. [CrossRef] [PubMed]
45. Rady, M.; Lampin, Q.; Barthel, D.; Watteyne, T. g6TiSCH: Generalized 6TiSCH for Agile Multi-PHY Wireless Networking. *IEEE Access* **2021**, *1*, 84465–84479. [CrossRef]
46. Rady, M.; Lampin, Q.; Barthel, D.; Watteyne, T. 6DYN: 6TiSCH with Heterogeneous Slot Durations. *Sensors* **2021**, *21*, 1611. [CrossRef] [PubMed]
47. Iridium. Datasheet Detailing the Specifications of the Iridium 9603 Transceiver. 2020. Available online: https://www.cls-telemetry.com/wp-content/uploads/2018/12/Iridium-9603-9603N_Developers-Guide.pdf (accessed on 20 July 2022).
48. Tuset, P.; Vilajosana, X.; Watteyne, T. OpenMote+: A Range-Agile Multi-Radio Mote. In Proceedings of the International Conference on Embedded Wireless Systems and Networks (EWSN), Graz, Austria, 15–17 February 2016; pp. 333–334.
49. Industrial Shields. Open Mote B User Guide. 2019. Available online: https://openmote.com/wp-content/uploads/2021/06/content.pdf (accessed on 20 July 2022).
50. Texas Instruments. Datasheet: CC2538 Powerful Wireless Microcontroller System-On-Chip for 2.4-GHz IEEE 802.15.4, 6LoWPAN, and ZigBee Applications. 2015. Available online: https://www.ti.com/product/CC2538 (accessed on 20 July 2022).

51. Atmel. Atmel AT86RF215 Device Family Datasheet. 2016. Available online: https://www.alldatasheet.com/view.jsp?Searchword=At86rf215%20datasheet&gclid=CjwKCAjwhNWZBhB_EiwAPzlhNl_tgjCml2-2kS3A-bt-j5LtPM-gELvaFy8Gn-b-VnGuJWTNWYsNEhoCONYQAvD_BwE (accessed on 20 July 2022).
52. Dobkin, D.M. *RF Engineering for Wireless Networks: Hardware, Antennas, and Propagation (Communications Engineering)*; Newnes: London, UK, 2004.

sensors

MDPI

Article

ParcEMon: IoT Platform for Real-Time Parcel Level Last-Mile Delivery Greenhouse Gas Emissions Reporting and Management

Ali Yavari [1,*], Hamid Bagha [1], Harindu Korala [2], Irfan Mirza [1], Hussein Dia [3], Paul Scifleet [4], Jason Sargent [4] and Mahnaz Shafiei [1]

1 School of Science, Computing and Engineering Technologies, Swinburne University of Technology, Melbourne, VIC 3122, Australia
2 Institute of Railway Technology, Monash University, Melbourne, VIC 3800, Australia
3 Department of Civil and Construction Engineering, Swinburne University of Technology, Melbourne, VIC 3122, Australia
4 School of Business, Law and Entrepreneurship, Swinburne University of Technology, Melbourne, VIC 3122, Australia
* Correspondence: mail@aliyavari.com

Abstract: Transport is Australia's third-largest source of greenhouse gases accounting for around 17% of emissions. In recent times, and particularly as a result of the global pandemic, the rapid growth within the e-commerce sector has contributed to last-mile delivery becoming one of the main emission sources. Delivery vehicles operating at the last-mile travel long routes to deliver to customers an array of consignment parcels in varying numbers and weights, and therefore these vehicles play a major role in increasing emissions and air pollutants. The work reported in this paper aims to address these challenges by developing an IoT platform to measure and report on real-world last-mile delivery emissions. Such evaluations help to understand the factors contributing to freight emissions so that appropriate mitigation measures are implemented. Unlike previous research that was completed in controlled laboratory settings, the data collected in this research were from a delivery vehicle under real-world traffic and driving conditions. The IoT platform was tested to provide contextualised reporting by taking into account three main contexts including vehicle, environment and driving behaviours. This approach to data collection enabled the analysis of parcel level emissions and correlation of the vehicle characteristics, road conditions, ambient temperature and other environmental factors and driving behaviour that have an impact on emissions. The raw data collected from the sensors were analysed in real-time in the IoT platform, and the results showed a trade-off between parcel weight and total distance travelled which must be considered when selecting the best delivery order for reducing emissions. Overall, the study demonstrated the feasibility of the IoT platform in collecting the desired levels of data and providing detailed analysis of emissions at the parcel level. This type of micro-level understanding provides an important knowledge base for the enhancement of delivery processes and reduction of last-mile delivery emissions.

Keywords: IoT; greenhouse gas; sustainable logistics; last-mile emission; supply chain

Citation: Yavari, A.; Bagha, H.; Korala, H.; Mirza, I.; Dia, H.; Scifleet, P.; Sargent, J.; Shafiei, M. ParcEMon: IoT Platform for Real-Time Parcel Level Last-Mile Delivery Greenhouse Gas Emissions Reporting and Management. *Sensors* **2022**, 22, 7380. https://doi.org/10.3390/s22197380

Academic Editors: Liliana Anchidin, Adrian I. Petrariu and Alexandru Lavric

Received: 2 September 2022
Accepted: 24 September 2022
Published: 28 September 2022

1. Introduction

Fossil fuel combustion has made vehicles one of the main sources of greenhouse gas (GHG) emissions. In Australia, transport is the second main sources of GHG emissions after electricity, and this emission rate is on the increase. In 2017, the per capita CO_2 emission produced by transport sector in Austerlia was 32.7 tons. More importantly, road transport in Australia contributes to 85% of transport sector pollution and GHG emissions, which is more than global average [1]. One of the main transport sectors which contributes to GHG emissions is last-mile delivery. In particular, rapid growth of e-commerce has significantly increased the number of last-mile deliveries in the last decade [2–4]. Last-mile delivery emissions have been discussed in previous studies [5–8]. However, research to date has concentrated on estimating the last-mile delivery emissions based on theoretical data about vehicle emission and have not conducted field experiments to determine last-mile delivery emissions.

This paper utilises the Internet of Things (IoT) to deploy multiple sensors on a delivery van to analyse the emissions at parcel level in real-time. The resulting data provide comprehensive information on how different factors can impact delivery and how the delivery process can be enhanced to reduce the last-mile delivery emission.

Vehicle emissions are an inevitable consequence of fossil fuel combustion, but there are certain factors which impact vehicle fuel combustion and emission rates that can be monitored and addressed. These factors can be categorised into several major categories including road condition, driving style, vehicle condition, vehicle mass and weather condition [9].

In fuel consumption analysis research, driving style is categorised into two main categories which are aggressive driving and eco-driving. Aggressive driving refers to high acceleration and deceleration and high speed with sudden breaking patterns. Eco-driving, on the other hand, refers to smooth acceleration and deceleration, optimal gear shifting, and driving with optimal speed. Vehicle condition is another category that impacts vehicle emissions and includes multiple factors such as lubrication, tyres condition, engine tune and air filter. Another aspect impacting vehicle emissions is weather conditions, with rain, snow and ambient temperature effecting fuel consumption [9]. While each of these categories has been a focal point of other research aiming to determine their impact on fuel consumption and vehicle emissions, most of these studies have been conducted in a laboratory environment on chassis dynamometers. Studies show that the results of such tests in controlled laboratory environments are remarkably different from actual vehicles emissions in the real world [10–12]. Weiss et al. [10] performed emissions testing in both laboratory and real-world contexts. They argue that, even though that laboratory testing can be used to perform repetitive tests in identical conditions to compare the acquired results, the laboratory environment fails to capture all the factors which impact fuel consumption and vehicle emissions in real-world. Therefore, they argue that there is need for data collection under ordinary operating condition on the road to complement the laboratory data and obtain accurate information to find the correlation between different factors and vehicle emissions.

These studies demonstrate that, in order to obtain comprehensive and accurate data regarding last-mile delivery emissions and their correlation with internal and external factors, there is need to perform field evaluation in real-world context. Although obtaining fuel consumption and vehicle emissions data in a real-world environment would be beneficial to better determine various emission factors impact, but performing such tests has certain challenges. One of the challenges is that laboratory devices are designed to be in fixed positions and are usually connected to the vehicle using wired communication technologies. In addition, gas analyser devices are mostly designed to collect data from cars in stationary mode. These facts hinder deployment of sensors and gas analysers on a moving car. In order to capture live data from vehicles in real-world environment the devices which are used in laboratory must be modified to transfer live data to cloud via wireless communication. Technological advancement and emergence of the IoT has provided substantial advantages to address similar challenges in capturing and processing live and heterogeneous data from multiple sensors in several real-world applications such as precision agriculture, smart cities, healthcare, environmental monitoring and so forth [13–18].

The IoT enables the automation of data collection with different types of sensors and integration of various data types into a single data model without human interaction. In addition, through edge computing, the IoT enables primary data processing in the same location where data are collected by sensors. The data are then transferred from edge device to the IoT cloud for real-time visualisation and data analysis. To the best of our knowledge, no field experiment to determine last-mile delivery emission at a parcel level has been conducted before. Such research can provide micro level understanding of last-mile delivery vehicles emission and can provide the information required to improve the efficiency and effectiveness of last-mile delivery procedures.

The main contribution of this research is the design, implementation, and evaluation of an IoT-based emission monitoring platform (referred to as ParcEMon) which enables parcel level emission analysis of last-mile delivery vehicles in real-world contexts.

The rest of this paper is organised as follows: Section 2 discusses the related work, Section 3 presents the methodology including platform architecture, platform implementation and data collection process. Section 4 presents the results of the research. Section 5 concludes this paper.

2. Related Work

Different modes of transport such as road, rail, aviation and shipping all result in the emission of GHG and air pollutants through fossil fuel combustion. However, the amount of emission across each transport mode is different with road transport being the most prolific producer of emissions in this sector. Road transport causes around 12% of entire global GHG emissions [19] with light duty vehicles contributing around 72% of this value [20]. Due to the fact that light duty vehicles such as vans are among the main vehicles used in last-mile deliveries, last-mile deliveries must be considered a contributor to GHG emissions. Although emission factors are highly intertwined, research identifies the most prominent emission factors to include driving behaviour, vehicle condition, vehicle mass, aerodynamics, road condition and weather condition.

One of the main factors which impact vehicle emission is driving behaviour. Merkisz et al. [21] conducted a research to measure driving style influence on CO_2 emission in the real environment. They characterised driving behaviour into three different eco, normal and aggressive styles. Their research shows that eco-driving results in 4.5% less fuel consumption compared to normal driving style and 12.4% less compared to aggressive driving style. Allison et al. [22] argue that eco-driving training has a short-term impact on drivers' behaviour and after a short period they return to their normal habit of driving. They mention that there is a need for a constant feedback mechanism to continuously inform drivers about the financial and environmental benefits of their eco-driving behaviour. The IoT can play a major role in implementing such real-time feedback systems to continuously encourage the drivers to follow eco-driving behaviour.

Different aspects of vehicle conditions such as lubricants, tyres and engine maintenance can impact fuel consumption. Around 25% of vehicle energy is used to overcome friction in different components of vehicles and using low viscosity lubricants can improve energy consumption. Lowering internal friction in an engine using a suitable motor oil can reduce fuel consumption by 2.5%. Tyres directly impact vehicle resistance which in turn impacts fuel consumption. Low resistance tyres can reduce fuel consumption by 3%. Misaligned wheels and poorly tuned engines are among other vehicle conditions which adversely impact fuel consumption. Flaws like this in vehicles can increase fuel consumption by 3.5% [9]. Vehicle weight is another factor which directly impacts fuel consumption. Zervas et al. [23] argue that, in order to control future gas emission, not only the efficiency of cars in fuel consumption must increase but also their weight must not exceed certain upper limits to reduce fuel consumption. They conducted research on passengers cars and analysed the CO_2 emission reduction when the weight of cars decreases. The result shows that, when compared to the reference 1600 kg weight limit, cars with 1400 kg weight generate 9% less CO_2, cars with 1200 kg weight generate 16% less CO_2 and cars with 1000 kg weight generate 28% less CO_2.

Studies show that driving uphill can increase fuel consumption by 13%. Road roughness and unevenness also can increase fuel consumption by 2.7% [9]. Zabaar and Chatti [24] conducted research to analyse road roughness on fuel consumption. They used five vehicles in field trials with different weights including a medium car, SUV, van, light truck and heavy truck. They argue that the impact of roughness is intertwined with other factors including vehicle weight, aerodynamics, temperature and road grade. Although several road factors would impact fuel consumption and emission rate, but the most impactful road characteristics on fuel consumption is from traffic congestion, based on the number of vehicles on the road and traffic status. Vehicles on congested roads with heavy traffic condition require repeated decreases and increases in the speed over a long period of time and such fluctuations in speed increase fuel consumption. Greenwood et al. [25] conducted a test to analyse the impact of traffic congestion on fuel consumption and emissions. They found out that fuel consumed when compared to steady speed consumption for a real-life section of motorway increased by around 13% over a 24 h period. Similarly, various vehicle emissions increased by as much as 25%.

Different weather conditions such as precipitation, temperature and air density also impact fuel consumption. Rain and snow both impact the rolling resistance of a car as well as the road surface characteristics. For a depth of one mm, rain can increase fuel consumption by 30%. Temperature is another factor which impacts tyres and also engines due to a cold start. A study shows that temperature between 0 to 20 can increase fuel consumption by 10% [9]. Saboohi and Farzaneh [26] developed a model for eco-driving based on least fuel consumption. They argue that air resistance has the highest impact on fuel consumption compared to other weather-related emission factors. In addition to direct impacts on vehicle and road conditions, various weather conditions can also increase fuel consumption and vehicle emissions by forcing the driver to regularly decelerate and accelerate and prevent eco-driving style.

Laboratory, simulation and real-world context are the environments where data analysis is performed to determine vehicle emissions and air pollutants.

Pelkmans and Debal [11] have performed data collection using a chassis dynamometer to compare laboratory emission with on-road emissions. The data which were captured include speed, relative positive acceleration (RPA), fuel consumption and CO_2, CO, NOx, THC and PM emissions. Based on their findings, the emissions rate in laboratory is 10–20% less compared to on-road emission. Weiss et al. [10] used a portable emission measurement system (PEMS) to analyse on-road emission of light duty vehicles in Italy. They installed gas analyser and other components including GPS, humidity, temperature and pressure sensors inside the test vehicles and the data were stored locally. The test was conducted using 12 light duty vehicles on four different routes with different characteristics representing rural, urban, uphill/downhill and motorway driving. They argue that their test provides accurate data which indicate on-road NOx emissions of light duty diesel vehicles exceeds Euro 3–5 emission limits, whereas on-road CO and THC emissions generally remain below the Euro 3–5 emission limits.

Miles et al. [27] propose an IoT-based decision support system for monitoring and mitigating pollution in smart cities. They used the IoT capability to integrate data from multiple sources to determine the factors which impact vehicle emission. They integrated vehicle, weather and traffic data using an IoT platform to improve their decision-making process and select appropriate mitigation strategies such as signal optimisation, heavy vehicle ban, parking regulations and road closure or diversion. However, the data collection from vehicle emissions is not IoT-based and standardised vehicle emission models were integrated into the IoT platform to simulate the data.

Much of the prior research conducted to analyse how various emission factors impact fuel consumption and vehicle emissions has acquired data for analysis either from laboratory or from simulation applications. Even though such data can reveal significant information regarding correlation between emission factors and fuel consumption, they cannot fully implement real on-road contexts and characteristics [11]. Therefore, there is always a certain data gap between laboratory data and on-road data. Moreover, to the best of our knowledge, no research has been conducted so far to integrate the IoT capabilities with an advanced industrial gas analyser to analyse vehicle on-road emission in real-time.

As a result, the focus of this research is to utilise the IoT capabilities to capture data from multiple emission factors in real on-road contexts in order to analyse parcel level emissions in last-mile delivery processes. The architecture, implementation and data collection of the ParcEMON IoT-based emission monitoring system are addressed in the following section.

3. Methodology

In this research, contextual data [28] regarding three different areas of impact on emissions are gathered. These contexts include environment, vehicle and driver behaviour contexts. Each of these contexts have certain parameters which can effect the GHG emission rate. Figure 1 shows the impacting contexts and related parameters.

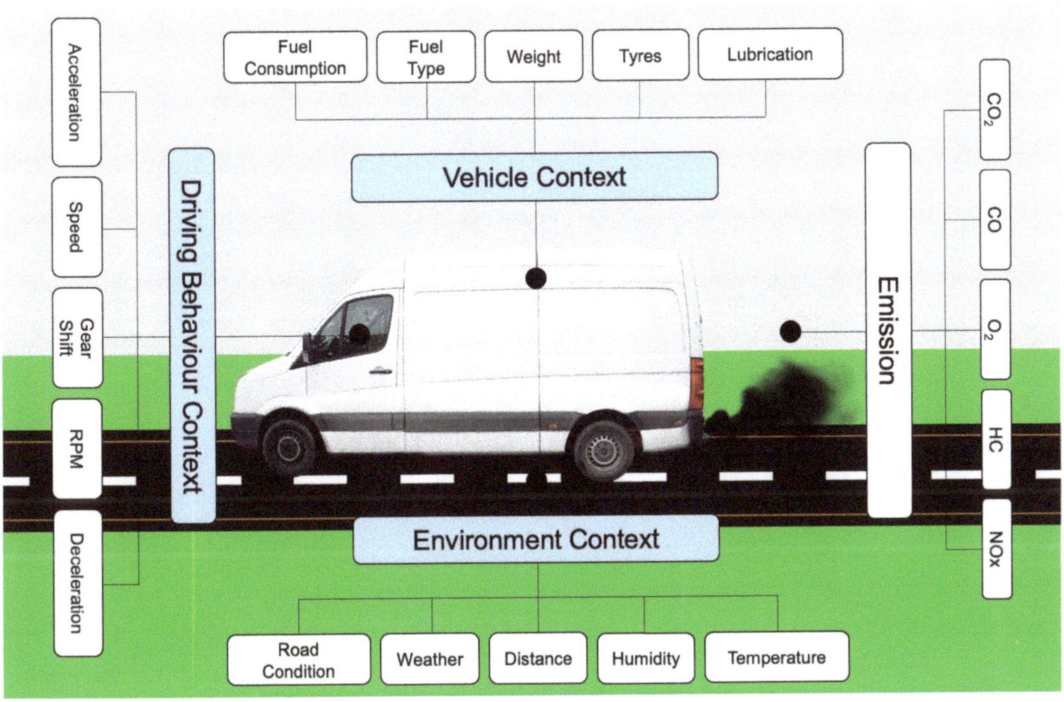

Figure 1. Emission impacting contexts and related parameters.

Data collection from the discussed contexts in Figure 1 was conducted using several IoT-based sensors from a van in last-mile deliveries. Data collection was conducted during the autumn season where van emissions were monitored in real time via the proposed ParcEMon platform. The van loaded the goods to be delivered and started the deliveries to different locations from 7:00 a.m. Different data including speed, acceleration, deceleration, gear shift, RPM, external temperature, road condition and vehicle path were captured. The data were then transferred to the IoT cloud using an edge device installed in the van for further data analysis and identification of their correlation with vehicle emissions. In Section 3.1, platform architecture is addressed, and, in Section 3.2, platform implementation is discussed.

3.1. Platform Architecture

In this research, an IoT testbed was developed to enable data capturing of a last-mile delivery van in a real-world context. The developed platform is based on a three layer IoT architecture including device, edge and application layers. The data between these layers can be exchanged through a wide range of wired and wireless communication protocols. The IoT architecture and the communication protocols are depicted in Figure 2.

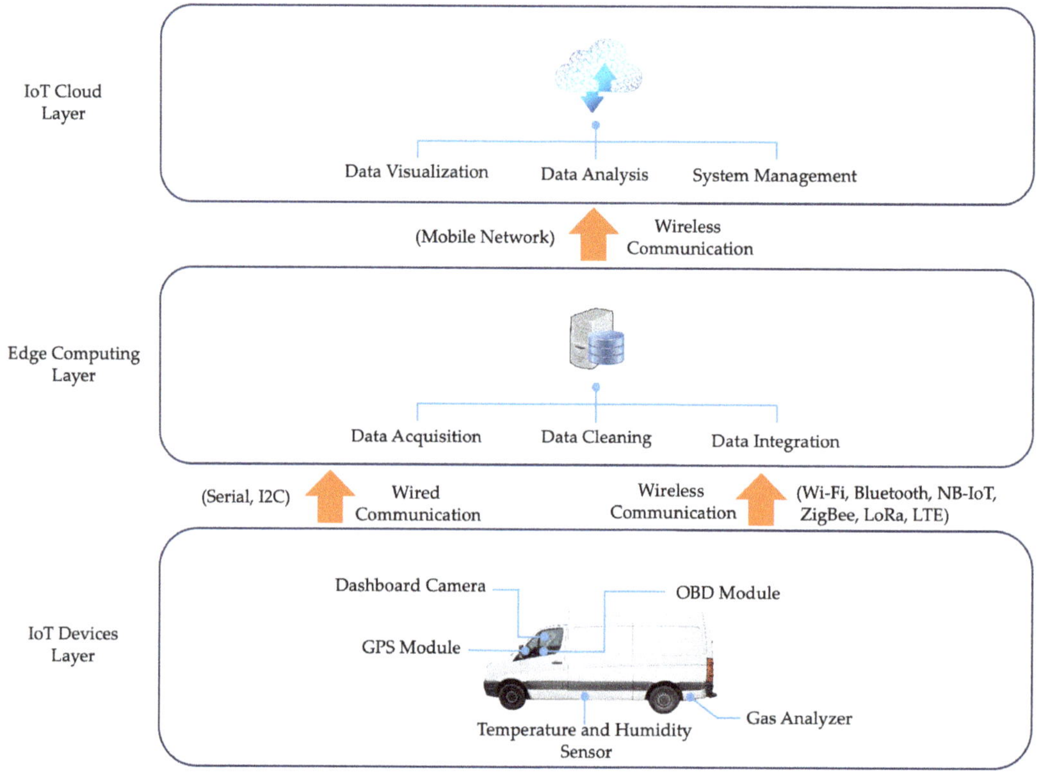

Figure 2. ParcEMon platform architecture.

The device layer includes the physical components of the IoT-based emission monitoring platform. This includes the sensors which are used to capture data regarding environment, vehicle and driving behaviour contexts. The data from the device layer were transferred to the edge layer using both wired and wireless communications. In the edge layer, three data processing operations including data acquisition, data cleaning and data integration were conducted using a Raspberry PI single board computer (SBC). Data acquisition refers to the process of sampling the signals transferred from the IoT device and converting them into numeric values to be manipulated by the data analysis software. Data acquisition is followed by data integration where data from multiple sensors with different data types are integrated into a single structured database for further processing. Data cleaning is also performed on the edge layer where incorrect, corrupted and incomplete data are removed from the data set. The data are then transferred to the application layer using a mobile network for data visualisation and data analysis. For data visualisation, an interactive dashboard was designed to present the acquired data from sensors in the IoT device layer. The dashboard also enables the management of the emission monitoring platform by providing an interactive interface for the system administrator. This enables different data sets to be filtered and generate different type of reports with different levels of granularity for data analysis purposes. In addition, the system management component enables additional data, such as the log of parcel size and delivery times, to be added into the database and integrated with the sensor readings from the IoT device layer. A detailed list of sensors, communication technologies and dashboard functionalities are discussed in Section 3.2.

3.2. Platform Implementation

Since the gas analyser receives and reads toxic gases form the vehicle emissions, it was necessary to design and install all sensor components outside of the vehicle. The gas analyser and its components were installed on the exterior chassis of the vehicle in such a way as to prevent any object or road bump damage to the devices. The temperature sensor was also installed underneath the van to capture ambient temperature. Other sensors including OBD, dashboard camera, GPS and the edge device were installed inside the driver cabin.

The implemented platform included multiple hardware and software components which are as follows.

- On-Board Diagnostics (OBD) Module: An OBD module was used to read vehicle performance data throughout the deliveries. The data which were captured by OBD included speed, acceleration, deceleration, and gear shift. The data were transferred to the edge device using Bluetooth communication. The OBD module used in this paper supported OBD version 2 with the frequency range of 2402 to 2480 MHz, supply voltage of 12 V, and data transmission range of up to 10 m.
- Dashboard Camera: A dashboard camera was used to record an entire trip in HD quality. The data were used to check the road condition and special events in case of sudden changes in OBD data to better perceive the impact of road condition, special events and driver behaviour on vehicle emission. The dashboard camera data were locally stored on an SD card with time and GPS location integrated in the video. Furthermore, a low resolution video was sent to the dashboard to show the real-time situation of the vehicle.
- GPS Module: A GPS module was integrated into the platform to continuously record the vehicle location. The data were used to determine the vehicle path and distance travelled and also to determine whether vehicle was doing deliveries in rural or urban areas. The GPS data were transferred to the edge device via USB communication. The GPS module used in this paper was GlobalSat BU-353-S4 utilising SiRF STAR IV GSD4e GPS chipset.
- Temperature and Humidity Sensor: A temperature sensor was installed on an Arduino microcontroller to continuously record the ambient temperature. The sensor was deployed underneath the van on the opposite side of the exhaust to prevent exhaust heat from impacting the sensor readings. The data were then transferred to the edge device using USB communication. The ambient temperature was used to analyse the impact of temperature on fuel consumption and vehicle emission in particular the impact of early morning cold weather on engine performance. The ambient condition sensor used in this paper was an SHT31 weather-proof temperature and humidity sensor.
- Gas Analyser: The vehicle gas emission (i.e., CO_2) was measured using an industrial gas analyser. The gas analyser used in this paper was able to capture five gases including CO, CO_2, O_2, HC and NOx. However, in this paper, we only report CO_2, which is the most prolific GHG. The gas analyser included a nozzle which was fixed inside the exhaust and transferred the exhaust emissions to the gas analyser module using a pipe. The data were then transferred from a gas analyser to the edge device using serial communication. The specification of the gas analyser used in this paper is shown in Table 1.
- Edge device: The edge device was used to collect data from all the sensors and transferred the data to the IoT cloud. The edge device was designed using a Raspberry Pi single board computer. The data were captured using different communication protocols including USB, serial and Bluetooth based on each sensor specification. The data were transferred to the IoT cloud using Wi-Fi communication. In addition, the edge device was equipped with a 60,000 mAh battery, which was the power source

of all sensors except the OBD which is directly powered via the vehicle OBD port. At the end of the experiment, 70% of the battery power was used by the deployed components. The components of the edge device are described in Table 2.

Table 1. Gas analyser specifications.

Gases Measured	CO (Carbon Monoxide) HC (HydroCarbons—Hexane (Gasoline) Propane (LPG), or Methane (CNG or LNG) CO_2 (Carbon Dioxide) O_2 (Oxygen) NO (NOx, Nitric Oxide)
Analysis Method	CO, HC, CO_2: NDIR (Non Dispersive Infra-Red) O_2, NO: Electro-Chemical Sensor
Reporting Ranges	CO: 0–10.00% HC (Hexane and Propane): 0–9999 ppm HC (Methane) 0.000–5.000% CO_2: 0–20.0% O_2: 0.0–25.0% NO: 0–5000 ppm
Resolution	CO: 0.01% HC (Hexane, Propane): 1 ppm HC (Methane): 0.001% CO_2: 0.1% O_2: 0.1% NO: 1 ppm
Accuracy	All gas channels +/−5% relative to gas reading
Repeatability	All gas channels +/−3% full scale
Response Time	Less than 8 s to 90% final value.
Warm-Up Time	30 s to 10% accuracy 5 min to full accuracy. (Constant ambient conditions)
Gas Sample Rate	350 mL/min typical. (Flow control pneumatics system).

Table 2. Edge device components.

Components	Description
Raspberry Pi 4	Broadcom BCM2711, Quad core Cortex-A72 (ARM v8) 64-bit SoC @ 1.5 GHz, 4 GB LPDDR4-3200 SDRAM 2.4 GHz and 5.0 GHz IEEE 802.11 ac wireless, Bluetooth 5.0, BLE, Gigabit Ethernet Raspberry Pi standard 40 pin GPIO header Micro-SD card slot for loading operating system and data storage 5V DC via USB-C connector
Power Bank	Battery Type: Li-Polymer Capacity: 64,000 mAh DC Input: 19 V 2 AAC Output: 220 V 50 HZ 130 W TYPE-C Output: 5 V/9 V/12 V 3 A USB Output: 5 V/9 V/12 V 3 A Recharging Time: 8–10 h
7 inches Touchscreen Display	TFT Display Screen Dimensions: 194 mm × 110 mm × 20 mm Viewable screen size: 155 mm × 86 mm Screen Resolution 800 × 480 pixels 10 finger capacitive touch Connects to the Raspberry Pi board using a ribbon cable connected to the DSI port.
Huawei 4G USB E3372	FDD: DD800/900/1800/2100/2600 UMTS: 900/2100 GSM: 850/900/1800/1900 LTE FDD: Cat4 DL: 150 Mbps/UL: 50 Mbps @20 M BW UMTS: DCHSPA+: 42/5.76 Mbps; 21 M/5.76 Mbps; 14 M/5.76 M HSUPA: 7.2 M/5.76 M 2G: EDGE packet data service of up to 236.8 kbps

Figure 3 illustrates the platform implementation and the IoT components.

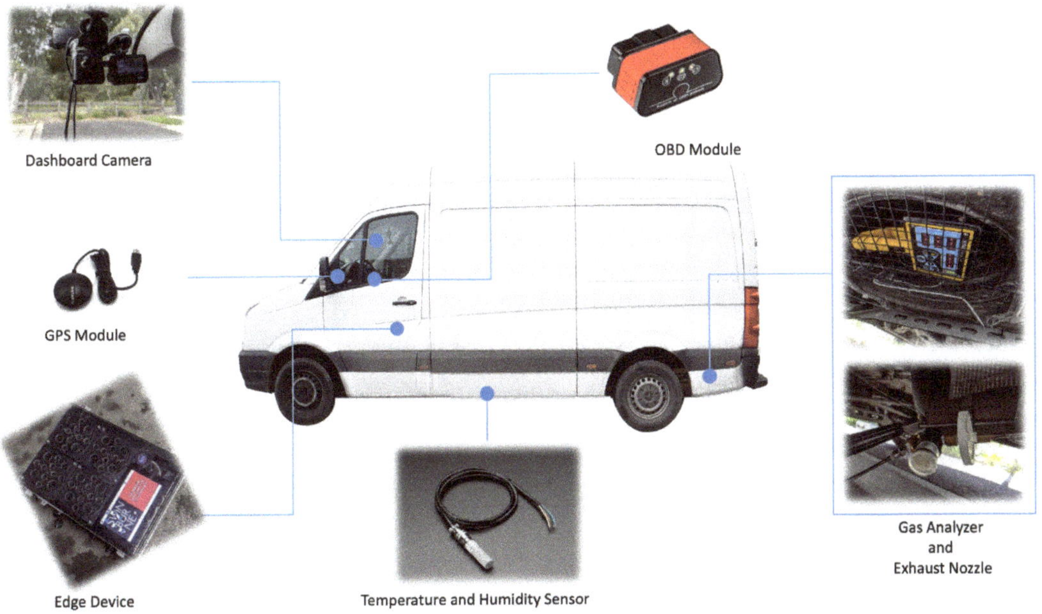

Figure 3. ParcEMon platform implementation.

The acquired data from the sensors were transferred to the IoT cloud platform for real-time visualisation and data analysis.

The dashboard presented in Figure 4 is a web-based online application that enables monitoring of the ParcEMon in real-time to ensure the platform, and its components are operational during the field-trial. In addition, it reports and records the vehicle location and status. The dashboard also generates several emission reports from the collected data with a preferred granularity level. Furthermore, this dashboard can be extended to facilitate lower emission route suggestion based on the historical and real-time data.

3.3. Data Collection

In this research, the IoT sensors, including a gas analyser, were deployed to capture live data from a delivery van. The van was working for an Australian household chain in the city of Melbourne and was handling last-mile deliveries. The acquired data covered a wide range of parameters which can be divided into two categories. The first category relates to environmental parameters including weather condition, temperature, humidity and road condition. The second category relates to car parameters including engine's revolution per minute (RPM), acceleration, deceleration, gear shift, Global Positioning System (GPS) coordinates, distance travelled and the travel path. The data from the gas analyser and other sensors were transferred to an edge device installed in the van via different communication protocols and subsequently the integrated data were transferred to the IoT cloud. The last-mile delivery vehicle emission and the data related to other contexts described in Section 3 were recorded from the vehicle departure point from the warehouse until the last parcel delivery. As it is shown in Figure 5, the warehouse is represented using the warehouse icon and each parcel, and its order of delivery is represented in circle icons. In total, data for 13 delivery points were recorded. The delivery points were in different suburbs with different road congestion and road conditions.

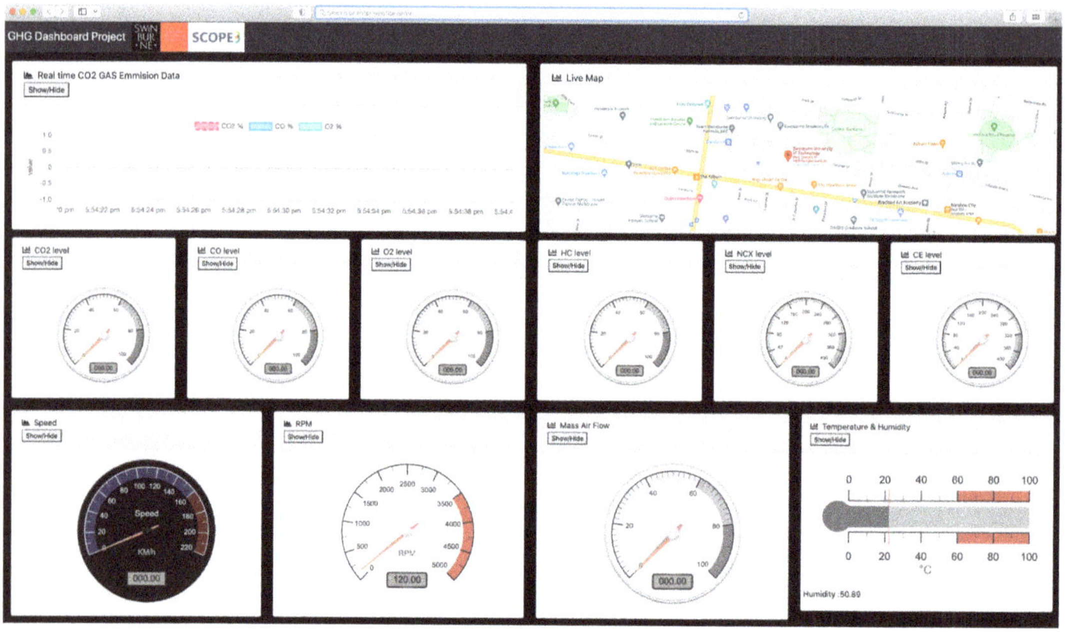

Figure 4. ParcEMon online dashboard.

Figure 5. The warehouse location and the order of 13 last mile parcel deliveries.

The data collection was captured during the autumn season with a temperature range between around 17 °C to 26 °C. The ambient temperature and humidity were recorded to ensure that the range of these two factors is within the operational range of the gas analyser sensor. Figure 6 shows the ambient temperature and sensor thorough the delivery process.

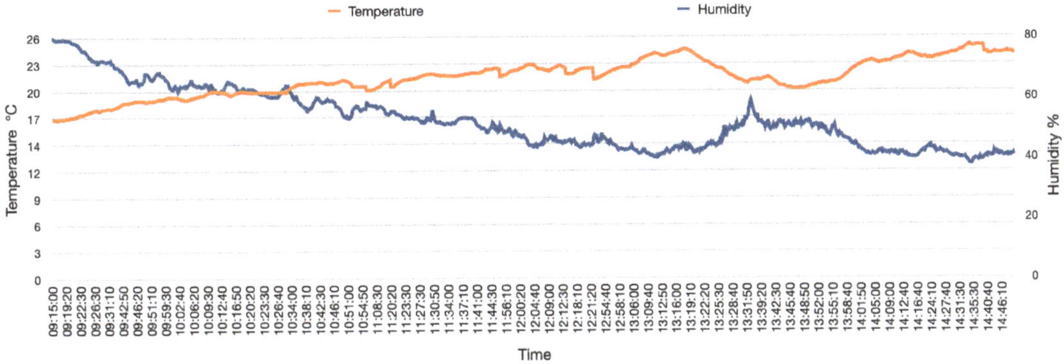

Figure 6. Ambient temperature and humidity during the field-trial.

The gas analyser temperature shows a sudden increase at the beginning of the delivery process. However, after an hour, the gas analyser temperature increase rate reduces and shows a mild increase around 5 °C for the rest of the delivery day. This gas sensor temperature increase is due to exhaust gas heat generated throughout the delivery process. Figure 7 shows the gas analyser sensor temperature throughout the delivery process.

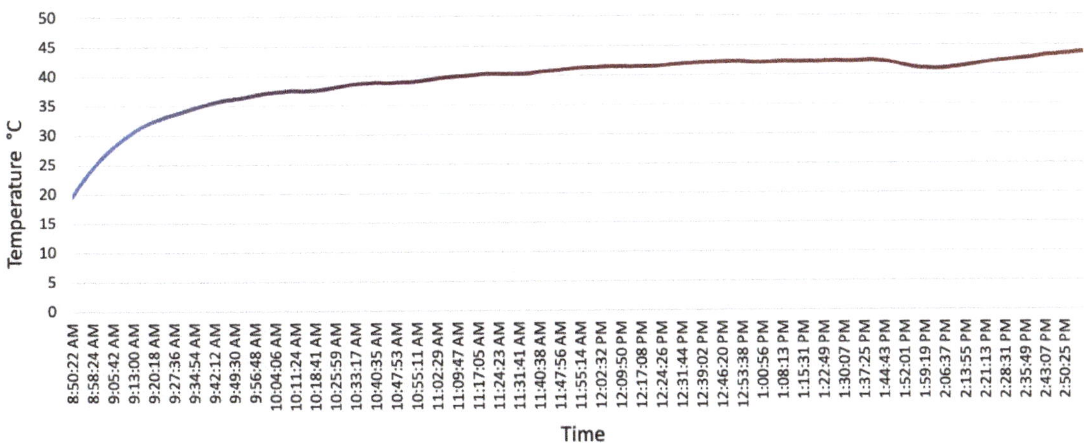

Figure 7. Gas analyser operating temperature during the field-trial.

For each parcel, the weight and the time of delivery were recorded. In addition, the consignment type for each parcel, which represents the size category of the parcel, is registered. The consignment types include economy, small and medium parcels, which represent the parcel weight from light to heavy. The lightest parcel was parcel No. 6 with a weight of 1 kg, and the heaviest parcel was parcel No. 1 with the weight of 194 kg. Other information including the delivery address, suburb and the postcode are also recorded to identify the suburb changes during the last-mile deliveries. The distance from each delivery point to the next delivery point is also recorded. These data can be used to analyse how many emissions have been released for each parcel based on the distance travelled, and this can be beneficial for analysing the best possible path to reduce the emission for all deliveries in a single day. Table 3 shows the recorded data for last-mile delivery on a single day which includes 13 deliveries.

Table 3. Consignment delivered during the field-trial.

No	Consignment Type	Weight (kg)	Delivery Time	Approximate Delivery Street	Delivery Suburb	Delivery Postcode
1	Medium	194	09:37	Rye Street	Mitcham	3132
2	Medium	179	09:55	Mahoneys Road	Forest Hill	3131
3	Medium	58	10:23	Manningham Road	Bulleen	3105
4	Medium	176	11:01	Ballantyne Street	Burwood East	3151
5	Small	16	11:18	Burwood Highway	Burwood East	3151
6	Economy Parcel	1	11:36	Ringwood Square Shopping Centre	Ringwood	3134
7	Economy Parcel	20	11:48	Lockhart Road	Ringwood North	3134
8	Medium	119	12:27	Ferntree Gully	Knoxfield	3180
9	Economy Parcel	10	13:02	Koornang Road	Scoresby	3179
10	Small	5	13:33	Caroline Street	Selby	3159
11	Medium	50	14:21	Fairbank Road	Clayton	3168
12	Small	29	14:39	Jarrah Court	Glen Waverley	3150
13	Small	30	14:48	Lincoln Avenue	Glen Waverley	3150

4. Results

The acquired data of this research are analysed from four different perspectives to determine the factors which impact each parcel delivery's CO_2 emission. These perspectives include van weight at each delivery point, distance travelled, interruption and driver behaviour. The driver behaviour is determined by analysing the vehicle Revolution per Minute (RPM), speed, acceleration and deceleration data. These factors impact the fuel consumption and subsequently the CO_2 emission. Figure 8 shows the RPM fluctuation during the last-mile deliveries. The deliveries which have resulted in higher RPM than average can contribute to more fuel consumption. The data also enable determining the road condition such as road gradient or traffic congestion.

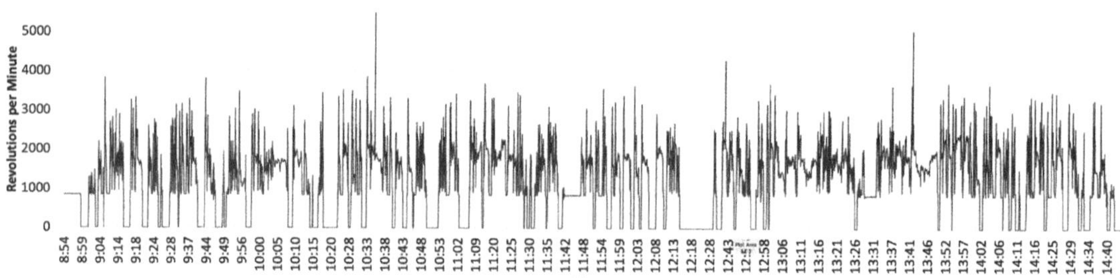

Figure 8. Engine's RPM data collected during the field-trial.

Similarly, the speed chart represented in Figure 9 enables analysis of driver behaviour in different conditions. In Victoria, the default speed limit for built-up areas is 50 km/h and on the highway it is 100 km/h. Using the speed chart, it can be determined in which time of the day the delivery van has been on the highway or in built-up areas. For example, the data points on 10:11 and 10:40 show that the vehicle has been on the highway to deliver parcel No. 3. This can be used to analyse the impact of constant speed on parcel level emission in comparison to the condition where there has been constant halts in vehicle movement such as traffic condition in built-up areas. This can provide beneficial data to determine the best possible route to reduce the deliveries emission.

The other factors which reflect driver behaviour are acceleration and deceleration. In order to show the impact of acceleration and deceleration on CO_2 emission, a five minute segment of driving and the CO_2 emission has been shown in Figures 10 and 11. Figure 11 shows the acceleration and deceleration, and Figure 10 shows the CO_2 emission percentage which results a few seconds after acceleration and deceleration. Figure 11 shows that the acceleration and deceleration can slightly impact the CO_2 percentage in exhaust gas;

however, most of the time the CO_2 percentage is 15%. This value is an ideal fuel combustion level for a diesel engine vehicle such as the delivery van engine examined in this research.

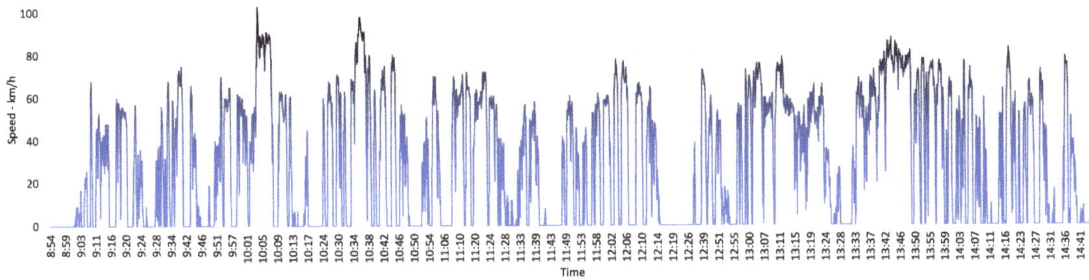

Figure 9. Vehicle's speed data collected during the field-trial.

Figure 10. CO_2 percentage in emitted gas from the tailpipe.

Figure 11. Acceleration and deceleration of the vehicle.

Another factor which must also be considered is interruptions during the deliveries. The interruptions during delivery such as a driver's stop for breakfast and lunch have been removed from the data to correctly acquire the emission ratio compared to duration for each parcel delivery. Figure 12 shows the duration for each delivery when the van has been on, and also the duration when the van has been moving (having speed more than 0 km) in comparison to the engine running time.

As it can be seen in Figure 12, parcel No. 5 has the highest engine off duration which indicates an interruption in the delivery. Parcel No. 10 has the highest engine on duration and also the vehicle movement in that period. These data can be used to show how long the delivery van has been utilised in each parcel delivery. In addition to driver behaviour and interruptions, parcel weight and the distance travelled also impact the emission level for each parcel delivery. The parcel weight has been addressed in Table 3. Table 4 integrates distance and interruption factors along with previously discussed driver behaviour for each parcel.

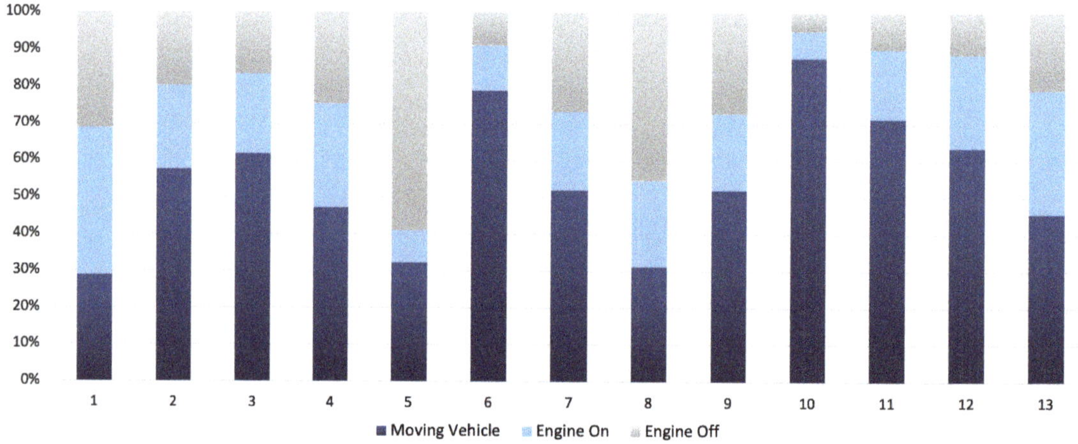

Figure 12. Engine status and vehicle movement (i.e., speed greater than zero) during each delivery in the field-trial.

Table 4. Delivery segments during the field-trial.

Segment	Round per Min Avg	Round per Min Max	Speed (km/h) Avg	Speed (km/h) Max	Start	End	Duration hh:mm	Max Acceleration km/s	Max Deceleration km/s	Distance km	Interruption (Engine Off)
Depot → 1	1064	3821	13	68	8:54	9:40	00:46	11	15	7	31%
1 → 2	1297	3784	26	75	9:40	9:56	00:16	12	20	6	20%
2 → 3	1267	3486	38	103	9:56	10:26	00:30	11	11	16	17%
3 → 4	1188	5824	33	98	10:26	11:05	00:39	13	15	16	25%
4 → 5	1515	3424	28	70	11:06	11:20	00:14	10	10	2	59%
5 → 6	1663	3682	42	72	11:20	11:38	00:18	12	9	11	9%
6 → 7	1310	3085	24	58	11:38	11:49	00:11	8	8	4	27%
7 → 8	1172	3642	26	78	11:49	12:50	01:01	20	15	14	45%
8 → 9	1243	4288	23	73	12:50	1:05	00:15	9	14	4	27%
9 → 10	1624	3673	46	79	13:05	13:33	00:28	13	16	21	5%
10 → 11	1544	5056	43	88	13:33	14:23	00:50	15	15	32	10%
11 → 12	1436	3476	31	83	14:23	14:40	00:17	12	11	8	11%
12 → 13	993	3280	19	79	14:40	14:49	00:09	12	10	2	21%

Each row presents the data from the previous delivery point starting from warehouse to delivery 1 in row No. 1. In total, 143 km has been travelled by the last-mile delivery van. Parcel 3 delivery has the highest maximum speed which shows the driver has been driving on the highway. This parcel has taken 30 min to be delivered over a 16 km distance. The longest path travelled for a parcel belongs to parcel No. 11 where 32 km has been travelled to make the parcel delivery.

The CO_2 emission for each parcel has been calculated based on the parcel contribution percentage to overall CO_2 emission. These data are presented in Figure 13.

As it can be noticed from Figure 13, parcel No. 13 with a weight of 30 kg and a distance of 2 km have the lowest emission of 1.81% from overall emission. On the other hand, parcel No. 11 with weight of 50 kg and the distance of 32 km has the highest CO_2 emission percentage. This parcel has contributed to 20.42% of overall CO_2 emission during the delivery day. However, it must be also be taken into consideration that the weight of parcels which have not yet been delivered including parcels 12 and 13 can contribute to this number as they are still being carried by the delivery van during parcel 11 delivery. As a result, the cumulative weight of the van through entire delivery has been taken into consideration to determine the CO_2 emission of each parcel by considering the weight of other existing parcels in the van at each delivery point. Figure 13 shows the cumulative weight of the van at each delivery point. The weight of the van was 2000 kg, which is also added to the overall parcel weight to have the cumulative weight at each delivery point.

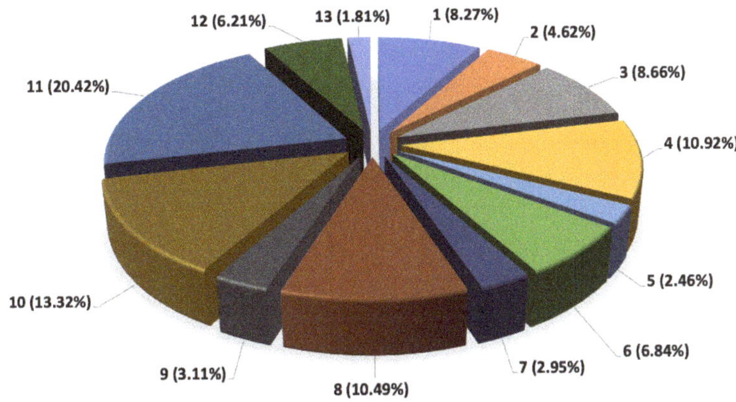

Figure 13. CO_2 emission percentage per delivery.

Figure 14 shows the emission of each parcel based on the cumulative weight of the vehicle.

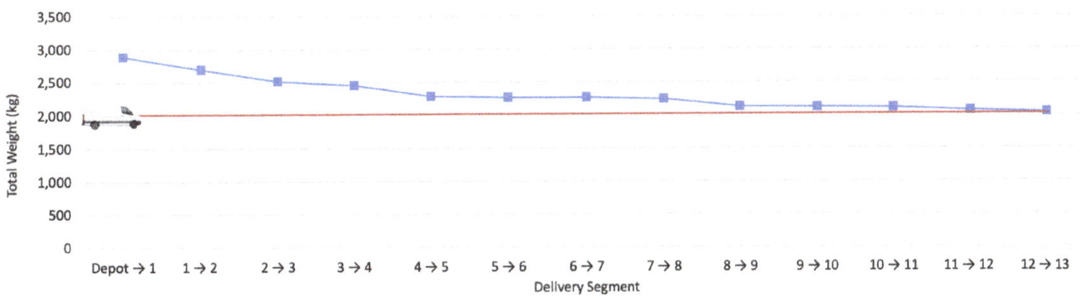

Figure 14. Van and parcels weight during the field-trial.

As shown in Figure 15 after considering the cumulative weight of the van, the parcels' contribution to overall CO_2 changes considerably. Parcel No. 8 with weight of 199 kg contributes to 21.88% of overall CO_2 emission. This parcel has been in the van for 76 km before it is delivered. Considering the fact that this parcel is a heavy weight parcel, it has resulted in high CO_2 emission over a long distance being on the move by delivery van. Parcel No. 11, which showed the highest CO_2 emission percentage when considered individually, after cumulative weight analysis is dropped to the third highest contributing parcel after parcel No. 8 and No. 4. In cumulative analysis, parcel No.11 has contributed to 16.31% of CO_2 contribution. Parcel No. 11 has been in the van for a distance of 133 km, but since its weight is 50 kg, which is much less than parcel No. 8, it therefore has continued to be less than CO_2 percentage compared to parcel No. 8. Parcel No. 4, despite the fact that it has only been in the van for 45 km, due to its heavy weight of 176 kg has resulted in 17.9% of overall CO_2 emission. Parcel No. 13, which had only 1.81% contribution to CO_2 emission before cumulative weight analysis, shows 10.74% contribution to overall CO_2 emission due to being in the vehicle until the end of the delivery day.

The results of this research show that the cumulative weight and the distance travelled can remarkably impact the parcel level CO_2 emission. Delivering heavy weight parcels at the beginning of delivery day can reduce the CO_2 emission reporting. However, it must also be considered that delivering the heavy weight parcels at the beginning of delivery day can increase the total path travelled by the delivery van. Therefore, although the CO_2 emission from a weight perspective could be reduced, the CO_2 emission from a distance travelled perspective would increase. This shows that there is a trade-off between parcel weight and total distance which must be considered when selecting the best delivery order for reducing CO_2 emission as much as possible.

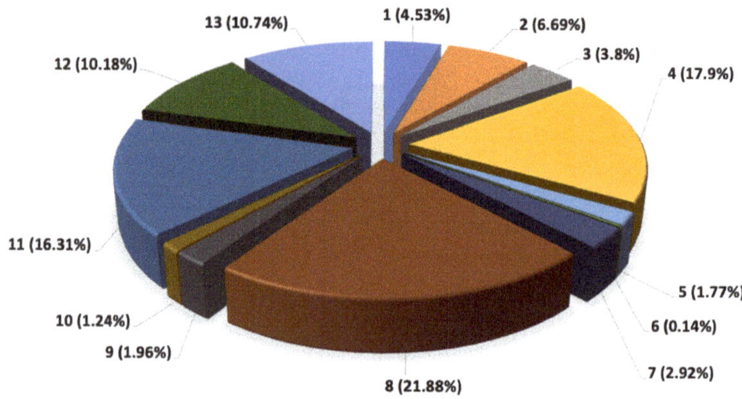

Figure 15. CO_2 cumulative emission percentage per consignment.

During this research, several limitations were identified which impacted the platform implementation and data analysis process. One of the limitations was that the gas analyser functionality could be impacted by environmental factors such as road bumps, mud and water splash. In addition, the gas analyser could not be located in a sealed enclosure since such enclosure prevents air flow and impacts the gas analyser reading. Moreover, the gas analyser performs periodic self calibration which results in a few seconds of data loss after each calibration cycle. Another limitation was that the weight of the driver and other possible passengers and materials inside the van other than the parcels were not considered in this research findings.

Future research directions in this topic should consider instrumentation and measurement of a large number of delivery vehicles representing the Australian vehicle fleet. The trial should also be for longer duration and cover different ambient conditions to capture the range of representative real-life conditions that are typical for Australian parcel delivery operations and road network conditions.

5. Conclusions

The transition to a low carbon and low emissions future depends on mitigating climate change in all sectors of the economy. Within the transport sector, freight transport networks, supply chains and last-mile deliveries present particular challenges to successful emissions mitigation. This paper presented the design, implementation and evaluation of an IoT platform for real-time parcel level last-mile delivery emissions reporting. The study demonstrated the feasibility of the technology, which comprised multiple IoT sensors, in measuring real-time emissions per parcel in real-world last-mile delivery vehicles. The results showed that the cumulative vehicle weight and the total distance travelled have substantial impacts on parcel level emissions. The findings present an approach for moving away from static models of carbon emissions assessment, to a more detailed analysis of on road data, including driving conditions and carbon intensity measures across the last mile in the logistics chain. These findings provide freight delivery operators with valuable insights in optimising their delivery schedules such that heavy-weight parcels are delivered at the start of the vehicle's journey to minimise the total distance travelled with heavy loads, which would result in substantial reduction of the vehicle's carbon footprint during those deliveries. However, the results also showed that there is a trade-off between delivery of heavy items first, and the total distance travelled by the vehicle during the day, in situations where the heavy parcels are located further away from the distribution centres. Data from the IoT platform developed in this study can help in ameliorating the negative impacts of such trade-offs by providing micro-level data that can be used to determine optimal locations for distributions' centres around the city, which would result in reducing the distances travelled and subsequently the levels of emissions emitted by each delivery vehicle.

Author Contributions: Conceptualisation, A.Y., H.D., P.S., J.S. and M.S.; Methodology, A.Y. and H.B.; Software, A.Y., H.K. and I.M.; Investigation, A.Y., H.D., P.S., J.S. and M.S.; Writing—Original Draft Preparation, A.Y. and H.B.; Writing— Review & Editing, H.D., P.S., J.S. and M.S. ; Visualisation, A.Y. and H.B. All authors have read and agreed to the published version of the manuscript.

Funding: This research was co-funded by Scope 3 Pty Ltd and Swinburne University of Technology's Research Office.

Institutional Review Board Statement: The study was ethics approved at Swinburne University of Technology, No. 20204101-5592.

Acknowledgments: The authors acknowledge Michael Blake and Gerry Walsh from EmissionsIQ for their support. This project utilised Swinburne Sensor Research Lab's facility funded by Australian Renewable Energy Agency as part of ARENA's Research and Development Program - Renewable Hydrogen for Export (Contract No. 2018/RND012). The authors acknowledge Chris Harrison from Swinburne University of Technology for his assistance.

Conflicts of Interest: The authors declare no conflict of interest.

References

1. Stock, P.; Steffen, W.; Bourne, G.; Brailsford, L. *Waiting for the Green Light: Transport Solutions to Climate Change;* Climate Council: Surry Hills, NSW, Australia, 2018.
2. Siikavirta, H.; Punakivi, M.; Kärkkäinen, M.; Linnanen, L. Effects of e-commerce on greenhouse gas emissions: A case study of grocery home delivery in Finland. *J. Ind. Ecol.* **2002**, *6*, 83–97. [CrossRef]
3. Laghaei, J.; Faghri, A.; Li, M. Impacts of home shopping on vehicle operations and greenhouse gas emissions: Multi-year regional study. *Int. J. Sustain. Dev. World Ecol.* **2016**, *23*, 381–391. [CrossRef]
4. Jiang, L.; Chang, H.; Zhao, S.; Dong, J.; Lu, W. A travelling salesman problem with carbon emission reduction in the last mile delivery. *IEEE Access* **2019**, *7*, 61620–61627. [CrossRef]
5. Brown, J.R.; Guiffrida, A.L. Carbon emissions comparison of last mile delivery versus customer pickup. *Int. J. Logist. Res. Appl.* **2014**, *17*, 503–521. [CrossRef]
6. Song, L.; Guan, W.; Cherrett, T.; Li, B. Quantifying the greenhouse gas emissions of local collection-and-delivery points for last-mile deliveries. *Transp. Res. Rec.* **2013**, *2340*, 66–73. [CrossRef]
7. Wygonik, E.; Goodchild, A.V. Urban form and last-mile goods movement: Factors affecting vehicle miles travelled and emissions. *Transp. Res. Part D Transp. Environ.* **2018**, *61*, 217–229. [CrossRef]
8. Li, L.; He, X.; Keoleian, G.A.; Kim, H.C.; De Kleine, R.; Wallington, T.J.; Kemp, N.J. Life cycle greenhouse gas emissions for last-mile parcel delivery by automated vehicles and robots. *Environ. Sci. Technol.* **2021**, *55*, 11360–11367. [CrossRef]
9. Zacharof, N.; Fontaras, G.; Ciuffo, B.; Tsiakmakis, S.; Anagnostopoulos, K.; Marotta, A.; Pavlovic, J. *Review of in Use Factors Affecting the Fuel Consumption and CO_2 Emissions of Passenger Cars;* European Commission: Brussels, Belgium, 2016.
10. Weiss, M.; Bonnel, P.; Hummel, R.; Manfredi, U.; Colombo, R.; Lanappe, G.; Le Lijour, P.; Sculati, M. *Analyzing On-Road Emissions of Light-Duty Vehicles with Portable Emission Measurement Systems (PEMS);* JRC Scientific and Technical Reports, EUR; Publications Office of the European Union: Luxembourg, 2011.
11. Pelkmans, L.; Debal, P. Comparison of on-road emissions with emissions measured on chassis dynamometer test cycles. *Transp. Res. Part D Transp. Environ.* **2006**, *11*, 233–241. [CrossRef]
12. Ježek, I.; Drinovec, L.; Ferrero, L.; Carriero, M.; Močnik, G. Determination of car on-road black carbon and particle number emission factors and comparison between mobile and stationary measurements. *Atmos. Meas. Tech.* **2015**, *8*, 43–55. [CrossRef]
13. Bagha, H.; Yavari, A.; Georgakopoulos, D. Hybrid Sensing Platform for IoT-Based Precision Agriculture. *Future Internet* **2022**, *14*, 233. [CrossRef]
14. Yavari, A.; Jayaraman, P.P.; Georgakopoulos, D.; Nepal, S. ConTaaS: An approach to internet-scale contextualisation for developing efficient internet of things applications. In Proceedings of the 50th Hawaii International Conference on System Sciences, Hilton Waikoloa Village, HI, USA, 4–7 January 2017.

15. Gaire, R.; Sriharsha, C.; Puthal, D.; Wijaya, H.; Kim, J.; Keshari, P.; Ranjan, R.; Buyya, R.; Ghosh, R.K.; Shyamasundar, R.; et al. Internet of Things (IoT) and cloud computing enabled disaster management. In *Handbook of Integration of Cloud Computing, Cyber Physical Systems and Internet of Things*; Springer: Cham, Switzerland, 2020; pp. 273–298.
16. Yavari, A.; Georgakopoulos, D.; Stoddart, P.R.; Shafiei, M. Internet of Things-based hydrocarbon sensing for real-time environmental monitoring. In Proceedings of the 2019 IEEE 5th World Forum on Internet of Things (WF-IoT), Limerick, Ireland, 15–18 April 2019; pp. 729–732.
17. Forkan, A.R.M.; Montori, F.; Georgakopoulos, D.; Jayaraman, P.P.; Yavari, A.; Morshed, A. An industrial IoT solution for evaluating workers' performance via activity recognition. In Proceedings of the 2019 IEEE 39th International Conference on Distributed Computing Systems (ICDCS), Dallas, TX, USA, 7–10 July 2019; pp. 1393–1403.
18. Yavari, A.; Georgakopoulos, D.; Agrawal, H.; Korala, H.; Jayaraman, P.P.; Milovac, J.K. Internet of Things milk spectrum profiling for industry 4.0 dairy and milk manufacturing. In Proceedings of the 2020 International Conference on Information Networking (ICOIN), Barcelona, Spain, 7–10 January 2020; pp. 342–347.
19. Ritchie, H.; Roser, M.; Rosado, P. *CO_2 and Greenhouse Gas Emissions*; Our World in Data: Oxford, UK, 2020.
20. CO_2 Emissions from Cars: Facts and Figures (Infographics). *European Parliament*, 15 June 2022.
21. Merkisz, J.; Andrzejewski, M.; Merkisz-Guranowska, A.; Jacyna-Gołda, I. The influence of the driving style on the CO_2 emissions from a passenger car. *J. KONES* **2014**, *21*, 219–226. [CrossRef]
22. Allison, C.K.; Stanton, N.A. Eco-driving: The role of feedback in reducing emissions from everyday driving behaviours. *Theor. Issues Ergon. Sci.* **2019**, *20*, 85–104. [CrossRef]
23. Zervas, E.; Lazarou, C. Influence of European passenger cars weight to exhaust CO_2 emissions. *Energy Policy* **2008**, *36*, 248–257. [CrossRef]
24. Zaabar, I.; Chatti, K. Calibration of HDM-4 models for estimating the effect of pavement roughness on fuel consumption for US conditions. *Transp. Res. Rec.* **2010**, *2155*, 105–116. [CrossRef]
25. Greenwood, I.; Dunn, R.; Raine, R. Estimating the effects of traffic congestion on fuel consumption and vehicle emissions based on acceleration noise. *J. Transp. Eng.* **2007**, *133*, 96–104. [CrossRef]
26. Saboohi, Y.; Farzaneh, H. Model for developing an eco-driving strategy of a passenger vehicle based on the least fuel consumption. *Appl. Energy* **2009**, *86*, 1925–1932. [CrossRef]
27. Miles, A.; Zaslavsky, A.; Browne, C. IoT-based decision support system for monitoring and mitigating atmospheric pollution in smart cities. *J. Decis. Syst.* **2018**, *27*, 56–67. [CrossRef]
28. Yavari, A. Internet of Things Data Contextualisation for Scalable Information Processing, Security, and Privacy. Ph.D. Thesis, RMIT University, Melbourne, VIC, Australia, 2019.

sensors

Article

Cost-Efficient Coverage of Wastewater Networks by IoT Monitoring Devices †

Arkadiusz Sikorski [1], Fernando Solano Donado [2] and Stanisław Kozdrowski [1,*]

1 Institute of Computer Science, Warsaw University of Technology, Nowowiejska 15/19, 00-665 Warsaw, Poland
2 Institute of Telecommunications, Warsaw University of Technology, Nowowiejska 15/19, 00-665 Warsaw, Poland
* Correspondence: stanislaw.kozdrowski@pw.edu.pl; Tel.: +48-22-234-5048

† This paper is an extended version of our paper published in International Conference on Software, Telecommunications and Computer Networks (SoftCOM), Split, Hvar, Croatia, 23–25 September 2021, https://doi.org/10.23919/SoftCOM52868.2021.9559098.

Abstract: Wireless sensor networks are fundamental for technologies related to the Internet of Things. This technology has been constantly evolving in recent times. In this paper, we consider the problem of minimising the cost function of covering a sewer network. The cost function includes the acquisition and installation of electronic components such as sensors, batteries, and the devices on which these components are installed. The problem of sensor coverage in the sewer network or a part of it is presented in the form of a mixed-integer programming model. This method guarantees that we obtain an optimal solution to this problem. A model was proposed that can take into account either only partial or complete coverage of the considered sewer network. The CPLEX solver was used to solve this problem. The study was carried out for a practically relevant network under selected scenarios determined by artificial and realistic datasets.

Keywords: sewer network; wireless sensor network; internet of things; combinatorial optimization; mixed integer programming

Citation: Sikorski, A.; Donado, F.S.; Kozdrowski, S. Cost-Efficient Coverage of Wastewater Networks by IoT Monitoring Devices. *Sensors* **2022**, *22*, 6854. https://doi.org/10.3390/s22186854

Academic Editor: Stefano Mariani

Received: 27 July 2022
Accepted: 7 September 2022
Published: 10 September 2022

1. Introduction

Wastewater networks are a critical infrastructure: an asset essential for the functioning of society and the economy. Its proper functioning can be impaired by several threats, such as sewage pipe leaks or ruptures, malfunctioning of the wastewater treatment plant (WWTP), etc.

One of the most important threats for its correct functioning in an urban environment relates to the illegal disposal of harsh chemicals in the sewer network. These chemicals may spread beyond the sewer network, and since the capacity of the sewage network and of the WWTP is limited, these chemicals may leak and contaminate groundwater reservoirs, or damage the wastewater treatment plants and render it offline. Examples of unlawful activities of industrial organizations in the sewage network are discharges of: (a) sulfuric acid (H_2SO_4), resulting from the etching of semiconductors, accumulator acid, or the production of organic chemical substances [1]; (b) sodium hydroxide (NaOH), resulting from cleaning of surfaces in metal processing in industrial applications [2]; (c) sodium sulfate (Na_2SO_4), resulting from regeneration of cation exchange resins, which are used for softening of water in industrial water treatment [3]. Illegal discharges of such dangerous harsh industrial waste into sewage networks could be harmful for the biological stage of WWTP, its personnel, sewer pipes, and the general public.

Detecting illegal discharges of any of three substances mentioned above can be performed by sampling the wastewater with commercial pH and Electrical Conductivity (EC) sensors. Nevertheless, due to wastewater dilution and mixing effects in sewer pipes throughout the sewage network, the concentration of such substances may be below the minimum detection threshold of such sensors several hundred meters downstream in a populated sub-catchment area. Therefore, it is important to monitor the wastewater composition at multiple points in the sub-catchment area.

As a result, several portable Internet of Things (IoT) systems for monitoring wastewater composition have been proposed in recent years [4–18]. These IoT systems are adapted for working at manholes or main sewer lines, and usually comprise a set of sensors (electrochemical sensors, optical sensors, mass spectrometry, ion spectrometry, etc.) for detecting the presence or concentration of specific marker pollutants.

One of such IoT systems is the Micromole system [4,5]. The Micromole system consists of one or more battery-operated devices mounted at sewer main lines. Each device is equipped with pH and Electrical Conductivity (EC) sensors, specially designed for its operation in flowing wastewater [19]. The micromole device is composed of several detachable replaceable modules. In Figure 1 a micromole device comprising five of such modules can be observed. Some of these modules contain batteries, while others contain sensor electronics.

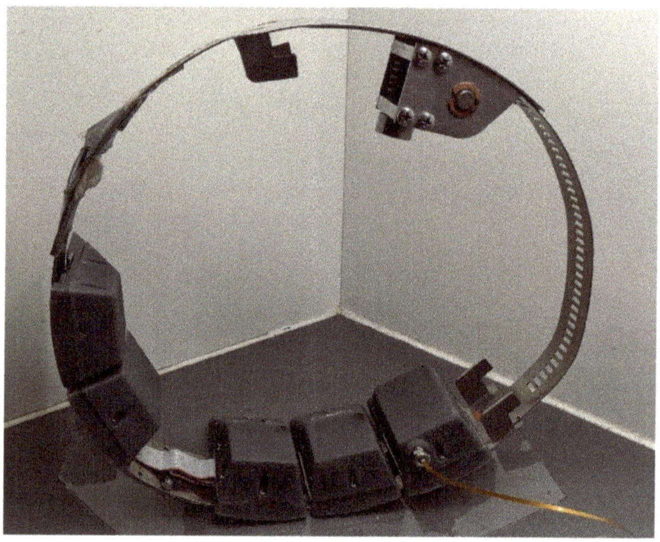

Figure 1. Micromole ring with five modules attached for measuring sewage wastewater physical parameters. From left to right, the attached modules are: battery module, wireless communication module, pH sensor module, Electrical Conductivity sensor module, and a Water Level sensor module.

This articles focuses on the planning of an cost-effective positioning of a network of IoT devices monitoring a sewage network. Below we provide an overview of the most recent methods proposed in the literature for the planning of monitoring devices in the sewage network.

This paper is organized as follows. Section 2 presents a description of the most relevant works on the subject. In Section 3, the problem is described and the model is presented with a brief explanation of the dispersion phenomena in wastewater networks. In Section 4 we describe a set of numerical experiments realized within a sewage network in the sub-catchment area of an European city. Section 5 provides the conclusions of our findings.

2. Related Work

The SIMONA project [20] has as one of its main goals proposing methods and algorithms for the planning of water quality monitoring stations in sewer systems. Banik et al. propose a set of solutions [21–24], all of which share the following approach. First, the authors consider as input to the problem a set of time-series of measurements, where one time-series consists of the measurements that would be observed at a given point in the sewage network if one potential source in the network makes a discharge. The measurements provide an indication of the quality of the wastewater, e.g., Electrical Conductivity, following certain given hydraulic conditions. Each measurement of the time-series is then quantized in discrete steps: rounding each measurement to its nearest value in the new scale. As a result, the number of potential different input values is constrained. Next, Banik et al. calculate the information entropy, or information content, of each time-series. After the previously described procedure for pre-processing is executed, Banik et al. consider a dual-objective optimization problem for the placement of the sensor devices. The objective function and meta-heuristic used for finding these solutions vary among Banik et al. contributions, which we summarise below.

In Ref. [21], the two objectives are: (1) maximum information content attained by a group of monitoring stations and (2) minimum the dependency among the monitoring stations. The first objective is achieved by maximizing the joint entropy of the selected monitoring stations, while the second one is attained by minimizing the total correlation of the chosen solution subset of monitoring stations. The set of Pareto optimal solutions is found by using an NSGA-II heuristic. According to Ref. [22], the final decision of selecting the set of monitoring stations from this Pareto front is made by maximizing the amount of information gained by a set of monitors, maintaining the consistency of the selected set of monitors for both variables (concentration and detection time) and having minimum total correlation within a set. The information theory approach taken by Banik et al. has been previously used in related areas [25,26].

In Refs. [22,23] Banik et al. extend their study by considering two additional objectives: detection time of an anomaly and reliability of the solution. The objective related to the detection time aims at minimizing the elapsed time from the discharge event until its detection, when using a fixed number of sensors. The objective related to reliability is related to the number of contamination scenarios that could be potentially correctly detected. Solution to this multi-objective optimization problem were found using the greedy algorithm proposed by Alfonso et al. in Ref. [27], originally designed for other applications.

Our previous work [28] presented the problem of optimising the number of IoT devices in a sewer network, while considering a fixed battery capacity for our sensors in a way that any potential illegal discharge in the sewage network could be detected. In this article we, instead, consider a partial network coverage and include the limitations imposed by sewage physical dimensions on the allocated battery capacities and sensor sampling rates. To the best of our knowledge, this article is the first one in the literature tackling such a problem.

Even though there are design methods in the literature—e.g., Genetic Algorithms [29,30], or Particle Swarm Optimization Algorithms [31]—for solving network coverage problems using Wireless sensor networks (WSN), none of them exploit the flow propagation properties and hydraulic dilution phenomena, as discussed in this article, in their solutions.

3. Related Background Knowledge and Proposed Methods

In this manuscript, we consider the problem of optimising the positioning of a wireless sensor network for monitoring the sewer network. In addition, the tackled problem also considers the appropriate allocation of the battery capacity of each sensor device, while considering energy requirements.

Two important requirements for the design of such sensor devices are: (1) to allow its placement in sewer mainline pipes of at least 250 mm of diameter without blocking the flow of sewage, and (2) ease of sensor and battery replacement. Micromole devices fulfil the first requirement by adopting a ring mechanical structure, as shown in Figure 1. Micromole devices fulfil the second requirement by housing electronics into a set of interchangeable modules, each of which share the same dimensions and electronic interconnections. These modules are mechanically and electronically interconnected through the ring mechanical structure, as shown in Figure 1. Since all modules have the same volume, the energy capacity that can be stored using batteries is the same for each module. Nevertheless, even though the energy capacity provided by any module is the same, the number of modules that can be attached to a Micromole device varies and largely depends on the circumference of the ring and, hence, is limited by the pipe diameter where it will be installed: wider pipes allow for a placement of more battery modules for a single device.

The energy consumption of the Micromole device is mostly dependent on the sampling frequency used by its sensors. The sampling frequency shall be set as to avoid situations where the device fails to notice a short discharge, due to its proximity to the source, fast flow speed, or short discharge time. Mitigating such situations can be achieved by assuming that the sampling frequency is dependent on the sewage flow velocity: fast flowing sewage requires high sampling frequency.

In this article we consider that the overall cost of a sensor device comprises the cost of the sensor electrodes themselves—which we consider as a fix cost per sensor device unit—and the cost of the chosen number of allocated battery units.

3.1. Pollution Detection and Sensor Localisation

In this article we assume that there is only a single polluting source at a time in the monitored sewer network. This is motivated by the fact that illegal discharges or wastewater pollution is a rare event. Nevertheless, the location of the polluting source, if present, is unknown.

The concentration of an injected pollutant fluctuates from pipe to pipe and, over time, due to the dispersion and dilution effects caused by the mixing of inflows in the sub-catchment area. This effect can be observed in Figure 2, where the EC of wastewater is shown for 82 measuring points downwards a polluting source, from which 50 L of sulphuric acid were disposed.

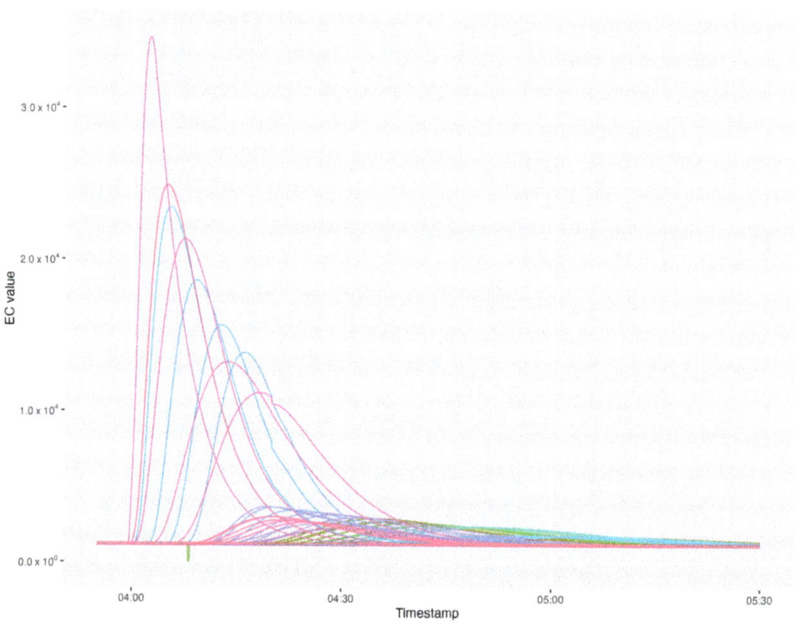

Figure 2. EC broadening and flattening caused by dispersion as seen at different measuring points, when 50 litres of sulphuric acid are discharged 81 manholes upstream from the sink point of the network shown in Figure 3 in low wastewater flow conditions.

A similar effect can be observed when measuring the amount of the diluted pollutant at the same pipe at different points in time during the day: as social and industrial activities demand more usage of water at certain hours, the amount of total flow in a pipe increases and so does the dilution factor of the pollutant. We refer to *flow conditions* as the amount of flow on every pipe at a given point of time.

Figure 3. Sewage network used for numerical experiments.

Due to the dilution effects and limited sensitivity of the sensor devices, the pollutant can only be detected in those pipes where the diluted amount of the substance exceeds the minimum limit of detection of the sensor. We say that a sensor located at pipe e *covers* a potential pollution source s_i when considering flow conditions f, if the sensor can detect the injection of a pollutant with an anomaly detection method using its collected time-series of sensor measurements. For the purpose of this study, we use a simple threshold criteria as our anomaly detection method: if a measured value exceeds a predefined threshold Q, then the sensor can detect the injection of the pollutant. The usage of a simple threshold as an anomaly detection method does not exclude the usage of more complex methods for anomaly detection—such as those based on pattern matching or Artificial Intelligence [32], for instance, or data fusion [33,34].

As a consequence, and given that the flows of wastewater is acyclic in a sewage network, the set of pipes where the pollution from a particular source can be detected form a directed acyclic sub-graph, $\mathcal{G}(s_i)$, of the sewage network. It shall be noted that for two polluting sources s_i and s_j, the corresponding sub-graphs, $\mathcal{G}(s_i)$ and $\mathcal{G}(s_j)$, may have edges in common. If a sensor device is installed in a common edge, it is not possible to discern whether the detected pollutant originates from either s_i or s_j, by only using our threshold criteria.

3.2. Model Description

Made assumptions in terms of domain language. These are expressed mathematically in the next sub-section.

Nodes:

- a set of nodes denoted as $\mathcal{V}$ is defined, each node of this set represents a sewer manhole;
- a few nodes are distinguished as outlet nodes of the given sewer network;
- a set $\mathcal{V}_s \subset \mathcal{V}$ is defined and represents nodes that can be sources of undesirable substances.

Edges:

- a set of directed edges is defined, each edge represents a pipe in the sewer network;
- any two nodes can have at most only one *direct* connection;
- edges can be marked as private or public. In Figure 4, dashed lines represent private pipes and solid lines represent public ones;
- each of these edges is characterized by a parameter that determines the size/flow capacity of water in each pipe;
- each pipe has a limited cross-area section, which limits the number of slots that can be used for attaching sensors and batteries in a single device. Sensors and batteries can only be installed on a ring device. Such a ring has a fixed cost.

Sensors:

- sensors detect undesired substances in the sewage, they are to be installed in the edges of the graph;
- only public edges are eligible for sensor installation while the private ones are not;
- each sensor has a fixed cost of installation;
- it is not known a priori how many sensors are required;
- sensors can only detect the contamination if the concentration in the pipe is not too low, since each sensor has a detection threshold. Each potential source of contamination is associated with a subgraph where the contaminant will be effectively detected and only there it makes sense to install sensors;
- discharge of undesired substances is a rare event and there can be only one at a time; there is no need to install sensors in a way that several sources can be distinguished;
- each sensor can sample the sewage at a given frequency—the bigger the flow, the more sensors will be needed to sample the flowing sewage—linearly more (one sensor is enough to sample the sewage having velocity 1 m/s but flow having the velocity 2 m/s requires two sensors).

Battery:

- sensors require batteries to run;
- the number of batteries required by each sensor depends on where the sensor will be installed;
- the number of batteries depends functionally on the sampling frequency, which depends on the flow rate and size of the pipe where the sensor will be installed;
- each battery has a fixed cost;

- size of the pipe limits the number of batteries that can be installed in the pipe as well as the number of sensors; the number of sensors and batteries combined is limited by the number of slots on the ring;
- the installed batteries deplete linearly, all at once.

Coverage of the sources:

- all potential pollution sources in the sewage network should be covered, i.e., any contamination discharge should be detectable by at least one sensor;
- one sensor can cover several sources since discharge from only one of them can happen at the time and there is no need to distinguish them;
- definition of coverage: for each source node $s \in \mathcal{V}_s$ there is defined a subgraph $\mathcal{G}_s$ where it makes sense to install sensors. If there is at least one sensor in each such subgraph, we satisfy the coverage condition;
- any solutions where any pollution source is not covered is not approvable;
- the coverage constraint is satisfied in Figure 4—the sensor covers both pollution sources that are denoted as red triangles. There is no need to put a sensor in the second leg of the network.

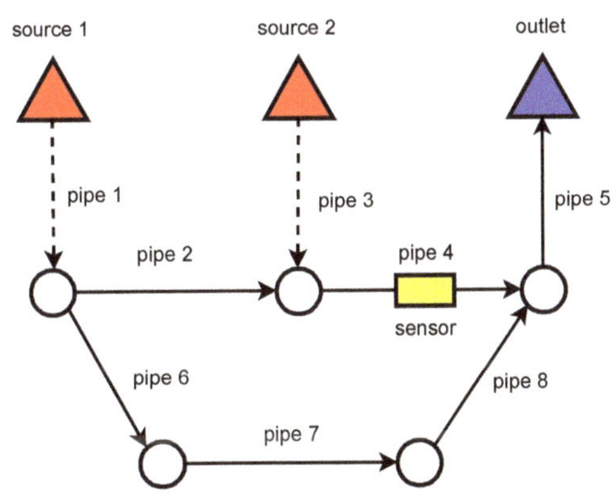

Figure 4. Network coverage. Assuming that $\mathcal{V}_1 = 2,4,6,7$ and $\mathcal{V}_2 = 4,5$, the sensor installed in pipe 4 is enough to cover both sources. Installing it in pipe 5 would cover only the second source.

Objective:

- Minimise the total cost of installing the sensors together with the cost of purchasing batteries for each sensor;
- We are interested in covering all or parts of the network, so that there is no potential source of contamination that is not detected by at least one sensor.

3.3. *Mixed Integer Programming Model*

The Mixed Integer Programming (MIP) method is proposed to solve the presented problem [35]. The advantage of this method is that it guarantees an optimal solution, as it searches the entire space of admissible solutions to the given problem. In general, the disadvantage of this method is that it often takes a long time to calculate the optimum [36]. In this case, in the problem under consideration, the MIP method performs quite well, even for networks with a large number of nodes.

The following will present the proposed model in mathematical terms. We will describe the definitions of the indices, sets, constants, and variables that appear in this model before the objective function and the necessary constraints are presented. We will operate with the indices e and s. The former refers to the edges and the latter to the nodes, which are the sources of pollution in the network under consideration. The sets, variables, and constants, on the other hand, are presented in Tables 1, 2 and 3, respectively.

Table 1. Sets description.

Set	Description
$\mathcal{V}$	Each vertex $s \in \mathcal{V}$ of the graph $\mathcal{G}$ represents a manhole
$\mathcal{E}$	The set of directed edges of graph $\mathcal{G}$, which represent the sewage pipes over which the sewage flows
$\mathcal{V}_s$	The set of vertices that could be potential sources of pollution; $\mathcal{V}_s \subset \mathcal{V}$
$\mathcal{E}_s$	The set of edges at which the concentration of pollutants allows effective detection of harmful substances after they have been emitted from the vertex s, also known as proximity; $\mathcal{E}_s \subset \mathcal{E}$.

Table 2. Variables description.

Variable	Description
α_e	$\alpha_e \in \mathcal{N}$; variable indicating how many sensors have been installed at edge $e \in \mathcal{E}$
β_e	$\beta_e \in \mathcal{N}$; variable indicating how many batteries have been installed at edge $e \in \mathcal{E}$
γ_e	Binary; equal 1 if the ring is installed on the edge $e \in \mathcal{E}$; 0 otherwise
δ_s	Binary; equal 1 if the source $s \in \mathcal{V}_s$ is covered; 0 otherwise.

Table 3. Constants description.

Constant	Description
Λ_e	Number of slots in the ring installed on the edge $e \in \mathcal{E}$
Γ_e	The cost of installing a ring on the edge $e \in \mathcal{E}$
A	The cost of one sensor
B	The cost of one battery
Ω_e	Total battery life at the edge $e \in \mathcal{E}$; expressed in sec; We assume that the sensor samples continuously; an example value is 10^6 s.
Φ_e	Sampling frequency of the sensor at the edge $e \in \mathcal{E}$; e.g., once per minute, then $\Phi_e = 1/60$
Θ	Capacity of one battery; expressed in the number of samples made, e.g., $\Theta = 10^5$, assuming that the batteries are the same on each edge $e \in \mathcal{E}$
Π	Percentage of source coverage.

Objective:

$$\min\left\{\sum_{e\in\mathcal{E}}(A\cdot\alpha_e + B\cdot\beta_e) + \Gamma_e\sum_{e\in\mathcal{E}}\gamma_e\right\} \tag{1}$$

Constraints:

$$\alpha_e + \beta_e \le \Lambda_e\cdot\gamma_e \qquad \forall_{e\in\mathcal{E}} \tag{2}$$

$$\sum_{e\in\mathcal{E}_s}\alpha_e \ge \delta_s \qquad \forall_{s\in\mathcal{V}_s} \tag{3}$$

$$\sum_{e\in\mathcal{E}_s}\gamma_e \ge \delta_s \qquad \forall_{s\in\mathcal{V}_s} \tag{4}$$

$$\beta_e\cdot\Theta \ge \Omega_e\cdot\Phi_e\cdot\alpha_e \qquad \forall_{e\in\mathcal{E}} \tag{5}$$

$$\sum_{s\in\mathcal{V}_s}\delta_s \ge \lceil\Pi\cdot|\mathcal{V}_s|\rceil \tag{6}$$

Formula (1) represents the cost function of the presented problem, which is subject to minimisation. Constraint (2) guarantees us that the number of slots in the ring installed on edge e does not exceed the available number of slots. Then, constraint (3) means that each potential source is covered by at least one sensor. Constraint (4) tells us that at least one ring must be installed on each edge where the concentration allows detection of harmful substances, while constraint (5) ensures that the capacity of all batteries must be greater than the lifetime and sampling frequency of the edge e. Finally, constraint (6) indicates the percentage of sources to be covered.

4. Experimental Results and Discussion

The proposed mathematical model was tested with two different datasets, each of which was derived from the same sewage network, which is depicted in Figure 3. The sewage network consists of 3297 manholes, 3343 pipes, and 1315 sources of pollution.

The first dataset uses a sub-graph of the base network and consists of 1124 pipes and 402 pollution sources while the second one uses the whole network.

Sections 4.1 and 4.2 describe how $\mathcal{E}_s$ sets were created—using discharge simulations and a simplified dispersion model respectively. Section 4.3 describes how sampling frequencies were pre-computed for both datasets. The following two subsections provide results and discussion of the actual cost optimization process using the linear model.

4.1. Dataset 1: Simulated Discharges and Dispersion Modelling

All flow and discharge simulations were performed using the software package ++SYSTEM Isar [37], which capabilities were extended by a reaction and transport model based on the concept of total alkalinity in the course of the Micromole project [4].

Due to computational constraints of the ++SYSTEM Isar system, it was not possible to simulate a discharge from every single building in the sub-catchment area. Instead, a subset of 402 buildings were chosen as potential sources of pollution. From every single potential source of pollution, we simulated discharges of 50 L of sulphuric acid, with pH 1 and EC 1400 mS/cm, with low flow conditions and with high flow conditions. Low flow conditions—f_L—represent the amount of flow found in this sewage network at 03 h 00 m, while high flow conditions—f_H—represent the amount of flow found in this sewage network at 08 h 00 m during a normal work day.

For establishing the sensor coverage for every particular pipe, we set a threshold for the EC value. In our experiments, we evaluated three different threshold values for EC: Q_1 = 2 mS/cm, Q_2 = 3 mS/cm, and Q_3 = 4 mS/cm, where the normal EC value of wastewater is nearly 1.3 mS/cm. As a result, the combination of the two flow conditions and the three EC threshold values results in six different scenarios that we evaluate below.

4.2. Dataset 2: Simplified Dispersion Model

Since discharge simulation is a heavy computational task, an inherited method of proximity generation was introduced to provide test data for a greater number of pollution sources. The algorithm of generating $\mathcal{E}_s$ sets is presented as Algorithm 1.

Algorithm 1 Simplified generation of proximities

```
function GENERATEPROXIMITIES(𝒢, k)
    𝒱_s ← findSourceNodes(𝒢)
    for s ∈ 𝒱_s do
        d ← findNearestDrainNode(s, 𝒢)
        p ← findShortestPathBetween(s, d, 𝒢)
        ℰ_s ← takeEdgesFromPath(p, k)
    return {ℰ_s ∀s ∈ 𝒱_s}
```

The above pseudocode requires some commentary:

1. All source nodes should be found or defined at the beginning; a source node has exactly one outcoming edge and no incoming edges;
2. For each source node s the shortest path between s and the closest drain node d needs to be found. It is the shortest in the terms of lowest number of edges;
3. Each shortest path is shortened and only the first k edges are taken. We assume that k pipes is enough for a pollutant to become undetectable by a sensor. This simplification is precise enough since pipes in the neighbourhood of each source have comparable lengths. k is chosen based on simulated data. We decided to test cases for k = 10, 20, 30, 40 since the average and the median length of a path in simulations was about 20 edges.

This method does not require dispersion simulation, which is computationally challenging. Instead, it uses simple graph algorithms, such as shortest path finding. The paths are limited to a length obtained from the simulations run using the smaller network.

4.3. Determining Sampling Frequencies for Both Datasets

Sampling frequencies in each pipe had to be calculated for both datasets. The sampling frequency in pipe e is affected by two factors:

1. The volume of sewage flowing through the pipe denoted as u_e. The greater the quantity of sewage in the pipe, the greater sampling frequency needs to be;
2. The area of the pipe's section, denoted as Ψ_e, calculated using a standard formula for disk area. The greater the section's area, the slower the flow in the pipe, so the sampling frequency can be lower.

Assuming that each source s continuously adds 1 discrete flow unit of sewage to the network, the flow values are generated as follows (see Figure 5):

1. For each e: set flow value $u_e = 0$;
2. For each source s: find the shortest path between s and the closest drain node d;
3. For each path p: for each edge e belonging the path p, increase flow value u_e by 1 unit.

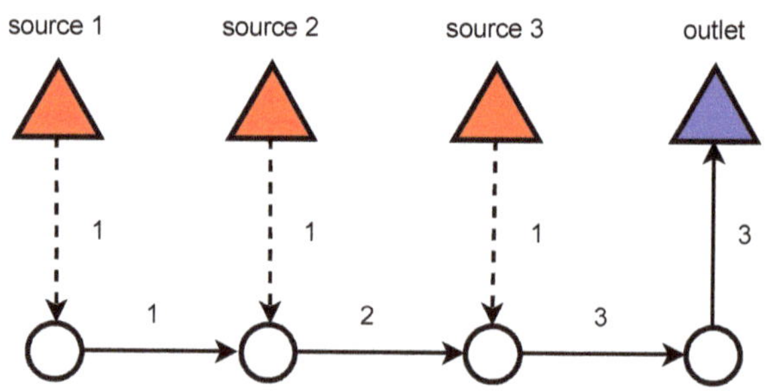

Figure 5. Flow units propagating through the network. The number over the edge is the number of flow units in the pipe. The greater number of flow units next to the outlet node means that a bigger volume of sewage flows in that part of the network when compared to pipes next to the sources.

Finally, sampling frequencies can be determined using the formula $\Phi_e = (\Phi_b + \Phi_c u_e) \cdot \Psi_e^{-1}$. Φ_b is the base frequency and Φ_c is the scaling factor of how much sampling frequency needs to be increased per each flow unit.

Values of sampling frequency determined by the described method are presented in Figure 6 as a histogram.

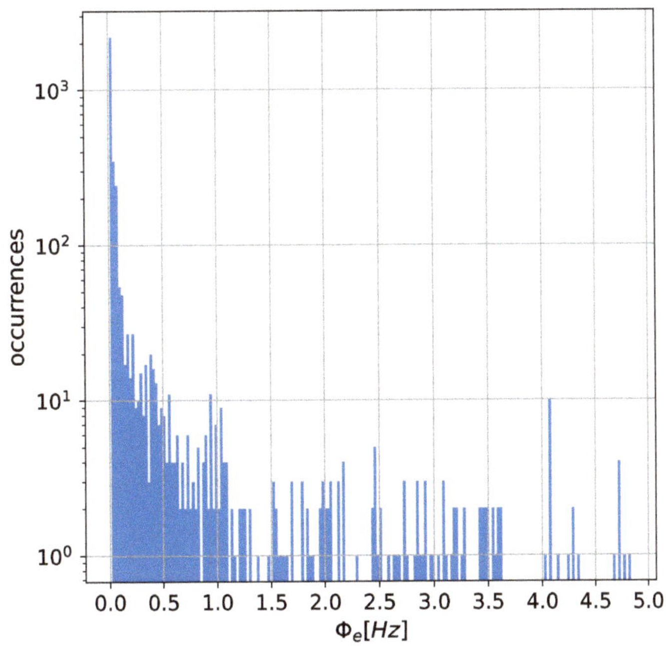

Figure 6. Histogram of sampling frequencies in the network.

4.4. Experiments

This section presents results of numerical experiments obtained with MIP solver and constant parameters presented in Table 4. Our experiments were divided into two cases:

- Case A—simplified dispersion model data—as explained in Section 4.2—with sampling depending on flow and pipe size;
- Case B—dispersion model data based on simulated discharges—as explained in Section 4.1—with sampling depending on flow and pipe size.

Table 4. Values of the used parameters.

Parameter	Value
A	7
B	3
Γ_e	5
Ω_e	10^6
Φ_b	1/60
Φ_c	1/60
Θ	10^6

Each case was tested with $\Pi = 0.1, 0.2, \ldots, 0.9, 1.0$ to determine how the cost changes when the constraint on how many pollution sources have to be covered is changed. The obtained results are presented in Table 5 and in Figure 7 for dataset 1 and in Table 6 and Figure 8 for dataset 2.

Table 5. Cost function values for the test scenarios of dataset 1.

Coverage [%]	Cost [Cost Units] EC 2000 3:00	EC 3000 3:00	EC 4000 3:00	EC 2000 8:00	EC 3000 8:00	EC 4000 8:00
10	15	15	15	15	15	15
20	15	15	15	15	15	30
30	15	30	30	15	30	45
40	30	30	30	30	45	60
50	30	45	45	30	60	90
60	30	60	60	45	75	120
70	45	75	90	60	120	165
80	60	90	135	90	183	255
90	75	135	213	120	303	393
100	168	303	471	250	600	750

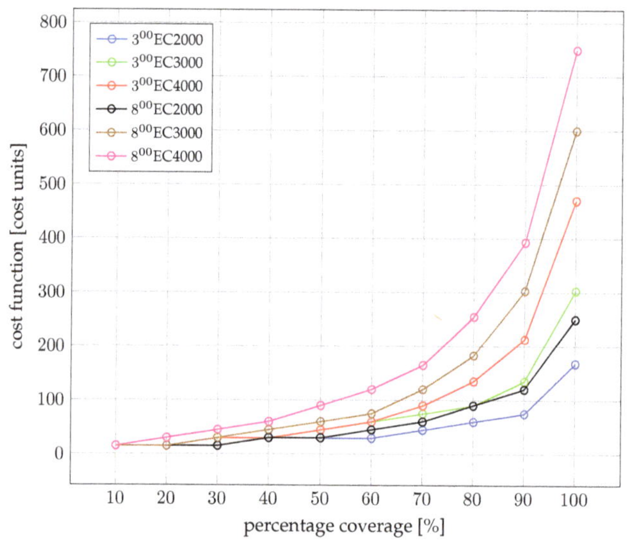

Figure 7. Optimal cost of IoT equipment deployment for dataset 1. Sampling frequency in a given pipe depends on the flow and the size of the pipe.

Table 6. Cost function values for the test scenarios of dataset 2.

Coverage [%]	Cost [Cost Units] $k = 10$	$k = 20$	$k = 30$	$k = 40$
10	36	24	21	21
20	87	48	36	33
30	144	69	54	48
40	210	102	69	69
50	285	132	93	87
60	369	171	126	111
70	480	222	168	144
80	645	285	213	186
90	915	378	285	261
100	1563	744	597	597

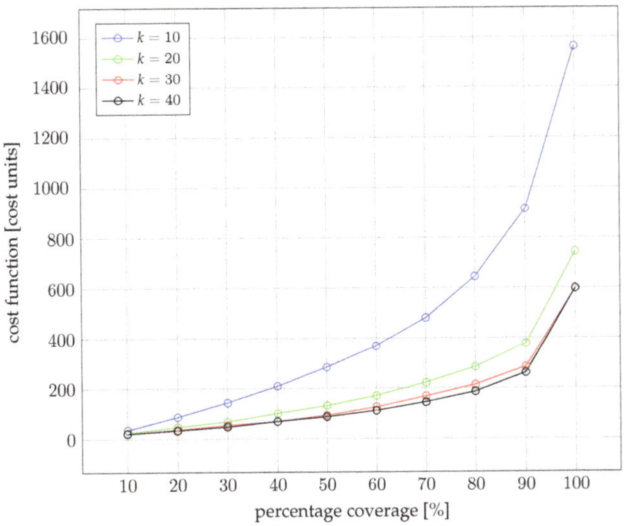

Figure 8. Optimal cost of IoT equipment deployment for dataset 2. Sampling frequency in a given pipe depends on the flow and the size of the pipe.

Both Figures 7 and 8 show an exponential increase of the cost for an increase in the demanded percentage of the sub-catchment area coverage. For instance, for the scenario when the threshold is set to Q_3 = 4 mS/cm and there are low flow conditions (f_L), a reduction of the cost of 47.6%—i.e., from 750 cost units to 393 cost units—can be achieved when relaxing the covered area from 100% to 90%. Similar relative cost reductions can be achieved at 90% coverage for all other five evaluated scenarios in each case.

Such results demonstrate that a wide area coverage is economically feasible for end-users—Law Enforcement Agencies and Environmental Agencies (LEAEA)—interested in monitoring an urban area, if the requirement of covering the whole sub-catchment area is relaxed. From these results, we conjecture that end-users may attempt to select for omission in the planning 10% of sources with a low probability of illegal discharges with the aim of reducing the cost of deployment by almost one half. This conjecture shall be studied in further work.

Figures 9 and 10 show the computational efficiency of the proposed method. Figure 9 shows the time as a function of the percentage coverage of the network for a representative case of the experiment shown in Figure 8. It should be emphasised that the computational time is satisfactory, with the cases between 40% and 80% coverage taking the most computational time.

On the other hand, Figure 10 shows convergence curve as a function of gap and the number of iterations. The gap reflects the difference between the best known bound and the objective value of the best solution produced by a particular algorithm.

Some statistical results concerning space utilization in the edges for both data-set scenarios are also presented in Appendix A.

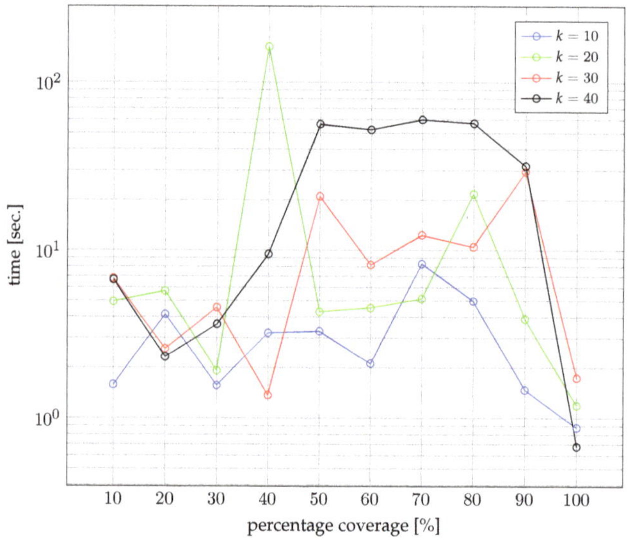

Figure 9. Computation time. Sampling frequency in a given pipe depends on flow and the size of the pipe.

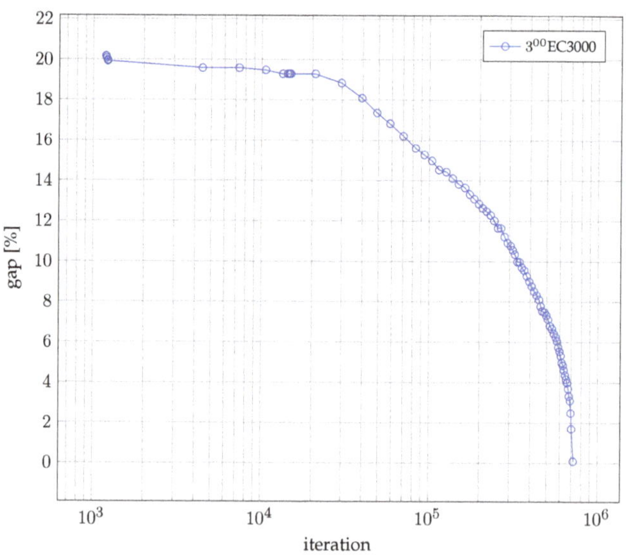

Figure 10. Convergence of the MIP method.

5. Conclusions

This work has addressed the problem of coverage in the sewage network. A model is proposed that provides a coverage problem in a sewer network and at the same time optimises network infrastructure resources such as Micromole rings with modules including sensors and batteries. We proposed the mixed integer programming method, which guarantees to find an optimal solution. In the experiments we used an example of a wide-ranging realistic sewage network from a big-sized city. The method we proposed proved to be effective, giving optimal results in a reasonable computational time.

The convergence curves show an exponential increase in cost for an increase in the desired percentage of coverage of the sub-catchment area. These results show that a wide range of coverage is economically feasible for end users. Based on these results, we conjecture that end-users may try to select up to a dozen percent of sources with low probability of illicit discharges for omission in planning in order to reduce the cost of deployment by almost half. This idea will be the subject of our further research in this area. We plan to develop a model and cost function to locate a potential source of pollutant discharge in the sewer network. We also plan to use evolutionary and bee algorithms if the computation time is long.

Author Contributions: Conceptualization, S.K., F.S.D. and A.S.; methodology, S.K.; software, A.S.; validation, S.K., F.S.D. and A.S.; formal analysis, S.K.; data curation, F.S.D.; writing—original draft preparation, S.K.; writing—review and editing, F.S.D., S.K. and A.S.; visualization, A.S.; supervision, S.K. All authors have read and agreed to the published version of the manuscript.

Funding: This work was partially supported by the H2020 SYSTEM project, which had received funding from the European Union's Horizon 2020 research and innovation programme under Grant Agreement No. 787128.

Institutional Review Board Statement: Not applicable.

Informed Consent Statement: Not applicable.

Data Availability Statement: Not applicable.

Acknowledgments: The authors would like to thanks the contribution of Steffen Krausse and Omar Shehata from Bundeswehr University Munich, Germany, for the discharge simulation performed by the ++SYSTEM Isar in Section 4.

Conflicts of Interest: The authors declare no conflict of interest.

Abbreviations

The following abbreviations are used in this manuscript:

WSN	Wireless Sensor Networks
IoT	Internet of Things
MIP	Mixed Integer Programming
WWTP	WasteWater Treatment Plants
EC	Electrical Conductivity
NSGA	Non-Dominated Sorting Genetic Algorithm
SBN	Simulated Binary Crossover
AI	Artificial Intelligence
DFS	Depth-First Search
gap	Difference between current best integer solution and optimal value of LP relaxation
LEAEA	Law Enforcement Agencies and Environmental Agencies
CPLEX	Mixed Integer Programming solver.

Appendix A

In the appendix, aggregated statistics of cross-sectional area utilization of pipes by sensors and batteries, or simply *edge space utilization*, per test scenario are included. Only edges with $\gamma_e = 1$ were considered in the statistics. In all cases $\alpha_e = 1$, so statistics of α_e were omitted in the tables. Edge utilization is measured as the ratio between the number of slots used by batteries and sensors and the total number of slots available in the given edge. Edge utilization means the number of edges with $\gamma_e = 1$.

For dataset 1 it can be concluded that for cases with hour 8:00, edge utilization is greater than for cases with hour 3:00. Space utilization is lower for 8:00, however. The statistics are presented in Table A1. For dataset 2 it can be concluded that the greater the k value is, the lower the edge utilization is. The same observation can be made for average slot (space) utilization—the greater the k value, the lower the space utilization. In addition, the greater the coverage percentage, the greater the space utilization is. The statistics are presented in Table A2. For both datasets it can be observed that the greater the coverage percentage is, the greater the edge utilization is.

Table A1. Aggregated statistics of space utilization in the edges of dataset 1 test scenarios.

Scenario	Percentage Coverage [%]	Utilized Edges	β_e Min	Max	Mean	Space Utilization [%] Min	Max	Mean
EC2000 03:00	10	1	1	1	1.00	13.33	13.33	13.33
	20	1	1	1	1.00	13.33	13.33	13.33
	30	1	1	1	1.00	13.33	13.33	13.33
	40	2	1	1	1.00	13.33	13.33	13.33
	50	2	1	1	1.00	13.33	13.33	13.33
	60	2	1	1	1.00	13.33	13.33	13.33
	70	3	1	1	1.00	13.33	13.33	13.33
	80	4	1	1	1.00	13.33	13.33	13.33
	90	5	1	1	1.00	13.33	13.33	13.33
	100	11	1	2	1.09	13.33	20.00	13.94
EC2000 08:00	10	1	1	1	1.00	13.33	13.33	13.33
	20	1	1	1	1.00	13.33	13.33	13.33
	30	1	1	1	1.00	13.33	13.33	13.33
	40	2	1	1	1.00	13.33	13.33	13.33
	50	2	1	1	1.00	13.33	13.33	13.33
	60	3	1	1	1.00	13.33	13.33	13.33
	70	4	1	1	1.00	13.33	13.33	13.33
	80	6	1	1	1.00	13.33	13.33	13.33
	90	8	1	1	1.00	13.33	13.33	13.33
	100	9	1	1	1.00	13.33	13.33	13.33
EC3000 03:00	10	1	1	1	1.00	13.33	13.33	13.33
	20	1	1	1	1.00	13.33	13.33	13.33
	30	2	1	1	1.00	13.33	13.33	13.33
	40	2	1	1	1.00	13.33	13.33	13.33
	50	3	1	1	1.00	13.33	13.33	13.33
	60	4	1	1	1.00	13.33	13.33	13.33
	70	5	1	1	1.00	13.33	13.33	13.33
	80	6	1	1	1.00	13.33	13.33	13.33
	90	9	1	1	1.00	13.33	13.33	13.33
	100	20	1	2	1.05	13.33	40.00	15.00
EC3000 08:00	10	1	1	1	1.00	13.33	13.33	13.33
	20	1	1	1	1.00	13.33	13.33	13.33
	30	2	1	1	1.00	13.33	13.33	13.33
	40	3	1	1	1.00	13.33	13.33	13.33
	50	4	1	1	1.00	13.33	13.33	13.33
	60	5	1	1	1.00	13.33	13.33	13.33
	70	8	1	1	1.00	13.33	13.33	13.33
	80	12	1	2	1.08	13.33	20.00	13.89
	90	20	1	2	1.05	13.33	20.00	13.67
	100	23	1	2	1.05	13.33	20.00	13.67
EC4000 03:00	10	1	1	1	1.00	13.33	13.33	13.33
	20	1	1	1	1.00	13.33	13.33	13.33
	30	2	1	1	1.00	13.33	13.33	13.33
	40	2	1	1	1.00	13.33	13.33	13.33
	50	3	1	1	1.00	13.33	13.33	13.33
	60	4	1	1	1.00	13.33	13.33	13.33
	70	6	1	1	1.00	13.33	13.33	13.33
	80	9	1	1	1.00	13.33	13.33	13.33
	90	14	1	2	1.07	13.33	20.00	13.81
	100	31	1	2	1.06	13.33	40.00	15.48

Table A1. *Cont.*

Scenario	Percentage Coverage [%]	Utilized Edges	β_e Min	β_e Max	β_e Mean	Space Utilization [%] Min	Space Utilization [%] Max	Space Utilization [%] Mean
EC4000 08:00	10	1	1	1	1.00	13.33	13.33	13.33
	20	2	1	1	1.00	13.33	13.33	13.33
	30	3	1	1	1.00	13.33	13.33	13.33
	40	4	1	1	1.00	13.33	13.33	13.33
	50	6	1	1	1.00	13.33	13.33	13.33
	60	8	1	1	1.00	13.33	13.33	13.33
	70	11	1	1	1.00	13.33	13.33	13.33
	80	17	1	1	1.00	13.33	13.33	13.33
	90	26	1	2	1.04	13.33	20.00	13.59
	100	31	1	2	1.04	13.33	20.00	13.59

Table A2. Aggregated statistics of space utilization in the edges of dataset 2 test scenarios.

Scenario	Percentage Coverage [%]	Utilized Edges	β_e Min	β_e Max	β_e Mean	Space Utilization [%] Min	Space Utilization [%] Max	Space Utilization [%] Mean
$k = 10$	10	2	1	2	2.00	20.00	20.00	20.00
	20	5	1	3	1.80	13.33	26.67	18.67
	30	7	1	7	2.86	13.33	53.33	25.71
	40	11	1	7	2.36	13.33	53.33	22.42
	50	15	1	7	2.33	13.33	53.33	22.22
	60	20	1	7	2.15	13.33	53.33	21.00
	70	26	1	7	2.15	13.33	85.71	22.78
	80	35	1	7	2.14	13.33	100.00	24.16
	90	48	1	7	2.35	13.33	100.00	30.26
	100	87	1	7	1.99	13.33	100.00	31.26
$k = 20$	10	1	4	4	4.00	33.33	33.33	33.33
	20	2	1	7	4.00	13.33	53.33	33.33
	30	3	1	7	3.67	13.33	53.33	31.11
	40	5	1	4	2.80	13.33	33.33	25.33
	50	6	1	7	3.33	13.33	53.33	28.89
	60	8	1	7	3.12	13.33	53.33	27.50
	70	11	1	7	2.73	13.33	53.33	24.85
	80	14	1	7	2.79	13.33	53.33	25.24
	90	20	1	7	2.30	13.33	53.33	22.00
	100	39	1	7	2.36	13.33	85.71	29.80
$k = 30$	10	1	3	3	3.00	26.67	26.67	26.67
	20	2	1	3	2.00	13.33	26.67	20.00
	30	2	3	7	5.00	26.67	53.33	40.00
	40	3	1	7	3.67	13.33	53.33	31.11
	50	4	1	7	3.75	13.33	53.33	31.67
	60	6	1	7	3.00	13.33	53.33	26.67
	70	8	1	7	3.00	13.33	53.33	26.67
	80	10	1	7	3.10	13.33	53.33	27.33
	90	15	1	7	2.33	13.33	53.33	22.22
	100	33	1	7	2.03	13.33	66.67	27.58
$k = 40$	10	1	3	3	3.00	26.67	26.67	26.67
	20	1	7	7	7.00	53.33	53.33	53.33
	30	2	1	7	4.00	13.33	53.33	33.33
	40	3	1	7	3.67	13.33	53.33	31.11
	50	4	1	7	3.25	13.33	53.33	28.33
	60	5	1	7	3.40	13.33	53.33	29.33
	70	7	1	7	2.86	13.33	53.33	25.71
	80	9	1	7	2.89	13.33	53.33	25.93
	90	14	1	7	2.21	13.33	53.33	21.43
	100	33	1	7	2.03	13.33	66.67	27.58

References

1. Hauser, F.M.; Hulshof, J.W.; Rößler, T.; Zimmermann, R.; Pütz, M. Characterisation of aqueous waste produced during the clandestine production of amphetamine following the Leuckart route utilising solid-phase extraction gas chromatography-mass spectrometry and capillary electrophoresis with contactless conductivity detection. *Drug Test. Anal.* **2018**, *10*, 1368–1382. [PubMed]
2. Cormier, G.J. Alkaline Cleaning. In *Surface Engineering*; Cotell, C.M., Sprague, J.A., Smidt, F.A., Eds.; ASM International: Materials Park, OH, USA, 1994; pp. 18–20. [CrossRef]
3. Ghasemipanah, K. Treatment of ion-exchange resins regeneration wastewater using reverse osmosis method for reuse. *Desalin. Water Treat.* **2013**, *51*, 5179–5183. [CrossRef]
4. Consortium, H.M. Micromole—Sewage Monitoring System for Tracking Synthetic Drug Laboratories. Available online: http://www.micromole.eu (accessed on 17 October 2019).
5. Consortium, H.S. H2020 SYSTEM—Synergy of Integrated Sensors and Technologies for Urban Secured Environment. Fact Sheet Available at EC Website. Under Project ID 787128. Available online: https://cordis.europa.eu/project/rcn/220304/factsheet/en (accessed on 6 September 2022).
6. De Vito, S.; Fattoruso, G.; Esposito, E.; Salvato, M.; Agresta, A.; Panico, M.; Leopardi, A.; Formisano, F.; Buonanno, A.; Delli Veneri, P.; et al. A Distributed Sensor Network for Waste Water Management Plant Protection. In *Convegno Nazionale Sensori*; Andò, B., Baldini, F., Di Natale, C., Marrazza, G., Siciliano, P., Eds.; Springer International Publishing: Cham, Switzerland, 2018; pp. 303–314.
7. Lepot, M.; Makris, K.F.; Clemens, F.H. Detection and quantification of lateral, illicit connections and infiltration in sewers with Infra-Red camera: Conclusions after a wide experimental plan. *Water Res.* **2017**, *122*, 678–691. [CrossRef] [PubMed]
8. Tan, F.H.S.; Park, J.R.; Jung, K.; Lee, J.S.; Kang, D.K. Cascade of One Class Classifiers for Water Level Anomaly Detection. *Electronics* **2020**, *9*, 1012. [CrossRef]
9. Tashman, Z.; Gorder, C.; Parthasarathy, S.; Nasr-Azadani, M.M.; Webre, R. Anomaly Detection System for Water Networks in Northern Ethiopia Using Bayesian Inference. *Sustainability* **2020**, *12*, 2897. [CrossRef]
10. Zhang, D.; Heery, B.; O'Neil, M.; Little, S.; O'Connor, N.E.; Regan, F. A Low-Cost Smart Sensor Network for Catchment Monitoring. *Sensors* **2019**, *19*, 2278. [CrossRef] [PubMed]
11. Perfido, D.; Messervey, T.; Zanotti, C.; Raciti, M.; Costa, A. Automated Leak Detection System for the Improvement of Water Network Management. *Proceedings* **2016**, *1*, 28. [CrossRef]
12. Rojek, I.; Studzinski, J. Detection and Localization of Water Leaks in Water Nets Supported by an ICT System with Artificial Intelligence Methods as a Way Forward for Smart Cities. *Sustainability* **2019**, *11*, 518. [CrossRef]
13. Ji, H.; Yoo, S.; Lee, B.J.; Koo, D.; Kang, J.H. Measurement of Wastewater Discharge in Sewer Pipes Using Image Analysis. *Water* **2020**, *12*, 1771. [CrossRef]
14. Kuchmenko, T.A.; Lvova, L.B. A Perspective on Recent Advances in Piezoelectric Chemical Sensors for Environmental Monitoring and Foodstuffs Analysis. *Chemosensors* **2019**, *7*, 39. [CrossRef]
15. Pisa, I.; Santín, I.; Vicario, J.; Morell, A.; Vilanova, R. ANN-Based Soft Sensor to Predict Effluent Violations in Wastewater Treatment Plants. *Sensors* **2019**, *19*, 1280. [CrossRef]
16. Drenoyanis, A.; Raad, R.; Wady, I.; Krogh, C. Implementation of an IoT Based Radar Sensor Network for Wastewater Management. *Sensors* **2019**, *19*, 254. [CrossRef]
17. Ma, J.; Meng, F.; Zhou, Y.; Wang, Y.; Shi, P. Distributed Water Pollution Source Localization with Mobile UV-Visible Spectrometer Probes in Wireless Sensor Networks. *Sensors* **2018**, *18*, 606. [CrossRef]
18. Desmet, C.; Degiuli, A.; Ferrari, C.; Romolo, F.; Blum, L.; Marquette, C. Electrochemical Sensor for Explosives Precursors' Detection in Water. *Challenges* **2017**, *8*, 10. [CrossRef]
19. Solano, F.; Krause, S.; Wöllgens, C. An Internet-of-Things Enabled Smart System for Wastewater Monitoring. *IEEE Access* **2022**, *10*, 4666–4685. [CrossRef]
20. Consortium, S. SIMONA—Sistema Integrato di Competenze per il MONitoraggio, la Protezione e il Controllo Delle Infrastrutture Idriche, Fognarie e Ambientali. Available online: http://www.progettosimona.it (accessed on 15 March 2021).
21. Banik, B.; Alfonso, L.; Torres, A.; Mynett, A.; Di Cristo, C.; Leopardi, A. Optimal Placement of Water Quality Monitoring Stations in Sewer Systems: An Information Theory Approach. *Procedia Eng.* **2015**, *119*, 1308–1317. [CrossRef]
22. Banik, B.; Alfonso, L.; Di Cristo, C.; Leopardi, A. Greedy Algorithms for Sensor Location in Sewer Systems. *Water* **2017**, *9*, 856. [CrossRef]
23. Banik, B.K.; Alfonso, L.; Cristo, C.D.; Leopardi, A.; Mynett, A. Evaluation of Different Formulations to Optimally Locate Sensors in Sewer Systems. *J. Water Resour. Plan. Manag.* **2017**, *143*, 04017026. [CrossRef]
24. Ali, H.; Osmani, S.; Banik, B. Integrating fuzzy logic with Pearson correlation to optimize contaminant detection in water distribution system with uncertainty analyses. *Environ. Monit. Assess.* **2019**, *191*, 441. [CrossRef]
25. Alfonso, L.; He, L.; Lobbrecht, A.; Price, R. Information theory applied to evaluate the discharge monitoring network of the Magdalena River. *J. Hydroinformatics* **2013**, *15*, 211–228. [CrossRef]
26. Alfonso, L.; Lobbrecht, A.; Price, R. Information theory–based approach for location of monitoring water level gauges in polders. *Water Resour. Res.* **2010**, *46*, 3. [CrossRef]

27. Alfonso, L.; Lobbrecht, A.; Price, R. Optimization of water level monitoring network in polder systems using information theory. *Water Resour. Res.* **2010**, *46*, 12. [CrossRef]
28. Sikorski, A.; Kozdrowski, S.; Donado, F.S. IoT Device Deployment for Optimal Wastewater Network Coverage. In Proceedings of the 2021 International Conference on Software, Telecommunications and Computer Networks (SoftCOM), Split, Hvar, Croatia, 23–25 September 2021; pp. 1–6. [CrossRef]
29. Tossa, F.; Abdou, W.; Ezin, E.C.; Gouton, P. Improving Coverage Area in Sensor Deployment Using Genetic Algorithm. In *Computational Science—ICCS 2020*; Krzhizhanovskaya, V.V., Závodszky, G., Lees, M.H., Dongarra, J.J., Sloot, P.M.A., Brissos, S., Teixeira, J., Eds.; Springer International Publishing: Cham, Switzerland, 2020; pp. 398–408.
30. Hanh, N.T.; Binh, H.T.T.; Hoai, N.X.; Palaniswami, M.S. An efficient genetic algorithm for maximizing area coverage in wireless sensor networks. *Inf. Sci.* **2019**, *488*, 58–75. [CrossRef]
31. Hanh, N.T.; Nam, N.H.; Binh, H.T.T. Particle Swarm Optimization Algorithms for Maximizing Area Coverage in Wireless Sensor Networks. In *Proceedings of the SAI Intelligent Systems Conference (IntelliSys) 2016*; Bi, Y., Kapoor, S., Bhatia, R., Eds.; Springer International Publishing: Cham, Switzerland, 2018; pp. 893–904.
32. Buras, M.P.; Solano Donado, F. Identifying and Estimating the Location of Sources of Industrial Pollution in the Sewage Network. *Sensors* **2021**, *21*, 3426. [CrossRef]
33. Chachuła, K.; Nowak, R.; Solano, F. Pollution Source Localization in Wastewater Networks. *Sensors* **2021**, *21*, 826. [CrossRef]
34. Chachuła, K.; Słojewski, T.M.; Nowak, R. Multisensor Data Fusion for Localization of Pollution Sources in Wastewater Networks. *Sensors* **2022**, *22*, 387. [CrossRef]
35. Fourer, R.; Gay, D.M.; Kernighan, B.W. AMPL: A Mathematical Programing Language. In *Algorithms and Model Formulations in Mathematical Programming*; Wallace, S.W., Ed.; Springer: Berlin/Heidelberg, Germany, 1989; pp. 150–151.
36. Kozdrowski, S.; Żotkiewicz, M.; Wnuk, K.; Sikorski, A.; Sujecki, S. A Comparative Evaluation of Nature Inspired Algorithms for Telecommunication Network Design. *Appl. Sci.* **2020**, *10*, 6840. [CrossRef]
37. Tandler.com. Software for Water Management ++ Systems Isar. Available online: www.tandler.com (accessed on 21 November 2021).

sensors

Article

Planning and Optimization of Software-Defined and Virtualized IoT Gateway Deployment for Smart Campuses

Divino Ferreira, Jr. [1,2], João Lucas Oliveira [2], Carlos Santos [3], Tércio Filho [4], Maria Ribeiro [5], Leandro Alexandre Freitas [6], Waldir Moreira [7] and Antonio Oliveira-Jr [2,7,*]

1 Campus Senador Canedo, Federal Institute of Education, Science and Technology of Goiás (IFG), Senador Canedo 75250-000, Brazil; divino.alves@ifg.edu.br
2 Institute of Informatics (INF), Federal University of Goiás (UFG), Goiânia 74690-900, Brazil; joaooliveira@discente.ufg.br
3 Campus Palmas, Federal Institute of Education, Science and Technology of Tocantins (IFTO), Palmas 77021-090, Brazil; carlosedu@ifto.edu.br
4 Institute of Biotechnology (IBiotec), Federal University of Catalão (UFCAT), Catalão 75705-220, Brazil; tercioas@ufg.br
5 Institute for Systems and Computer Engineering, Technology and Science (INESC-TEC), 4200-465 Porto, Portugal; maria.r.ribeiro@inesctec.pt
6 Campus Inhumas, Federal Institute of Education, Science and Technology of Goiás (IFG), Inhumas 75402-556, Brazil; leandro.freitas@ifg.edu.br
7 Fraunhofer Portugal AICOS, 4200-135 Porto, Portugal; waldir.junior@fraunhofer.pt
* Correspondence: antoniojr@ufg.br

Citation: Ferreira, D., Jr.; Oliveira, J.L.; Santos, C.; Filho, T.; Ribeiro, M.; Freitas, L.A.; Moreita, W.; Oliveira-Jr, A. Planning and Optimization of Software-Defined and Virtualized IoT Gateway Deployment for Smart Campuses. *Sensors* **2022**, *22*, 4710. https://doi.org/10.3390/s22134710

Academic Editors: Alexandru Lavric, Liliana Anchidin, Adrian I. Petrariu and Pal Varga

Received: 8 April 2022
Accepted: 20 June 2022
Published: 22 June 2022

Abstract: The Internet of Things (IoT) is based on objects or "things" that have the ability to communicate and transfer data. Due to the large number of connected objects and devices, there has been a rapid growth in the amount of data that are transferred over the Internet. To support this increase, the heterogeneity of devices and their geographical distributions, there is a need for IoT gateways that can cope with this demand. The SOFTWAY4IoT project, which was funded by the National Education and Research Network (RNP), has developed a software-defined and virtualized IoT gateway that supports multiple wireless communication technologies and fog/cloud environment integration. In this work, we propose a planning method that uses optimization models for the deployment of IoT gateways in smart campuses. The presented models aimed to quantify the minimum number of IoT gateways that is necessary to cover the desired area and their positions and to distribute IoT devices to the respective gateways. For this purpose, the communication technology range and the data link consumption were defined as the parameters for the optimization models. Three models are presented, which use LoRa, Wi-Fi, and BLE communication technologies. The gateway deployment problem was solved in two steps: first, the gateways were quantified using a linear programming model; second, the gateway positions and the distribution of IoT devices were calculated using the classical K-means clustering algorithm and the metaheuristic particle swarm optimization. Case studies and experiments were conducted at the Samambaia Campus of the Federal University of Goiás as an example. Finally, an analysis of the three models was performed, using metrics such as the silhouette coefficient. Non-parametric hypothesis tests were also applied to the performed experiments to verify that the proposed models did not produce results using the same population.

Keywords: Internet of Things (IoT); smart campus; IoT gateway; cluster; optimization

1. Introduction

Information technology has become essential in the daily lives of people and businesses and the Internet of Things (IoT) concept directly contributes to changes in everyday life [1–7]. The IoT is a communication paradigm in which objects communicate with each other and with users via network communication technologies, mostly wireless networks [8].

This paradigm enables interaction between several devices (called things), such as electronic appliances, home appliances, vehicles, hospital equipment, sensors, surveillance cameras, etc. The technologies that are used in the IoT context directly contribute to the evolution of services that are associated with smart cities and smart campuses. Hence, smart campuses are integrated work, study, and living environments that are based on the Internet of Things [9]. A multitude of data types that correspond to distinct applications are transmitted over the IoT infrastructure and need to be processed in different units [10].

Two of the main challenges that are analyzed and faced by IoT are interoperability and heterogeneity [11]. One solution to these challenges is the deployment of IoT gateways that support multiple communication technologies [12]. Deploying IoT gateways in smart campuses requires prior planning. Consequently, it is necessary to deploy IoT gateways that address the communication needs of IoT devices in university campuses, such as bandwidth, communication technologies, energy efficiency, etc. The SOFTWAY4IoT project was funded by the National Education and Research Network (RNP) and developed an IoT gateway that can establish communication within an Internet of Things environment. This gateway is virtualized, utilizes software-defined networking (SDN) and fog computing that is integrated with cloud computing, and supports multiple wireless communication technologies [13]. The area coverage of wireless networks is also a challenging problem. The challenges start with the choice of the communication technology that is to be used through to the architectural characteristics of the environment, the number of things or people that access the network simultaneously, the mobility of the people or things that connect to the network, etc. It is possible to find several works that have focused on area coverage in the literature.

We found several works in the literature that focused on area coverage [14–19]. Services that require communication infrastructures need a guarantee of the quality of the communication. Distributing the gateways within these environments is a big challenge comprising many factors that are relevant to the positioning and quantification of the devices that aim to meet the presented demand.

In this context, this work proposes a method for the planning and optimization of the deployment of IoT gateways, with the SOFTWAY4IoT IoT gateway as its motivation. Hence, this paper presents three optimization methods for gateway infrastructure planning that aim to minimize the number of required gateways and maximize the coverage area, considering the communication capacity of the data link and the range of the communication technology. A comparison of the models is also presented in order to evaluate the obtained results. The study focused on the LoRa, BLE, and Wi-Fi communication technologies.

The LoRa technology was chosen because it has a long communication range that increases the coverage area, as well as a low energy consumption. On the other hand, the BLE technology opposes LoRa technology in terms of range, as it covers a small area (personal area) but has a similarly low power consumption. Finally, we chose Wi-Fi because it is one of the most popular communication technologies within wireless networking. It has an area coverage that has a larger range than the BLE but a smaller range than the LoRa and has a high power consumption.

However, the determination of how many IoT gateways to use and where to place them in order to maximize the coverage area and minimize the number of gateways is an NP-hard problem, called the WMN node placement problem. The complexity of this problem grows exponentially with small changes in the size of the problem. Therefore, we tested three different hybrid methods to find near optimal solutions [20].

The presented optimization model aimed to minimize the number of required IoT gateways and to position them within the area to be covered, taking into account the size of the area, the range of the communication technology, and the location in which the devices are deployed. The gateway deployment problem was split in two steps: (a) the determination of the number of required gateways by solving a linear programming problem (LPP), which aimed to minimize the gateways quantities; (b) the K-means and particle swarm optimization (PSO) algorithms were employed to establish the gateway positioning.

The use case scenario was a university campus and the chosen areas and the size and characteristics of the environment (indoor/outdoor) are presented in Table 1. We tested larger and smaller areas with different architectural characteristics.

Table 1. The measurements of the areas that were used in the use case.

Desired Area	Size of Area (m^2)	Type of Environment
INF	2.425	Indoor
Samambaia Campus	761.380	Outdoor
Academic Blocks	26.664	Indoor

The contributions of this article can be summarized as follows:

- The proposal of an optimization model that uses linear programming for the quantification and positioning of *gateways* to minimize network deployment costs;
- The proposal of an algorithm that uses the K-*means* clustering method to define the positioning of *gateways* within a (predefined) area;
- The proposal of an algorithm that uses the PSO clustering method to define the positioning of gateways within a (predefined) area via two different initialization approaches;
- A comparative evaluation of the methods to establish which method produces the best results for area coverage, using the silhouette coefficient as the metric [21];
- The employment of non-parametric hypothesis testing to verify that the different metaheuristics did not produce results using the same population [22].

This paper is organized as follows. Section 2 presents the related work. Section 3 presents the modeling of the optimization methods that were applied in this study. Section 4 presents the application scenario and Section 5 presents the evaluation of the results. Finally, the final considerations and directions for future work are presented in Section 6.

2. Related Works

We considered studies regarding the Internet of Things, intelligent environments, optimization models, clustering methods, and signal propagation, among others, as relevant to the development of this work [23–28]. Some studies have analyzed area coverage using equipment for different communication technologies; however, our research considered an intelligent environment using IoT gateways for heterogeneous networks.

The related works were found using the keywords "optimization", "coverage", "IoT gateway", "placement", and "planning". These keywords were used to search for works that were related to area coverage and the placement of IoT gateways. We also searched for papers that were related to network signal propagation and smart environments (smart campuses, smart cities, etc.). Initially, the search focused on papers that were published from 2015 onward, but relevant papers were found also from earlier periods. As a search tool, we first used Google Scholar and then we mostly used the IEEE and Springer databases. The searches for related papers were based on the topics of wireless network area planning and coverage, with a greater focus on IoT networks, although this was not restrictive.

A proposal for multi-hop network planning that aimed to minimize hardware (gateways) and operational costs was presented by [14]. Path loss was also considered as an operational cost. The work presented a mathematical optimization model and used three evolutionary algorithms that were based on swarm intelligence [14]. Although the work had similar goals and metrics to this research, there was no application in real space. Our work presents results that were obtained directly from the study environment using three communication technologies and three optimization models.

The model presented by [29] aimed to minimize the latency of communication between fog nodes and gateways. To meet this objective, ref. [29] developed a mathematical formulation that considered the activation of a minimum number of fog nodes. Compared to the proposal presented in this paper, the model in question did not present any study of applied scenarios and adopted empirical parameters in the experiments. There was no quantification of the number of gateways, only a comparison of the executions that varied the numbers.

A study developed by [15] presented a model for application in IoT systems, with a focus on smart cities. The study aimed to reduce the costs of the deployment of an IoT system by considering the use of two gateway models for communication. One of the gateway models communicated with the Internet and the other did not. The presented study used three technologies (Wi-Fi, ZigBee, and RFID) to establish communication between the systems. The work also developed an algorithm to minimize the number of gateways, with the objective of reducing the costs of deployment, and a mechanism to tolerate communication failure.

A smart city scenario integrates several applications of different technologies and may have hundreds of networks using different domains. Each network is coordinated by a coordination device (CD) and each CD needs to transmit its data to the Internet via its own connection that is established using a gateway. To reduce costs, this work proposes two gateway models: an IGW (IoT gateway), which establishes communication using the Internet, and an SSGW (solution-specific gateway), which has no direct communication with the Internet. The gateway presented in our study has features such as SDN usage, virtualization, and integration between the edge and the cloud. These features allow all gateways to communicate with the Internet and can use anything from machines with low processing power, such as a Raspberry, to more robust machines, such as servers with high processing power.

A study on multi-objective planning in WLAN networks was conducted in [16]. The study developed a planning tool that was capable of finding the best position for an access point (AP) and load balance within a WLAN network, which minimized the interference between access points. In this work, a multi-objective evolutionary algorithm with a greedy heuristic was used. The IEEE 802.11 standard was also used, but the study did not address other communication technologies and was restricted to the communication that was established by the access points. For IoT scenarios, heterogeneity is a latent challenge and thus, a study that focuses on IoT needs to consider multi-technology communication. In this regard, our work presents an access point that can integrate multiple communication technologies, i.e., a gateway with multiple communication technologies.

The determination of gateway placements within a network to connect IoT devices is a crucial point when it comes to deployment costs. To solve this problem, ref. [10] presented a solution using integer linear programming (ILP) that minimized the total network costs in relation to deployed devices, while taking into account mandatory quality of service (QoS) requirements. A gateway could be placed anywhere within a given area, but an initial set of candidate positions was considered. Communication took place over multiple hops. To obtain the lowest deployment costs and guarantee the QoS of the fixed transmission range, the specific data rates, end device costs, gateway costs, generated traffic, and the distance between the nodes were used as data. In order to provide QoS, the capacity of the links had to be sufficient to handle multiple simultaneous transmissions.

An optimization approach for gateway deployment in heterogeneous sensor networks was presented by [17]. This study focused on ILP-based optimization and wireless gateway locations. The goal was to minimize the installation costs and maximize the energy efficiency of the wireless sensor network, considering multi-hop coverage and connectivity constraints. Although this study addressed the overheads of sensors that are considered to be critical, multiple hops could demand more time before the message reached its destination.

A gateway placement approach was presented by [18]. This approach aimed to optimize the number of gateways, the average number of mesh routers, and the variations in gateway loads within wireless mesh networks (WMNs). Minimizing the average hop count of the network was one of the objectives since long paths reduce the throughput. The proposed work used two stages to achieve their objectives.

The works of [10,17,18] took into consideration communication using multiple hops. Direct connections between devices and gateways may present higher costs, but the message delivery time tends to be shorter, thereby improving network performance. The particularities of each deployment environment have to be considered, along with issues regarding device density, which involve denser environments or more spacious environments, and other factors. The proposed work considered an intelligent scenario that focused on indoor and outdoor environments and direct communication using software-defined and virtualized gateways. Our optimization model aimed to minimize the number of required gateways as a function of the number of deployed devices to improve the range of the communication technology and the communication capacity of the gateways.

A study developed by [19] addressed the optimal deployment of IoT gateways in smart home environments (smart homes). The work in question solved the optimization problem using the branch and bound method with the goal of minimizing the gateway deployment costs, subject to the constraint that all service areas of the home must be covered. The smart home environment is relatively small compared to a smart campus or smart city. A small environment assumes the use of a single gateway and few devices. For a small environment, the model in question proved to be effective; however, its application in larger environments is necessary to evaluate the effectiveness of the model since there are large areas with high densities of IoT devices.

A literature review article presented by [20] surveyed the optimization approaches that have been implemented to solve the node placement problem in WMNs. In the literature, several WMN node positioning approaches have been proposed. This paper presented a classification that was based on the type of method that was used. The classification was split into four categories: methods that are based on exact approaches, methods that use heuristics, methods that use metaheuristics, and methods that apply hybrid approaches. Additionally, their paper presented a case study using the greedy algorithm (GA), simulated annealing (SA), particle swarm optimization (PSO), and the firefly algorithm (FA) to investigate the impacts of varying the number of mesh clients, the number of mesh routers, and the coverage radius.

Table 2 presents a comparison of the related work, according to network planning. There were works that did not have a focus on IoT networks and others that did not have a focus on gateways, but all of them proposed an optimization model for network planning. All of the works presented map positioning for the devices that established the network communication, although each paper had its own particularities and metrics for mapping. Hence, in this work (and contrary to the related work), our goal was to quantify and position IoT gateways (considering the range of the communication technology that was employed) and the maximum number of devices that could be supported by the gateways.

Section 3 presents the optimization models that were employed in this work and shows the particularities that were adopted in each model.

Table 2. Comparison of network planning related works.

Author	Objective	Method of Optimization	Technology Comunication	Quantifies
Gravalos, Ilias, et al. [10]	Minimize coast/ QoS	LP	-	Yes
Ali, Hafiz Munsub, et al. [14]	Minimize coast	evolutionary algorithms	-	No
Maiti, Prasenjit, et al. [29]	Minimize latency	randomized, greedy, k-median, K-means and simulated annealing	-	No
Karthikeya, Surabhi Abhimithra, J. K. Vijeth, and C. Siva Ram Murthy [15]	Minimize coast	Heuristics	Wi-Fi, ZigBee, RFID	Yes
Matni, Nagib, et al. [30]	Minimize coast/ QoS	Fuzzy C-means	Lora	Yes
Lima, Marlon Paolo, Eduardo G. Carrano, and Ricardo HC Takahashi [16]	Minimize AP/ load	Genetic. Algorithm	Wi-Fi	Yes
Capone, Antonio, et al. [17]	Minimize coast/ Max. energy efficiency	LP	-	Yes
Wu, Wenjia, Junzhou Luo, and Ming Yang [18]	Minimize gateways/ position	LP/ Heuristics	-	No
Lin, Po-Chiang [19]	Minimize coast	LP	-	No

3. Optimization Model for IoT Gateway Planning, Coverage, and Positioning

Several wireless communication technologies are available that can meet the heterogeneous characteristics that guide the Internet of Things paradigm. The application scenarios of this paradigm may include residences, offices, stores, hospitals, industries, universities, cities, etc. The implementation of systems within the context of the Internet of Things demands prior study. In the first instance, this study aimed to analyze the coverage area, the devices that are to be deployed, and the communication technology that would best meet the needs of the environment.

The focus of this work was the planning of IoT gateway deployment within intelligent environments. This research evaluated the signal coverage of three communication technologies (Wi-Fi, LoRa, and BLE) in a smart campus scenario. LoRa is an emerging communication technology with a long range and low power consumption. Wi-Fi is a mid-range technology with a higher power consumption; however, it is widespread in communications and has great relevance to indoor environments. BLE is a short-range communication technology with a low power consumption and personal area coverage.

The proposed work was split into two steps. The first step defined the number of gateways using an optimization model with linear programming. In the second step, the linear programming model, the K-means clustering method, and the PSO method were used to define the positions of the IoT gateways, considering the number of gateways that was defined in the first step. Finally, a comparison was made between the three gateway positioning models, which had the aim of evaluating the results of each model. Figure 1 shows the flowchart of the optimization model.

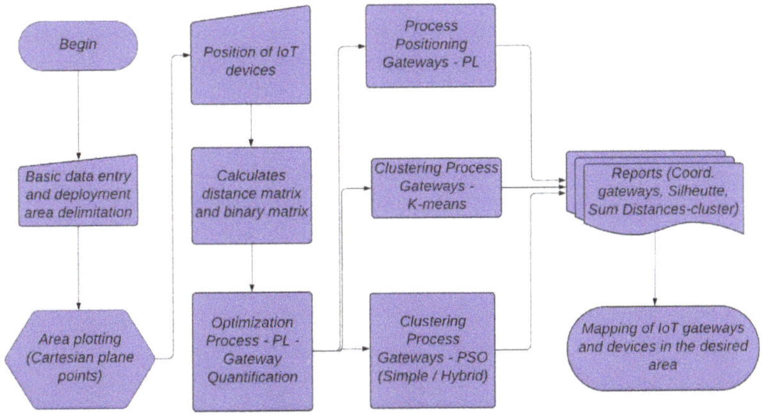

Figure 1. The flowchart of the optimization model.

The candidate points matrix was implemented with the goal of determining the possible installation sites for gateways and IoT devices. We considered the matrix to be a key point for area coverage as its implementation reproduced the floor plan of the scenario, which allowed the algorithm to check each point within the desired area.

The linear programming model developed in this work had a relevance to quantification. The constraints that were associated with the objective allowed for reductions in the deployment costs. The methods, such as K-means and PSO, were used based on the proposed locations of the gateways within the search space that was delimited by the floor plan of the scenario, which is reproduced in this paper in the form of points on a Cartesian plane. These details allowed the quantification and positioning to come closer to the real circumstances of the deployment.

The advantages of the linear programming model over the K-means and PSO models involved the gateway positioning being restricted by the range of the technology; in this case, no device could connect to a gateway when the distance between the points exceeded the range of the technology. On the other hand, the dispersion of devices in relation to the gateways was greater. In the clustering models, there was no restriction on range; however, the degree of dispersion was lower because the devices were positioned considering proximity to the gateways.

The presented models enabled the association of the devices and the experiments were implemented in a real environment with distinct architectures. In this way, it was possible to evaluate their applicability as close to reality as possible. It is noteworthy that the range of the technology needed to consider the study of signal propagation within the deployment scenario. Although this study did not consider the actual coordinates of the deployment of devices, the models were developed with this objective; therefore, when we had the coordinates of the installation site of a device, we inserted them as an input parameter and obtained a result that was even more realistic.

3.1. Gateway Quantification and Positioning Using Linear Programming

The planning and optimization of IoT gateway deployment for communication networks aimed to calculate the minimum gateway quantity that is required to cover a defined area and the number of IoT devices per gateway. For the quantification, the range of the employed wireless communication technology and the gateway communication link capacity were considered. The ultimate goal was to minimize the number of gateways while maximizing the coverage for devices within the desired area.

The area coverage was calculated based on the signal propagation of the communication technology. The signal propagation of each communication technology had distinct characteristics and also varied with respect to the propagation environment (indoor/outdoor). The signal propagation models allowed the range of the communication technology to be abstracted as a function of path loss.

The coverage area was an input parameter for the optimization model. Using Google Earth as a tool (available online: https://www.google.com.br/intl/pt-BR/earth/ (accessed on 10 May 2021)), it was possible to ascertain the measurements (dimensions) of the desired area. This area needed to be delimited and the possible locations of the devices and gateways needed to be plotted. In the optimization model, the points were plotted according to the coordinates of a Cartesian plane, as shown in Figure 2. In this case, we obtained an area of 2425 m^2 that was considered for the possible device locations. Each m^2 was one positioned point (i.e., a possible point in the desired area), totaling 2425 possible IoT device installation points.

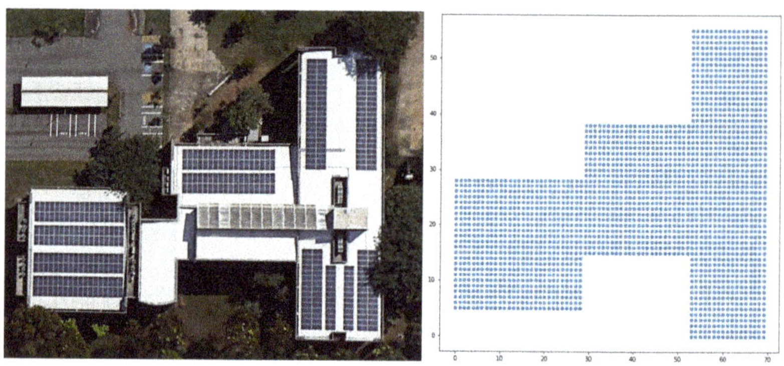

Figure 2. An example of area mapping using the Institute of Informatics, UFG.

Each pair of coordinates for the plotted points in the specific area was used as an input parameter for the linear programming-based optimization model. The proposed model chose the installation points of the IoT devices randomly using a seed, but they could also be defined according to the reality of their deployment (i.e., their real positions). When using the real positions, it was necessary to have the coordinates of the location in which the device was to be deployed.

As input parameters, we also used the range of the communication technology, the number of devices, the gateway data link capacity, and the IoT device link demands. The presented metaheuristic quantified the IoT gateways within the desired area, considering the coverage of the devices that were deployed in that area. Table 3 presents the parameters that were used for the optimization model that was based on linear programming. Considering the goal of minimizing the number of gateways, the decision variables were defined first. Then, we obtained:

- $X_{i,j}$ as the matrix that associated a device to a gateway;
- Y_i as the vector that received the gateways that were activated to meet the demand of the devices.

Given the coordinates of the IoT devices ($D_{i,j}$), the optimization model associated each device with a gateway (Y_i). The gateway had to be within the range of the technology (r). This association occurred via the decision variables. Each device was assigned to a gateway but each defined gateway could have multiple devices associated with it. The model used these constraints to make associations that obeyed the necessary criteria in order to meet the proposed objective. The objective of the optimization model is represented by Equation (1):

$$min \sum_{i=1}^{n} Y_i \tag{1}$$

where Y_i represents the active gateways. Equation (1) represents the objective of minimizing the number of gateways (Y_i). Note that each $D_{i,j}$ is a candidate gateway point. Using this logic, a gateway could always be activated at one of the points $D_{i,j}$, which was equivalent to an IoT device point.

Table 3. The variables that were used for the LP optimization model, including the input parameters and data that were generated from the input parameters.

Variable	Description
r	Range of the communication technology
n	Number of deployed devices
c	Device bandwidth consumption
cb	Total link communication capacity
$A_{i,j}$	All possible points within the desired area
$D_{i,j}$	Coordinates of the deployed devices
$Mdist_{i,j}$	Distance matrix (measured between the point of a device and the points of the other deployed devices)
$Mativ_{i,j}$	Binary matrix (where 0 indicates that the distance between the devices is not within range and 1 indicates that the distance between the devices is within the range of the technology)

Constraints were defined to achieve the goal of ensuring coverage for the IoT devices using the lowest number of gateways. As constraints, we used:

$$\sum_{j=1}^{n} X_{i,j} = 1, \quad i = 1, \ldots, n. \tag{2}$$

$$X_{i,j} \leq Mativ_{i,j} * Y_i \qquad i,j = 1, \ldots, n \tag{3}$$

$$\sum_{j=1}^{n} X_{i,j} * c - cb \leq 0 \tag{4}$$

$$X_{i,j} \geq 0, \quad i,j = 1, \ldots, n \tag{5}$$

$$Y_i \geq 0, \quad i,j = 1, \ldots, n \tag{6}$$

The restrictions caused by Equations (2) and (3) aimed to associate a device with a single gateway. Equation (4) limited the number of devices per gateway, considering the link transmission capacity, and did not allow for the exceedance of the defined maximum transmission capacity. In this case, when the maximum capacity was exceeded, another gateway was activated to meet the required demands. The constraints caused by Equations (5) and (6) were non-negative constraints.

Running the optimization model produced the active gateways (Y_i) and the associations between the devices and their corresponding gateways ($X_{i,j}$) as solutions . From those results, a graph could be plotted using the plant model represented by Figure 2, which showed the positions of the gateways and the respective IoT devices, thus forming a cluster.

3.2. Gateway Placement Using the K-Means Clustering Algorithm

The K-means clustering method is used in data mining [31]. This method aims to partition n observations into k clusters. In this research, the K-means algorithm was used to define the gateway placements and the distribution of devices per gateway.

The definition of a cluster quantity in the K-means approach was used as an input parameter. The input parameters that were used in this model are presented in Table 4.

Table 4. The variables that were used in the K-means clustering model, including the input parameters.

Variable	Description
k	Number of gateways/devices to be deployed equals the number of centroids
n	Number of deployed devices
$D_{i,j}$	Coordinates of the deployed devices
$A_{i,j}$	All coordinates of the possible points within the desired area

The K-means method uses the distance between points for clustering and attempts to separate samples into n clusters of equal variance by minimizing a criterion, which is known as inertia, or the sum of squares within the cluster according to Equation (7):

$$\sum_{j=1}^{n} (\| x_j - \mu \|^2) \tag{7}$$

The proposed algorithm divided a set of n IoT devices into k groups. The center of the groups (centroid) represents the position of the IoT gateway and $D_{i,j}$ represents the coordinates/position of the IoT devices within the defined area $A_{i,j}$. In the performed experiments, the devices ($D_{i,j}$) were defined randomly using points within the desired area ($A_{i,j}$). It is worth pointing out that they were defined randomly because it was an experiment. The points of a device must have its position defined as coordinates on a Cartesian plane in a real situation.

As the number of gateways was defined by the results (C.f. Figure 3) of the first stage of execution, it was understood that the areas and positions of the devices had to be the same for the execution of the second stage (clustering). In order to retain the same position of the devices, the same seed was used for all execution steps. In this way, the devices were randomly chosen based on the same seed.

The K-means algorithm selected the position of the centroids randomly. The input K-means parameter from the scikit-learn library was applied to initialize the centroids more intelligently in order to speed up convergence. Using this parameter, the K-means algorithm randomly positioned a first centroid, then the other centroids were positioned so that they were as far away from each other as possible. This initial positioning helped the algorithm to converge faster.

Algorithm 1 positioned all gateways (c) and assigned to them the nearest devices ($D_{i,j}$), so that all devices were associated with a gateway. The process of assigning and reallocating centroids was repeated until the positions of the centroids were stable (convergence). The goal was make the sum of the distances between the devices ($D_{i,j}$) and the gateways (c) as small as possible.

Algorithm 1: The K-means pseudocode.

```
input  : N, k, A_{i,j}, D_{i,j}
begin
    Define number of clusters (cluster = k)
    Initialize centroids - (c ← init = KMeans++)
    Receives data to be clustered - (.fit(D_{i,j}))
    repeat
        assign each D_{i,j} to c - smallest distance between points criterion
        reposition c by minimizing the distance between D_{i,j} and c;
    until centroids remain stable;
output: Grouping the devices to the respective gateway forming the clusters
output: Reports with metrics related to the clustering process
```

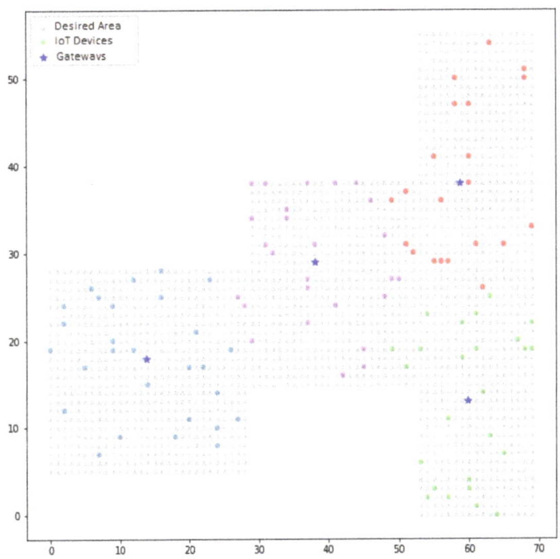

Figure 3. An example of the results of device and gateway placement using the K-means method.

3.3. Positioning of Gateways Using the PSO Optimization Algorithm

The particle swarm optimization model is so named because it is an evolutionary algorithm that arose from studies on algorithms that modeled the social behavior of animals (e.g., bird flocking and bee swarming behavior).

It had the same parameters as the K-means method: the number of gateways (c), the number of devices (n), the desired area ($A_{i,j}$), and the device positions ($D_{i,j}$). The number of particles (p) was a unique parameter for the PSO model. The number of particles was then defined as an input parameter. Each particle was a vector that contained the same number of positions as gateways, so we obtained $P_i = (P_1, P_2, \ldots, P_c)$.

Each particle in the population represented one possible solution. The particles moved around the search space, looking for the position that best met the objective function (fitness function). These particles moved at a certain speed (v) and had a "memory" that saved their best position ($pbest$). A "collective memory" was considered for the swarm and represented the best global position that was reached ($gbest$). Table 5 presents a summary of the variables that were applied to the PSO model that was developed in this research.

Table 5. The variables that were used for the PSO model.

Variable	Description
n	Number of deployed devices
g	Number of gateways/centroids
$A_{i,j}$	Desired area/all possible points
$D_{i,j}$	Coordinates of the deployed devices
w	Inertia factor ($w_{initial}/w_{final}$)
mi	Maximum iterations
$c1, c2$	Constants (equivalent to the cognitive and social coefficients)
$r1, r2$	Random constants (generated with values between 0 and 1)
$pbest_i$	Best position of each particle
$gbest$	Best global position
p_i	Position vector of particle i within the search space
v_i	Velocity vector of particle i

The initialization of the particles took place randomly within the search area. Once initialized, the $gbest$ was updated with the positions of the particles. This model positioned the gateways considering the best global position of the particles. Following the flow, the particle velocity was updated, as presented in Equations (8) and (9), which updated the particle positions within a given iteration:

$$v_i^{(t+1)} = w^{(t)}v_i^{(t)} + c_1r_1(pbest_i^{(t)} - p_i^{(t)}) + c_2r_2 * (gbest_i^{(t)} - p_i^{(t)}) \quad (8)$$

where $v_i^{(t+1)}$ is the velocity to be updated, $w^{(t)}$ is the inertia factor at iteration t, c_1 and c_2 are the cognitive and social coefficients, respectively, r_1 and r_2 are the random values between 0 and 1, and $pbest_i^{(t)}$ is the best local point and best global point at iteration t.

$$p_i^{(t+1)} = p_i^{(t)} + v_i^{(t+1)} \quad (9)$$

where $p_i^{(t+1)}$ represents the new position of the particle, $p_i^{(t)}$ is the current position of the particle, and $v_i^{(t+1)}$ is the new velocity, as calculated by Equation (8).

The inertia factor (w) contributes to particle convergence. Larger values of w contribute to a global search that explores new areas of the search space. As values of w decrease, they favor local searches, which is interesting when the particles are close to a good solution. The presented PSO model applied a linear variation of the inertia factor over the number of iterations, as shown in Equation (10):

$$w^{(t)} = w_{initial} - (w_{initial} - w_{final})\frac{t}{mi} \quad (10)$$

where $w(t)$ is the inertia at iteration t, $w_{initial}$ is the initial inertia, w_{final} is the inertia for the last iteration, t is the current iteration, and mi is the maximum number of iterations.

The coefficients c_1, c_2, and $w_{initial}/w_{final}$ were input parameters, while the coefficients r_1 and r_2 were random values between 0 and 1 that were calculated by the system.

The PSO updated the positions of the particles until it reached the maximum number of iterations, as presented in Algorithm 2. The particles converged to a point that was considered to be the best global position.

Algorithm 2: The PSO pseudocode.

Input $g, mi, A_{i,j}, D_{i,j}, c_1, c_2, w_{initial}, w_{final}$ **begin**
  Starts the population randomly (p_i, v_i)
  Evaluates all particles and updates $pbest_i$
  repeat
    updates velocities
    updates particle position
    Evaluates and updates $pbest_i$ and $gbest$
  until *up to maximum iterations (mi)*;
output: Grouping of the devices to the respective gateway forming the clusters
output: Reports with metrics related to the clustering process

4. Description of the Application Environment

The proposed models were applied in a case study in order to evaluate their results. The models allowed for the quantitative measurement of gateway positioning within the coverage area.

From the point of view of the deployment of IoT devices in a smart campus, several possibilities are envisioned to meet diverse demands, such as: monitoring the environment, humidity, gas levels, and temperature; monitoring parking lots; access monitoring via video cameras; and intelligent lighting control. All of these examples contribute toward facilitating daily activities and improving the quality of life of the academic community.

This section discusses the case study scenario and presents the characteristics and parameters that were explored.

Scenario Description

Smart campus deployment is marked by the heterogeneity of services, assuming the heterogeneity of the technologies that can be used to deploy services within the environment. The application of the optimization model to three communication technologies was proposed in this work: one with a personal area coverage with low power consumption (BLE); one with local and commonly known area coverage (Wi-Fi); and one with long range areas with low power consumption (LoRa).

Each technology has a different range. It has also been observed in published studies in the literature that each environment has objects and structures that interfere with the propagation of wireless network signals. For the case study, we searched the literature for signal propagation studies regarding BLE, Wi-Fi, and LoRa technologies. The signal propagation studies were based on the received signal strength indication (RSSI). From the measurement of the RSSI and the application of signal propagation models, it was possible to determine the range of the technology. The RSSI for the range radii of the tested technologies was below −90 dBm. In this work, we adopted the range measurements according to Table 6.

Table 6. The ranges of the communication technologies that were adopted in the use case.

Technology	Range (m)	Environment	Reference
BLE	5	Indoor	[32]
Wi-Fi	25	Indoor	[33]
LoRa	70	Indoor	[34]

The case study was applied in a smart campus scenario. As an outdoor environment, we could use the whole area of the Samambaia Campus at the Universidade Federeal de Goiás (UFG). In this work, we considered an area of 761,380 m^2, excluding the woods and unbuilt areas (Figure 4a). The indoor environment was the INF with 2425 m^2 (Figure 4b) and an area of academic blocks with 26,664 m^2 (Figure 4c), comprising the Institute of

Biological Sciences (ICB I and II), Institute of Chemistry (IQ), Institute of Physics (IF), Faculty of Philosophy (FAFI), Faculty of History (FH), and Faculty of Communication and Librarianship (FACOMB). The measurements were taken using Google Earth.

(**a**)

(**b**)

(**c**)

Figure 4. The Samambaia Campus and departments of UFG: (**a**) the Samambaia Campus; (**b**) the Institute of Informatics; (**c**) the academic blocks.

The presented optimization model recognized the desired area by defining points on a Cartesian plane (x and y coordinates). These coordinates indicated where an IoT device or gateway could be deployed. Figure 5 shows an example of plotting the points on a Cartesian plane, with an area of 100 m^2 containing 10 deployed IoT devices. The points in gray are all possible deployment points, while the points in red represent the deployed IoT devices.

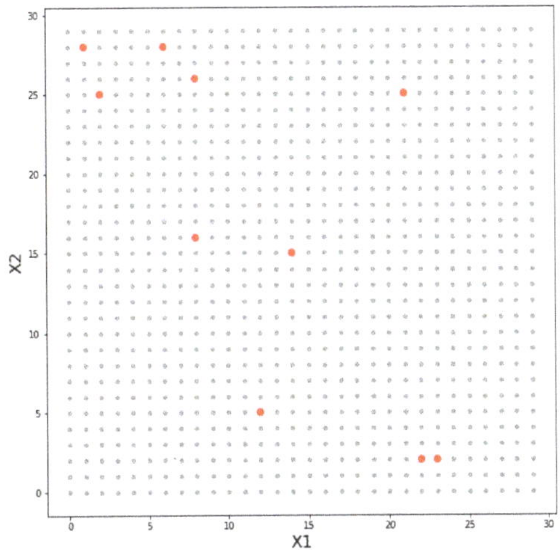

Figure 5. An example of how IoT points and devices are plotted within the desired area by generating an *x*,*y* coordinate matrix.

A real application of the optimization model should have the coordinate pair of each device as an input parameter. For the purposes of these experiments, the coordinates of the IoT devices were set randomly among the points in the desired area. Once the device coordinates were defined, a distance matrix between the device points (see Table 7) was calculated, which was the basis for the generation of the gateway activation matrix (see Table 8).

The distance matrix was generated from the calculation of the Euclidean distance between one point and all of the other points. In the example presented in Tables 7 and 8, 10 IoT devices and a range of 25 m were considered. The activation matrix was binary and was generated from the distance matrix. When the distance was less than the range, it was assigned the value of 1 (one); when the distance was greater than the range, it was assigned the value of 0 (zero).

The parameters that were used in the proposed models were presented in Section 3. In particular, the PSO model used the following values as the initial values for the execution of the experiments: $w_{initial} = 0.9$, $w_{final} = 0.3$, $np = 20$, $mi = 150$, $c1 = 2.1$, and $c2 = 1.9$.

In Section 5, the results of the experiments and an analysis of the comparison of the three metaheuristics are presented.

Table 7. The matrix of the distances between devices, which was used by the LP model to generate the activation matrix.

	0	1	2	3	4	5	6	7	8	9
0	0.0	25.5	10.2	10.44	21.38	11.7	11.4	21.93	23.77	22.36
1	25.5	0.0	18.38	33.42	7.28	13.89	34.06	20.22	5.0	3.16
2	10.2	18.38	0.0	15.26	12.53	6.08	15.81	12.21	15.26	15.62
3	10.44	33.42	15.26	0.0	27.78	19.8	1.0	23.02	30.53	30.48
4	21.38	7.28	12.53	27.78	0.0	10.0	28.3	13.04	2.83	6.08
5	11.7	13.89	6.08	19.8	10.0	0.0	20.52	15.81	12.17	10.82
6	11.4	34.06	15.81	1.0	28.3	20.52	0.0	23.09	31.06	31.14
7	21.93	20.22	12.21	23.02	13.04	15.81	23.09	0.0	15.3	19.0
8	23.77	5.0	15.26	30.53	2.83	12.17	31.06	15.3	0.0	5.0
9	22.36	3.16	15.62	30.48	6.08	10.82	31.14	19.0	5.0	0.0

Table 8. The matrix that was used by the LP model to generate the gateway activation.

	0	1	2	3	4	5	6	7	8	9
0	1	0	1	1	1	1	1	1	1	1
1	0	1	1	0	1	1	0	1	1	1
2	1	1	1	1	1	1	1	1	1	1
3	1	0	1	1	0	1	1	1	0	0
4	1	1	1	0	1	1	0	1	1	1
5	1	1	1	1	1	1	1	1	1	1
6	1	0	1	1	0	1	1	1	0	0
7	1	1	1	1	1	1	1	1	1	1
8	1	1	1	0	1	1	0	1	1	1
9	1	1	1	0	1	1	0	1	1	1

5. Presentation and Evaluation of Results

In order to achieve the proposed goal, the research aimed to present three algorithms that focused on planning and optimizing the deployment of IoT gateways. The first was based on linear programming, which aimed to quantify and position the gateways. The second and the third employed clustering using the K-means and PSO algorithms, which aimed to position the gateways.

The research relied on the case study to refine the results; then, the results were compared to evaluate the three methods that were applied to the case study. The initial parameters were set. A floor plan of the studied environment with the plots of the relevant points on a Cartesian plane and the definition of the points of the installed IoT devices were used as input parameters, as well as the range of the technologies and the capacity of the data links. The same parameters were applied to the three metaheuristics. After the experiments, the results were compared to analyze the positioning of the gateways and the distribution of the devices. Three environments were used in the case study (indoor and outdoor environments) to evaluate the behavior of the metaheuristics and the obtained results.

5.1. LP Optimization Model: Gateway Quantification and Positioning

The experiments presented in this section aimed to evaluate the optimization model that was based on linear programming. Table 9 presents the results of the run, considering the defined values.

Table 9. The number of required gateways considering the INF area and bandwidth consumption variations.

Tech Comms	Quantity Devices	Range (m)	Gateways (Demand 3%)	Silhouette (Demand 3%)	Gateways (Demand 0%)	Silhouette (Demand 0%)
BLE	100	5	35	0.35794	35	0.35794
BLE	300	5	37	0.28183	37	0.29547
Wi-Fi	100	25	4	0.10635	3	0.35984
Wi-Fi	300	25	10	−0.07833	3	0.33021
LoRa	100	70	4	−0.23305	1	-
LoRa	300	70	10	−0.15928	1	-

The first experiment took into account the deployment of 100 or 300 devices for each communication technology. It was observed that for the BLE technology, with a demand of 0% or 3% of the data link for each device, we obtained the same results of 35 and 37 gateways, respectively. The value remained the same because the range was the parameter that prevailed in the definition of the quantity of gateways. The opposite was observed with the Wi-Fi and LoRa technologies. When Wi-Fi technology consumed 30% of the link demand, it was necessary to have 4 gateways to serve 100 devices and 10 gateways to serve 300 devices; the same occurred with LoRa. This equality between LoRa and Wi-Fi occurred because the same bandwidth consumption was defined for both technologies. Thus, when analyzing the same technologies with 0% bandwidth consumption, it was observed that Wi-Fi communication required 1 less gateway to serve 100 devices and 7 less gateways to serve 300 devices. It was also found that LoRa demanded 3 fewer gateways to serve 100 devices and 9 fewer gateways to serve 300 devices. This variation in the number of required gateways proved that the presented optimization model considered the range and the demand for the data links.

The main goal of this optimization model was to quantify the gateways. The model was proposed based on gateway activation criteria as a function of the Euclidean distance between the IoT devices and the corresponding gateway and the transmission capacity of the communication link to the gateway. When the gateway was activated, it was assigned the devices that were within its range, while obeying the established constraints. The process was repeated until the smallest possible number of gateways was found to serve all of the deployed devices.

The association between the devices and the gateway allowed for a map of their positions to be plotted, as shown in Figure 6. In the LP model, the association was defined considering the range of the communication technology, whereas in the clustering process, the association was based on the proximity of the devices to the gateway.

The association between the devices and the corresponding gateway is represented by the alternating colors of the devices in the graph. These figures were equivalent to the execution, the results of which are shown in Table 9. A device could be at different ends of the defined area, but when it was within the range of coverage, it could be associated with the gateway. There are two columns called "Silhouette" in Table 9: this metric is a coefficient that indicates how cohesive a data cluster is. This coefficient ranged from −1 to 1 in this study. The closer to 1 the coefficient, the more cohesive the data; the closer to −1, the wider the dispersion.

As mentioned above, the distribution of the gateway devices that is represented by Figure 6e had the lowest Silhouette coefficient, indicating that the distribution was the least cohesive. This coefficient proved that dispersion when distributing the devices was significant.

Although a graphical representation of the positions of the devices/gateways with a bandwidth consumption that was equal to zero is not presented, it was observed that the distribution was more cohesive, according to the silhouette coefficients that are seen in Table 9. The silhouette coefficient can only be calculated when there are two or more clusters, thus justifying the absence of this coefficient from some of the presented results.

The optimization model could be applied to different areas, each with its own architectural characteristics. In this work, it was applied in three areas: two indoor areas (INF and the UFG academic blocks) and one outdoor area (the entire Samambaia Campus area of the UFG). The measurements of these areas are presented in Table 1 and the ranges of the technologies are presented in Table 6.

The experiments that were performed in other areas produced the results that are shown in Figure 7a,b. In these experiments, we tested the Samambaia Campus with 300 IoT devices and 31 gateways to serve a 100-m range (Wi-Fi outdoor environment) with a 3% bandwidth consumption and we tested 300 IoT devices and 19 gateways in the academic blocks of the Samambaia Campus with the same bandwidth consumption and a 25-m range (indoor environment).

The LoRa communication technology has as a remarkable long-distance range. Taking this characteristic into account, we conducted the experiments at the Samambaia Campus (outdoor environment). We considered a range of 500 m for the LoRa technology in the outdoor environment. The distribution of the gateways and IoT devices are presented in Figure 8. In the scenario in question, 7 gateways were required for 200 devices with 3% bandwidth consumption, as shown in Figure 8a. Figure 8b presents the results for the same scenario but for 300 devices with 3% bandwidth consumption, which demanded 10 gateways.

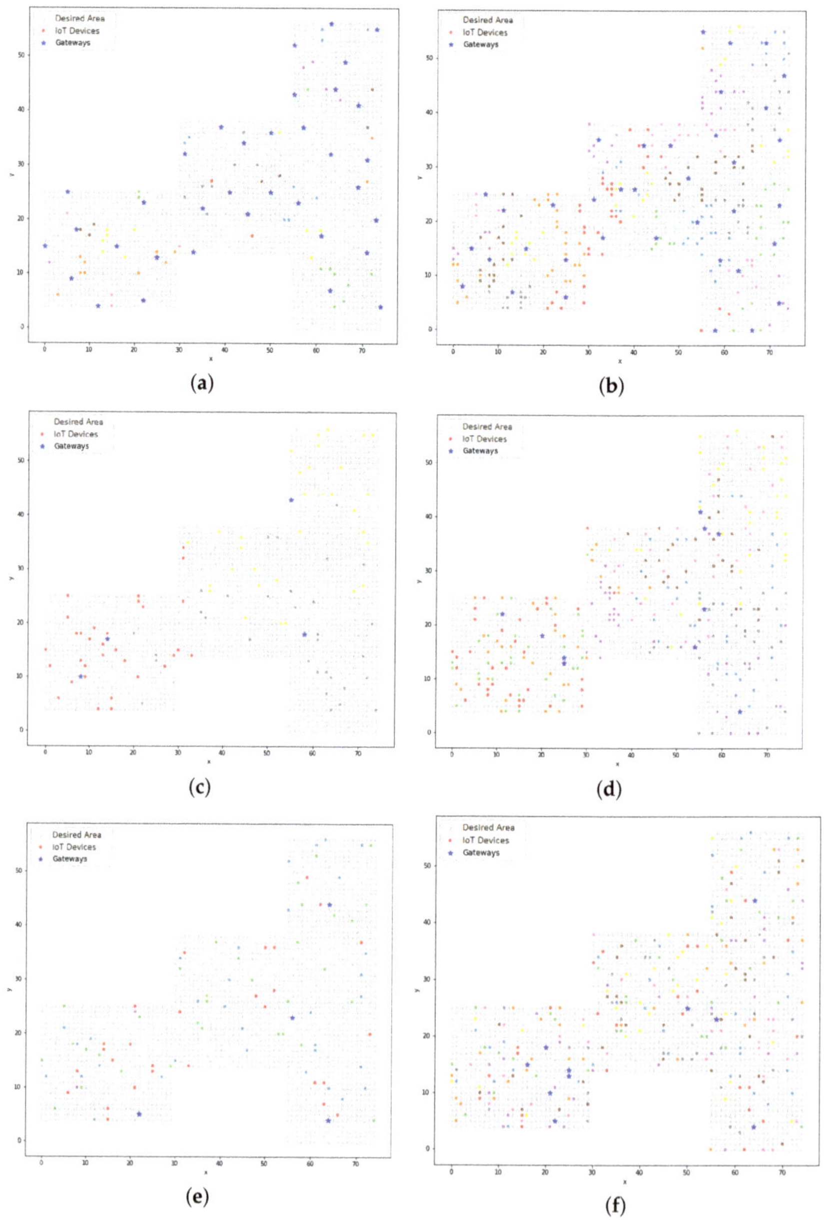

Figure 6. The plots of the devices and their respective gateways with a3% data link consumption (LP method): (**a**) BLE technology with 100 devices and 35 gateways; (**b**) BLE technology with 300 devices and 37 gateways; (**c**) Wi-Fi technology with 100 devices and 4 gateways; (**d**) Wi-Fi technology with 300 devices and 10 gateways; (**e**) LoRA technology with 100 devices and 4 gateways; (**f**) LoRA technology with 300 devices and 10 gateways.

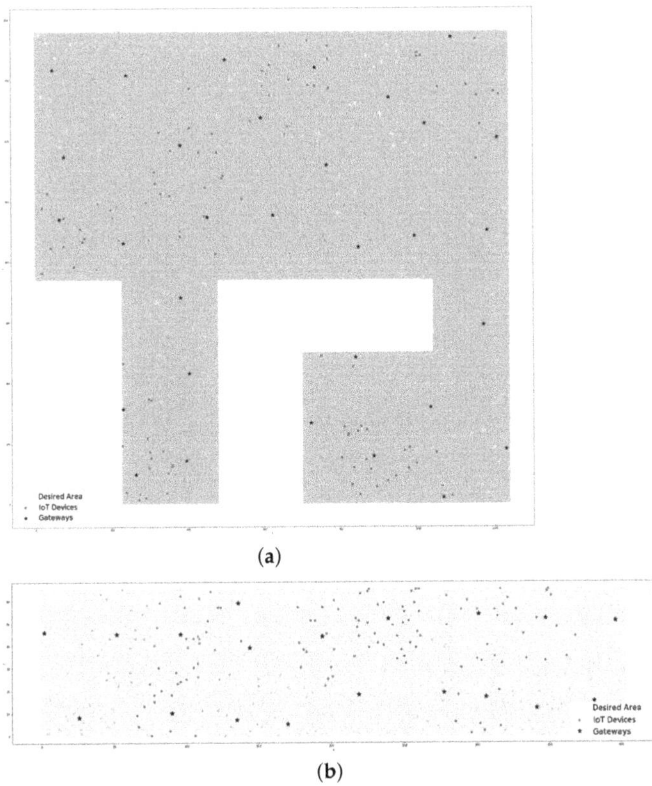

Figure 7. The plots of the devices and their gateways (Wi-Fi technology): (**a**) the Samambaia Campus with 300 devices (outdoor environment); (**b**) the academic blocks with 300 devices (indoor environment).

Table 10 presents a report of the execution of the optimization model when applied to the academic blocks of the Samambaia Campus. Table 10 shows the number of IoT devices that were served by each gateway. Taking into account the bandwidth consumption of 3%, each gateway could serve a maximum of 33 devices. In the presented results, the number of clusters/gateways that served the most IoT devices was the 17, serving 31 devices. The second column of the table shows the largest calculated distance between a device and the corresponding gateway within the respective cluster. It can be seen that no value in that column exceeded the range of the technology (25 m).

The results that are presented in these three scenarios show that the optimization model could be applied in different scenarios by changing the parameters according to the desired situation.

Some of the experiments that were applied to the INF scenario considered 100 devices that were communicating via Wi-Fi technology with varying demands for link capacity. The results are presented in Table 11. It can be seen that an increase in demand for bandwidth consumption caused an increase in the number of gateways, so the dispersion of the devices within the clusters was noticeable. Evidence for the dispersion can also be seen in the variations in the silhouette coefficient. Figure 9a shows the increases in the bandwidth demand and the number of gateways and Figure 9b shows the decrease in the coefficient.

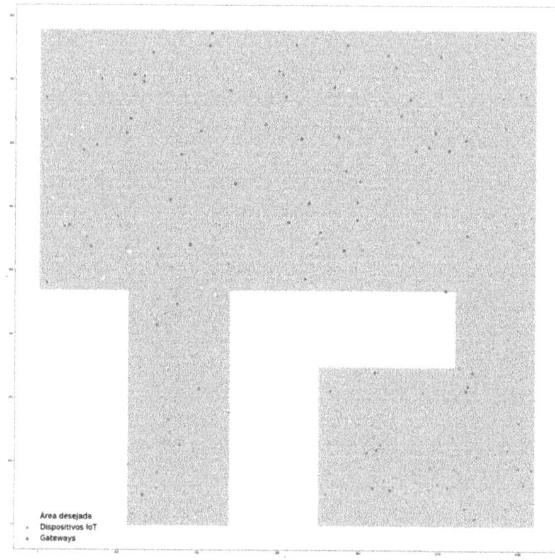

(**a**)

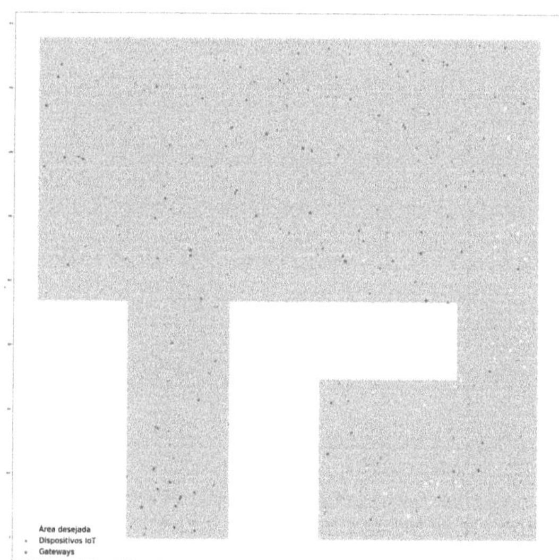

(**b**)

Figure 8. The plots of the devices and their gateways (LoRa technology; outdoor environment): (**a**) the Samambaia Campus with 200 devices and 7 gateways; (**b**) the Samambaia Campus 300 devices and 10 gateways.

Table 10. The result of the gateway quantification experiments in the academic blocks scenario with 300 devices that were communicating via Wi-Fi technology, as presented in Figure 7b.

Quantity of Devices/Gateways	Increased Distance Between Device/Gateway	Sum of the Distance Between Device/Gateway	Cluster	Silhouette Coefficient
21	25	338.62	0	
19	25	286.12	1	
21	24.7	361.36	2	
17	24.02	205.54	3	
16	24.7	263.59	4	
20	25	283.56	5	
10	22.56	153.04	6	
11	24.35	143.4	7	
17	24.08	231.6	8	
18	24.17	242.7	9	0.31363
11	24.76	179.06	10	
15	25	268.54	11	
13	24.33	185.21	12	
15	23.41	230.15	13	
13	23.19	197.4	14	
16	24.02	247.65	15	
10	23.71	155.17	16	
31	24.7	499.38	17	
6	22.47	107.08	18	
	Total Distance:	4579.17		

Table 11. The results of the experiments with variations in the demand for data link consumption (LP method).

Area	Technology	Quantity of Devices	Demand	Quantity of Gateways	Silhouette Coefficient
INF	Wi-Fi	100	0	3	0.35984
INF	Wi-Fi	100	2	3	0.34473
INF	Wi-Fi	100	3	4	0.09411
INF	Wi-Fi	100	4	4	0.05009
INF	Wi-Fi	100	5	5	0.06368
INF	Wi-Fi	100	10	10	−0.12747
INF	Wi-Fi	100	15	17	−0.20703
INF	Wi-Fi	100	20	20	−0.21001
INF	Wi-Fi	100	30	34	−0.38046

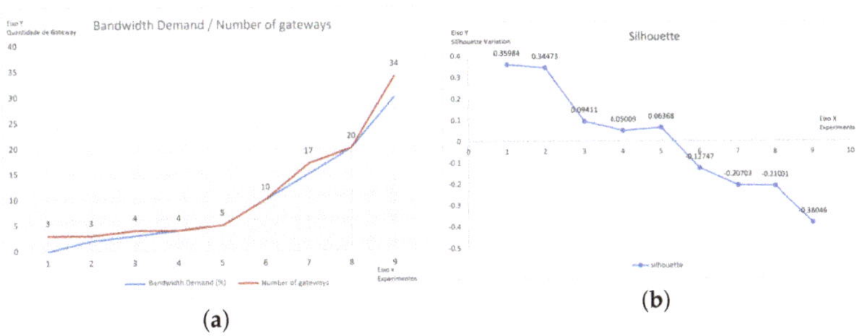

Figure 9. Graphs showing the results of the experiments with variations in the demand for data link consumption: (**a**) the data link consumption × quantity of gateways; (**b**) the silhouette coefficient.

The obtained results confirmed that varying the range of the communication technology and the demand for data link consumption directly influenced the required number of gateways.

It was also observed that the distribution of gateways and devices using this optimization model was not achieved using a clustering technique. In Sections 5.2 and 5.3, the results from the placement of gateways using the K-means clustering model and the PSO model are presented, respectively.

5.2. K-Means Model: Clustering and Gateway Placement

The K-means clustering model uses Euclidean distance as a metric to divide the devices into n groups. The center of each group (centroid) is considered to be a gateway. This section presents the results of the application of the clustering model, taking into account the parameters presented in Table 9, which show the comparison between the three communication technologies (Wi-Fi, BLE, and LoRa) when varying the number of devices that were distributed in the area.

Table 12 presents the silhouette coefficients after running the K-means model. The results shown by the silhouette coefficients were good, proving the cohesion of the clustering. A comparative analysis between the silhouette coefficients from the K-means model and those from the other clustering models is presented in Section 5.4.

Table 12. The silhouette coefficients after applying the K-means model, considering the number of gateways that resulted from the application of the LP model.

Communication Technology	Number of Devices	Range (m)	Number of Gateways (Demand 3%)	Silhouette Coefficient (Demand 3%)
BLE	100	5	35	0.42388
BLE	300	5	37	0.38236
Wi-Fi	100	25	4	0.50848
Wi-Fi	300	25	10	0.37390
LoRa	100	70	4	0.50848
LoRa	300	70	10	0.37390

Graphically, it can be observed in Figure 10 that the clusters were formed by the devices that were close to the centroids, unlike the dispersion that was observed in Figure 6. The dispersion was verified by the silhouette coefficient. Analyzing the resulting positions from the K-means model using the LP model, it could be seen that the K-means model treated the gateways as the center of the cluster and, in this case, the radius of the technology was not a limiting factor in the clustering process (unlike the treatment from the LP model). In Section 5.3, the PSO model is addressed and in Section 5.4, a comparative analysis between the proposed models is presented, which found that in some cases with the clustering model, the distance between a device and a gateway could be greater than the range of the technology.

5.3. PSO Model: Clustering and Gateway Placement

The PSO positioning model was inspired by the collective movement of animals, such as a flock of birds or a school of fish. The model proposed in this research had two approaches: a simple PSO and a hybrid PSO. The difference between the approaches was in the initialization of the particles. In the simple approach, the particles initialized randomly within the search space and, depending on their initial position, moved during each iteration in search of the best global position. In the hybrid approach, the initialization of the particles occurred using the initialization model that was adopted in the K-means algorithm, called K-means++. The results of the two approaches are presented in Sections 5.3.1 and 5.3.2, respectively.

5.3.1. Simple PSO

As with the K-means clustering model, Table 13 demonstrates the silhouette coefficient values from the simple PSO model. The variations in these coefficients came from variations in the adopted parameters, but they still showed good results regarding the cohesion of the clusters.

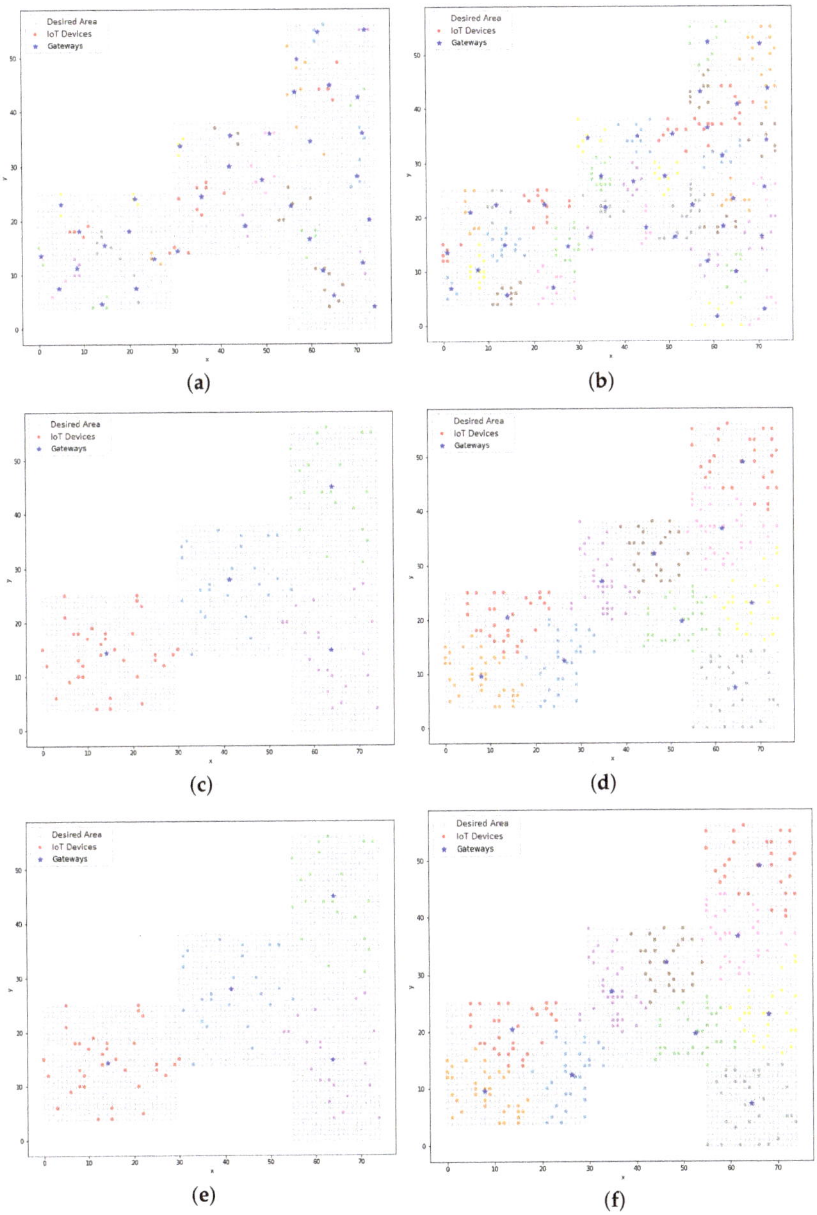

Figure 10. The clustering of IoT devices and their respective gateways using the K-means model: (**a**) BLE technology with 100 devices and 35 gateways; (**b**) BLE technology with 300 devices and 37 gateways; (**c**) Wi-Fi technology with 100 devices and 4 gateways; (**d**) Wi-Fi technology with 300 devices and 10 gateways; (**e**) LoRa technology with 100 devices and 4 gateways; (**f**) LoRa technology with 300 devices and 10 gateways.

Table 13. The silhouette coefficients after applying the PSO positioning model (simple approach), considering the number of gateways that resulted from the application of the LP model.

Communication Technology	Number of Devices	Range (m)	Number of Gateways (Demand 3%)	Silhouette Coefficient (Demand 3%)
BLE	100	5	35	0.23778
BLE	300	5	37	0.24597
Wi-Fi	100	25	4	0.45798
Wi-Fi	300	25	10	0.31159
LoRa	100	70	4	0.45150
LoRa	300	70	10	0.31159

Figure 11 shows the distribution of gateways and IoT devices that was achieved using the PSO model (simple approach). In this model, it could be seen that the distribution of gateways allowed for a higher concentration in certain regions of the search area as well as other sparser regions, as shown in Figure 11b. This occurred because of the randomness of the initial positions of the particles, which interfered with the positioning results across the iterations. The initial positions could be concentrated in one region and by updating the positions of the particles using the velocity and the cognitive and social factors, this positioning was achieved.

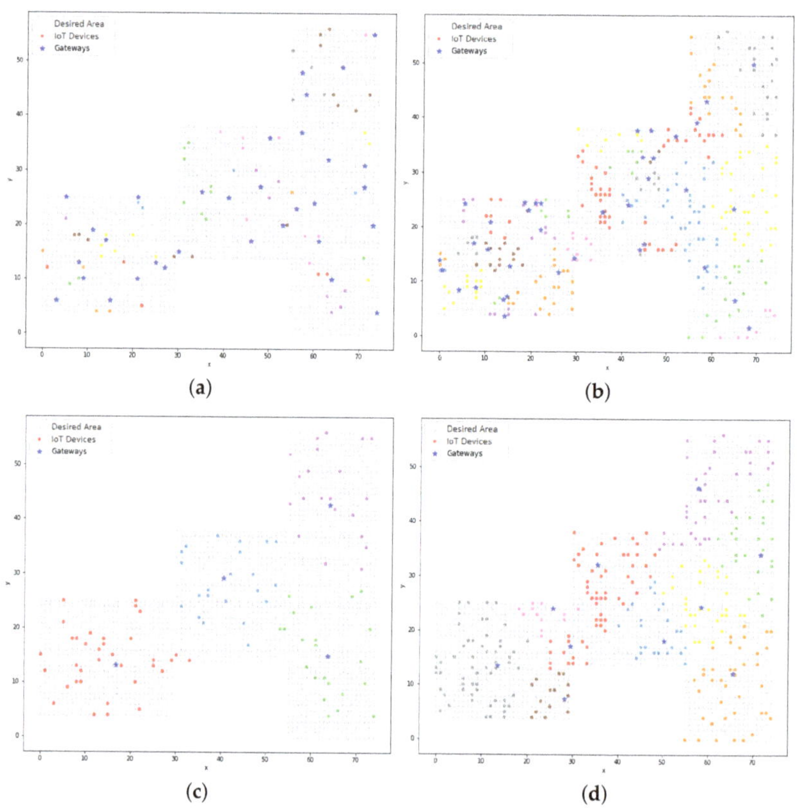

Figure 11. *Cont.*

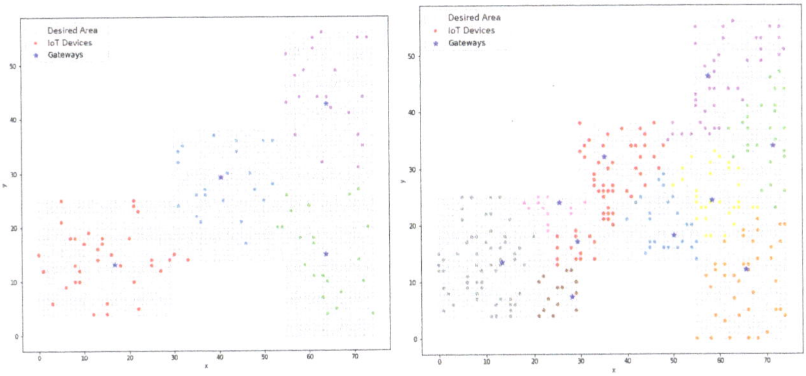

(e) (f)

Figure 11. The clustering of IoT devices and their respective gateways using the PSO model (simple approach): (**a**) BLE technology with 100 devices and 35 gateways; (**b**) BLE technology with 300 devices and 37 gateways; (**c**) Wi-Fi technology with 100 devices and 4 gateways; (**d**) Wi-Fi technology with 300 devices and 10 gateways; (**e**) LoRa technology with 100 devices and 4 gateways; (**f**) LoRa technology with 300 devices and 10 gateways.

5.3.2. Hybrid PSO

The hybrid PSO approach was a clustering model that aimed to improve the initial positioning of the particles using the initialization method that was adopted in the K-means model. The initialization was used in the first iteration and forced the centroids to be further away from each other.

When comparing the silhouette coefficients from the simple PSO model (Table 13) to those from the hybrid PSO model (Table 14), it could be seen that the larger coefficients were always from the hybrid approach, which confirmed that the better initial positioning of the particles resulted in better final positions.

Table 14. The silhouette coefficients after applying the hybrid PSO model, considering the number of gateways that resulted from the application of the LP model.

Communication Technology	Number of Devices	Range (m)	Number of Gateways (Demand 3%)	Silhouette Coefficient (Demand 3%)
BLE	100	5	35	0.36686
BLE	300	5	37	0.37331
Wi-Fi	100	25	4	0.50842
Wi-Fi	300	25	10	0.37530
LoRa	100	70	4	0.50842
LoRa	300	70	10	0.37530

The distribution of IoT devices and their corresponding gateways that resulted from the hybrid PSO model is presented in Figure 12. When comparing these results to those from the simple PSO model, it could be visually observed that the gateway positioning in the hybrid PSO model always maintained a better distribution within the desired area, especially when looking at Figures 11b and 12b.

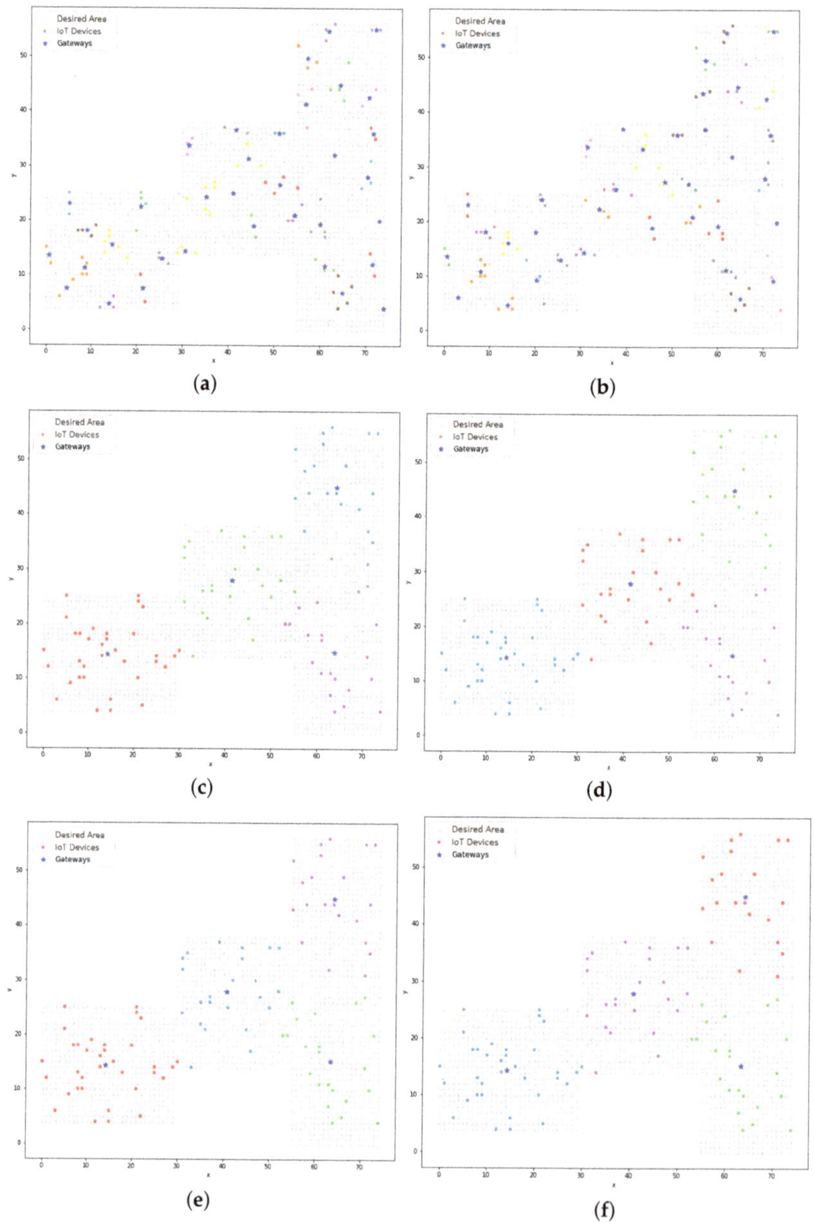

Figure 12. the clustering of IoT devices and their respective gateways using the PSO model (hybrid approach): (**a**) BLE technology with 100 devices and 35 gateways; (**b**) BLE technology with 300 devices and 37 gateways; (**c**) Wi-Fi technology with 100 devices and 4 gateways; (**d**) Wi-Fi technology with 300 devices and 10 gateways; (**e**) LoRa technology with 100 devices and 4 gateways; (**f**) LoRa technology with 300 devices and 10 gateways.

Figure 13 presents a comparison of the two PSO approaches. The comparison was carried out by summing the calculated Euclidean distances between the IoT devices and their respective gateways. The larger the sum of the distances, the less cohesive the cluster.

In all of the presented runs, the hybrid PSO model had smaller sums of the distances, thereby proving that the clustering in the hybrid model was better.

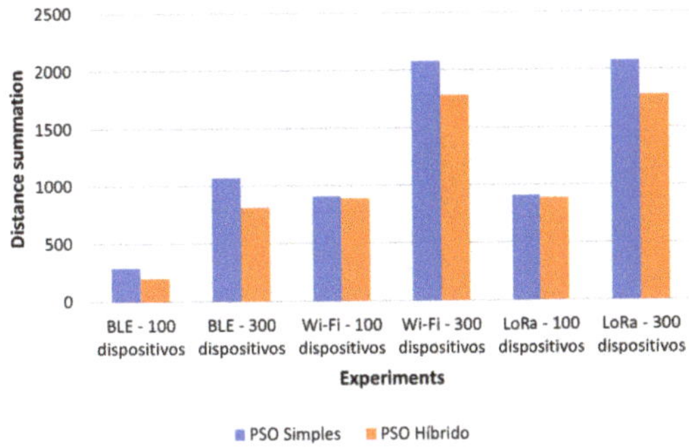

Figure 13. The comparison of the total distances between the IoT devices and their gateways using the simple and hybrid PSO models.

The experiments presented in Sections 5.1–5.3 were mainly aimed at testing how well the models worked. From those experiments, it was proved that it was possible to quantify, assign, and cluster the IoT devices and their gateways. A comparative analysis was performed to evaluate the four models: LP, K-means, simple PSO, and hybrid PSO. The analysis is described in Section 5.4.

5.4. Discussion: Comparative Analysis and Evaluation of Presented Results

The present work proposed three optimization models. The LP model aimed to quantify and position gateways by assigning IoT devices to a respective gateway within the desired area as a function of the radius and data link demand. The K-means model and the PSO model positioned the gateways according to the number that was defined by the LP model and distributed the IoT devices by considering the Euclidean distance between them.

For the three models in question, metrics could be abstracted for a comparative analysis. The silhouette coefficient was one of the metrics considered. Another factor taken into consideration was the sum of the Euclidean distances between the devices and their respective gateways. This result was calculated for each cluster and then totaled.

The models could be applied in several scenarios. In this work, we considered three scenarios: INF, the Samambaia Campus, and the academic blocks. The architecture of the buildings does not represent a regular geometric figure, which aroused interest in analyzing the behavior of the models.

Since we tested different scenarios, Table 15 presents the comparisons between the models within the same scenario. It can be seen that the total distances and the best silhouette coefficients were found by the K-means and hybrid PSO models. The K-means model defined the position of the centroids and, at each iteration, repositioned them to the center of the cluster. In this way, the clusters always had their gateways in the center. The hybrid PSO model defined the initial positions of the particles using the K-means initialization technique and led the particles to better initial positions, which was reflected throughout each iteration, thereby proving that once initialized well, particle movements tend to finalize well. The LP model showed good summations and silhouette coefficients, proving that they were cohesive. The simple PSO model did not perform as well as the hybrid approach in terms of clustering. When there was a good silhouette coefficient, it could be observed that the summation always tended to be lower because the devices were more cohesive with the centroid.

Table 15. The distance summations and silhouette coefficients of the experiments that were performed in the three scenarios, considering 300 devices and Wi-Fi communication technology.

Model	Area	Range (m)	Number of Gateways	Summed Distance	Silhouette Coefficient	Increased Distance Between Devices and Gateways
LP	Samambaia Campus	100	31	17,359.73	0.38812	95.82
Simple PSO	Samambaia Campus	100	31	20,554.47	0.29233	120.49
Hybrid PSO	Samambaia Campus	100	31	14,796.27	0.40764	92.98
K-means	Samambaia Campus	100	31	14,965.47	0.40943	114.85
LP	Academic block	25	19	4604.6	0.30675	24.74
Simple PSO	Academic Blocks	25	19	5384.96	0.27088	24.52
Hybrid PSO	Academic Blocks	25	19	3982.1	0.39224	29.57
K-means	Academic Blocks	25	19	3974.86	0.38964	27.01

In relation to the greatest distance that was found between a device and a gateway (Table 15), it was observed that in some cases the greatest distance was greater than the range of the technology. In the LP model, this distance was always less than the radius, while in the K-means model and the simple PSO and hybrid PSO models, values that were greater than the range were found, which occurred because these models did not adopt the range as a restriction to the clustering process. Out of the proposed models, the only method that guaranteed the range was the LP model; on the other hand, this model was the only one to present a negative silhouette coefficient, as shown in Table 9. In the LP model, the gateways were activated to serve the devices that were in range. So, the larger the range, the more likely the silhouette coefficient was to be bad. Although the clustering models did not use the range as a criterion, they clustered the devices by considering the best position; in this case, it was a clustering process, not a gateway activation process.

In the presented clustering models, the number of gateways was informed based on the results of the LP model. It was possible to enter a larger number of gateways and obtain the greatest distance within the range of the technology. It is worth noting that the idea was to quantify the smallest possible number of gateways and the model achieved the expected result, as did the clustering process.

Throughout Sections 5.1–5.3, the experiments were described considering the same parameters for each proposed model. It was concluded that the models corresponded to expectations. The experiments with the LP model only had one sample. In the performed tests, it was found that repetition with several samples and the same parameters returned the same result, matching the search for the best result. The clustering models presented different results for each sample because a random initial position was defined for each sample, which interfered with the final results. For evaluation purposes, 32 samples were run and the average of the silhouette coefficients and the total distances within the clusters were calculated to compare the models.

Table 16 presents the parameters that were applied in the experiments. In order to compare the results, six experiments were performed, all of which were applied in the INF scenario with an area of 2425 m^2 in an indoor environment.

Table 16. The parameters that were applied in the experiments.

Experiment	Communication Technology	Range (m)	Number of Devices	Number of Gateways (Demand 3%)
1	BLE	5	100	35
2	BLE	5	300	37
3	Wi-Fi	25	100	4
4	Wi-Fi	25	300	10
5	LoRa	70	100	4
6	LoRa	70	300	10

The results presented in Tables 17 and 18 show a comparison between the six experiments, using the total distance and the silhouette coefficient as metrics. The higher the total distance, the greater the distance between the devices and their respective gateways. The silhouette coefficient represented the cohesion of the clusters.

Table 17. The comparison of the total distance metric (average), considering the experiments that were performed using the same parameters.

Experiment	K-Means Summed Distance	LP Summed Distance	Simple PSO Summed Distance	Hybrid PSO Summed Distance
1	196.80823	220.14000	315.47147	196.06500
2	809.21433	939.62000	1154.32925	1779.13501
3	885.48468	1289.35000	1056.12834	885.49252
4	1776.56160	4261.13000	2206.50263	1775.86463
5	885.50771	1289.35000	1050.06652	885.48468
6	1776.56160	8916.60000	2206.50263	1775.86463

Table 18. The comparison of the silhouette coefficient metric (average), considering the experiments that were performed using the same parameters.

Experiment	K-Means Silhouette Coefficient	LP Silhouette Coefficient	Simple PSO Silhouette Coefficient	Hybrid PSO Silhouette Coefficient
1	0.42334	0.35794	0.23778	0.42535
2	0.38212	0.28193	0.24597	0.37331
3	0.50842	0.09411	0.42984	0.50845
4	0.37390	−0.07833	0.31159	0.37530
5	0.50848	0.09411	0.45149	0.50842
6	0.37390	−0.15928	0.31159	0.37530

When analyzing the results, the smallest distance found was from the hybrid PSO model in Experiment 1 and the largest distance was from the LP model in Experiment 6. In almost all of the experiments, the largest distance found was from the LP model.

When comparing the total distances in each experiment, it was found that the hybrid PSO model stood out as having the smallest distance in four experiments; the K-means model was second in this ranking. Regarding the greatest distance, the LP model achieved the greatest total distance in four experiments, followed in the ranking sequence by the simple and hybrid PSO models.

The same evaluation was conducted for the silhouette coefficient. The lowest silhouette coefficient out of the experiments was verified using the LP model in Experiment 6, which was equivalent to the highest distance summation that was found. As can be observed in Figure 6f, there was no cohesion between the devices and the gateways as the negative coefficient that was ascertained by the model characterized a high degree of dispersion. The K-means model showed the best silhouette coefficient in Experiment 5, with a value above 0.5; the closer to 1, the more cohesive the cluster. There were other results of above 0.5: Experiment 3 with the hybrid PSO model ranked second; Experiment 3 with the K-means model and Experiment 5 with the hybrid PSO model ranked in joint third position. When only comparing the models within the same experiment, it could be seen that the hybrid PSO model stood out in four experiments (1, 3, 4, and 6) and the K-means model stood out in two (2 and 5). The smallest silhouette coefficients were from the LP model, which achieved the smallest results in four experiments (3, 4, 5 and 6), followed by the simple PSO model with the smallest results in Experiments 1 and 2.

It was possible to observe that each model had its own contribution that could be improved by inserting other parameters or constraints. The K-means and hybrid PSO models were the best models for clustering, but the linear programming model was the only one that restricted the devices within a range.

A statistical analysis using the Friedman test is presented in Section 5.5.

5.5. The Friedman Test

The Friedman test is a non-parametric statistical test that was developed by Milton Friedman, who was an economist, statistician, and writer [35]. This test is used to detect differences between treatments in various experiments, allowing a choice to be made between two or more hypotheses using the data from a given experiment. The objective of this test is to determine whether there are at least two samples that represent the populations of distinct means out of a set of n samples ($n \geq 2$). In this way, it is possible to detect significant differences between the behaviors of two or more metaheuristics [36].

The Friedman test was applied to our experiments in order to evaluate the optimization models. The arithmetic means of the silhouette coefficients and the summation of the distances were used as metrics.

The Friedman test is based on ranking the data, so the lowest ranking value is assigned to the best performing algorithm. This test returned a p-value that allowed the similarities between the proposed models to be evaluated. The p-value was compared to an α value, which represented the significance of the test. $\alpha = 0.05$ was adopted; when the p-value $< \alpha$, it was concluded that the algorithms were different. The degree of confidence was equal to $1 - \alpha$. For $\alpha = 0.05$, we obtained a confidence level of 0.95 (or 95%).

Although the results showed whether or not there was a difference between the algorithms, it was not possible to know what was different. In view of this, the Friedman test compared the models and evaluated the differences between them. This comparison is called a post hoc test. Each comparison returned a value of p, which represented the similarity between the compared algorithms.

The tests were applied using the Keel (Knowledge Extraction based on Evolutionary Learning) software, which is a free software that was developed in Java by a group of researchers from Spain and the UK and is capable of performing various experiments involving data mining, including the Friedman test [36].

The test was applied using the data from the experiments presented in Tables 17 and 18. The results are shown in Tables 19–22. The standard deviations and the arithmetic means and medians of the silhouette coefficients were considered, along with the total distances between the gateways and the IoT devices.

Table 19. The results from the Friedman test: the ranking applied to the arithmetic mean values and the sums of the distances between the gateways and the IoT devices.

Model	Ranking
LP	1.50000
Simple PSO	1.83330
K-means	3.33330
Hybrid PSO	3.33330
p-value	0.01694

Table 20. The results from the Friedman test: the comparison between each metaheuristic using the arithmetic means of the sums of the distances between the gateways and the IoT devices.

Comparison	p-Value With $\alpha = 0.05$
K-means vs. LP	0.01391
LP vs. Hybrid PSO	0.01391
K-means vs. Simple PSO	0.04417
Simple PSO vs. Hybrid PSO	0.04417
LP vs. Simple PSO	0.65472
K-means vs. Hybrid PSO	1.00000

Table 21. The results from the Friedman test: the ranking applied to the arithmetic mean values of the silhouette coefficients.

Models	Ranking
Hybrid PSO	1.33330
K-means	1.66670
Simple PSO	3.33330
LP	3.66670
p-value	0.00200

Table 22. The results from the Friedman test: the comparison between each metaheuristic using the arithmetic means of the silhouette coefficients.

Comparison	p-Value With $\alpha = 0.05$
LP vs. Hybrid PSO	0.00175
Simple PSO vs. Hybrid PSO	0.00729
K-means vs. LP	0.00729
K-means vs. Simple PSO	0.02535
K-means vs. Hybrid PSO	0.65472
LP vs. Simple PSO	0.65472

When analyzing the ranking results presented in Tables 19 and 21, it could be observed that the models all had a degree of similarity of less than 0.05, indicating that the optimization models were different. When analyzing the rankings, the hybrid PSO model performed the best in terms of the silhouette coefficient metric. In terms of the sum of the distances between the devices and the gateways, we obtained a different result for each metric.

When evaluating the post hoc tests, which aimed to compare the models to each other, it was observed that the K-means and hybrid PSO models had a p-value of greater than 0.05 in all of the presented cases, which confirmed that these metaheuristics were similar. It is noteworthy that the hybrid PSO model initialized the particles using the same initialization technique as the K-means model, which justified the similarity that was found in the Friedman test. The LP and simple PSO models showed a similarity when evaluating the arithmetic means of the silhouette coefficients and the sums of the distances between the gateways and the IoT devices.

The four models presented in this paper contributed to the planning and deployment of IoT gateways in a smart campus environment. Based on the results, it was possible to choose which model best fit a specific case. In Section 6, the final considerations of this work are presented.

6. Conclusions

This paper presented an approach for planning and deploying IoT gateways in smart campus environments, using the minimum number gateways that was required for the desired area coverage. The application scenario for the experiments was the Samambaia Campus of the UFG. In this environment, some departments were chosen for indoor and outdoor experiments. The heterogeneity of IoT devices led us to think about environments with multiple communication technologies, considering the characteristics of each technology. Thus, the proposed model focused on the LoRa, Wi-Fi, and BLE technologies.

The linear programming-based model considered the desired area, the range of the technology, the consumption capacity of the data link, and the deployed IoT devices to return the lowest possible number of gateways and to subsequently position them. For positioning, we presented the linear programming, K-means, and PSO models.

The experiments were applied to four scenarios (INF, academic blocks, Samambaia Campus, and IFTO Palmas) and achieved considerable results for both the quantification model and the gateway positioning models. The problems that were addressed by the model have been considered as difficult to solve and as the number of devices increased, the size of the area and the demand for processing and memory also increased.

With the results of these experiments, a comparative analysis was conducted that allowed for the evaluation of the behavior of each model and the determination of their advantages. It was concluded that the results were satisfactory and proved the efficiency of the models in relation to the proposed objective. The IoT gateway quantification was accurate and respected the established range and data demands. The positioning models defined the positions of the gateways and created clusters of devices that were associated with a gateway.

With the presented optimization models, it would be possible to plan the required area coverage for establishing wireless communication technologies. This planning would reveal the amount of communication equipment that is needed and would define the best positions for that equipment to be installed.

This work offers contributions to the field of IoT gateway planning and deployment through the optimization model. Throughout the research, we envisioned studies that could be developed in the future.

For future work, we intend to improve the optimization models, both from a quantification and positioning point of view by:

- Considering other parameters and constraints that may contribute to gateway quantification, with the goal of obtaining results that are even closer to the existing design;
- Associating weights with the objective functions or processes that could improve gateway positioning using the clustering models;
- Working on a model that can consider all three communication technologies simultaneously for gateway quantification.

Author Contributions: Conceptualization, D.F.J., C.S. and A.O.-J.; methodology, D.F.J., J.L.O., C.S., T.F., M.R., L.A.F., W.M. and A.O.-J.; software, D.F.J. and J.L.O.; validation, D.F.J., J.L.O., C.S., L.A.F. and A.O.-J.; formal analysis, D.F.J., C.S., L.A.F. and A.O.-J.; investigation, D.F.J., J.L.O. and C.S.; resources, W.M. and A.O.-J.; data curation, D.F.J.; writing—original draft preparation, D.F.J.; writing—review and editing, D.F.J., J.L.O., C.S., T.F., M.R., L.A.F., W.M. and A.O.-J.; visualization, D.F.J., J.L.O., C.S., T.F., M.R., L.A.F., W.M. and A.O.-J.; supervision, C.S., T.F., M.R., L.A.F., W.M. and A.O.-J.; project administration, A.O.-J.; funding acquisition, W.M. and A.O.-J. All authors have read and agreed to the published version of the manuscript.

Funding: This research was partially funded by Fraunhofer Portugal (AICOS) and partially by the Federal University of Goiás (UFG). M.R. acknowledges financial support from the Fundação para a Ciência e a Tecnologia (FCT) under scholarship number SFRH/BD/138302/2018.

Institutional Review Board Statement: Not applicable.

Informed Consent Statement: Not applicable.

Data Availability Statement: Not applicable.

Conflicts of Interest: The authors declare no conflict of interest.

References

1. Kumar, M.; Kumar, S.; Kashyap, P.K.; Aggarwal, G.; Rathore, R.S.; Kaiwartya, O.; Lloret, J. Green Communication in Internet of Things: A Hybrid Bio-Inspired Intelligent Approach. *Sensors* **2022**, *22*, 3910. [CrossRef] [PubMed]
2. Spadaccino, P.; Crinó, F.G.; Cuomo, F. LoRaWAN Behaviour Analysis through Dataset Traffic Investigation. *Sensors* **2022**, *22*, 2470. [CrossRef] [PubMed]
3. Almarzoqi, S.A.; Yahya, A.; Matar, Z.; Gomaa, I. Re-Learning EXP3 Multi-Armed Bandit Algorithm for Enhancing the Massive IoT-LoRaWAN Network Performance. *Sensors* **2022**, *22*, 1603. [CrossRef] [PubMed]
4. Qadri, Y.A.; Nauman, A.; Zikria, Y.B.; Vasilakos, A.V.; Kim, S.W. The future of healthcare internet of things: A survey of emerging technologies. *IEEE Commun. Surv. Tutor.* **2020**, *22*, 1121–1167. [CrossRef]
5. Dang, L.M.; Piran, M.J.; Han, D.; Min, K.; Moon, H. A survey on internet of things and cloud computing for healthcare. *Electronics* **2019**, *8*, 768. [CrossRef]
6. Ghorbani, H.R.; Ahmadzadegan, M.H. Security challenges in internet of things: Survey. In Proceedings of the 2017 IEEE Conference on Wireless Sensors (ICWiSe), Kuching, Malaysia, 13–14 November 2017; IEEE: Piscataway, NJ, USA, 2017; pp. 1–6.
7. Said, O.; Masud, M. Towards internet of things: Survey and future vision. *Int. J. Comput. Netw.* **2013**, *5*, 1–17.
8. Atzori, L.; Iera, A.; Morabito, G. The internet of things: A survey. *Comput. Netw.* **2010**, *54*, 2787–2805. [CrossRef]
9. Liu, X. A study on smart campus model in the era of big data. In Proceedings of the 2016 2nd International Conference on Economics, Management Engineering and Education Technology (ICEMEET 2016), Sanya, China, 12–13 November 2016; Atlantis Press: Dordrecht, The Netherlands, 2017.
10. Gravalos, I.; Makris, P.; Christodoulopoulos, K.; Varvarigos, E.A. Efficient gateways placement for internet of things with QoS constraints. In Proceedings of the 2016 IEEE Global Communications Conference (GLOBECOM), Washington, DC, USA, 4–8 December 2016; IEEE: Piscataway, NJ, USA, 2016; pp. 1–6.
11. Venkatesan, V.P.; Devi, C.P.; Sivaranjani, M. Design of a smart gateway solution based on the exploration of specific challenges in IoT. In Proceedings of the 2017 International Conference on I-SMAC (IoT in Social, Mobile, Analytics and Cloud)(I-SMAC), Palladam, India, 10–11 February 2017; IEEE: Piscataway, NJ, USA, 2017; pp. 22–31.
12. Qiu, T.; Chen, N.; Li, K.; Atiquzzaman, M.; Zhao, W. How can heterogeneous internet of things build our future: A survey. *IEEE Commun. Surv. Tutor.* **2018**, *20*, 2011–2027. [CrossRef]
13. Cardoso, K.; Jr, A.O.; Correa, S. SOFTWAY4IoT: SOFTware-defined gateWAY and fog computing for IoT (Internet of Things). In Proceedings of the WRNP 2019, Gramado, Brazil, 6–7 May 2019; Volume 1.
14. Ali, H.M.; Liu, J.; Bukhari, S.A.C.; Rauf, H.T. Planning a secure and reliable IoT-enabled FOG-assisted computing infrastructure for healthcare. *Clust. Comput.* **2022**, 25, 2143–2161. [CrossRef]
15. Karthikeya, S.A.; Vijeth, J.; Murthy, C.S.R. Leveraging solution-specific gateways for cost-effective and fault-tolerant IoT networking. In Proceedings of the 2016 IEEE Wireless Communications and Networking Conference, Doha, Qatar, 3–6 April 2016; IEEE: Piscataway, NJ, USA, 2016; pp. 1–6.
16. Lima, M.P.; Carrano, E.G.; Takahashi, R.H. Multiobjective planning of wireless local area networks (WLAN) using genetic algorithms. In Proceedings of the 2012 IEEE Congress on Evolutionary Computation, Brisbane, Australia, 10–15 June 2012; IEEE: Piscataway, NJ, USA, 2012; pp. 1–8.
17. Capone, A.; Cesana, M.; Donno, D.D.; Filippini, I. Deploying multiple interconnected gateways in heterogeneous wireless sensor networks: An optimization approach. *Comput. Commun.* **2010**, *33*, 1151–1161. [CrossRef]
18. Wu, W.; Luo, J.; Yang, M. Gateway placement optimization for load balancing in wireless mesh networks. In Proceedings of the 2009 13th International Conference on Computer Supported Cooperative Work in Design, Santiago, Chile, 22–24 April 2009; IEEE: Piscataway, NJ, USA, 2009; pp. 408–413.
19. Lin, P.C. Optimal smart gateway deployment for the Internet of Things in smart home environments. In Proceedings of the 2015 IEEE 4th Global Conference on Consumer Electronics (GCCE), Osaka, Japan, 27–30 October 2015; IEEE: Piscataway, NJ, USA, 2015; pp. 273–274.
20. Taleb, S.M.; Meraihi, Y.; Gabis, A.B.; Mirjalili, S.; Ramdane-Cherif, A. Nodes placement in wireless mesh networks using optimization approaches: A survey. *Neural Comput. Appl.* **2022**, 34, 5283–5319. [CrossRef]
21. Rousseeuw, P.J. Silhouettes: A graphical aid to the interpretation and validation of cluster analysis. *J. Comput. Appl. Math.* **1987**, *20*, 53–65. [CrossRef]
22. Friedman, M. The use of ranks to avoid the assumption of normality implicit in the analysis of variance. *J. Am. Stat. Assoc.* **1937**, *32*, 675–701. [CrossRef]
23. Jinaporn, N.; Saengudomlert, P. Impact of Gateway Placement and Energy Consumption for Data Processing on Lifetime of IoT Networks. In Proceedings of the 2021 18th International Conference on Electrical Engineering/Electronics, Computer, Telecommunications and Information Technology (ECTI-CON), Chiang Mai, Thailand, 19–22 May 2021; IEEE: Piscataway, NJ, USA, 2021; pp. 90–93.
24. Rathod, N.; Sundaresan, R. Relay Placement Algorithms for IoT Connectivity and Coverage in an Outdoor Heterogeneous Propagation Environment. *IEEE Access* **2022**, *10*, 13270–13289. [CrossRef]
25. Zhao, D.; Zou, Q.; Boshkani Zadeh, M. A QoS-Aware IoT Service Placement Mechanism in Fog Computing Based on Open-Source Development Model. *J. Grid Comput.* **2022**, *20*, 1–29. [CrossRef]

26. Petroni, A.; Biagi, M. Interference Mitigation and Decoding Through Gateway Diversity in LoRaWAN. *IEEE Trans. Wirel. Commun.* **2022**. [CrossRef]
27. Majid, M.; Habib, S.; Javed, A.R.; Rizwan, M.; Srivastava, G.; Gadekallu, T.R.; Lin, J.C.W. Applications of Wireless Sensor Networks and Internet of Things Frameworks in the Industry Revolution 4.0: A Systematic Literature Review. *Sensors* **2022**, *22*, 2087. [CrossRef] [PubMed]
28. Matni, N.; Moraes, J.; Oliveira, H.; Rosário, D.; Cerqueira, E. LoRaWAN Gateway Placement Model for Dynamic Internet of Things Scenarios. *Sensors* **2020**, *20*, 4336. [CrossRef] [PubMed]
29. Maiti, P.; Apat, H.K.; Sahoo, B.; Turuk, A.K. An effective approach of latency-aware fog smart gateways deployment for IoT services. *Internet Things* **2019**, *8*, 100091. [CrossRef]
30. Matni, N.; Moraes, J.; Rosário, D.; Cerqueira, E.; Neto, A. Optimal gateway placement based on fuzzy C-means for low power wide area networks. In Proceedings of the 2019 IEEE Latin-American Conference on Communications (LATINCOM), Salvador, Brazil, 11–13 November 2019; IEEE: Piscataway, NJ, USA, 2019; pp. 1–6.
31. MacQueen, J. Some methods for classification and analysis of multivariate observations. In Proceedings of the Fifth Berkeley Symposium on Mathematical Statistics and Probability, Oakland, CA, USA, 1 January 1967; Volume 1, pp. 281–297.
32. Thaljaoui, A.; Val, T.; Nasri, N.; Brulin, D. BLE localization using RSSI measurements and iRingLA. In Proceedings of the 2015 IEEE International Conference on Industrial Technology (ICIT), Seville, Spain, 17–19 March 2015; IEEE: Piscataway, NJ, USA, 2015; pp. 2178–2183.
33. Najnudel, M. Estudo de Propagação em Ambientes Fechados para o Planejamento de WLANs. Master's Thesis, Pontifícia Universidade Católica do Rio de Janeiro, Departamento de Engenharia Elétrica, Rio de Janeiro, Brazil, 2004.
34. Xu, W.; Kim, J.Y.; Huang, W.; Kanhere, S.S.; Jha, S.K.; Hu, W. Measurement, Characterization, and Modeling of LoRa Technology in Multifloor Buildings. *IEEE Internet Things J.* **2020**, *7*, 298–310. [CrossRef]
35. Butler, E. *Milton Friedman*; Leya: Alfragide, Portugal, 2012.
36. MESQUITA, E.d.M. Estudo Comparativo de Metaheurísticas Aplicadas ao Controle Preditivo Baseado em Modelo. Master's Thesis, Universidade de Brasília, Brasília, Brazil, 2018.

sensors

MDPI

Article

Self-Sovereignty Identity Management Model for Smart Healthcare System

Pinky Bai [1], Sushil Kumar [1], Geetika Aggarwal [2], Mufti Mahmud [3,4,5], Omprakash Kaiwartya [3,4,*] and Jaime Lloret [6]

1 School of Computer and Systems Sciences, Jawaharlal Nehru University, New Delhi 110067, India; pinky82_scs@jnu.ac.in (P.B.); skdohare@mail.jnu.ac.in (S.K.)
2 Department of Engineering, Nottingham Trent University, Nottingham NG11 8NS, UK; geetika.aggarwal@ntu.ac.uk
3 Department of Computer Science, Nottingham Trent University, Nottingham NG11 8NS, UK; mufti.mahmud@ntu.ac.uk
4 Computing and Informatics Research Centre, Nottingham Trent University, Nottingham NG11 8NS, UK
5 Medical Technologies Innovation Facility, Nottingham Trent University, Nottingham NG11 8NS, UK
6 Department of Communications, Universitat Politècnica de València, 46022 Valencia, Spain; jlloret@dcom.upv.es
* Correspondence: omprakash.kaiwartya@ntu.ac.uk

Abstract: An identity management system is essential in any organisation to provide quality services to each authenticated user. The smart healthcare system should use reliable identity management to ensure timely service to authorised users. Traditional healthcare uses a paper-based identity system which is converted into centralised identity management in a smart healthcare system. Centralised identity management has security issues such as denial of service attacks, single-point failure, information breaches of patients, and many privacy issues. Decentralisedidentity management can be a robust solution to these security and privacy issues. We proposed a Self-Sovereign identity management system for the smart healthcare system (SSI-SHS), which manages the identity of each stakeholder, including medical devices or sensors, in a decentralisedmanner in the Internet of Medical Things (IoMT) Environment. The proposed system gives the user complete control of their data at each point. Further, we analysed the proposed identity management system against Allen and Cameron's identity management guidelines. We also present the performance analysis of SSI as compared to the state-of-the-art techniques.

Keywords: internet of things; blockchain; self-sovereign identity; IoMT; security; privacy

Citation: Bai, P.; Kumar, S.; Aggarwal, G.; Mahmud, M.; Kaiwartya, O.; Lloret, J. Self-Sovereignty Identity Management Model for Smart Healthcare System. *Sensors* **2022**, *22*, 4714. https://doi.org/10.3390/s22134714

Academic Editors: Alexandru Lavric, Liliana Anchidin and Adrian I. Petrariu

Received: 30 May 2022
Accepted: 16 June 2022
Published: 22 June 2022

Publisher's Note: MDPI stays neutral with regard to jurisdictional claims in published maps and institutional affiliations.

1. Introduction

Blockchain plays a crucial role in healthcare applications, from improving medical record management, enhancing insurance claim processes, and accelerating clinical/biomedical research to advancing healthcare data by recording on the ledger. Blockchain technology can provide feasible and secure solutions to healthcare applications. The blockchain's main characteristics, i.e., decentraliseddata management, data provenance, immutable audit trails, high availability, and, most importantly, security and privacy, increase the usability of blockchains in healthcare applications compared to traditional databases [1].

In smart healthcare applications, patients are implanted with wearable biosensors on their bodies and non-wearable sensors in nearby environments. These wearable biosensors and non-wearable sensors collect vital and biological data (e.g., cardiac activity, pulse rate, blood pressure, temperature, etc.). Biological data and personal patient profiles are addressed as an Electronics Health Record (EHR). The security and privacy requirements for an EHR have become more difficult and necessary as the movement to an EHR is one click away from being across the world. Challenges emerge as more health data is collected from wearable devices and Electronic Health Record (EHR) systems. The current centralized smart healthcare system has data isolation, data ownership, accountability, security, and privacy issues. Further, patients do not have control over their health data; the self-sovereignty concept is an excellent way to deal with these privacy issues. The current centralized concept is better for the scalability and mobility of the system. However, it is not good in terms of privacy, security, usability, single-point failure, and system complexity [2].

Identification is essential for public health management and quality delivery of health services to the end-user. Patients should be uniquely identified in the smart healthcare system to access the appropriate medical treatment and services. The service providers should also ensure that they provide consistent and correct services to the right person. The unique identification of patients helps researchers and administrators analyze records in order to generate statistics and other data planning, pandemic management, treatment improvements, tracking a patient in case of spreading diseases like covid, emergency response, and many more. Further, health insurance companies must also identify a patient to ensure the correct claims are submitted and provide insurance money based on the patient's treatment history [3]. Smart healthcare consists of smart medical equipment, wearable sensors, or the internet of medical things, making identity management difficult for health service providers [4]. We can summarize that the healthcare system needs a secure, inclusive identity management system to provide quality health services.

The existing centralized identity management system for smart healthcare faces security, privacy, single point of failure, and interoperability issues. Further, individuals are given fewer or no options to control their health data and data transactions encompassing how, where, when, by whom, to whom, by what time, and which specific data is shared. The right of users to control and rectify personal information, including health information, has decreased in the digital era [4,5]. To solve the issues of a centralized identity model and patient privacy, the Self-Sovereign Identity Model (SSI) is an emerging concept of identity management. An SSI is a decentralised and owner-centric identity model that can solve the identification issues of a smart healthcare system.

In this paper, we are proposing a decentralised identity management SSI for smart healthcare to provide patients control over their EHR. The proposed identity model covers the IoMT identification and gives control of device data to the device owner. The smart healthcare system has IoMT devices or sensors at different stages, like sensors installed with patients for remote monitoring, wearable devices, patient motion detection, and at hospitals to measure different health parameters. In the proposed identity model, the owner of the device (mostly the patient) has complete control of the sensors or the IoMT devices that collect data, and the owner chooses to share the information.

The motivation of this research is to consider the IoMT device as an essential identity in the smart healthcare system. The identity management system aims to ensure that the service provider provides services to the trusted user based on the trusting relationship with an identity provider. However, there is no limit on the IoMT for providers to offer their services to any requestor. Traditional identity management systems focus only on real users' identities and negate the end-users like the application, the IoMT. Researchers were motivated to do this research to provide solutions to these limitations.

The contribution of this paper can be summarized as follow:

- First, a system architecture is presented for a Self-Sovereign Identity Model for smart healthcare, including the IoMT network. The IoMT network is integrated with the smart healthcare distributed network.
- Second, the registration and authentication process of stakeholders in smart healthcare is presented along with the smart device installed or the patient's collected EHR, registration, and authentication in the smart healthcare system.
- Third, we have implemented a prototype for the proposed SSI model using the permission blockchain, Hyperledger Indy, to collect the results for performance analysis.
- Finally, the proposed identity model is analysed with respect to the Allen identity model rules. Further performance analysis with respect to the execution time and

storage is presented. The proposed distributed identity model gives complete control of personal data to the data owner. The patient and other stakeholders can choose the limited disclosure of personal information.

The organisationof the paper is as follows. Section 2 describes Preliminaries on Smart Healthcare and Identity Management concepts. In Section 3, we reviewed the existing identity model based on blockchains and without blockchains for smart healthcare. Section 4 describes the proposed identity model with architecture, communication flow, and process. The experimental implementation, results and their analyses are discussed in Sections 5 and 6, respectively. Section 7 concludes the research paper and discusses future direction.

2. Preliminaries on Smart Healthcare and Identity Management

2.1. IoT Enabled Smart Healthcare Model

The internet of things (IoT) has been used as a potential solution to reduce the pressure on the healthcare system and provide healthcare services to everyone, anytime and anyplace. A large amount of research focuses on this direction. It shows the considerable use of the IoT in healthcare, such as remote monitoring of specific conditions, aiding rehabilitation through constant monitoring of a patient's progress, constant monitoring of patients using wearable devices, and many more. Baker et al. presented a range of uses for the IoT in healthcare and proposed a unique identity model for future IoT-based healthcare systems [5].

Figure 1 presents the IoT-enabled healthcare system. Figure 1 captures all stakeholders of smart healthcare. The healthcare system comprises many stakeholders such as doctors, hospitals, clinics, pharmaceuticals, insurance companies, researchers, and core healthcare "patients." The second important factor in the IoT-enabled healthcare is that the sensors are deployed with patients, and the generated large data is sent to the storage location. The storage server, like cloud, blockchain, or any database, stores this large health data, and the end-user applications access this data for further analysis and provide the services. Further, machine learning, deep learning, and soft computing or other computation techniques are used in the analysis to get specific results.

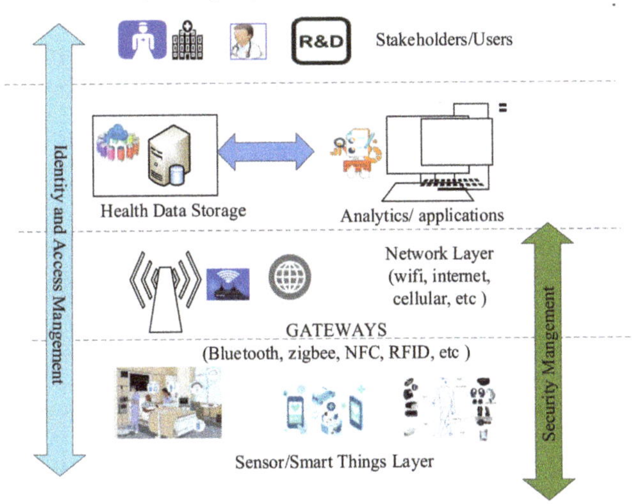

Figure 1. IoT enabled healthcare system.

In this work, we present the four key players as follows:

Healthcare Consumers: People or patients who receive healthcare services, treatment, or care. Patients access their healthcare records and share the same with doctors or hospitals, or other healthcare providers party to get the health services.

Healthcare Regulators: Government institutes or public health departments that regulate health services among consumers. Health regulators monitors and fame the policies related to healthcare services. They aggregate the health data and process it to make new healthcare frameworks or policies.

Healthcare Providers: Any entity that provides healthcare services to patients. It can be doctors, hospitals, nurses, ambulances, clinics, and others who are responsible for delivering health services. They collect the health data to provide health services.

Industry Representative: Includes pharmaceutical firms, insurance companies, drug manufacturers, and medical device companies. They help operate the healthcare system and provide the latest and advanced solutions for health services. They collect the health data to provide good solutions to advance the health system.

2.2. Digital Identity Management System

An identity management system (IDMS) or digital identity management system contains a set of rules and conditions for authentication, authorization, and the system's access control. The IDMS ensures that only authorized entities can access the services in an organization. The core entities of any IDMS are the user, the identity issuer, and the service provider. In most IDMSs, a single centralized authority, like an organization, controls and owns the digital identities of specific organizations or systems [6].

Mainly, three types of IDMS models have been present since the internet's beginning: Centralized Identity, Federated Identity, and user-centric Identity/decentralised identity model. The service provider authenticates users in the centralized Identity model before providing a service. Here, the service provider controls the identities and provides the credentials to access each service and time. The centralized system is the model we have been using for a long time: government ID, license card, college identity card, voter ID, Facebook, Twitter login, and so on. The identity issuer (mainly government and service provider) issues the identifiers and credentials to the user in the centralized identity model [7,8].

In the federated identity model, the identity provider manages the identities of more than one service provider. Users register for identity providers and can access the services from a federation. There are three popular federated identity protocols available: SAML, OAuth, and OpenID since 2005. Using protocols like OpenID Connect, social login buttons from Facebook, Google, Twitter, LinkedIn, etc., are now a standard feature on many consumer-facing websites [8,9].

In the user-centric identity model, the user controls its identifiers and defines a policy to share the attributes with the service provider to access the service. The decentralised identity model is based on peer-to-peer connection and does not have a centralized authority to manage the identity of a system. uPort, ShoCard, BitID, and soverin are examples of decentralised identity models. Further, the self-sovereign identity (SSI) model is a decentralised model that facilitates the recording and exchange of identity attributes and the propagation of trust among participating entities [9,10].

2.3. Self-Sovereign Identity (SSI)

In February 2012, a developer, Moxie Marlinspike, first wrote about the "Sovereign Source Authority" and mentioned that "individuals have a right to an "Identity"" [10]. Simultaneously, In March 2012, Patrick Deegan also started working on an open-source framework that gives the control of a digital identity to the user [11]. SSI introduces a layer of flexibility and security in distributed identity management systems. SSI is the concept where organizations and individuals have whole ownership of their identities along with self-defined attributes and identifiers, while the distributed identified management system (DIMS) uses the user's already existing trusted credentials like PAN, Voter ID, Passport, etc [7].

In the SSI model, there is no central authority that holds user data and passes data on to other parties on a request. The user holds his/her own data. The cryptography and distributed ledger technology allow users to present claims about identity, and others can verify it with cryptographic certainty [11].

2.4. Architecture of Self Sovereign Identity Model

The SSI model uses the core concept of identity management, blockchain or distributed ledger technology, distributed computing, and cryptography and provides a user-centric identity model. These concepts have been well established for a long time, and SSI put them together to create a more secure, persistent, and interoperable identity model. Figure 2 represents the sequence flow along with the important component of the SSI model. The conceptual architecture of the SSI model has four layers: Identifiers and Keys (DID), Secure communication and Interfaces, Verifiable Credentials, and Governance. Further, these four layers need seven building blocks to achieve the user-centric identity management goal [12,13]. The seven building blocks are:

- The trust triangle (issuer, holder, and verifier): Issuers are the source of credentials. The holder saves credentials issued by the issuers in its digital wallet and presents proof of claims when a verifier requests. The verifier verifies the credentials presented by the holder.
- Verifiable credentials or digital credentials: The digital equivalent of physical credentials are the verifiable credentials to prove the identity. The subject of the credentials creates a set of claims, and the verifiable credentials contain those claims. The issuer in the SSI model issues the verifiable credentials.
- Digital wallets: Digital wallets store credentials and other sensitive data and work with digital agents to securely exchange credentials among peers.
- Digital Agent: Digital agent is a software on the digital wallet that provides security to the digital wallet, participates in secure credentials exchange, and forms connections via a decentralised, secure message protocol. Edge Agents and cloud agents are two general categories of the digital agent.
- Decentralised Identifiers (DIDs): DIDs are decentralised, cryptographically verifiable, resolvable, and unique identifiers. DIDs are combinations of the private and public keys of a user. DIDs are decentralised by the nature that makes credentials available at all times for verifications. DIDs create a secure, unique, and private peer-to-peer connection between two parties who agree to connect with each other based on their requirements. The identity owner has complete control of the DIDs.
- Verifiable Data registries: A DID can be registered with any type of decentralised network, verifiable data registry, or even exchanged peer-to-peer. Blockchain can be a vital choice for verifiable data registry because a blockchain is a highly tamper-resistant transactional distributed database that no single party controls.
- Trust Framework: The trust framework contains the set of business, legal, and technical rules to use the SSI infrastructure and enables interoperable digital trust ecosystems of any size and scale.

The basic steps of information flow in the SSI model are:

- The issuer issues the verifiable credentials to the identity owner/holder. The VC includes the claims and attention.
- The user/holder stores this information himself. Users and holders can be the same sometimes. Furthermore, the VC holders have complete control of the VCs.
- When the user wants to access any service, he/she presents its VC to the verifier.
- The verifiers verify the VC without connecting the issuers. The verifier connects with a distributed registry (blockchain), verifies the user, and grants authorized services.
- The distributed verifiable registry has the VC schemas and DID, which helps in user verification.

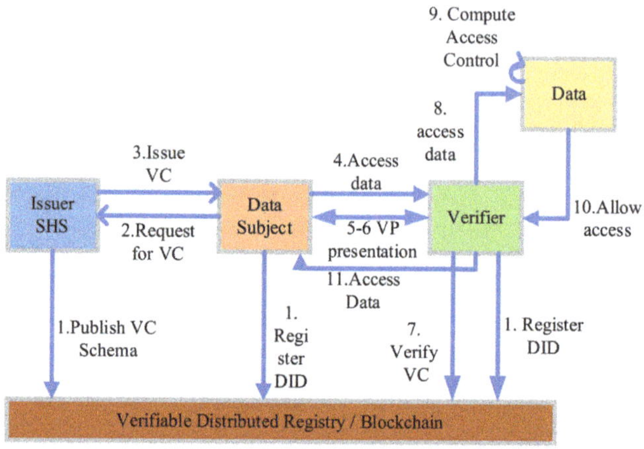

Figure 2. SSI communication Sequence.

The SSI model is an advancement of DIMS and user-centric identity. The main focus of the SSI model is that the user must be the controller of the identity, and the identity must come with interoperability across multiple locations with the user's permission. There is vast literature available that writes about Self-Sovereign Identity Models. Here, we can summarize the ten principles of SSI. These principles focused on the user's identity control, system transparency and fairness, and interoperability. Table 1 defines the ten principles of SSI and categorizes them based on focus area [13,14].

Table 1. Principle of SSI.

User's Control	Security	Portability
Users must have control of their data like which information can be seen the other	Keep identity information secure	Users can move anywhere without being tied to a provider
Existence	Protection	Access
Control	Persistence	Transparency
Consent	Minimization	Interoperability
Persistence		

Alex Preukschat and Drummond Reed analysed and listed the major features and benefits of the SSI model. The SSI model can help in the following areas: fraud detection, reducing customer onboarding cost, auto authentication, auto authorization, automated workflow, data security, privacy, protection, portability, and much more [15].

Limitations:

This part addresses the challenges and limitations of developing the fully decentralised identity system or SSI system. The SSI is a distributed identity model and relies on the DID and Blockchain system, so blockchain performance directly affects the system. The issue faced in blockchain implementation, like storage limitations, scalability, predefined set of users, and so on, is directly accepted in the SSI model. Key storage is also an essential part of the implementation of SSI. When DID is created, the private keys and verifiable credentials are stored in secure storage with the user. The challenges come when the user loses the secure storage for any reason. In that case, it is like the user losing their access to identity.

3. Literature Survey

In this section, we have reviewed and analysed the literature on centralized and decentralised identity management systems. As the literature on SSI and distributed identity in smart healthcare is scarce, we consider access control and identity management in smart healthcare, e-health, and traditional healthcare. Some surveys are presented in the open literature on identity management based on blockchain technology. X, Zhu et al. presented the survey on blockchain-based identity solutions for the internet of things. The survey analysed the identity solution for IoT digital identity management and studied the recent increment in blockchain-based SSI solutions for IDMS [15]. Further, Kuperberg surveyed essential aspects of blockchain-based identity systems like compliance and liability, regulations, standards, integration, and user-friendliness [16].

We divide the literature analysis into a centralized identity model and a decentralised or SSI identity model for smart healthcare.

3.1. Centralised Identity Model

The health information system contains individuals' personal information and critical health data. Bouras et al. stated that the current centralized identity management system is good in terms of scalability and mobility. However, the centralized identity model is poor regarding security, privacy, usability, single point of failure, and complex ecosystem [17].

Aghili et al. proposed lightweight authentication and an ownership transfer protocol (LACO), a secure and energy-efficient protocol that provides authentication and key agreement. The proposed work also covers the access control of health data and preserves doctor and patient privacy. The author designed a threat model for IoT and analysed the proposed protocol against around eleven security attacks. Further, the authors presented a comparison between the proposed model and the ZZTL [18] protocol [19].

Yang et al. proposed a big data storage system with self-adaptive access control to preserve privacy in smart IoT-based healthcare. The proposed work aims to provide emergency and normal access control and prevent duplication from saving space. Further, the system supports sharing encrypted medical files from IoT networks to different domains by applying the cross-domain sharing policy [20]. PASH, privacy-centered access control for the health system, is proposed. The proposed system revealed that only attribute names of access policies and attribute values are encrypted and stored in records because the attribute values have sensitive information, not the attribute names. Further, the security analysis shows that PASH is also secure as per the standard model [21]. Farid et al. present identity management solutions for IoT and cloud computing-based personal healthcare systems. The solution uses biometrics to perform the authentication in the system. However, the framework does not have user consent and is a combination of federated and centralized identity management. The proposed work does not present an end-to-end security analysis [22].

3.2. Decentralised Identity Model

As we have already discussed the properties and technical aspects of the Self-Sovereign Identity Model, the SSI model is a robust solution for protecting the data owner's privacy as SSI gives its owner control of identity. Further, Houtan et al. analysed identity projects based on the SSI model, such as uPort, Soverin, evernym, ShoCard, TheKey, and other projects based on blockchain technology for the patient identity system [22].

Augot et al. developed a zero-knowledge proof-based solution for identity management. However, the proposed framework has two significant drawbacks. First, the authentication is not free. As the authentication is encoded with bitcoin transactions, the users have to pay a transaction fee to the miners. Second, the bitcoin transactions are public, so the user's privacy is at risk [23]. Liang et al. presented personal health data management using blockchain and Intel SGX. The intel SGX stores the healthcare records; these records are trusted timestamping and free from redundancy, preserving both availability and accountability. However, the research does not include the identity management of the IoMT device [24].

AU et al. also proposed user-centric and privacy-preserving identity management for the distributed e-health system. Healthcare consumers maintain pseudonymous identifiers for use in different healthcare systems. However, this work does not include the implementation of the proposed architecture in an e-health system controlled environment and particle deployment in an e-health system [25]. Shuaib et al. explore the applicability of blockchain-based SSI solutions for healthcare, their advantages, and their requirements. Further, they proposed a model to demonstrate the use case of SSI. However, this work did not present the proposed model's formal implementation and performance analysis [26]. Mikula et al. proposed identity and access management for EHR in the healthcare system to support authentication and authorization (use this in comparison). This paper does not cover IoMT device authentication and authorization [27]. Further, Zoho et al. applied sovereign identity claims to provide the distributed data vending system for the personal healthcare system [28]. Buzachis et al. also used uPORT, a self-sovereign identity platform, to identify patients in healthcare [29,30]. In recent times, researchers have included new technologies along with blockchain to secure the health system. Neelakandan et al. used deep learning with blockchain to secure the healthcare and diagnostic model [31]. Kamalraj et al. applied an interpretable filter-based convolutional neural network in the healthcare system for glucose prediction and further analysis [32]. Harshavardhan et al. proposed an optimization model for healthcare systems using LSGDM with biogeography-based optimization [33]. It is clear from the literature survey and the surveys mentioned earlier that no single distributed solution is fully distributed, covering users' consent, privacy, and compliance with privacy standards [34,35]. The proposed research has the potential to be used in a smart city environment, such as smart traffic management [36–38]. Smart cities can be connected with smart healthcare to reduce the insurance claim process time and prevent fraud in the traffic-centric health insurance process [39–42].

4. Proposed Framework

We proposed an SSI model to manage the digital identities for a smart healthcare system. The proposed model includes the IoMT devices and provides digital identity to all stakeholders in smart healthcare. Further, the model gives complete control to the data owner at the time of sharing to PII and PHI. Whenever a user wants to access another user's data, the requester's data must authenticate himself to the requestee user. This provides the security and privacy of personal data and users' health data. We proposed a distributed blockchain-based SSI that does not require a central authority to control the identity lifecycle. The following section will explain the whole SSI model in detail.

4.1. High-Level System Architecture of SSI Model for Smart Healthcare

In this section, we present the high-level system architecture of the proposed model for the smart healthcare system. Figure 3 explains the main components of the SSI model of smart healthcare system (SSI-SHS) with reference to the SSI architecture presented in Section 2. The three main access-based roles of the SSI model are Subject (Identity Holder), Issuer (Identity issuer), and Verifier (Identity Verifier) in the context of smart healthcare as follows: the smart healthcare system (SHS) issues the verifiable credentials to all stakeholders (patients, doctors, labs, researchers, and others) based on DID. The subjects who hold the identity can be any stakeholders, and we include IoMT devices to cover the end-to-end information flow of smart healthcare. The verifier is the entity that provides any kind of service to others. For example, if a doctor wants to access the data of any medical device, in this case, a patient who owns the device verifies the doctor's identity directly without any help from SHS (the issuer).

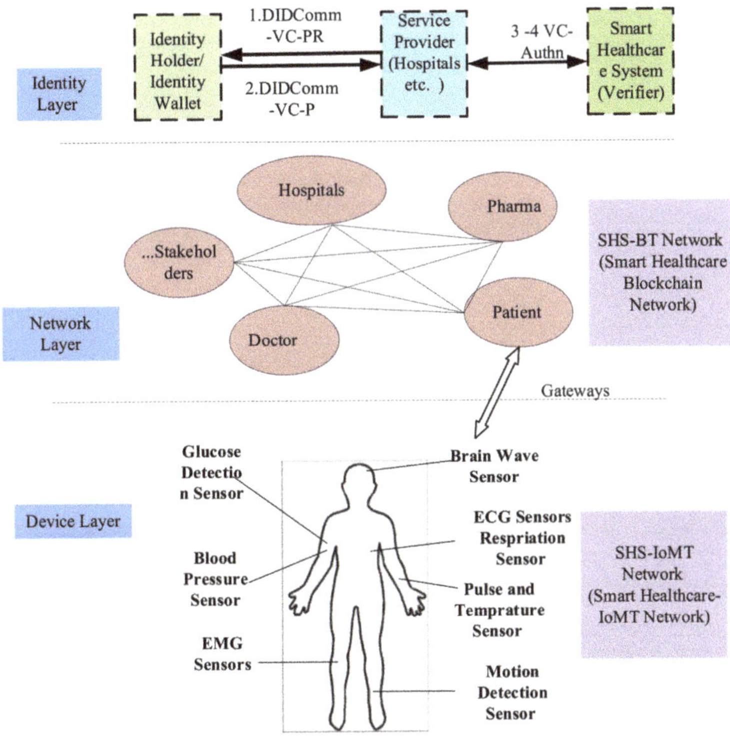

Figure 3. SSI-SHS architecture.

Further, all the stakeholders are in the same blockchain network (SHS-BT), and any entity that wants to access health services must register on the SHS-BT network. For the transactions, we will discuss the generation of transactions (identity management related) in the coming sections, stored at the SHS-BT blockchain distributed ledger (Li). The owners of IoMT devices register their devices on SHS-BT by providing the DID of devices along with their own DID. The SSI-SHS uses blockchain for verifiable data registry (VDR) also.

VDR sets the rules in the distributed system for entities to create identifiers as per their own rules. VDR is a role or system that mediates the creation and verification of identifiers, verifiable credentials schema, keys, and other relevant data, such as public keys, revocation registries and so on, which are required in the verification of verifiable credentials.

4.2. Communication Sequence Flow

In this section, we will describe the flow in SSI-SHS. The backbone of communication is DID communication between the Edge Agents of respective users. A user contains the user's Edge Agent, front-end DID wallet, secure element, and micro ledger. The URL consists of a community resolver, a driver for DID methods, and a cache. We choose Sovrin [34] to demonstrate the interaction of DID among the users. A steward, a DID syntax checker, cache and resolution result constructor, and serialization validators are part of Sovrin. The VDR is any blockchain network.

The system uses an agent that is a delegated entity by the DID subject. The agent controls the agent-to-agent DID communication, DID wallet cryptography-based operations, and sharing of credentials to authorized agents as per the relationships. Agents are categorized as Edge Agents and cloud agents. The Edge Agent resides within the wallet software locally. The cloud agent resides in the cloud and has extended features like identity wallet backup to the cloud, 24/7 DID communication when an Edge Agent is offline, data storage in the cloud, and key management. We use an Edge Agent (EA) in the proposed system.

To explain the architecture, we take the most common communication in smart healthcare, where a patient wants to talk with a doctor, share health data, and device readings to get medical service. To establish the communication, the patient's Edge Agent first queries the doctor's DID from the Edge Agent to the community resolver within the universal resolver (UR). Then DID methods return the DDO of the doctor's EA to the UR through VDR interaction. Now, the patient's EA retrieves the DDO from UR. After that, the patient's EA establishes the DID communication as per the data present in DDO.

The whole smart healthcare identity architecture can be discussed in two parts: high-level user interaction named "SSI-SHS: SSI for smart healthcare system"; and the second part, "SSI-SHS-IoMT: Interaction of IoMT to SSI-SHS", is a network among sensor and patient named as IoMT network. The term "IoMT" covers all types of medical devices, sensors, and other smart medical devices with the patient, as elaborated in Figure 4, the high-level SSI-SHS architecture.

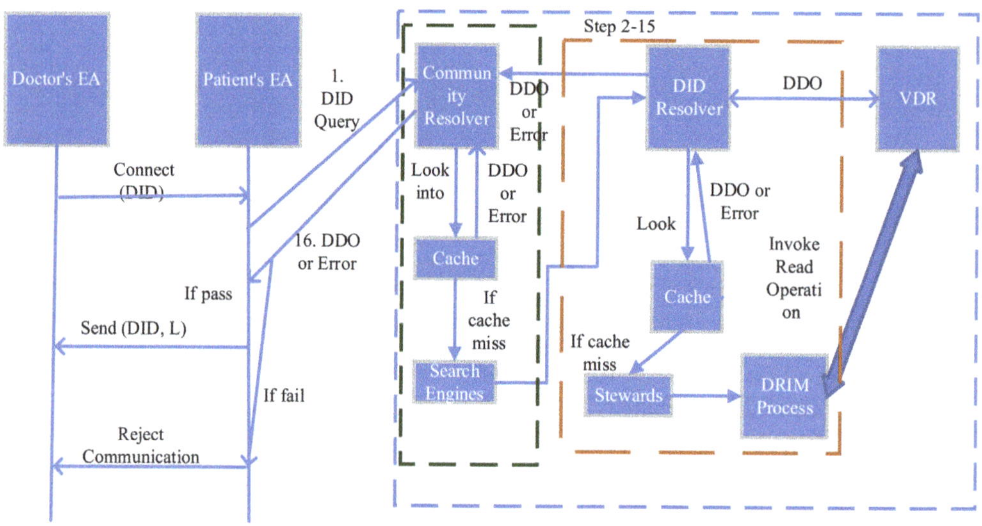

Figure 4. Authentication Process.

Part 1: SSI-SHS

Phase 1: Registration: Identity Wallet and Agent Installation

In the registration phase, the shareholders create their own DID with the help of a digital wallet and Edge Agent. We described the registration of patients on the SHS-BT network. The following steps describe the process of getting the verifiable credentials from the SHS and registration:

Step 1: The patient installs the digital wallet software and initiates the creation of the first Edge Agent. The user uses the wallet to receive credentials from various entities and presents these credentials to prove himself on the system.

Step 2: The Edge Agent (EA) creates a DID for agent communication, credentials for secure element and link secret (link secret is used in DID relationship establishment through a blinded commitment).

$$EA_{patient} \rightarrow DID_{EA}, Cred_{SE}, linksecret \tag{1}$$

Step 3: The Edge Agent requests the Secret Element to create verifiable credentials (VC) and a VC presentation (VCP) for specific credentials schema (CS). The EA requests different types of keys like DID keys for signing and verification, Agent Policy (AP) keys, and encryption and decryption keys for the wallet.

$$EA_{patient}\ Req(VC, VCP, CS) \rightarrow SE \tag{2}$$

Step 4: The Secret Element (SE) stores the following: VC, signing key of DID, decryption key of wallet, and AP keys.

$$SE \leftarrow stores\ VC, PR_{patient}, D.Key_{wallet}, Key_{AP} \tag{3}$$

Step 5: The SE returns the following to the front-end wallet: VCP for CS, DID verification keys, and decryption keys for wallet backup.

$$SE\ return\big(D.Key_{wallet}, VCP, PR_{patient}\big) \rightarrow DigitalWallet_{frontend} \tag{4}$$

Step 6: The EA asks to store the following in the front-end wallet: agent IDs, CS registry address P, and a link secret. The address P denotes the storage location of CS in the public ledger. After that, CS establishes the authorization level for each agent based on each different credentials. The newly added agent stores VCP at address P in the PROVE section of CS.

$$DigitalWallet_{frontend} \leftarrow store\big(EA_{patient}ID, linksecret, P_{AP}\big) \tag{5}$$

The result of the above process achieves the SSI and ensures that the privacy of the identity system is preserved via control and confidentiality. The system provides minimum controllable disclosure of the proof to achieve control of identity. The system stores the user credentials in the decentralised key management system identity wallet to satisfy confidentiality.

Further, the EA proves the authorization using VCP without disclosing the secret value defined in Step 6. The CS Address Commitment (CSAC) can also be generated via the VCP with CS address to achieve herd privacy in the system.

The DID keys are composed of a signing and verification key, as defined in steps 3 and 5. The signing key is based on the Edwards curve Digital Signature Algorithm using SHA-2 and Curve 25519 (ED25519). Next, when a new relationship is started, the DID and verification keys are shared with other parties. Lastly, CA is created for backup purposes.

Phase 2: Authentication using DID Method

After installing the agent and identity wallet, if the doctor wanted to access the patient's data, the doctor would have to send a DID communication request to the patient. The EA of the patient gets the invitation for a DID connection from a remote doctor. The process flow is as follows:

$$EA_{doctor}connect(DID_{doctor}) \rightarrow EA_{patient} \tag{6}$$

Step 1: The EA of the patient asks the query to the Community resolver (CR) of the Universal resolver (UR).

$$EA_{patient}query(DID_{doctor})$$

Step 2: The CR checks the cache first after receiving the DID query.

Step 3: The CR returns the stored DDO (DID Document) immediately if the DID query hits the cache. The DDO is passed with other metadata like DDO metadata, and DID resolution metadata.

$$CR\ return(DDO, DID_{doctor}) \rightarrow EA_{patient} \tag{7}$$

If the cache miss in Step 3, the CR invokes the resolution process (RP). The CR first chooses the driver that is similar to the DID method, as received in DID. The proposed system uses the DID method as defined in the sovran for demonstration and implementation purposes.

$$CR\ invoke(RP, method_{DID})$$

Step 4: The driver passes the DID and DID Resolution Input Metadata (DRIM) to RP to get the DID method. After getting the DID method, the DRIM is passed to the DDO.

Step 5: The steward hits the cache to check that the DDO is present in the cache for the provided DID.

Step 6: If the cache hits, the steward returns the DDO immediately. If the cache misses, the steward resolves the DID query.

Step 7: the steward first checks the DID format; the input DID should be in a standard format. The steward throws an error on any syntax or semantics error.

Step 8: The steward passes the DID method to invoke the read operation to VDR. If the DID has a public DID URL, then the DID URL has to be dereferenced to get more information regarding the specific resources to be targeted in the DDO. The DDO resources are identified with the help of the DID URL components like query, path, and fragments.

Step 9: In the first VDR operation, the DDO is returned to the steward after processing the read operation on VDR.

Step 10: The serialization validator validates the DDO format as per defined in the DID. The serialization format can be JSON, JSONLD, and CBOR as per the DID core data model standard. If the serialization validator gives an error on a query, then the steward forwards the error to the requestor (driver from UR).

Step 11: The requested DDO is sent to the steward and to the resolution constructor. The resolution result constructor makes the representation form of the DID resolution result.

Step 12: The resolution result constructor sends the DDO to the serialization validator, and the cache updates with DDO.

Step 13: The serialization validator sends this DDO to the steward.

Step 14: The steward sends the DDO to the driver of UR.

Step 15: The driver of UR returns the DDO to CR and caches updates with the DDO.

Step 16: The CR passes the DDO to the patient's EA.

$$CRsend(DDO, DID_{doctor}) \rightarrow EA_{patient} \tag{8}$$

Step 17: After verifying the DDO, if the patient is satisfied and ready to connect, then the EA of the patient sends a DID communication message along with a delta of micro ledger.

$$EA_{patient}\ send(DID_{commn}, encrypt_{PR_{EA}}(\nabla L)) \tag{9}$$

The EAs use a message-based protocol to communicate and exchange a series of messages with each other. The delta of the micro ledger consists of the record of DID events from each EA. The delta represents the updates of the micro ledger, and it is organized in the Merkle tree. The EA exchanges the delta via authenticated encryption using the verification key of the EA.

Step 18: The doctor's EA stores the DID events in the relationship between patient and doctor. The micro ledger of the doctor stores the delta of the micro ledger sent by the patient to make sure that the patient and doctor have the same copy of the DID events.

Step 19: The change of state of the doctor's micro ledger is sent in response to the recording operation.

Step 20: The doctor's EA sends a reply of DID communication message and a copy of the micro ledger delta of the doctor to the EA of the patient.

Step 21: The EA of the patient stores the DID events in the micro ledger and stores the delta of the micro ledger sent by the doctor.

Step 22: The micro ledger sends a state change response to the EA of the patient.

Step 23: The EA of the patient checks that the DID events are in sync with the doctor's DID events.

Step 24: The identity ledger (Li) stores the transaction as either failed or passed. The blockchain transactions include the EA of the requester and provider along with the status (pass or fail) and a hash of the ledger with a timestamp as below.

Tx (EApatient, EAdoctor, status, timestamp, hash(L))

Figure 4 describes the authentication process of a doctor by a patient without any centralized system or hospital authority. These two phrases describe the registration and authentication of high-end users and stakeholders. The privacy and secrecy concerns in smart healthcare increase with the number of IoT devices in the system, and smart healthcare is loaded with lots of medical devices, and these devices are exposed and vulnerable to the outside world.

Part 2 SSI-SHS-IoMT: Interaction of IoMT to Healthcare SSI

Here, we present the second part, where a medical device connected with a patient registers itself and authenticates in a smart healthcare system. We assume that all the medical devices/sensors (IoMT) are connected to the internet and pass the information to a remote doctor/hospital to analyze the patient's health and provide health services. Now the patient owns his DID and VC and has some medical sensors and devices which send data to smart healthcare.

We design smart contracts for authentication named "DIDMaster" that provides the information related to the device after supplying the DID: (diddev.H(docdev). URIdocdev .stadiddev). The "AccessData" smart contract provides access to the data of the patient's devices. These smart contracts are deployed on VDR, and the node in the SHS-BT network calls the smart contract by passing the correct parameter, as explained in the next part.

Phase 1: Registration of IoMT devices

Here, we consider that all the devices are bootstrapped in the patient's environment. Now, the device registration of the device is with the patient. The patient is an onsite registration authority for devices in SSI healthcare. In smart healthcare, a device is owned by a single patient, and a single patient can own many devices. After receiving the VC from smart healthcare, the patients register their IoMT device and bind the ownership of the device.

The Bootstrapping process:

1. Once the device is active, the patients send their own DID to the device.
2. The device creates an authentication token that includes the patient's DID, signs this token with a private key (AuthToken1), and sends the token back to the patient.
3. The client creates another token (AuthToken2), including AuthToken1, and signs this with its own private key.
4. The patient calls the "DIDRegister" smart contract as a message sender and passes the device address. "DIDRegister" registers the assignment between the device and registrar with a tentative state.
5. The patient submits the AuthToken2 to a smart health system node which is a server application connected to the blockchain.
6. The SHS checks the validity of AuthToken2 and the registration status of "DIDRegister" (step 4). If both are valid, the identity provider node proofs the assignment of "DIDRegister". Afterward, the "DIDRegister" changes the state to active.
7. The SHS node generates an AuthToken3 with a confirmation about the assignment, signs it with its private key, and sends it to the patient.
8. The client forwards AuthToken3 to the device.
9. The device verifies the signature of the SHS node with its built-in list (in a secured environment) and, if ok, adds the patient to its trust list.

Phase 2: Authentication of IoMT on network

Here, the patient who is the owner of the medical device has complete control over the data collected by their medical/IoT device. A detailed description of how the doctor accesses the data from IoMT with user consent:

Step 1: The patient provides the DID of a device to the doctor to access the data; after receiving the DID of the device, the doctor calls the "DIDMaster" smart contract bypassing the DID of the device and the patient.

Step 2: The smart contract returns the details of the device: hash of the DID device, hash of DDO of the device, status of the device, and URI of the device.

Step 3: If the status of the device is active, the doctor sends the DID to the verifier to verify the DID.

Step 4: The EA of the doctor gets the address of "AccessData" from step 3.

Step 5: The EA of the doctor calls "AcesssData" by passing his own parameters: didoc, reqdoc; where diddoc is the doctor's DID and reqdoc denotes the data set that doctor would like to access. Moreover, the smart contract AcesssData also checks whether the data access request reqdoc is valid. After the successful execution of AcesssData, a data request event, which contains the doctor's request parameters, is emitted on the blockchain.

Step 6: Successful execution of the "AcesssData" smart contract saves on the blockchain.

Step 7: The EA of the patient, on receiving a new data request, sends the DID of the doctor to the UR to resolve. Moreover, the further process is the same as "Phase 2: Authentication using DID Method".

Step 8: If the doctor's data request passes all the validations, the patient's EA invokes the smart contract AccessData with parameters diddoc, reqdoc, and Sigskpatient (Addrs doc. Reqdoc), which grants the doctor's data access request, and saves the data access granted txn on the identity ledger (Li).

$$\text{Tx}\ (\text{EA}_{\text{patient}}, \text{DID}_{\text{dev}}, \text{EA}_{\text{doctor}}, \text{status}_{\text{dev}}, \text{timestamp}, \text{hash(L)}) \tag{10}$$

Step 9: When a doctor's data access request is granted, he obtains the access token Sig*sk*patient (*addr*doc, reqdoc) from the saved txn event and requests to download data from the service endpoint specified in the docdev.

Step 10: To identify that the data requester is a doctor, the patient sends a random challenge *r* to the doctor. Then, the doctor generates a tuple ⟨*addr*dctr, reqdctr, *pk*dctr, Sig*sk*patient (*addr*dctr, reqdctr), Sig*sk*dctr (*r*, *pk*dctr)⟩ as the response.

Step 11: The patient first checks that the access token Sig*sk*patient (*addr*dctr, reqdctr) is valid, which indicates patient authorization for the doctor making the reqdctr query on data collected by the patient's IoT device.

Step 12: The patient then checks that *addr*dctr is derived from *pk*dctr. Finally, it verifies the signature Sig*sk*dctr(*r*, *pk*dctr) to confirm that the response actually comes from the doctor.

Step 13: If the doctor's response passes all the above validations, the patient generates a download link and sends it back to the doctor. Otherwise, the data retrieval request is rejected.

We have described the registration and authentication process of all entities of smart healthcare in the proposed identity model SSI-SHS. Figure 5 presents the complete process of SSI-SHS; the most common scenario starts with a doctor who wants to connect a patient to a doctor who gets access to the IoMT data based on the request. Other communication also follows the same process. Any nodes who want to participate in smart healthcare first get the identity and then present the identity proof to get the healthcare services.

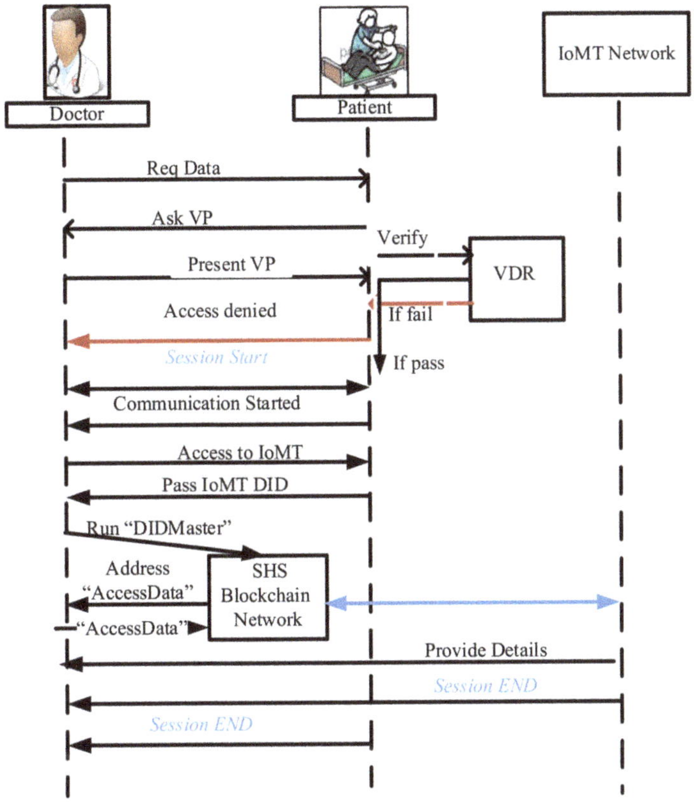

Figure 5. SSI-SHS process flow scenario: The doctor access the IoMT data.

5. Implementation

A prototype of the proposed SSI-SHS identity model is implemented. The blockchain network (SHS-BT) is designed on Hyperledger Aries blockchain, a private network using four nodes: doctor, patient, hospital, and laboratory. Further, the patient's IoMT device is a health band that measures the patient's heartbeat, BP, and sugar.

Terminology:

Identity Ledger (Li): The ledger is the verifiable directory that stores identity records and transactions. The public data like public keys, service endpoints, credentials schema, credentials definition, etc. define the identity record. The relationship between identity record and DID is 1:1, meaning each record has one DID. The DID is unique and resolvable via an identity ledger without needing any third-party centralized authority.

Trust Anchor: A trust anchor (person or organization) that the ledger already knows, bootstrap others. In the smart healthcare system, we can think that an organisation(SHS), hospital, doctors, and other stakeholders must trust anchor roles that bootstrap other entities into the process.

DID Creation: The DID creation is defined in the DID method, and according to the DID method, the node passes the minimum information for DID creation. The W3C organisationmaintains the DID specification registries that include all implemented DID specifications.

Here, we took the minimum input parameter and generated the DID using the "genrateDID" function. The IoMT device generates cryptographically verifiable public and private keys in a trusted execution environment. The device stores the Private key in its secure environment and passes the public key to generate the DID.

$$genrateDID(\text{"}0x5576E95935566Ebd2637D9171E4C92e60543fg10\text{"}, \\ \text{"}8806157fdcbcae265667576fa72d88568db7f9ca8b36tydfe3755ae80457eaf5\text{"}, \\ \text{"}user : password@tcp(example\_connection\_string : 3106)/\text{"})$$

This "genrateDID" function returns the DID for the subject in a format:

$$did : abc : \ H3C2AVvLMv6gmMNam3uVAjZpfkcJCwDwnZn6z3wXmqPV$$

Here, "*did*" is scheme, "*abc*" is DID method, and the remaining part is method specific identifier.

DID Documents (DDO): The DID document, as per the W3C following parameter, must be present in DDO while creating a new DID for a node. The DDO expresses the cryptographical equations, verification methods, services, and controls. The services enable secure and trusted interaction of DID subjects with others. The verification method defines the verification of the DID subject by the verifiers. The DID resolver resolves the DID into DDO. We used hyperledger aries blockchain to implement the prototype. Listing 1 presents the DID and DDO used in a prototype implementation.

Verifiable credentials: Verifiable credentials represent statements made by an issuer in a tamper-evident and privacy-respecting manner. When an organisationissues verifiable credentials, they attach their public DID to the credential, and the verifier can verify the same without contacting the issuing authority. The verification method is presented in the DID document along with other attributes. The issuer cryptographically signs the VC. The VC includes proofs and claims for the subject. The sample of VC from the implementation environment is presented by Listing 2.

Listing 1: Example of blockchain based security implementation DDO.


```
  "@context": "https://w3id.org/did/v1",
  "id": "did:sov:123456789abcdefghij",
  "publicKey": [{
    "id": "did:sov:123456789abcdefghij#keys-1",
    "type": "RsaVerificationKey2018",
    "controller": "did:sov:123456789abcdefghij",
    "publicKeyPem": "-----BEGIN PUBLIC KEY...END PUBLIC KEY-----\r\n"
  }],
  "authentication": [{
    "type": "RsaSignatureAuthentication2018",
    "publicKey": "did:sov:123456789abcdefghi#keys-1"
  }],
  "service": [{
    "id": "did:sov:123456789abcdefghij;exam_svc",
    "type": "ExampleService",
    "serviceEndpoint": "https://example.com/endpoint/8377464"
  }],
  "created": "2018-02-08T16:03:00Z",
  "proof": {
    "type": "LinkedDataSignature2015",
    "created": "2018-02-08T16:02:20Z",
    "creator": "did:sov:8uQhQMGzWxR8vw5P3UWH1ja#keys-1",
    "signatureValue": "QNB13Y7Q9...1tzjn4w=="
  }
}
```

Listing 2: Example of blockchain based security implementation credential schema.

```
"@context": [
  "https://www.w3.org/2018/credentials/v1",
    "https://www.w3.org/2018/credentials/examples/v1"
  ],
  // specify the identifier for the credential
  "id":z6MkjBWPPa1njEKygyr3LR3pRK-
kqv714vyTkfnUdP6ToFSH5#z6Mkn9kQbVXdeDW3h3GvYUV5BzTQDw5oh26CGDqS7sZq3kBN
  // the credential types, which declare what data to expect in the credential
  "type": ["VerifiableCredential", "PatientCredential"],
  // the entity that issued the credential
  "issuer": "https://smarthealth.in/issuers/366079",
  // when the credential was issued
  "issuanceDate": "2021-09-01T19:73:24Z",
  // claims about the subjects of the credential
  "credentialSubject": {
    // identifier for the only subject of the credential
    "id":"did:abc:z6MkjBWPPa1njEKygyr3LR3pRK-
kqv714vyTkfnUdP6ToFSH5#z6Mkn9kQbVXdeDW3h3GvYUV5BzTQDw5oh26CGDqS7sZq3kBN",
    // assertion about the only subject of the credential
    "patientOf": {
      "id": "did:abc:c276e12ec21ebfeb1f712ebc6f1",
      "name": [{
        "value": "patient1",
        "lang": "en"
      }]     }},
  // digital proof that makes the credential tamper-evident
  "proof": {
    // the cryptographic signature suite that was used to generate the signature
    "type": "RsaSignature2018",
    // the date the signature was created
    "created": "2017-06-18T21:19:10Z",
    // purpose of this proof
    "proofPurpose": "assertionMethod",
    // the identifier of the public key that can verify the signature
    "verificationMethod": "https://smarthealth.in/issuers/keys/1",
    // the digital signature value
    "jws":      "eyJhbGciOiJSUzI1NiIsImI2NCI6ZmFsc2UsImNyaXQiOlsiYjY0Il19..TCYt5X      sITJX1CxPCT8yAV-
TVkIEq_PbChOMqsLfRoPsnsgw5WEuts01mq-pQy7UJiN5mgRxD-WUc
      X16dUEMGlv50aqzpqh4Qktb3rk-BuQy72IFLOqV0G_zS245-kronKb78cPN25DGlcTwLtj"
  }
}
```

The Credential Schema is a document that is used to guarantee the structure and, by extension, the semantics of the set of claims comprising a Verifiable Credential. A shared Credential Schema allows all parties to reference data in a known way.

The ledger stores a number of different types of transactions. The transactions that:

- Write a new DID and DDO to the ledger.
- Update existing DDO such as rotating keys
- Define a new Schema name, version, and list of attributes for new credentials
- Define a revocation registry for specific credentials
- Update the revocation registry when the issuer issues or revokes the credentials.

Write the public key from a generated pair of signature algorithms for a specific credentials schema.

6. Result and Analysis

6.1. Identity Framework Evaluation and Result Analysis

The SSI Identity Model for smart healthcare, including IoMT, has been described in detail in the previous section. Here, the SSI-SHS framework is evaluated and analyses the results derived from the prototype experiment.

SSI-SHS Identity Model Evaluation

There are no standard criteria available on how to evaluate an SSI system. Allen proposed the SSI requirements focusing on personal data control, security, and privacy [12]. In the same direction, Cameron also presented "Seven Laws of Identity" [14]. This is a well-established framework for the digital identity system. NIST has a standard for "digital identity." We reference both guidelines to evaluate the SSI-SSH. We have modified and deleted some rules per the framework requirement and the practicality of implementing the guidelines. For example, the interoperability of identities is designed within the smart healthcare system, but the interoperability outside the smart healthcare system needs a standardized format and procedure.

The requirements and guidelines are divided into three groups: User control, Security and Privacy, and Portability.

User Control: This group includes "Existence", "Control", "Consent", and "protection".

1. Existence: NIST defines that every digital identity must have a non-digital existence that manages and represents the online identity. In the proposed architecture, the device and the stakeholder generate their public and private key pair and register themselves on smart healthcare. The main focus is on the patient and his own devices.
2. Control: "Control" of the owner on their identity is proposed by Allen and Cameron. This principle defines that users must have control over their identity and be able to decide which part of their identity they want to share. They should be able to decide which data they share with others, for how long, and be able to refer to, update or hide the identity. In the proposed framework, multiple DID can be derived with a single key pair with different DDOs.
3. Consent: The use of the user's identity should always be with the user's agreement. The user should decide which information and with whom it is shared. Further, the user should decide the time; this also means what time the other party can have access to this information.

Security and Privacy: This group includes "Access," "Transparency," and "Minimization."

1. Protection: To preserve the freedom of the user and to keep the balance in the system, a censorship-resistant, independent, and force-resilient algorithm needs to be run in a decentralised manner.
2. Minimization: This law describes that the closure of credentials should be as minimal as possible. The minimum disclosure protects the privacy of the user. The proposed framework uses zero-knowledge proofs (ZKP) as verifiable credentials presentation. The ZKP allows cryptographically proven claims without sharing the actual information. The claims and proofs are present on the identity ledger, where a verifier can verify the specific claims.
3. Persistence: The lifetime of the digital address of identity/identifiers should be under the user's control. The identifier should exist till the user wants it. In the proposed system, the revocation of the DID is covered, which fulfils this requirement.
4. One further principle could be privacy-preserving. Even though this is already partly integrated without explicitly saying it, the privacy-preserving design of services plays a key role in Self-Sovereign Identity. Reselling user-related information is a large business on the internet.

Portability

1. Access: Access to the user's identity should be accessible to the user at any time. No intermediaries should prevent the user from accessing their identity. The distribution and access of data or identity should be accessible to the authorized parties only. In the proposed framework, only public information is available on the ledger, and the stakeholders have their personal information (or PII) on local storage. An Access Control List (ACL) is also designed on blockchain to prevent unauthorized access. If any party (doctor, hospital, pharmaceuticals, and other stakeholders) wants to access others' information, they must first authenticate themselves in the system.
2. Transparency: The identity system must be transparent to each stakeholder. This leads to high trust and continuous improvement. Further, the participants can control the actions of each other and prevent and detect malicious actions from happening. The proposed framework is designed on a blockchain distributed network. Blockchain is the solution for transparency and trust.
3. Interoperability: The identities should be usable for many services; they should not be limited to a single service.

We proposed SSI for smart healthcare, and smart healthcare has many services with a large number of stakeholders. The identity information should be accessible by services in a standardized way. This should be created.

6.2. Security Analysis of SSI-SHS

In the proposed solution, private information does not store on blockchain. The PII is stored within the wallet (mobile wallet in implantation) in encrypted mode, and the wallet is secured using a fingerprint (or PIN or biometric feature the vendor provides). The proposed system defines the requirement of a secure interface while accessing the wallet and considers that vendors provide their implementation, so the security analysis of these implementations is generally challenging. Here, we suppose the vendor provides a secure way to protect the wallet and its data.

Even though the data in rest compiled the privacy of the design concept, at some point, the private information must be processed by the service provider's server. At the time of personal information processing, data privacy depends on the trust level of implementation and the segregation of the component. This implies that the highest level of privacy may only be achieved if a separate organisationoperates each component and there is a process set up to ensure the implementation does not store data that are supposed to be transient. Both of these may be done by regular organizational audits.

The security analysis of the issuance and authentication process can tell the security strength of any system. In the proposed framework, the issuance of the entity's credentials and authentication is crucial. The analysis will be done for the security strength of both processes.

6.2.1. Issuance Process Security Analysis

The issuance of VCs to the user consists of many steps, starting from getting the DID from the user and passing the VC to the user storing these VCs in a secure wallet. The identity requestor/holder and issuer communicate via a secure communication channel. The following threats can happen while issuing the credentials:

- The attacker gets the exchanged data between the issuer and holder;
- Man in the middle attacks over DID communication;
- Key Exposure attack;
- DDO forgery attack.

In the proposed SSI-SHS framework, the communication between the issuer and the requestor is protected at several levels. The certificate pinning is where the public key is associated with its host and is recommended and used in a prototype implementation. The framework considers the exchange of critical encrypted data and is implemented in the prototype. The certificate pinning prevents the man-in-middle attacks.

Most of the attacks are rooted in key vulnerability exposure attacks. In the proposed system, the keys, VC, and secret links are secured elements (TEE). Moreover, the keys are maintained through DKMS. This method can prevent key exposure.

Further, the wallet stores all data in encrypted form. For the DDO forgery, the attacker needs a trapdoor key with the help of the key exposure technique. Moreover, we have already discussed that the proposed framework is resistant to key exposure, so DDO forgery can also be prevented using the proposed framework.

6.2.2. Authentication Process Security Analysis:

The authentication process can have the following attacks:

- Wallet attacks;
- Man in the middle attacks on DID communication;
- By passing an authentication attack.

All wallets at the user's end need strong encryption. The encryption algorithm must be strong and searchable, and it should not depend on the storage technology of data in the wallet. The encryption technique should be able to hide the data pattern in the encrypted data and rotate the key to protect the wallet without having to re-encrypt the whole data. This would not be possible with the help of a trivial encryption algorithm. Some strong encryption, like Ethereum, uses AES-128-CTR cipher with decrypt and MAC, and Indy wallet uses HMAC-SHA256 and other identity wallets implemented while keeping the above two requirements. We used a sovrin wallet to store the keys and PII of the user in a prototype implementation.

Two situations may threaten user personal information: the data are disclosed to the attacker during the authentication process, or the VC discloses more that the user consented to share. A common strategy mitigates the first threat by the solution: mandatory TLS usage with optional certificate pinning with additional application-level encryption. The second threat relies on Mobile Wallet implementation: it must ensure no data leaves the Mobile Wallet without the user's consent. The proposed framework considers that each function call to obtain data has an exact authorization procedure associated, and the authorization is applied to the whole identity and each item separately: the user enables access to his identity, but to access personal information, an additional step is needed. The verification processes differ for each VC technology. During the verification process of basic PKI credentials, personal information is exchanged as attributes linked to the VC after VC verification. During verification, data are exchanged. From this point of view, the verification approach is more privacy-friendly and provides a much better solution.

6.3. Performance Analysis:

Execution Time Analysis

We first run the primary operation of the SSI-SHS network, i.e., registration of new stakeholders, registration of IoMT devices with a patient, authentication of an entity on the network, credentials issuance, and credentials verification. We recorded the execution time of these primary operations to evaluate the performance of the proposed identity model. We analysed the model in terms of execution time with 50 peers and ten medical devices. We put data on medical devices directly (not real-time monitoring). The execution time depends on many factors like connectivity, hardware, and program complexity, so the execution time varies from network to network. Here, we captured the data as our local machine.

Figure 6a presents the time taken in the registration of a stakeholder and the medical device of a patient. It is clear from Figure 6a,b that IoMT device registration and authentication take more time than the registration and authentication of stakeholders. The reason is that the IoMT device lacks communication power and energy. Furthermore, device registration and authentication include smart contracts, unlike the direct registration and authentication of stakeholders. Figure 6b presents the authentication time for stakeholders and medical devices.

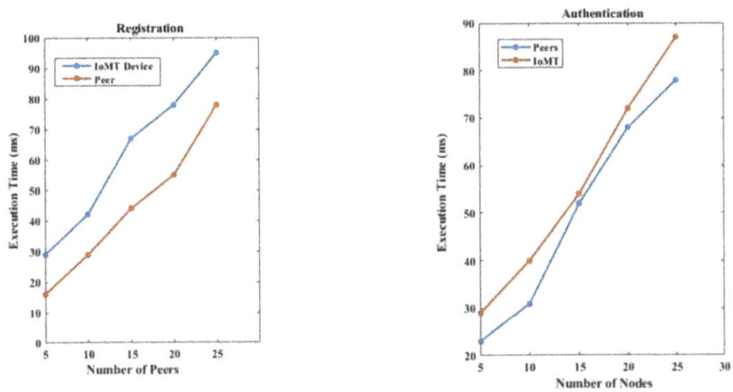

(a) (b)

Figure 6. (**a**) Registration and (**b**) authentication time of stakeholders and IoMT devices.

Further, we recorded the registration and authentication time with a varying number of transactions (50, 100, 150 txn) in the network and drew it into Figure 7a,b. Figure 7a presents the registration time when the number of transactions increases on a network scale, and Figure 7b presents the authentication time when the number of transactions increases on a network scale. The scalability of the network depends on the infrastructure resources used in the implementation. However, as the number of participants increases in the network, the number of transactions used in both the registration and authentication process increases. If we use high computation and large storage sources, then increased transactions can be handled in a timely manner.

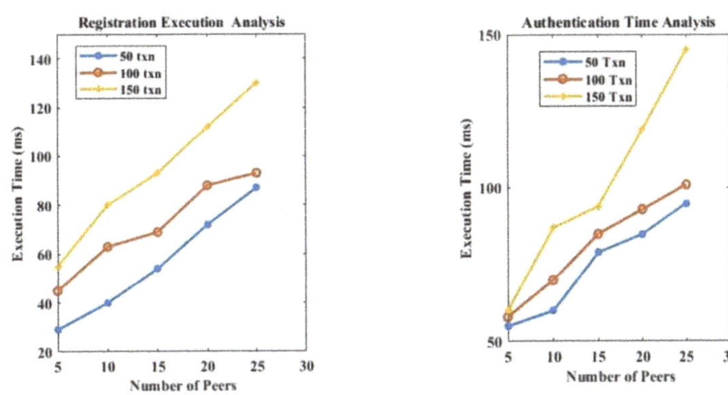

(a) (b)

Figure 7. (**a**) Registration time on network scale; (**b**) authentication time on network scale.

Figure 8 presents the time analysis and the contract deployment analysis for varying numbers of transactions and peers, and Figure 9 presents the execution time analysis of off-chain storage. The smart contract is deployed on the blockchain, and nodes trigger queries through smart contracts for different functionalities. As the network increased, the smart contract queries also increased. As shown in Figure 8, the execution time of a smart contract is increased with the number of nodes participating in the network. Further, the transactions storage also increases with the number of nodes in the network, as presented in Figure 9.

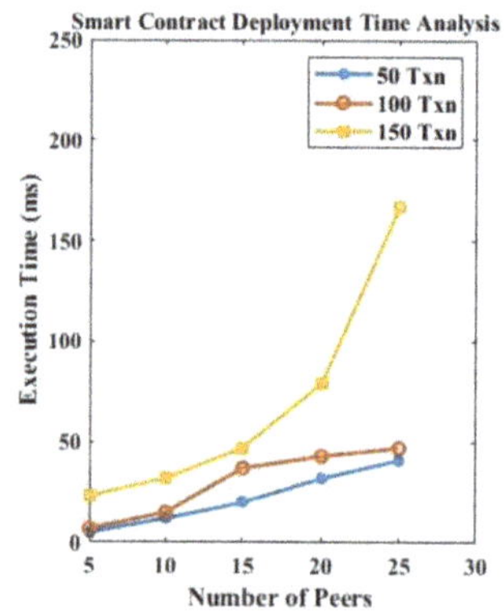

Figure 8. Contract deployment analysis.

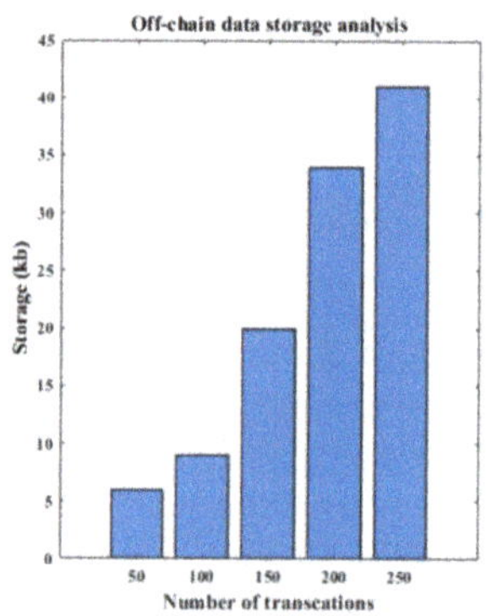

Figure 9. Execution time analysis of off-chain storage.

We compare our proposed identity model performance with the existing model proposed by Xueping Liang et al. [24] and Rafael Belchior et al. [30]. Our SSI-SHS provides better performance, as presented in Figure 10. SSI-SHS is better in both phases, i.e., registration and authentication. Further, we also include the authentication and registration of IoMT devices, which makes our approach much more robust to security issues.

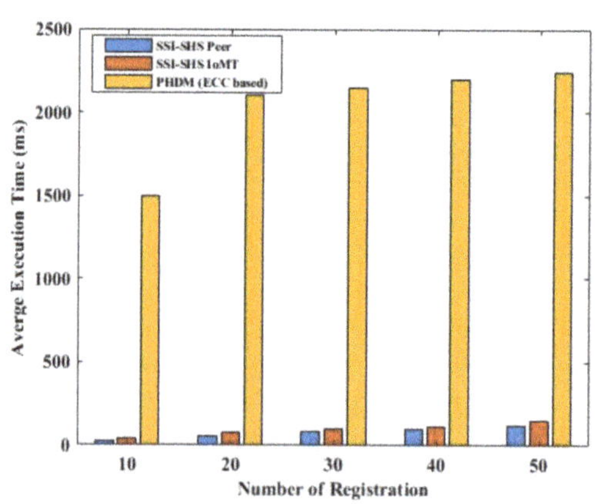

Figure 10. Performance comparison.

7. Conclusions

The SSI system is proposed for smart healthcare (SSI-SHS) to protect the user's privacy and give the owner control of health data. Smart contracts are designed to request the user's data and provide IoMT data access to other trusted parties with a time limit. First, we designed a distributed network of all stakeholders of smart healthcare on a permissioned blockchain to limit the participants at the healthcare level, which provides application-level security. In the proposed identity system, the SSI-SHS and IoMT-SSI are connected through the common participants in both networks, like patients. The IoMT-SSI manages the identity of IoMT devices through the device owner rather than the device's manufacturer. The device owner controls their health data even if the data is gathered via some smart medical devices. Further, the results and analysis show that SSI-SHS complies with the identity guidelines proposed by Allen and Cameron. For future research, the identity model can be expanded to make it interoperable with other smart parts of a smart city. For example, smart traffic management in smart cities can be connected with smart healthcare to reduce the insurance claim process time and prevent fraud in the health insurance process.

Author Contributions: Conceptualisation P.B.; Formal analysis G.A.; Investigation, P.B.; Methodology, P.B.; Resources, M.M., O.K.; Supervision, S.K., M.M., O.K., J.L.; Validation, G.A.; Writing, P.B.; Review and Editing, G.A., O.K., M.M. All authors have read and agreed to the published version of the manuscript.

Funding: This research is supported by the B11 unit of assessment, Center for Computing and Informatics Research Center, Department of Computer Science, Nottingham Trent University, UK.

Institutional Review Board Statement: Not applicable.

Informed Consent Statement: Not applicable.

Data Availability Statement: Research data will be available on individual requests to the corresponding author considering collaboration possibilities with the researcher or research team and with restrictions that the data will be used only for further research in the related literature progress.

Acknowledgments: The research work is scientifically supported by Nottingham Trent University, UK.

Conflicts of Interest: The authors declare no conflict of interest.

References

1. Rashid, M.M.; Choi, P.; Lee, S.-H.; Kwon, K.-R. Block-HPCT: Blockchain Enabled Digital Health Passports and Contact Tracing of Infectious Diseases like COVID-19. *Sensors* **2022**, *22*, 4256. [CrossRef] [PubMed]
2. McGhin, T.; Choo, K.K.R.; Liu, C.Z.; He, D. Blockchain in healthcare applications: Research challenges and opportunities. *J. Netw. Comput. Appl.* **2019**, *135*, 62–75. [CrossRef]
3. World Bank. *The Role of Digital Identification for Healthcare: The Emerging Use Cases*; World Bank: Washington, DC, USA, 2018.
4. Ullah, F.; Abdullah, A.H.; Kaiwartya, O.; Lloret, J.; Arshad, M.M. EETP-MAC: Energy efficient traffic prioritization for medium access control in wireless body area networks. *Telecommun. Syst.* **2020**, *75*, 181–203. [CrossRef]
5. Baker, S.B.; Xiang, W.; Atkinson, I. Internet of things for smart healthcare: Technologies, challenges, and opportunities. *IEEE Access* **2017**, *5*, 26521–26544. [CrossRef]
6. Windley, P.J. *Digital Identity: Unmasking Identity Management Architecture (IMA)*; O'Reilly Media, Inc.: Sebastopol, CA, USA, 2005.
7. Dunphy, P.; Petitcolas, F.A. A first look at identity management schemes on the blockchain. *IEEE Secur. Priv.* **2018**, *16*, 20–29. [CrossRef]
8. Ferdous, M.S.; Chowdhury, F.; Alassafi, M.O. In search of self-sovereign identity leveraging blockchain technology. *IEEE Access* **2019**, *7*, 103059–103079. [CrossRef]
9. Kim, B.G.; Cho, Y.S.; Kim, S.H.; Kim, H.; Woo, S.S. A Security Analysis of Blockchain-Based Did Services. *IEEE Access* **2021**, *9*, 22894–22913. [CrossRef]
10. Marlinspike, M. What Is 'Sovereign Source Authority'? The Moxie Tongue. 2012. Available online: http://www.moxytongue.com/2012/02/what-is-sovereign-source-authority.html (accessed on 1 February 2022).
11. Deegan, P. *Open Mustard Seed (OMS) Framework*; ID3: Cambridge, MA, USA, 2013.
12. Allen, C. The Path to Self-Sovereign Identity. Life with Alacrity. 2016. Available online: http://www.lifewithalacrity.com/2016/04/the-path-to-self-sovereign-identity.html (accessed on 1 February 2022).
13. Preukschat, A.; Reed, D. *Self-Sovereign Identity: Decentralised Digital Identity and Verifiable Credentials*; Simon and Schuster: New York, NY, USA, 2021.

14. Cameron, K. The laws of identity. *Microsoft Corp* **2005**, *12*, 8–11.
15. Zhu, X.; Badr, Y. Identity management systems for the internet of things: A survey towards blockchain solutions. *Sensors* **2018**, *18*, 4215. [CrossRef]
16. Kuperberg, M. Blockchain-based identity management: A survey from the enterprise and ecosystem perspective. *IEEE Trans. Eng. Manag.* **2019**, *67*, 1008–1027. [CrossRef]
17. Bouras, M.A.; Lu, Q.; Zhang, F.; Wan, Y.; Zhang, T.; Ning, H. Distributed ledger technology for eHealth identity privacy: State of the art and future perspective. *Sensors* **2020**, *20*, 483. [CrossRef] [PubMed]
18. Zhang, L.; Zhang, Y.; Tang, S.; Luo, H. Privacy protection for e-health systems by means of dynamic authentication and three-factor key agreement. *IEEE Trans. Ind. Electron.* **2018**, *65*, 2795–2805. [CrossRef]
19. Aghili, S.F.; Mala, H.; Shojafar, M.; Peris-Lopez, P. LACO: Lightweight three-factor authentication, access control and ownership transfer scheme for e-health systems in IoT. *Future Gener. Comput. Syst.* **2019**, *96*, 410–424. [CrossRef]
20. Yang, Y.; Zheng, X.; Guo, W.; Liu, X.; Chang, V. Privacy-preserving smart IoT-based healthcare big data storage and self-adaptive access control system. *Inf. Sci.* **2019**, *479*, 567–592. [CrossRef]
21. Zhang, Y.; Zheng, D.; Deng, R.H. Security and privacy in smart health: Efficient policy-hiding attribute-based access control. *IEEE Internet Things J.* **2018**, *5*, 2130–2145. [CrossRef]
22. Houtan, B.; Hafid, A.S.; Makrakis, D. A survey on blockchain-based self-sovereign patient identity in healthcare. *IEEE Access* **2020**, *8*, 90478–90494. [CrossRef]
23. Augot, D.; Chabanne, H.; Chenevier, T.; George, W.; Lambert, L. A user-centric system for verified identities on the bitcoin blockchain. In *Data Privacy Management, Cryptocurrencies and Blockchain Technology*; Springer: Cham, Switzerland, 2017; pp. 390–407.
24. Liang, X.; Shetty, S.; Zhao, J.; Bowden, D.; Li, D.; Liu, J. Towards Decentralised Accountability and Self-Sovereignty in Healthcare Systems. In Proceedings of the International Conference on Information and Communications Security, Beijing, China, 6–8 December 2017; Springer: Cham, Switzerland, 2017; pp. 387–398.
25. Au, R.; Croll, P. Consumer-Centric and Privacy-Preserving Identity Management for Distributed e-Health Systems. In Proceedings of the 41st Annual Hawaii International Conference on System Sciences (HICSS 2008), Waikoloa, HI, USA, 7–10 January 2008; IEEE: Piscataway, NJ, USA, 2008; p. 234.
26. Shuaib, M.; Alam, S.; Alam, M.S.; Nasir, M.S. Self-sovereign identity for healthcare using blockchain. *Mater. Today Proc.* 2021; *in press*. [CrossRef]
27. Mikula, T.; Jacobsen, R.H. Identity and Access Management with Blockchain in Electronic Healthcare Records. In Proceedings of the 2018 21st Euromicro Conference on Digital System Design (DSD), Prague, Czech Republic, 29–31 August 2018; IEEE: Piscataway, NJ, USA, 2018; pp. 699–706.
28. Zhou, J.; Tang, F.; Zhu, H.; Nan, N.; Zhou, Z. Distributed Data Vending on Blockchain. In Proceedings of the 2018 IEEE International Conference on Internet of Things (iThings) and IEEE Green Computing and Communications (GreenCom) and IEEE Cyber, Physical and Social Computing (CPSCom) and IEEE Smart Data (SmartData), Halifax, NS, Canada, 30 July–3 August 2018; IEEE: Piscataway, NJ, USA, 2018; pp. 1100–1107.
29. Buzachis, A.; Celesti, A.; Fazio, M.; Villari, M. On the Design of a Blockchain-as-a-Service-Based Health Information Exchange (Baas-Hie) System for Patient Monitoring. In Proceedings of the 2019 IEEE Symposium on Computers and Communications (ISCC), Barcelona, Spain, 29 June–3 July 2019; IEEE: Piscataway, NJ, USA, 2019; pp. 1–6.
30. Belchior, R.; Putz, B.; Pernul, G.; Correia, M.; Vasconcelos, A.; Guerreiro, S. SSIBAC: Self-Sovereign Identity Based Access Control. In Proceedings of the 2020 IEEE 19th International Conference on Trust, Security and Privacy in Computing and Communications (TrustCom), Guangzhou, China, 29 December 2020–1 January 2021; IEEE: Piscataway, NJ, USA, 2020; pp. 1935–1943.
31. Neelakandan, S.; Rene Beulah, J.; Prathiba, L.; Murthy, G.L.N.; Irudaya Raj, E.F.; Arulkumar, N. Blockchain with deep learning-enabled secure healthcare data transmission and diagnostic model. *Int. J. Model. Simul. Sci. Comput.* **2022**, *11*, 2241006. [CrossRef]
32. Kamalraj, R.; Neelakandan, S.; Kumar, M.R.; Rao, V.C.S.; Anand, R.; Singh, H. Interpretable filter based convolutional neural network (IF-CNN) for glucose prediction and classification using PD-SS algorithm. *Measurement* **2021**, *183*, 109804. [CrossRef]
33. Harshavardhan, A.; Boyapati, P.; Neelakandan, S.; Abdul-Rasheed Akeji, A.A.; Singh Pundir, A.K.; Walia, R. LSGDM with Biogeography-Based Optimization (BBO) Model for Healthcare Applications. *J. Healthc. Eng.* **2022**, *2022*, 2170839. [CrossRef]
34. Khovratovich, D.; Law, J. Sovrin: Digital identities in the blockchain era. *Github Commit Jasonalaw* **2017**, *17*, 38–99. Available online: https://sovrin.org/wp-content/uploads/AnonCred-RWC.pdf (accessed on 1 February 2022).
35. Verma, G.K.; Singh, B.B.; Kumar, N.; Kaiwartya, O.; Obaidat, M.S. PFCBAS: Pairing free and provable certificate-based aggregate signature scheme for the e-healthcare monitoring system. *IEEE Syst. J.* **2019**, *14*, 1704–1715. [CrossRef]
36. Kaiwartya, O.; Kumar, S. Cache agent-based geocasting in VANETs. *Int. J. Inf. Commun. Technol.* **2015**, *7*, 562–584. [CrossRef]
37. Cao, Y.; Kaiwartya, O.; Aslam, N.; Han, C.; Zhang, X.; Zhuang, Y.; Dianati, M. A trajectory-driven opportunistic routing protocol for VCPS. *IEEE Trans. Aerosp. Electron. Syst.* **2018**, *54*, 2628–2642. [CrossRef]
38. Kaiwartya, O.; Kumar, S. Geocast Routing: Recent Advances and Future Challenges in Vehicular Adhoc Networks. In Proceedings of the 2014 International Conference on Signal Processing and Integrated Networks (SPIN), Noida, India, 20–21 February 2014; IEEE: Piscataway, NJ, USA, 2014; pp. 291–296.

39. Kumar, S.; Singh, K.; Kumar, S.; Kaiwartya, O.; Cao, Y.; Zhou, H. Delimitated anti jammer scheme for Internet of vehicle: Machine learning based security approach. *IEEE Access* **2019**, *7*, 113311–113323. [CrossRef]
40. Khatri, A.; Kumar, S.; Kaiwartya, O.; Aslam, N.; Meena, N.; Abdullah, A.H. Towards green computing in wireless sensor networks: Controlled mobility–aided balanced tree approach. *Int. J. Commun. Syst.* **2018**, *31*, e3463. [CrossRef]
41. Kumar, S.; Kaiwartya, O.; Abdullah, A.H. Green computing for wireless sensor networks: Optimization and Huffman coding approach. *Peer-to-Peer Netw. Appl.* **2017**, *10*, 592–609.
42. Kaiwartya, O.; Kumar, S. Enhanced Caching for Geocast Routing in Vehicular Ad Hoc Network. In *Intelligent Computing, Networking, and Informatics*; Springer: New Delhi, India, 2014; pp. 213–220.

sensors

Review

A Smarter Health through the Internet of Surgical Things

Francesk Mulita [1,2,*], Georgios-Ioannis Verras [2], Christos-Nikolaos Anagnostopoulos [1] and Konstantinos Kotis [1,*]

[1] Intelligent Systems Lab, Department of Cultural Technology and Communication, University of the Aegean, 81100 Mytilene, Greece; canag@aegean.gr
[2] Department of Surgery, General University Hospital of Patras, 26504 Rio, Greece; georgiosverras@gmail.com
* Correspondence: cti21001@ct.aegean.gr (F.M.); kotis@aegean.gr (K.K.); Tel.: +30-6974822712 (K.K.)

Abstract: (1) Background: In the last few years, technological developments in the surgical field have been rapid and are continuously evolving. One of the most revolutionizing breakthroughs was the introduction of the IoT concept within surgical practice. Our systematic review aims to summarize the most important studies evaluating the IoT concept within surgical practice, focusing on Telesurgery and surgical Telementoring. (2) Methods: We conducted a systematic review of the current literature, focusing on the Internet of Surgical Things in Telesurgery and Telementoring. Forty-eight (48) studies were included in this review. As secondary research questions, we also included brief overviews of the use of IoT in image-guided surgery, and patient Telemonitoring, by systematically analyzing fourteen (14) and nineteen (19) studies, respectively. (3) Results: Data from 219 patients and 757 healthcare professionals were quantitively analyzed. Study designs were primarily observational or based on model development. Palpable advantages from the IoT incorporation mainly include less surgical hours, accessibility to high quality treatment, and safer and more effective surgical education. Despite the described technological advances, and proposed benefits of the systems presented, there are still identifiable gaps in the literature that need to be further explored in a systematic manner. (4) Conclusions: The use of the IoT concept within the surgery domain is a widely incorporated but less investigated concept. Advantages have become palpable over the past decade, yet further research is warranted.

Keywords: smart health; IoT; surgical practice; Internet of Surgical Things

Citation: Mulita, F.; Verras, G.-I.; Anagnostopoulos, C.-N.; Kotis, K. A Smarter Health through the Internet of Surgical Things. *Sensors* **2022**, *22*, 4577. https://doi.org/10.3390/s22124577

Academic Editors: Alexandru Lavric, Eugenio Santoro, Liliana Anchidin and Adrian I. Petrariu

Received: 3 April 2022
Accepted: 14 June 2022
Published: 17 June 2022

1. Introduction

Advances in surgical practice have been rapid and non-stop for the past few centuries. From the groundbreaking idea by Joseph Lister that postoperative deaths might be attributed to certain invisible pathogens that could be combated with antiseptic solutions, to the popularization of robotic surgery in many modern medical centers, progress has been continuous. Many related advances came in the form of technological milestones that changed the shape of surgical practice as we know it today. The latest in a series of breakthroughs in the smart health domain is the utilization of the Internet in everyday surgical practice, the role of which is ever-expanding.

To systematically study the technological advances in a particular sector, attributed to the utilization of the Internet, the term "Internet of Things" (IoT) was introduced. While not strictly defined, IoT describes a network of Internet-based connected things equipped with (embedded) sensing and actuating devices, with data production, processing, and consumption abilities. The utilization of the Internet and IoT in medical practice can take many shapes and forms. Ranging from the awe-inspiring telesurgical procedures [1,2] to complex AI machine learning applications that aid in medical decision making [3], to a simple email containing a preoperative CT scan, the Internet of Surgical Things (IoST) is here to stay. A representative example of the IoST is a smart ingestible sensor (pill) that is activated after being swallowed [4], "travels" in the body through the colon and sends data to outer devices such as computers and smartphones when it detects a threat for cancer. Such a device can be used instead of colonoscopy for people who cannot obtain a colonoscopy due to psychological and physiological problems. In the broader aspect of the IoST context, IoST entities are different types of connected entities that "live" in this smart setting, such as surgical things (e.g., a connected surgery tool) [5], organs (e.g., a connected colon, an artificial heart) [6], humans (e.g., a connected patient or doctor) [7], smart devices [8] (e.g., a connected heartbeat monitoring device, a smart pill, an ingestible sensor), services [3] (e.g., a connected telemonitoring service), data (e.g., a connected data steam of heart monitoring data), etc. In this extensive survey, we aim to present an overview of current uses of IoT-embedded surgical practice by focusing on Telesurgery and Surgical Telementoring. As a secondary research question, we also briefly review the latest advances in IoT-associated image guided surgery, and surgical patient telemonitoring by utilizing the IoT paradigm.

Current literature on the Internet of Medical Things (IoMT) includes a multitude of heterogenous reports of Internet-based applications within the medical/healthcare domain. Most of the related articles present network(s) of sensor arrays and data processing stations, with or without actuating devices, interconnected via the Internet infrastructure. There is, however, a lack of a systematic approach within the existing literature as to how the IoT concept has revolutionized the surgical world towards smarter health. To fill this gap in current knowledge, the present systematic literature review studies the Internet of Surgical Things (IoST) domain by focusing on the three most prominent areas of application: (a) image-guided surgery, (b) telesurgery and telementoring in surgery, and (c) surgical patient monitoring [3,5]. Between these three areas of applications within the surgical discipline, the applications of Telesurgery and Surgical Telementoring are undoubtedly the most influenced by the IoMT concept. Therefore, our systematic review will be centered around studies of these applications. Additionally, we will briefly discuss the current literature on patient monitoring and IoT applications in image-guidance.

In this review, we systematically discuss a novel concept (the IoST) that is being rapidly incorporated into surgical practice. This is one of the first systematic review manuscripts covering this area. By focusing on the three applications described above, we provide a thorough understanding of the feasibility and effectiveness of different IoST applications. In addition, this paper also summarizes known and emerging weak points in IoST ecosystems that should be the focus of future research efforts. Finally, by incorporating data from system development studies, we offer insight into promising future uses of the IoST that are yet to be popularized but have the potential to be groundbreaking advances.

While there is a lack of a universally accepted definition for the Internet of Things, for the purposes of this review, we considered studies looking into ecosystems of interconnected computing devices, digital screens, sensor-bearing instruments, robotic surgery systems, mobile phones, 5th generation mobile networks, and can even include people (in our case either operators or patients. We have focused our attention on systems operating within the surgical world, connecting two or more "Things" (as defined above), via a certain, defined, and dedicated network. Looking into our specified research questions, some examples of IoT applications would be, for instance, a network comprising a preoperative imaging modality (e.g., MRI scanner), a processing station, and software within a specialized robot or specialized augmented reality glasses that ultimately aim to facilitate a procedure by superimposing real-time image guidance. Telementoring/Telesurgery systems usually comprise specialized working stations connected to a user that are capable of transmitting audiovisual cues and/or controlling a surgical robot at a distant location. Finally, telemonitoring is carried out by an interconnected series of sensor-bearing devices that centripetally transmit patient data, either directly to the physician, or to a dedicated data-gathering station.

The structure of the paper is as follows: Section 2 presents the research methodology, Section 3 presents an overview of the main results/findings of this survey, Section 4 discusses open issues and challenges, and, finally, Section 5 concludes the paper.

2. Materials and Methods

To meticulously look through the current literature concerning IoT in surgical practice, we have formed one primary and two secondary research questions that allow us to produce a thorough literature review. These research questions are as follows:

Primary:

1. What is the latest experience of telesurgery and surgical telementoring with regard to the IoT concept? (See Supplemental Figure S1).

Secondary:

1. What are the current applications of IoT technology in image-guided surgery? (See Supplemental Figure S2).

2. How can the IoT be utilized for patient monitoring outside the operating room? (See Supplemental Figure S3).

The presented literature review was conducted using the PubMed (Medline) and Web of Science (Clarivate) directories. The search queries for each research question as well as universal inclusion and exclusion criteria are presented in Table 1. We have used relevant keywords/phrases that characterize the articles of interest, for each of the research questions. To collect as many published articles as possible, we have incorporated additional articles from less specific queries, which can be viewed in Table 1. This was necessary mainly due to the lack of terms "Internet" or "Internet of Things" within the related keywords. Additionally, due to most of the IoT-related applications involving specialized sensor data inputs that lead to data-driven action, we have added a separate relevant query for all questions, thus capturing any surgical-specific system that would have been missed by the other search queries. The PRISMA flowchart of this additional search is presented in Supplemental Figure S4.

Table 1. Search Queries.

Research Question	Queries Used in Medline	Queries Used in Web of Science
What are the current applications of IoT technology in image-guided surgery?	(image-guided surgery) and ((internet of things) or (internet))	(ALL = (image guided surgery) and (ALL = (internet of things) or ALL = (internet)))
What is the latest experience of telesurgery and surgical telementoring with regard to the IoT concept?	(telesurgery) and ((internet of things) or (internet)) + (telementoring) and (surgery) + (telesurgery)	(ALL = (telesurgery)) and ALL = (internet) + (ALL = (telementoring)) and ALL = (surgery) + ALL = (telesurgery)
How can the IoT network be utilized for patient monitoring outside the operating room?	((telemonitoring) and (surgery)) and (internet) + (surgery) and (internet of things)	((ALL = (telemonitoring)) and ALL = (surgery)) and ALL = (internet)
Supplemental Query for all Research Questions	(sensors) and (surgery) and ((internet) or (internet of things))	((ALL = (sensors)) and ALL = (surgery) and (ALL = (internet) or ALL = (Internet of Things)))
	Inclusion and Exclusion Criteria	
Inclusion Criteria	Articles that described clinical or feasibility studies of modalities incorporating the IoT framework. Articles that described a system in development, were able to demonstrate potential real-life application.	
Exclusion Criteria	Case reports of previously known modalities. Literature reviews. Opinion or Editorial articles. Lack of clarity regarding the utilization of an internet-based network. Articles focusing on technical developments rather than surgery-oriented application potential. Unavailability of text. Articles solely in non-English languages. Articles from pre-print servers or online-only publication platforms.	

The presented review includes articles dating from 2010 to 2021. Through these years, the development of the Internet was constant, and, as a direct consequence, so was the incurrence of IoST-related publications. We have decided to include all relevant publications, even when reported methods seemed outdated for today's standards (e.g., the use of dedicated land connections). By doing so, we have collected all reports of IoST-related concepts and showcased that the idea of an ecosystem of computing devices transferring data without the need for human interaction was present in the surgical sciences and practice long before high-speed or wireless Internet connections gave birth to IoT. Chronologically, we incorporated studies from 2010 onwards. This decision regarding the period was made largely to the fact that the IoT is a novel concept, mainly developed within the past decade.

Our search for related literature included articles published in peer-reviewed journals, as well as published presentations from conferences organized by notable scientific societies with a documented history of IoT expertise. We have included articles describing the clinical applications of the IoT concept specifically for surgical practice. Study designs that were considered for inclusion within the systematic review included clinical trials, case series, and animal trials. Several feasibility and prototype model studies were also incorporated if they presented clearly defined and well-supported potential for clinical application.

The exclusion criteria of the survey concern the following article types: opinion or editorial articles, articles presenting a singular case report of previously a published methodology, and literature reviews of a subject. Such articles were deemed as low-quality articles that would not aid in forming a subjective overview of the IoST. We also decided to exclude articles that failed to indicate the presence of Internet-based interconnected devices. Articles that focused on technical developments rather than clinical applications (e.g., latency-lowering methods, advances in the security of IoT networks, advances in specific hardware or software components of a network, etc.) were also not included due to the clinically oriented nature of the review. Lastly, we excluded articles that were published in non-peer-reviewed platforms such as several online-only journals or pre-print platforms, due to concerns regarding their methodological quality.

The reference management tool EndNote was used for detecting duplicates. The PRISMA flowchart for each research question was developed, as can be seen in the corresponding Figures S1–S4. In the first step of the selection process, our research team excluded detected duplicate entries and articles that were clearly labeled as one of the study designs incorporated within the exclusion criteria. Screening of the records in the second step of the elimination process was carried out by revising available abstracts for any obvious signs of mismatching with the inclusion criteria (technically oriented studies, singular case reports). All studies deemed appropriate for further evaluation were sought for full-text retrieval. The fourth and final step of the review process included the evaluation of the full manuscripts, where the more specific exclusion criteria were applied (lack of clarity, unavailability of potential clinical use, etc.). This review process was the same for each of the three research questions.

3. Results

Following a literature search and a gradual article elimination process, a total of 48 studies were selected for reviewing regarding our primary research question (Supplemental Figure S1). Of these studies, 36 were observational studies conducted within a clinical environment, evaluating either patients or healthcare professionals. Twelve studies were classified as feasibility or system development studies. In total, the studies included in this review incorporated data from 219 patients, and 757 healthcare professionals, and concern studies that use IoT in Telesurgery and Telementoring. The utility and main findings of each study is highlighted in the corresponding column of the supplemental tables. The studies in the field of IoST are rather few, and seemingly characterized by a high degree of heterogeneity regarding variables, study designs, and outcome types; therefore, it is not easy to identify a formal and systematic way to synthesize and report a cumulative and quantitative outcome. In that respect, in this paper we present a narrative review of the IoT in Telesurgery and Surgical Telementoring.

Regarding our secondary research questions on Image-Guided Surgery (IGS), and surgical patient Telemonitoring, we systematically gathered 14 and 19 studies, respectively. Studies regarding the IoT in IGS incorporated experimental data from 111 patients and 35 patient models in total. Of the included studies, 7 were clinically based, and another seven were system development/feasibility studies. As for the IoT in surgical patient Telemonitoring, of the 19 studies, 15 were clinical studies, and 4 discussed newly developed IoT-based telemonitoring systems.

3.1. The Internet of Telesurgery and Surgical Telementoring

Perhaps the most impressive advancement the technological applications were able to provide in the surgical world is the incurrence of the telesurgical procedures. When discussing the prospects of IoT applications within surgical theaters, there are two key concepts in existence to keep in mind: telesurgery, and telementoring. While telesurgery describes the performance of a surgical operation by a remotely located surgeon, utilizing a network of devices with the aim of transferring the surgeon's haptic commands to a robotic surgery system, telementoring is the performance of live surgery on-site, with the live assistance of a more experienced surgeon, located off-site [9–25]. In order to achieve the latter, a network of connected cameras, microphones, screens and computers is necessary [25–31]. A fundamental outline of the network and crosstalk required is depicted in Figure 1. This connection is mostly provided through the Internet nowadays, and with new and emerging technologies, such as the fifth generation of mobile networks, implementation of teleservices is expected to grow (Supplemental Table S2). A 2017 systematic review of the literature on telementoring in the operating room revealed that operating times and complication rates were the same when compared to on-site live mentoring of younger surgeons [32]. This showcases that telementoring possibly has the potential to supplement live surgical training and might allow trainees to even surpass their educational goals when mentored remotely.

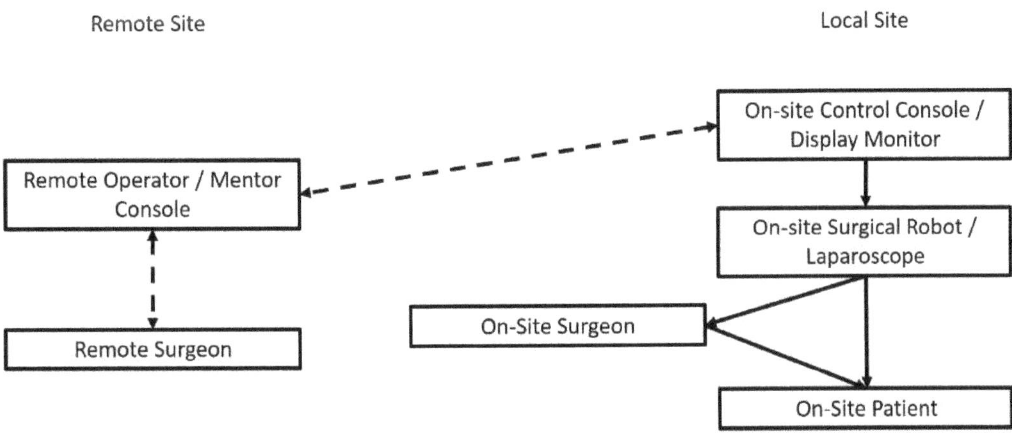

Figure 1. Connected IoST entities and workflow of an IoST-based Telementoring/Telesurgery System.

In addition, telementoring has the potential of long-distance consultation with senior and specialized surgeons who can provide live assistance in the operation rather than a simple preoperative consultation, or the costly option of transferring difficult to manage cases in specialized units [15–20]. Systems that are developed for surgical telementoring in live surgery can also be expanded with an array of biosensors for physiological parameter monitoring that is transmitted directly to the mentoring surgeon [27], allowing for the complete monitoring of the surgical patient, rather than just the surgical procedure. Studies evaluating the satisfaction outcomes of surgical trainees whenever live proctoring from a distance was used reveal universally high satisfaction rates [18–25] that were common to both mentors and mentees. Participants of said studies felt that telementorship sessions with more experienced surgeons would be a useful addition in surgical education curriculums [33–40]. Even when utilized for teleconferencing and simple audiovisual transmission of an operation for teaching purposes, participants rated positively to their experience more than 90% of the time [33,41–43]. This goes to show that the introduction of the Internet in the everyday surgical practice can have a pragmatic impact in the education of younger surgeons, particularly at low costs and without complex institutional requirements. Telementoring in surgery allows the safe transfer of knowledge and skill aptitude from a more experienced surgeon to a younger mentee.

Studies reporting telementoring techniques within the OR were split broadly in the two categories of the studied teaching method. The simplest form of surgical telementoring includes the live observation of a surgical procedure by a distant senior surgeon who can provide audiovisual guidance through an Internet connection, or a more advanced form of proctoring including the distant manipulation of a camera-bearing surgical robotic arm.

A study by Hinata et al. [28] compared in a systematic manner the perceived differences in live mentoring with telementoring in robotic surgery. Apart from requiring a stable and fast internet connection being the single drawback, all of the postoperative parameters of the surgical patients were the same between the differently mentored groups in almost all the included studies. In another comparative study of internet-based telementoring versus in-person telementoring, not only were the postoperative results the same between the two groups of mentees, but the mentors also showed a significant trend towards telementoring [27], while in some studies, mentees achieved better postoperative results when distance-mentored [43–48]. In a report on telementored trainees that performed bariatric surgery procedures, those that utilized Internet-based mentoring achieved fewer complications, shorter operative hours, as well as shorter hospitalization time [1]. In addition, Altieri et al. studied whether delivering a surgical skills course over the Internet would be less effective than live mentorship [29]. After the course was completed, the trainees were also found not to perform similarly in the post course assessment, but also showed non-inferior skill decline patterns for a few weeks after the course. These results prove that internet-based mentoring in surgery can be similarly effective in the long-term skill-honing process and is not confined in short-term positive outcomes. Additionally, they found that within every study describing an internet-based telementoring system, there were almost no unexpected intra-operative complications [47]. This proves that the use of the Internet as a mentoring tool can be a safe alternative, even when within a high-risk environment such as the operating room (OR).

With the incurrence of high-speed network connections, the handling of larger data transmissions at no expense of the perceived live mentoring was made possible. Utilizing such connections, several research teams were able to integrate AR within surgical telementoring. In these publications, authors describe the use of specialized AR glasses by the mentees [10,33–36,43–50]. The glasses were equipped with cameras providing live feeds of the surgical field that were transmitted through the Internet to a distant mentor. The mentoring station on the other end includes motion tracking sensors and cameras that capture the mentor's movements. These movements are superimposed on the mentee's field of view and on to the surgical field through the specialized glasses in order for the mentee to closely follow along. The trainees assigned to be mentored with this system achieved higher scores on operational evaluation and were prone to less mistakes [10,36,43–46,49–53]. In one clinical study, AR telementoring was compared with simple telementoring through audiovisual transmission and audio guidance [44,48]. People mentored with the AR system made less mistakes and were more accurate. However, in several similar studies, authors noted that mentees on such systems required more time to complete the simulated procedure [32,34,43,44]. Additionally, we must not forget the increased cost of such systems, as they require not only high-bandwidth internet connections, but sophisticated equipment for the mentor and mentee stations. Such equipment may not be readily available at most institutions. Andersen et al. [10] took the AR systems a step further by incorporating pre-recorded footage of similar operations that could be viewed directly onto the mentee's display. Operative results were encouraging; however, the individual variability in patients posed a drawback, since the pre-recorded operations could not account for variability.

More advanced telementoring systems include the ability of the remote mentor to assume control of part of the operating system, most commonly the laparoscope-containing arm of the surgical robot [40,42,45,47,48]. Advantages of this setup is the ability to provide direct feedback to the mentee and demonstrate the appropriate handling of the instruments, even though the mentor is located remotely. In our group of studies evaluating similar setups, all reported operations were carried out successfully by the mentored surgeons without intra-operative complications. Mentees also reported that they preferred the distant mentoring to on-site mentoring. In a study by Prince et al. on simulated surgical tasks [14], mentored surgeons that utilized the telementoring system who allowed for instrument control by the mentor scored higher on dexterity assessment tests than those who did not.

Telementoring with the use of Internet connections in surgical specialties, however, is not limited to the operating room. Authors have reported utilizing telementorship to organize skill stations, assess the work of mentee surgeons, broadcast live operations for teaching purposes, teach postgraduate courses, conduct virtual grand rounds, etc. [33–41,50–53]. In all of these activities, the mentees rated their experience almost universally positively. A recent endeavor by Greenberg et al. [39] evaluated a novel surgical skill simulation course that incorporated AR. Post-course evaluation revealed not only satisfied students, but increased aptitude as well. Suzuki et al., developed a surgical training system, using software that could provide an accurate virtual surgical case and robotic controllers that could be connected to the mentee's PC for a training session. In addition, the software, in addition to the patient models, was available as a cloud-based product, meaning that the exact same simulation can be accessed from anywhere in the world.

Indeed, distance mentoring with the use of the Internet also has certain drawbacks. Firstly, as it is often underlined by authors, all the systems described here require a stable, and, in many instances, high-bandwidth Internet connections. This is often an issue with smaller institutions, or institutions at developing countries. In addition, many of the systems included here, require much more than a webcam and a microphone. Complex setups for surgical telementoring often include robotic systems, expensive software, AR glasses, specialized simulation instruments and more. All the above can prove to be a substantial cost that reaches prohibitive status in certain healthcare systems. Trainees that make use of telementoring systems outside the operating room, have also reported that the experience, although positive, could never replace their presence within the OR [39]. In several studies, authors report one or more instance of technical difficulties with the remote mentoring systems, such as audio or video failure, latency times or interruption of connections [42]. Seeing that these studies are mostly dated prior to 2015, we can hypothesize that such difficulties were a result of inferior internet capabilities, an issue largely resolved in today's practice, where fast internet speeds are widely available. Scheduling of the mentoring sessions can also prove to be a minor issue when the mentor and mentee are in different time zones. However, this is not something that careful planning cannot address. Lastly, there is also an additional hindrance mentioned by authors in telementoring and telesurgery. This is none other than legal considerations that must be addressed. Although not expected to differ significantly than standard practice, there is still a gap of specific legislation around telemedicine in general that will need to be filled before telemedicine services are offered as a standard of care.

Telesurgery is closely related to another relatively recent advancement in the operating room: robotic surgery. In fact, robotic surgery (the next logical step to laparoscopic surgery) was crucial in planting the idea that if a surgeon can command a machine and perform a procedure from a few meters away, why not apply the same principle in larger distances, perhaps even transatlantic ones. In theory, by utilizing a complex internetworked system of cameras, video streaming, feedback data and data processing and transmission devices, the remotely situated surgeon is able to operate the robotic surgery system and provide real-life, quality surgical outcomes. In this scenario, the IoT concept is mainly applied in the collection, transmission and exploration of the collected data, as well as within the bidirectional transmission of signals. The data processed and transmitted include audio, video, and images. In some experimental systems, where remote control of robotic surgery modalities is involved, the data also include feedback related to the positioning of the robotic arms, the surgeon's hand positioning and movement. The major impending factor in achieving this, however, was none other than the delay, caused by relaying large amounts of data in a wireless manner over large distances. The first recorded instance of a long-distance surgical procedure was an experimental cholecystectomy performed by the team of Marescaux et al. on swine specimens [40]. In their experimental study, all operations were successful, and the measured latency between the command given by the remotely located surgeon, and the observed response was 155 ms, indicating future perspectives where such surgeries might be part of the everyday practice. The first validation that performing surgical procedures from a distance was feasible and safe came in 2002 from the same research team [47] that performed the first ever cholecystectomy on a real patient, utilizing a robot that was controlled overseas.

Since then, the telesurgery concept has come a long way, with one major milestone being added recently, namely, the incorporation of 5th generation mobile networks in surgical practice. Study groups evaluating the feasibility of remote-controlled robotic surgery systems have found that it is a realistic option for performing surgical operations, with no additional risk for the patient and comparable surgical outcomes [42–53]. The feasibility of remote surgery application is such that surgical teams have achieved successful operations with optimal postoperative results from a remote site of more than 3000 km away from the primary site where the operating suite is located. This means that patients are now able to receive specialized surgery without having to relocate to a specialized surgical center. Utilization of high-speed internet connections is usually in the form of fiber optic connections, or more recently 5G mobile networks, and allow minimal latency within surgical practice, meaning increased safety for the patient as the operator is confident of their movements within the surgical field [39]. Within the included studies, all of the authors report no intra-operative complications, in addition to no further time delays of the surgery. Therefore, we can only expect for direct telesurgery with remote handling of surgical robots to be further expanded in the future, making surgical care available for less developed areas in a safe and highly efficient manner, comparable to live surgery. However, before widespread implementation, key issues of telesurgical systems need to be addressed [41–50]. These include further limitation of latency between stations, addressing safety concerns regarding cyber security, as well as the newfound liability due to medical damage caused by remotely operated robotic surgery systems. Financial imbalances between populations and countries are perhaps the greatest obstacle to the popularization of long-distance surgery. However, as the years progress, the availability of stable and fast Internet connections is on the rise and the costs of medical infrastructure, while substantial, can be surpassed by the cost-effectiveness of such advances [48].

Authors have also compared the different options of Internet connections that were applied to telesurgical procedures, namely, land cable connection with satellite connection. No differences in operative parameters such as blood loss or total operational time were found, and the participants felt equally confident with both modalities. In one study, the satellite Internet connection produced significantly greater latency times between the operator and the surgical robot; however, they were not sufficient to constitute the operation as unsafe.

3.2. Image-Guided Surgery in the IoT Era

The basic principle of Image-Guided Surgery (IGS) is constituted by the utilization of a tracking device, alongside pre or even intra-operative imaging, to aid the surgeon in the spatial orientation during a surgical process. (Supplemental Table S1). Authors of relevant publications have used IoT networks to incorporate patient imaging, as well as preoperative planning, into the operating room.

The network-based sharing of data and the creation of a workflow through sequential data appraisals and data provision towards the next component makes IGS a prime example of an IoT application in surgery. An IGS system built around the IoT approach usually consists of input of preoperative imaging data of the surgical patient. These data are then used to make reconstructed models of the anatomical area of interest. Such models are then transmitted wirelessly to a modality of choice, ranging from augmented reality (AR) glasses to the viewing screen of a surgical robot [54–59]. A simplified schematic illustrating the workflow of these systems can be seen in Figure 2, encompassing the idea of the IoT concept defined as a network of inter-connected devices that process and exchange data.

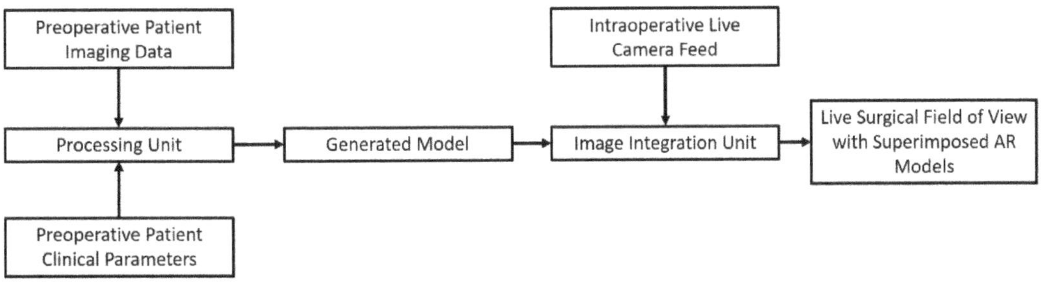

Figure 2. Connected IoST entities and workflow of an IoST-based Image-Guided Surgical System.

Current literature in IGS is heavily referred to as advances in the neurosurgical field. In this aspect, Internet-based IGS has a lot to offer by incorporating data from various sources and making them available to the surgeon in real-time. Augmented reality systems make use of the Internet to transmit preoperative renderings and patient imaging after registering these data with on-site patient images onto specialized smart glasses [54,55,60]. Coupling preoperative patient imaging with the image-guided system usually involves using certain sensors for the tracking of specific markers within the patient [55]. Studies on real-time surgery indicate that IGS has been very helpful in the identification and preservation of vital structures [58,59,61]. Almost all of the reports included here report that AR imaging systems allowed the preservation of vital structures in real-life patients [60–64], significant accuracy when utilized to assist in biopsies or electrode insertion [54–59] and real-time compensation for brain shift during neurological surgery [54], a novelty not achievable otherwise. In a study by Watanabe et al. [60], the intra-operative tablet devices, used to provide live camera views of the surgical field were also tracked in space by an overhead multi-angle camera system and special tracking spheres [60–64]. Internet use within the surgical suit enabled authors to utilize pre- and intra- operative patient imaging to create a live overlay of anatomical areas of interest (such as tumors or sensitive nearby structures) by connecting the preoperative image repository with an image processing software that returned the final image data either to a monitor or another specialized apparatus (e.g., smart glasses) [59].

Eftekhar et al. [63] took the AR integration within the surgical suite a step further by introducing a lesion-tracking smartphone app for mobile phones. The software will utilize the smartphone's camera to register anatomic landmarks of the patent's head surface that will be used to align the postoperative imaging with the live image. Therefore, we can see that the Internet can be used, not only as a "data highway" for short-term data handling, but as a connection to large servers, allowing direct access to a repository of patient data.

Related work by Guo et al. [8] and Li et al. [54] describe the integration of surgical robots within IGS systems. The coupling between the intra-operative MRI imaging, the surgical robot, the central processing unit and the end-display of the processed imaging data is achieved by utilizing wired or wireless Internet connections. Despite lacking trials on their performance in real-life surgery in humans, it is safe to say that the Internet in this case, has provided reliable interconnectivity with multiple appliances at once, and has the potential to become the unseen substrate of modern neurosurgical breakthroughs.

Ushimaru et al. [5] utilized RFID tracking tags placed on laparoscopic instruments in order to track usage in general, as well as the activation times of each tool. Data from the tags were transmitted onto RFID readers within the operating room. The RFID readers then transmitted the recorded data to specialized computer software that converted the electrical current readings to "on/off" indications and activation times. This experimental setup was successful in capturing the usage patterns of surgical instruments during cholecystectomy. This short proof of concept study opens the door towards incorporating network-based tracking systems within the everyday surgical practice that will capture surgical instrument use and possibly aid the surgeon in spatial navigation. Possible implications of this could include coupling a similar tracking system with an AI network that can aid in intra-operative, real-time decision making, or an in-hospital data accumulation system that will be able to measure surgical performance. A research group was recently able to combine the tracking of virtual laparoscopic instruments operated by a distant mentoring surgeon with the live feed image of the laparoscope and the real surgical instruments operated by a mentee [58]. Meier-Hein et al. also managed to construct a tracking algorithm for laparoscopic instruments by using sensor data and an Internet-based integration status. Their system was successful in producing an algorithm that could automatically detect and accurately annotate the laparoscopic surgical instruments, as well as their usage status in real-time operations.

On the other hand, the cost of implementation of such technological advances might prove to be restricting for certain institutions [54–58]. One of the biggest drawbacks of the systems described in the literature is that the vast majority of them report small patient cohorts and lack a direct comparison with other, more traditional surgical methods. Some authors even reported comparable accuracy when they compared novel IGS systems with their simpler, older counterparts [59]. Other reports that describe exciting new advances in the field of IGS, with promising results in simulations of surgical operations, lack a patient application altogether [8,60]. In that aspect, the precise effect the Internet has had on the evolution of IGS is harder to quantify. Still, we cannot deny that advances in the connectivity of devices achieved by an internet connection is a major component towards the modernization of medicine services in general, and surgery in particular.

3.3. *The Role of the IoT in Telemonitoring the Surgical Patient*

In order to complete our appraisal of the IoT concept in surgical specialties, we would be remiss if we did not include the potential uses outside the operating room (Supplemental Table S3). Medical telemonitoring usually consists of a specialized "smart" device that captures target parameters and transmits them through a wireless Internet connection, either directly to the referring physician, or to a centralized repository from which they can be accessed (Figure 3). The role of the Internet here is more straightforward: instead of being the network substrate that interconnects a variety of operating stations, data repositories and data processing modalities, here, it is used as a unidirectional "data highway" that runs towards the physician. In contrast to previous advances, telemonitoring has been widely implemented in some healthcare systems.

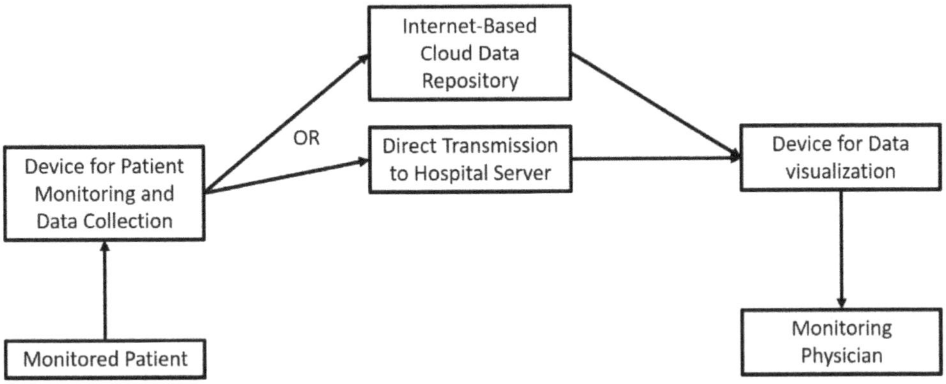

Figure 3. Connected IoST entities and workflow of an IoST-based Telemonitoring System.

Within the surgical patient subgroup, there have been a few clinical studies looking into the applicability of telemonitoring, usually in the postoperative period [3,65–80] (Supplemental Table S3). Patients enrolled in an at-home monitoring program after chest wall surgery were also monitored effectively by utilizing the Internet to input certain parameters in an online platform [65,67] Cardiac surgery patients were also studied in an IoST rehabilitation program that included wearable biomedical and motion tracking sensors [68]. The physician was therefore able to monitor the patients' activity levels and their performance in rehabilitation exercises. In a 2021 study by Cos et al., patients scheduled to undergo pancreatic surgery were monitored preoperatively by using a wearable smart device that was able to record heart rate, activity status, etc., and through an internet connection, transmit them to a central server. Not only did patients adhere to this novel concept, but the data that were automatically collected were of such quality that the research team developed an accurate predictive model for postoperative outcomes [3]. Biosensor-based systems are able to wirelessly transmit data on physiological parameters of the patients in order to assist with postoperative monitoring. Authors have reported the incorporation of pulse rate, blood pressure and activity tracking sensors as being successful in monitoring the rehabilitation process of surgical patients [69–77]. Reported advantages, include successful vital signs readings, short training period of nurses and patients alike, less unplanned office visits, and predicting unplanned postoperative complications by indirect monitoring of vital signs [8]. Kim et al. successfully developed a Doppler cuff that could be remotely monitored, allowing remote monitoring of the blood flow of skin flaps. This resulted in superior graft survivability rates [6]. Results such as these are indicative that we have come to a point at which smart devices with Internet connections can provide fast and accurate measurements of clinical parameters in a reproducible manner and have the potential to effectively substitute an in-office visit for routine monitoring. In addition to the universally observed accuracy of the requested parameter measurements, distance monitoring saves time for the patient and the physician alike, prevents missed appointments and is generally preferred by patients [78–80]. When studying the response of bariatric patients postoperatively, regarding the telemonitoring process, Vilallonga et al. found that patients themselves would prefer the telemonitoring option [71].

Postoperative monitoring of physiological parameters with the use of biosensors can be further expanded to include automated action after the received sensor signal. In an example by Wang et al. [75], patient-controlled analgesia was administered after indications from a biosensor feedback system that measured physiological responses to pain. Postoperative pain and nausea were reduced in the patients treated with this system. An ostomy alert sensor was developed by the team of Rouholiman et al. [80] that was capable of alerting nurses, patients, and physicians alike of the content status of the ostomy.

The vast array of remote monitoring systems described in the literature could very well be applied to a pre- or post-operative surgical patient, even if there are currently no studies for this specific cohort of patients. Systems used to monitor diseases such as heart failure, hypertension, pregnancy-related complications and more can easily be applied to the surgical patient in the future. A recently developed smartphone application could aid in the distant monitoring of COPD patients and could be useful in the detection of acute exacerbations and advise timely hospitalization [74]. All the above-mentioned modalities for long distance patient monitoring rely on the Internet for data transmission and could very well see their way in surgical patient monitoring or consultation in the not-so-distant future.

There are of course some hurdles still left in the way of universal internet-based patient monitoring. The biggest of which seems to be the reported difficulties elderly patients have with operating such systems [74–78]. Technological illiteracy is a persistent issue that seems harder to address than the technicalities of the systems. Patients of older age, of mental burden and patients without a reliable Internet connection that is readily accessible are in danger of being left out of such technological advancements, an observation reported in the majority of clinical studies. What is more, home-based distance monitoring relies entirely upon the adherence of the patient in data recording and the use of the instructed devices.

4. Discussing Open Issues and Challenges

The present literature review showcases current uses of the IoT paradigm within surgical practice, mainly by exploring the concept of telesurgery and surgical telementoring.

Telesurgery and surgical telementoring are undoubtedly the most impressive of the listed IoT applications within the surgical practice. Looking into the included particles of the presented survey, we can safely say that long-distance surgery on real-life patients is now feasible, although it seems to be scarcely performed. Authors do not mention any surgical safety compromises when IoT networks were utilized to perform telesurgery, which is most definitely the first hurdle that a newly emerging technique must overcome on its way to popularization. However, the most significant drawback of these applications, is the requirement of fast, stable Internet connections that will allow minimal latency and data loss. Paradoxically, such Internet connections may be lacking in the areas that are most in need of telesurgical and telementoring applications, such as rural and distant institutions that are not able to provide expert surgical consultations. Popularization of 5th generation (5G) mobile Internet networks is expected to be a big step towards that direction that will guarantee minimal latency and maximal connection stability. Particularly for telesurgical applications, there is also the issue of legal implications, as is underlined by several authors. The lack of specific legislature regarding long-distance surgery might prove to be grounds for liability of the providers, which is not previously described or covered by insurance. Due to the rapid implementation of such systems that is expected to follow, we must tackle such issues rapidly so that providers feel confident in partaking in long-distance operations or consultations. Financial costs are once again a key factor in play here. Teleconsultation or telementoring might not require much more than an audiovisual connection over the Internet; however, telesurgery itself requires surgical robots in order to transfer the instrument movements as instructed by the surgeon. Once again, these are not available everywhere on the globe, and rural centers in need of distant expert consultation are not likely to have robotic surgical systems available. From the above, it can be concluded that the feasibility and reliability of a new paradigm such as the IoST in telesurgery are not guarantees for its widespread application. There are still major logistic problems to overcome before telesurgery becomes part of the everyday surgical practice. Our research team aims to set up the first surgical telementoring system in Greece that will begin by providing a real-time audiovisual connection between an expert and a novice surgeon with live intraoperative guidance. This system will be evaluated against the traditional live mentoring of more inexperienced surgeons in order to provide a proof-of-concept. Further steps within our goals also include the introduction of more surgical centers to the said system in view of establishing a network of interconnected hospitals that provide regular surgical consultation over the Internet.

Image-guided surgery is one of the most developed forms of operative strategy due to the minimization of tissue damage, blood loss, operative hours and postoperative pain. By making use of serial interconnected data processing modules, researchers are able to construct IoT networks that greatly facilitate image-guidance in surgery. The real-time integration of preoperative imaging is the main goal in this case. The tracking of the imaging-specified patient anatomy in real-time surgical operations has proved highly significant in increasing surgical accuracy, and, in many cases, in assisting the prevention of accidental tissue damage. Despite encouraging results from proof-of-concept studies, however, these systems seem to be lacking adequate investigation in large-scale patient cohorts. As such, it is not safe to state with confidence that image-guided surgery incorporating the IoT paradigm today is widely accepted for clinical use. Still, we can safely conclude that IoT-based tool-tracking sensors will be proved valuable in the near future, especially for the surgeons that require maximum precision in instrument or anatomical landmark tracking. A major advantage of these systems, as already mentioned by a large number of authors, is the capability of IoST to account for interpersonal variability in anatomical structures in a real-time manner. Despite the lack of explicit reference in any of the included studies, acquisition of the necessary tool-tracking sensors and software is surely expected to be a major issue for several institutions. Therefore, the scientific community ought to aim for larger studies on surgical patients that will not only include image-guidance systems similar to those mentioned here but will randomize patients between image-guidance systems in order to better delineate the proposed advantages over older systems.

Long-distance monitoring using IoT is also one of the most common applications in medicine and healthcare. There is a limited number of articles including telemonitoring of the surgical patient specifically; however, such systems have proven their value in patient comfort and effective physiological parameter monitoring. The major challenge in the widespread implementation of said systems is the technological literacy of the patient population, as well as its Internet access. While improving Internet access is an ongoing global strife, researchers must focus on constructing adaptable and more intuitive user interfaces of such applications in order to appeal to older patients not comfortable with the everyday use of technological applications, wearable sensors and smart devices. Patient adherence is a challenge within these studies. The popularization of long-distance monitoring, in addition to patient education for these systems, will assist with the wider adoption of IoT-based monitoring of the surgical patient.

5. Conclusions

The aim of the present literature review is to collect and analyze the available knowledge on the most prominent fields that the IoT paradigm finds application to the surgical practice, i.e., the Internet of Surgical Things. Technological advances allow the incorporation of rendered preoperative data to the live surgical field, the valuable from-a-distance mentorship of younger from more experienced surgeons, the realization of a surgical procedure by remotely controlled robots, and the monitoring of surgical patients without the need for hospital visits. Despite the availability of reliable and fast Internet being a requirement for the actualization of these concepts in more areas of the world, the seamless incorporation of "smart" functions within the surgical world with the aid of the Internet is on a steady route to becoming a reality towards smarter and more efficient health services.

This study faces certain methodological limitations which arise from the narrative structure of most of the included literature. To begin with, this is a narrative systematic review, meaning that there are no statistical deductions to be made or pre-specified comparisons between different methodological approaches. The IoST is a concept, rather than a method, and therefore tangible comparisons can only be made in very specific applications. Current literature on IoST is lacking in comparative studies that produce results one can use to reach safe conclusions. Despite adhering to specific selection criteria, our study selection process was amenable to an unavoidable degree of bias, arising mainly from the lack of specific clinical applications in many of the screened publications. Finally, our inclusion of feasibility and animal model studies needs to be interpreted as a showcase of potential future applications, rather than everyday uses of the IoST concept. The review's findings can be summarized, per the research question, in the following paragraphs.

Telesurgical applications are undoubtedly the primary example of IoST systems. Their validity has been repeatedly evaluated over the years and the literature reveals that telesurgical networks are routinely incorporated in many surgical centers. Within such systems, there is a wide adoption of sensorial arrays that transmit data to distant locations. Telementoring is an even more inclusive concept in surgical education. Evaluation of distant teaching in surgery has revealed that it is a viable alternative to traditional teaching, which is at times preferred over in-person assistance. Recent literature also points out the incorporation of 5th generation cellular networks that are able to effectively eradicate latency times and connectivity issues.

IoST in image-guided surgery is currently an "under development" application that has produced tangible results in only a few clinical studies. Intraoperative use of IoST networks mainly focuses on superimposing preoperative imaging on live surgical camera feeds to assist surgeons in precision-requiring tasks. Despite being investigated mainly in neurosurgical procedures, these networks are predominantly software-dependent, thus making it possible to also be incorporated in more procedures in the years to come.

Patient telemonitoring involves the utilization of network-connected biosensors that track physiological patient parameters that are observed by a distant physician. Such sensorial ecosystems seem to be highly successful at detecting specified cutoff points and providing alerts to the physicians. Additionally, such systems can shorten hospital stays and lessen routine in-office patient visits, without compromising patient safety. Lastly, telemonitoring has relatively few requirements, considering the availability of biosensors in everyday technological products and "smart" wearable devices.

Supplementary Materials: The following supporting information can be downloaded at: https://www.mdpi.com/article/10.3390/s22124577/s1. PRISMA methodology (Supplemental Figures S1–S4), Supplemental Tables S1–S3.

Author Contributions: Conceptualization, K.K. and C.-N.A.; methodology, F.M. and G.-I.V.; software, C.-N.A.; validation, K.K., C.-N.A. and F.M.; formal analysis, F.M. and G.-I.V.; investigation, F.M. and G.-I.V.; resources, F.M.; data curation, F.M. and G.-I.V.; writing—original draft preparation, F.M. and G.-I.V.; writing—review and editing, K.K. and C.-N.A.; visualization, F.M.; supervision, C.-N.A.; project administration, K.K.; funding acquisition, C.-N.A. All authors have read and agreed to the published version of the manuscript.

Funding: This research received no external funding.

Institutional Review Board Statement: The present study required no prior ethical approval.

Informed Consent Statement: Not applicable.

Data Availability Statement: Not applicable.

Conflicts of Interest: The authors declare no conflict of interest.

References

1. Fuertes-Guiró, F.; Vitali-Erion, E.; Rodriguez-Franco, A. A Program of Telementoring in Laparoscopic Bariatric Surgery. *Minim. Invasive Ther. Allied Technol.* **2016**, *25*, 8–14. [CrossRef] [PubMed]
2. Mitsuno, D.; Hirota, Y.; Akamatsu, J.; Kino, H.; Okamoto, T.; Ueda, K. Telementoring Demonstration in Craniofacial Surgery With HoloLens, Skype, and Three-Layer Facial Models. *J. Craniofac. Surg.* **2019**, *30*, 28–32. [CrossRef] [PubMed]
3. Cos, H.; Li, D.W.; Williams, G.; Chininis, J.; Dai, R.X.; Zhang, J.W.; Srivastava, R.; Raper, L.; Sanford, D.; Hawkins, W.; et al. Predicting Outcomes in Patients Undergoing Pancreatectomy Using Wearable Technology and Machine Learning: Prospective Cohort Study. *J. Med. Internet Res.* **2021**, *23*, 11. [CrossRef] [PubMed]

4. Ohta, H.; Izumi, S.; Yoshimoto, M. A More Acceptable Endoluminal Implantation for Wirelessly Monitoring Vital Signals Using Ingestible Sensors Anchored to the Stomach Wall. In Proceedings of the Annual International Conference of the IEEE Engineering in Medicine and Biology Society (EMBC), Milano, Italy, 25–29 August 2015.
5. Ushimaru, Y.; Takahashi, T.; Souma, Y.; Yanagimoto, Y.; Nagase, H.; Tanaka, K.; Miyazaki, Y.; Makino, T.; Kurokawa, Y.; Yamasaki, M.; et al. Innovation in Surgery/Operating Room Driven by Internet of Things on Medical Devices. *Surg. Endosc.* **2019**, *33*, 3469–3477. [CrossRef] [PubMed]
6. Kim, S.H.; Shin, H.S.; Lee, S.H. 'Internet of Things' Real-Time Free Flap Monitoring. *J. Craniofac. Surg.* **2018**, *29*, e22–e25. [CrossRef]
7. Omboni, S. Connected Health in Hypertension Management. *Front. Cardiovasc. Med.* **2019**, *6*, 17. [CrossRef]
8. Guo, Z.; Dong, Z.; Lee, K.H.; Cheung, C.L.; Fu, H.C.; Ho, J.D.L.; He, H.; Poon, W.S.; Chan, D.T.M.; Kwok, K.W. Compact Design of a Hydraulic Driving Robot for Intraoperative MRI-Guided Bilateral Stereotactic Neurosurgery. *IEEE Robot. Autom. Lett.* **2018**, *3*, 2515–2522. [CrossRef]
9. Agrawal, R.; Mishra, S.K.; Mishra, A.; Chand, G.; Agarwal, G.; Agarwal, A.; Verma, A.K. Role of Telemedicine Technology in Endocrine Surgery Knowledge Sharing. *Telemed. J. E-Health* **2014**, *20*, 868–874. [CrossRef]
10. Andersen, D.; Popescu, V.; Cabrera, M.E.; Shanghavi, A.; Mullis, B.; Marley, S.; Gomez, G.; Wachs, J.P. An Augmented Reality-Based Approach for Surgical Telementoring in Austere Environments. *Mil. Med.* **2017**, *182*, 310–315. [CrossRef]
11. Anderson, S.M.; Kapp, B.B.; Angell, J.M.; Abd, T.T.; Thompson, N.J.; Ritenour, C.W.M.; Issa, M.M. Remote Monitoring and Supervision of Urology Residents Utilizing Integrated Endourology Suites—A Prospective Study of Patients' Opinions. *J. Endourol.* **2013**, *27*, 96–100. [CrossRef]
12. Andersen, D.; Popescu, V.; Cabrera, M.E.; Shanghavi, A.; Gomez, G.; Marley, S.; Mullis, B.; Wachs, J.P. Medical Telementoring Using an Augmented Reality Transparent Display. *Surgery* **2016**, *159*, 1646–1653. [CrossRef] [PubMed]
13. Artsen, A.M.; Burkett, L.S.; Duvvuri, U.; Bonidie, M. Surgeon Satisfaction and Outcomes of Tele-Proctoring for Robotic Gynecologic Surgery. *J. Robot. Surg.* **2022**, *16*, 563–568. [CrossRef] [PubMed]
14. Prince, S.W.; Kang, C.; Simonelli, J.; Lee, Y.H.; Gerber, M.J.; Lim, C.; Chu, K.; Dutson, E.P.; Tsao, T.C. A Robotic System for Telementoring and Training in Laparoscopic Surgery. *Int. J. Med. Robot.* **2020**, *16*, e2040. [CrossRef] [PubMed]
15. Patel, E.; Mascarenhas, A.; Subuhee, A.; Stirt, D.; Brady, I.; Perera, R.; Noël, J. Evaluating the Ability of Students to Learn and Utilize a Novel Telepresence Platform, Proximie. *J. Robot. Surg.* **2021**, *2021*, 1–7. [CrossRef]
16. Rojas-Muñoz, E.; Cabrera, M.E.; Lin, C.; Andersen, D.; Popescu, V.; Anderson, K.; Zarzaur, B.L.; Mullis, B.; Wachs, J.P. The System for Telementoring with Augmented Reality (STAR): A Head-Mounted Display to Improve Surgical Coaching and Confidence in Remote Areas. *Surgery* **2020**, *167*, 724–731. [CrossRef]
17. Rojas-Muñoz, E.; Cabrera, M.E.; Andersen, D.; Popescu, V.; Marley, S.; Mullis, B.; Zarzaur, B.; Wachs, J. Surgical Telementoring without Encumbrance: A Comparative Study of See-through Augmented Reality-Based Approaches. *Ann. Surg.* **2019**, *270*, 384–389. [CrossRef]
18. Rojas-Munõz, E.; Cabrera, M.E.; Lin, C.; Sánchez-Tamayo, N.; Andersen, D.; Popescu, V.; Anderson, K.; Zarzaur, B.; Mullis, B.; Wachs, J.P. Telementoring in Leg Fasciotomies via Mixed-Reality: Clinical Evaluation of the STAR Platform. *Mil. Med.* **2020**, *185*, 513–520. [CrossRef]
19. Rojas-Muñoz, E.; Lin, C.; Sanchez-Tamayo, N.; Cabrera, M.E.; Andersen, D.; Popescu, V.; Barragan, J.A.; Zarzaur, B.; Murphy, P.; Anderson, K.; et al. Evaluation of an Augmented Reality Platform for Austere Surgical Telementoring: A Randomized Controlled Crossover Study in Cricothyroidotomies. *NPJ Digit. Med.* **2020**, *3*, 75. [CrossRef]
20. Safir, I.J.; Shrewsberry, A.B.; Issa, I.M.; Ogan, K.; Ritenour, C.W.M.; Sullivan, J.; Issa, M.M. Impact of Remote Monitoring and Supervision on Resident Training Using New ACGME Milestone Criteria. *Can. J. Urol.* **2015**, *22*, 7959–7964.
21. Schlachta, C.M.; Lefebvre, K.L.; Sorsdahl, A.K.; Jayaraman, S. Mentoring and Telementoring Leads to Effective Incorporation of Laparoscopic Colon Surgery. *Surg. Endosc.* **2010**, *24*, 841–844. [CrossRef]
22. Talbot, M.; Harvey, E.J.; Berry, G.K.; Reindl, R.; Tien, H.; Stinner, D.J.; Slobogean, G. A Pilot Study of Surgical Telementoring for Leg Fasciotomy. *BMJ Mil. Health* **2018**, *164*, 83–86. [CrossRef] [PubMed]
23. Trujillo Loli, Y.; D'Carlo Trejo Huamán, M.; Campos Medina, S. Telementoring of In-Home Real-Time Laparoscopy Using Whatsapp Messenger: An Innovative Teaching Tool during the COVID-19 Pandemic. A Cohort Study. *Ann. Med. Surg.* **2021**, *62*, 481–484. [CrossRef] [PubMed]
24. Andersen, D.S.; Cabrera, M.E.; Rojas-Muñoz, E.J.; Popescu, V.S.; Gonzalez, G.T.; Mullis, B.; Marley, S.; Zarzaur, B.L.; Wachs, J.P. Augmented Reality Future Step Visualization for Robust Surgical Telementoring. *Simul. Healthc.* **2019**, *14*, 59–66. [CrossRef] [PubMed]
25. Dawe, P.; Kirkpatrick, A.; Talbot, M.; Beckett, A.; Garraway, N.; Wong, H.; Hameed, S.M. Telementored Damage-Control and Emergency Trauma Surgery: A Feasibility Study Using Live-Tissue Models. *Am. J. Surg.* **2018**, *215*, 927–929. [CrossRef] [PubMed]
26. DeKastle, R. Telesurgery: Providing Remote Surgical Observations for Students. *AORN J.* **2009**, *90*, 93–101. [CrossRef]
27. Din, N.; Chan, C.C.; Cohen, E.; Iovieno, A.; Dahan, A.; Rootman, D.S.; Litvin, G. Remote Surgeon Virtual Presence: A Novel Telementoring Method for Live Surgical Training. *Cornea* **2022**, *41*, 385–389. [CrossRef]
28. Hinata, N.; Miyake, H.; Kurahashi, T.; Ando, M.; Furukawa, J.; Ishimura, T.; Tanaka, K.; Fujisawa, M. Novel Telementoring System for Robot-Assisted Radical Prostatectomy: Impact on the Learning Curve. *Urology* **2014**, *83*, 1088–1092. [CrossRef]

29. Altieri, M.S.; Carmichael, H.; Jones, E.; Robinson, T.; Pryor, A.; Madani, A. Educational Value of Telementoring for a Simulation-Based Fundamental Use of Surgical EnergyTM (FUSE) Curriculum: A Randomized Controlled Trial in Surgical Trainees. *Surg. Endosc.* **2020**, *34*, 3650–3655. [CrossRef]
30. Glenn, I.C.; Bruns, N.E.; Hayek, D.; Hughes, T.; Ponsky, T.A. Rural Surgeons Would Embrace Surgical Telementoring for Help with Difficult Cases and Acquisition of New Skills. *Surg. Endosc.* **2017**, *31*, 1264–1268. [CrossRef]
31. Lenihan, J.; Brower, M. Web-Connected Surgery: Using the Internet for Teaching and Proctoring of Live Robotic Surgeries. *J. Robot. Surg.* **2012**, *6*, 47–52. [CrossRef]
32. Moore, A.M.; Carter, N.H.; Wagner, J.P.; Filipi, C.J.; Chen, D.C. Web-Based Video Assessments of Operative Performance for Remote Telementoring. *Surg. Technol. Int.* **2017**, *25*, 25–30.
33. Tel, A.; Bortuzzo, F.; Pascolo, P.; Costa, F.; Sembronio, S.; Bresadola, V.; Baldi, D.; Robiony, M. Maxillofacial Surgery 5.0: A New Paradigm in Telemedicine for Distance Surgery, Remote Assistance, and Webinars. *Minerva Stomatol.* **2020**, *69*, 191–202. [CrossRef] [PubMed]
34. Shin, D.H.; Dalag, L.; Azhar, R.A.; Santomauro, M.; Satkunasivam, R.; Metcalfe, C.; Dunn, M.; Berger, A.; Djaladat, H.; Nguyen, M.; et al. A Novel Interface for the Telementoring of Robotic Surgery. *BJU Int.* **2015**, *116*, 302–308. [CrossRef]
35. Kirkpatrick, A.W.; McKee, J.L.; Netzer, I.; McBeth, P.B.; D'Amours, S.; Kock, V.; Dobron, A.; Ball, C.G.; Glassberg, E. Transoceanic Telementoring of Tube Thoracostomy Insertion: A Randomized Controlled Trial of Telementored Versus Unmentored Insertion of Tube Thoracostomy by Military Medical Technicians. *Telemed. E-Health* **2019**, *25*, 730–739. [CrossRef]
36. Liu, P.; Li, C.; Xiao, C.; Zhang, Z.; Ma, J.; Gao, J.; Shao, P.; Valerio, I.; Pawlik, T.M.; Ding, C.; et al. A Wearable Augmented Reality Navigation System for Surgical Telementoring Based on Microsoft HoloLens. *Ann. Biomed. Eng.* **2021**, *49*, 287–298. [CrossRef] [PubMed]
37. Lacy, A.M.; Bravo, R.; Otero-Piñeiro, A.M.; Pena, R.; De Lacy, F.B.; Menchaca, R.; Balibrea, J.M. 5G-Assisted Telementored Surgery. *Br. J. Surg.* **2019**, *106*, 1576–1579. [CrossRef] [PubMed]
38. Netzer, I.; Kirkpatrick, A.W.; Nissan, M.; McKee, J.L.; McBeth, P.; Dobron, A.; Glassberg, E. Rubrum Coelis: The Contribution of Real-Time Telementoring in Acute Trauma Scenarios-A Randomized Controlled Trial. *Telemed. E-Health* **2019**, *25*, 1108–1114. [CrossRef]
39. Greenberg, J.A.; Schwarz, E.; Paige, J.; Dort, J.; Bachman, S. At-Home Hands-on Surgical Training during COVID19: Proof of Concept Using a Virtual Telementoring Platform. *Surg. Endosc.* **2021**, *35*, 1963–1969. [CrossRef]
40. Forgione, A.; Kislov, V.; Guraya, S.Y.; Kasakevich, E.; Pugliese, R. Safe Introduction of Laparoscopic Colorectal Surgery Even in Remote Areas of the World: The Value of a Comprehensive Telementoring Training Program. *J. Laparoendosc. Adv. Surg. Tech.* **2015**, *25*, 37–42. [CrossRef]
41. Chu, G.; Yang, X.; Luo, L.; Feng, W.; Jiao, W.; Zhang, X.; Wang, Y.; Yang, Z.; Wang, B.; Li, J.; et al. Improved Robot-Assisted Laparoscopic Telesurgery: Feasibility of Network Converged Communication. *Br. J. Surg.* **2021**, *108*, e377–e379. [CrossRef]
42. Wirz, R.; Torres, L.G.; Swaney, P.J.; Gilbert, H.; Alterovitz, R.; Webster, R.J.; Weaver, K.D.; Russell, P.T. An Experimental Feasibility Study on Robotic Endonasal Telesurgery. *Neurosurgery* **2015**, *76*, 479–484. [CrossRef] [PubMed]
43. Zheng, J.; Wang, Y.; Zhang, J.; Guo, W.; Yang, X.; Luo, L.; Jiao, W.; Hu, X.; Yu, Z.; Wang, C.; et al. 5G Ultra-Remote Robot-Assisted Laparoscopic Surgery in China. *Surg. Endosc.* **2020**, *34*, 5172–5180. [CrossRef] [PubMed]
44. Park, J.W.; Lee, D.H.; Kim, Y.W.; Lee, B.H.; Jo, Y.H. Lapabot: A Compact Telesurgical Robot System for Minimally Invasive Surgery: Part II. Telesurgery Evaluation. Minim. Invasive Ther. *Allied Technol.* **2012**, *21*, 195–200. [CrossRef]
45. Acemoglu, A.; Peretti, G.; Trimarchi, M.; Hysenbelli, J.; Krieglstein, J.; Geraldes, A.; Deshpande, N.; Ceysens, P.M.V.; Caldwell, D.G.; Delsanto, M.; et al. Operating From a Distance: Robotic Vocal Cord 5G Telesurgery on a Cadaver. *Ann. Intern. Med.* **2020**, *173*, 940–941. [CrossRef]
46. Tian, W.; Fan, M.; Zeng, C.; Liu, Y.; He, D.; Zhang, Q. Telerobotic Spinal Surgery Based on 5G Network: The First 12 Cases. *Neurospine* **2020**, *17*, 114–120. [CrossRef]
47. Morohashi, H.; Hakamada, K.; Kanno, T.; Kawashima, K.; Akasaka, H.; Ebihara, Y.; Oki, E.; Hirano, S.; Mori, M. Social Implementation of a Remote Surgery System in Japan: A Field Experiment Using a Newly Developed Surgical Robot via a Commercial Network. *Surg. Today* **2021**, *52*, 705–714. [CrossRef]
48. Huang, E.Y.; Knight, S.; Guetter, C.R.; Davis, C.H.; Moller, M.; Slama, E.; Crandall, M. Telemedicine and telementoring in the surgical specialties: A narrative review. *Am. J. Surg.* **2019**, *218*, 760–766. [CrossRef]
49. Sachdeva, N.; Klopukh, M.; Clair, R.S.; Hahn, W.E. Using conditional generative adversarial networks to reduce the effects of latency in robotic telesurgery. *J. Robot. Surg.* **2021**, *11*, 635–641. [CrossRef]
50. Shabir, D.; Abdurahiman, N.; Padhan, J.; Trinh, M.; Balakrishnan, S.; Kurer, M.; Ali, O.; Al-Ansari, A.; Yaacoub, E.; Deng, Z.; et al. Towards Development of a Telementoring Framework for Minimally Invasive Surgeries. *Int. J. Med. Robot.* **2021**, *17*, e2305. [CrossRef]
51. Nguyen, N.T.; Okrainec, A.; Anvari, M.; Smith, B.; Meireles, O.; Gee, D.; Moran-Atkin, E.; Baram-Clothier, E.; Camacho, D.R. Sleeve gastrectomy telementoring: A SAGES multi-institutional quality improvement initiative. *Surg. Endosc.* **2018**, *32*, 682–687. [CrossRef]
52. Snyderman, C.H.; Gardner, P.A.; Lanisnik, B.; Ravnik, J. Surgical telementoring: A new model for surgical training. *Laryngoscope* **2016**, *126*, 1334–1338. [CrossRef] [PubMed]

53. Kirkpatrick, A.W.; Tien, H.; LaPorta, A.T.; Lavell, K.; Keillor, J.; Beatty, H.E.W.L.; McKee, J.L.; Brien, S.; Robert, D.J.; Wong, J.; et al. The marriage of surgical simulation and telementoring for damage-control surgical training of operational first responders: A pilot study. *J. Trauma Acute Care Surg.* **2015**, *79*, 741–747. [CrossRef] [PubMed]
54. Li, G.; Su, H.; Cole, G.A.; Shang, W.; Harrington, K.; Camilo, A.; Pilitsis, J.G.; Fischer, G.S. Robotic System for MRI-Guided Stereotactic Neurosurgery. *IEEE Trans. Biomed. Eng.* **2015**, *62*, 1077–1088. [CrossRef] [PubMed]
55. Louis, R.G.; Steinberg, G.K.; Duma, C.; Britz, G.; Mehta, V.; Pace, J.; Selman, W.; Jean, W.C. Early Experience with Virtual and Synchronized Augmented Reality Platform for Preoperative Planning and Intraoperative Navigation: A Case Series. *Oper. Neurosurg.* **2021**, *21*, 189–196. [CrossRef]
56. Ivan, M.E.; Eichberg, D.G.; Di, L.; Shah, A.H.; Luther, E.M.; Lu, V.M.; Komotar, R.J.; Urakov, T.M. Augmented Reality Head-Mounted Display–Based Incision Planning in Cranial Neurosurgery: A Prospective Pilot Study. *Neurosurg. Focus* **2021**, *51*, E3. [CrossRef]
57. De Momi, E.; Ferrigno, G.; Bosoni, G.; Bassanini, P.; Blasi, P.; Casaceli, G.; Fuschillo, D.; Castana, L.; Cossu, M.; Lo Russo, G.; et al. A Method for the Assessment of Time-Varying Brain Shift during Navigated Epilepsy Surgery. *Int. J. Comput. Assist. Radiol. Surg.* **2016**, *11*, 473–481. [CrossRef]
58. Kiarostami, P.; Dennler, C.; Roner, S.; Sutter, R.; Fürnstahl, P.; Farshad, M.; Rahm, S.; Zingg, P.O. Augmented Reality-Guided Periacetabular Osteotomy—Proof of Concept. *J. Orthop. Surg. Res.* **2020**, *15*, 540. [CrossRef]
59. Padilla, J.B.; Arango, R.; García, H.F.; Cardona, H.D.V.; Orozco, Á.A.; Álvarez, M.A.; Guijarro, E. NEURONAV: A Tool for Image-Guided Surgery—Application to Parkinson's Disease. In *International Symposium on Visual Computing*; Springer: Cham, Switzerland, 2015; Volume 9474, pp. 349–358. [CrossRef]
60. Watanabe, E.; Satoh, M.; Konno, T.; Hirai, M.; Yamaguchi, T. The Trans-Visible Navigator: A See-Through Neuronavigation System Using Augmented Reality. *World Neurosurg.* **2016**, *87*, 399–405. [CrossRef]
61. Yoon, J.W.; Chen, R.E.; ReFaey, K.; Diaz, R.J.; Reimer, R.; Komotar, R.J.; Quinones-Hinojosa, A.; Brown, B.L.; Wharen, R.E. Technical Feasibility and Safety of Image-Guided Parieto-Occipital Ventricular Catheter Placement with the Assistance of a Wearable Head-up Display. *Int. J. Med. Robot. Comput. Assist. Surg.* **2017**, *13*, e1836. [CrossRef]
62. Fan, X.; Roberts, D.W.; Schaewe, T.J.; Ji, S.; Holton, L.H.; Simon, D.A.; Paulsen, K.D. Intraoperative Image Updating for Brain Shift Following Dural Opening. *J. Neurosurg.* **2017**, *126*, 1924–1933. [CrossRef]
63. Eftekhar, B. A Smartphone App to Assist Scalp Localization of Superficial Supratentorial Lesions–Technical Note. *World Neurosurg.* **2016**, *85*, 359–363. [CrossRef] [PubMed]
64. Hu, L.; Wang, M.; Song, Z. A Convenient Method of Video See-through Augmented Reality Based on Image-Guided Surgery System. In Proceedings of the 2013 Seventh International Conference on Internet Computing for Engineering and Science, Shanghai, China, 20–22 September 2013; Volume 2013, pp. 100–103. [CrossRef]
65. Wildemeersch, D.; D'Hondt, M.; Bernaerts, L.; Mertens, P.; Saldien, V.; Hendriks, J.M.; Walcarius, A.S.; Sterkens, L.; Hans, G.H. Implementation of an Enhanced Recovery Pathway for Minimally Invasive Pectus Surgery: A Population-Based Cohort Study Evaluating Short- and Long-Term Outcomes Using E-Health Technology. *JMIR Perioper. Med.* **2018**, *1*, e10996. [CrossRef] [PubMed]
66. Msayib, Y.; Gaydecki, P.; Callaghan, M.; Dale, N.; Ismail, S. An Intelligent Remote Monitoring System for Total Knee Arthroplasty Patients. *J. Med. Syst.* **2017**, *41*, 90. [CrossRef]
67. Peysson, L.; Gomez, C.; Giovannetti, P.; Coltey, B.; Dufeu, N.; Bregeon, E.; Gaubert, J.; Dutau, H.; Thomas, P.; Reynaud-Gaubert, M. Internet-Based Telemonitoring System of Daily Home Spirometry in Lung Transplant Recipients. *J. Heart Lung Transplant.* **2015**, *34*, S141. [CrossRef]
68. Jonker, L.T.; Lahr, M.M.H.; Festen, S.; Oonk, M.H.M.; de Bock, G.H.; van Leeuwen, B.L. Perioperative Telemonitoring of Older Adults with Cancer: Can We Connect Them All? *J. Geriatr. Oncol.* **2020**, *11*, 1244–1249. [CrossRef] [PubMed]
69. Cornelis, N.; Buys, R.; Dewit, T.; Benoit, D.; Claes, J.; Fourneau, I.; Cornelissen, V. Satisfaction and Acceptability of Telemonitored Home-Based Exercise in Patients with Intermittent Claudication: Pragmatic Observational Pilot Study. *JMIR Rehabil. Assist. Technol.* **2021**, *8*, e18739. [CrossRef]
70. Mullen-Fortino, M.; Rising, K.L.; Duckworth, J.; Gwynn, V.; Sites, F.D.; Hollander, J.E. Presurgical Assessment Using Telemedicine Technology: Impact on Efficiency, Effectiveness, and Patient Experience of Care. *Telemed. E-Health* **2019**, *25*, 137–142. [CrossRef]
71. Vilallonga, R.; Lecube, A.; Fort, J.M.; Boleko, M.A.; Hidalgo, M.; Armengol, M. Internet of Things and Bariatric Surgery Follow-up: Comparative Study of Standard and IoT Follow-Up. *Minim. Invasive Ther. Allied Technol.* **2013**, *22*, 304–311. [CrossRef]
72. Luo, M.; Yuan, R.; Sun, Z.; Li, T.; Xie, Q. A web-based computer aided system for liver surgery planning: Initial implementation on RayPlus. In Proceedings of the Medical Imaging 2016: Image-Guided Procedures, Robotic Interventions, and Modeling, San Diego, CA, USA, 18 March 2016; pp. 504–509. [CrossRef]
73. Yanni, R.M.T.; El-Bakry, H.M.; Riad, A.; El-Khamisy, N. Internet of Things for Surgery Process Using Raspberry Pi. *Int. J. Online Biomed. Eng.* **2020**, *16*, 96–115. [CrossRef]
74. Caggianese, G.; Calabrese, M.; Gallo, L.; Sannino, G.; Vecchione, C. Cardiac surgery rehabilitation system (CSRS) for a personalized support to patients. In Proceedings of the International IEEE Conference on Signal-Image Technologies and Internet-Based System, Jaipur, India, 4–7 December 2017; pp. 83–90.
75. Wang, R.; Wang, S.; Duan, N.; Wang, Q. From Patient-Controlled Analgesia to Artificial Intelligence-Assisted Patient-Controlled Analgesia: Practices and Perspectives. *Front. Med.* **2020**, *7*, 145. [CrossRef]

76. McGillion, M.; Ouellette, C.; Good, A.; Bird, M.; Henry, S.; Clyne, W.; Turner, A.; Ritvo, P.; Ritvo, S.; Dvirnik, N.; et al. Postoperative Remote Automated Monitoring and Virtual Hospital-to-Home Care System Following Cardiac and Major Vascular Surgery: User Testing Study. *J. Med. Internet Res.* **2020**, *22*, e15548. [CrossRef] [PubMed]
77. Colomina, J.; Drudis, R.; Torra, M.; Pallisó, F.; Massip, M.; Vargiu, E.; Nadal, N.; Funentes, A.; Bravo, M.O.; Miralles, F.; et al. Implementing mHealth-Enabled Integrated Care for Complex Chronic Patients With Osteoarthritis Undergoing Primary Hip or Knee Arthroplasty: Prospective, Two-Arm, Parallel Trial. *J. Med. Internet Res.* **2021**, *23*, e28320. [CrossRef] [PubMed]
78. Holmes, M.; Nieto, M.P.; Song, H.; Tonkin, E.; Grant, S.; Flach, P. Modelling Patient Behaviour Using IoT Sensor Data: A Case Study to Evaluate Techniques for Modelling Domestic Behaviour in Recovery from Total Hip Replacement Surgery. *J. Healthc. Inform. Res.* **2020**, *4*, 238–260. [CrossRef]
79. Kong, H.; Chen, J. Medical Monitoring and Management System of Mobile Thyroid Surgery Based on Internet of Things and Cloud Computing. *Wirel. Commun. Mob. Comput.* **2021**, *2021*, 7065910. [CrossRef]
80. Rouholiman, D.; Gamble, J.G.; Dobrota, S.D.; Encisco, E.M.; Shah, A.G.; Grajales, F.J., III; Chu, L.F. Improving Health-Related Quality of Life of Patients With an Ostomy Using a Novel Digital Wearable Device: Protocol for a Pilot Study. *JMIR Res. Protoc.* **2018**, *7*, e7470. [CrossRef] [PubMed]

sensors

MDPI

Article

IoT-Based Multi-Sensor Healthcare Architectures and a Lightweight-Based Privacy Scheme

Vassileios Aivaliotis, Kyriaki Tsantikidou and Nicolas Sklavos *

SCYTALE Group, Computer Engineering and Informatics Department, University of Patras, 26504 Patras, Greece; vaivaliotis@ceid.upatras.gr (V.A.); k.tsantikidou@upatras.gr (K.T.)
* Correspondence: nsklavos@upatras.gr

Abstract: Health 4.0 is a new promising addition to the healthcare industry that innovatively includes the Internet of Things (IoT) and its heterogeneous devices and sensors. The result is the creation of numerous smart health applications that can be more effective, reliable, scalable and cost-efficient while facilitating people with their everyday life and health conditions. Nevertheless, without proper guidance, the employment of IoT-based health systems can be complicated, especially with regard to security challenges such susceptible application displays. An appropriate comprehension of the structure and the security demands of IoT-based multi-sensor systems and healthcare infrastructures must first be achieved. Furthermore, new architectures that provide lightweight, easily implementable and efficient approaches must be introduced. In this paper, an overview of IoT integration within the healthcare domain as well as a methodical analysis of efficient smart health frameworks, which mainly employ multiple resource and energy-constrained devices and sensors, will be presented. An additional concern of this paper will be the security requirements of these key IoT components and especially of their wireless communications. As a solution, a lightweight-based security scheme, which utilizes the lightweight cryptographic primitive LEAIoT, will be introduced. The proposed hardware-based design displays exceptional results compared to the original CPU-based implementation, with a 99.9% increase in key generation speed and 96.2% increase in encryption/decryption speed. Finally, because of its lightweight and flexible implementation and high-speed keys' setup, it can compete with other common hardware-based cryptography architectures, where it achieves lower hardware utilization up to 87.9% with the lowest frequency and average throughput.

Keywords: healthcare architectures; IoT; Health 4.0; lightweight privacy; cryptography; sensors

Citation: Aivaliotis, V.; Tsantikidou, K.; Sklavos, N. IoT-Based Multi-Sensor Healthcare Architectures and a Lightweight-Based Privacy Scheme. *Sensors* **2022**, *22*, 4269. https://doi.org/10.3390/s22114269

Academic Editors: Alexandru Lavric, Liliana Anchidin and Adrian I. Petrariu

Received: 30 April 2022
Accepted: 1 June 2022
Published: 3 June 2022

1. Introduction

In recent years, the demand for healthcare services that are high quality, low cost, widely available and easily accessible from the comfort of the patients' homes is rapidly increasing. The result is the creation of notable problems that healthcare providers are struggling to resolve. These circumstances lead to the progressive inclusion of creative methodologies and technologies whose integration will greatly impact the healthcare industry. Some of these technologies are the Internet of Things (IoT), Internet of Services (IoS), Medical Cyber-Physical Systems, Cloud Computing, Fog, Big Data Analytics, 5G, Cryptography, Blockchain and Artificial Intelligence (AI).

Internet of Things (IoT) consists of heterogeneous devices that are interconnected with each other and with the Internet [1]. The utilization of IoT technology is rapidly growing with various systems emerging as efficient solutions to applicational restrictions [2]. It provides new services and intelligent capabilities by enabling the components to collect, store and analyze data from different resources while also retaining low-cost implementation. Specifically, the healthcare industry greatly benefits from these new services [3]. Health 4.0 introduces creative methods for the employment of innovative IoT technology [4].

Smart, cost-effective, reliable, scalable and easily accessible approaches, that facilitate multiple classes of people in various health-related situations, can be developed by following its main design principles, namely Interoperability, Virtualization, Decentralization, Real-Time Capability, Service Orientation and Modularity [5]. The first principle indicates the ability of the heterogeneous components to share data with each other and overall connect through Internet of Things (IoT) technology. Virtualization is another important principle that automates medical procedures by monitoring all healthcare elements and creating virtual copies of different processes. Decentralization requires all utilized healthcare devices to make decisions on their own and always execute their operations efficiently. Moreover, due to the critical nature of healthcare applications, the collection of data, decision making and healthcare functions must be executed in real-time without unnecessary delays, leading to the implementation of the Real-Time Capability principle. Service orientation implies that all operations and components of the Healthcare application are presented as services. These services can be accessible to all participants and they can be provided through various organizations. The last principle, Modularity, represents the flexibility of the system to constantly evolve and improve to fulfill new requirements and correspond to the scalability that characterizes IoT.

Nevertheless, the vision of Health 4.0, namely the integration of IoT technology and the healthcare industry, can add further difficulties to developers. The scalability and heterogeneous nature of IoT elements, the simultaneous utilization of numerous components and the delicate structure of healthcare systems can be a troublesome combination, especially without the proper comprehension of the architecture and operation of IoT-based healthcare infrastructures. Furthermore, security is an important factor in the development of IoT systems in healthcare [6]. The data that are collected by the sensors and are shared between the devices contain private information. Only authorized users must have access to the medical data, which must remain intact and efficiently encrypted throughout their transmission to the IoT smart health network. Their improper protection has legal and ethical implications because they can be exploited by attackers with the intention of possibly harming the user's well-being. Moreover, only data from trusted devices and components must be accepted by the system in order to prevent the falsification of medical history and enhance the accuracy of remote diagnosis. The system must also be able to constantly execute its main functions without being obstructed by attacks, because even a small delay in critical operation can threaten the user's life. For example, in smart health applications that require immediate responses for alerting medical services or providing first-aid assistance, specific types of attacks can delay these functionality processes and negate the system's purpose of facilitating the patient and healthcare providers. Therefore, the application and maintenance of the main security schemes, such as confidentiality, availability and integrity, are a major concern for the research community.

The primary security challenges of IoT employment for smart health originate from various threats and vulnerabilities and are divided into two categories, namely embedded and network challenges [2]. The embedded challenges refer to the hardware and software aspects of IoT-based systems. The IoT components are resource- and energy-constrained devices that operate remotely and autonomously. By their own nature, they cannot support security algorithms that are resource and computationally demanding. Lightweight cryptography can resolve this problem because of the ability to adequately secure the architecture while utilizing fewer hardware resources. Nevertheless, in many cases, specific approaches that are used as a replacement are not as reliable as the well-known heavy cryptographic primitives [7]. Moreover, most smart health providers do not spend their resources achieving the security concepts, viz confidentiality, availability and integrity [8]. They rather only focus on the application's healthcare functions and reduction in the implementational costs. Therefore, inept security practices and updates are adopted, resulting in an easily exploitable system. Another vulnerability originates from the scalable nature of IoT. Specifically, the devices are constantly being added to the system without the guaranty of their own integrity. An attacker can be granted access to the main system by compromising a small, unprotected device. Therefore, all devices, even the smallest ones, that interact with the central services must have a form of a lightweight security algorithm that protects them.

Network security challenges concern communication mechanisms and smooth data transmissions. All IoT devices continually connect and disconnect to a variety of communication networks that can either be private or unknown. In addition, the wireless networks utilized in IoT have many vulnerabilities on their own [9,10]. Their exploitation helps attackers intercept the network transmission and eavesdrop on the sensitive health data. Hence, they must implement flexible algorithms that are sufficiently capable to ensure the privacy and integrity of the collected data. This can be achieved by implementing encryption and decryption algorithms in the heterogeneous devices. However, the most common ciphers are resource-demanding and they cannot be utilized in the IoT environment. Furthermore, the scalability of IoT renders difficult the detection of malicious presences in the network, deeming necessary the employment of more complex data management schemes [11,12]. Despite the fact that the high production speed of the cipher key and quick encryption/decryption mechanism can alleviate some pressure on the transmission management, their achievement is challenging because of IoT hardware limitations. Lastly, a main challenge for IoT systems is the prevention and mitigation of Denial of Service (DoS) or Distributed Denial of Service (DDoS) attacks [2]. These attacks interfere with the communication processes and render the system unavailable to the end users by exploiting network vulnerabilities.

In recent years, many lightweight methods have been proposed for security, privacy and authenticity purposes in IoT-based healthcare applications. Some utilize new mechanisms such as Artificial Intelligence and Blockchain. However, these integrations add new vulnerabilities and implementational issues into the already security challenging IoT architecture [6]. Hence, they are not considered lightweight-efficient IoT-based security designs. Other implementations propose more flexible and appropriate approaches. The authors' previous works proposed exemplary frameworks for limiting the implementation vulnerabilities and preventing possible attacks [13,14]. Other proposed architectures offered a variety of efficient security approaches that employ flexible security mechanisms and interactive cryptographic algorithms with minimum hardware resources [15,16].

In this paper, as a continuation of the authors' research, an extensive study of healthcare frameworks and applications is presented. Afterwards, a basic architecture of an IoT-based multi-sensor healthcare infrastructure is demonstrated. The first contribution of this paper is this detailed analysis and calculated creation of an IoT-based healthcare structure which can benefit the research community by referencing this architecture and its properties in future related work. This general architecture has some security requirements and challenges regarding data transfers and encryption/decryption mechanisms. Specifically, the main security and functionality requirements are the following:

- A formal lightweight cryptographic primitive which can encrypt and decrypt the inputted messages with a considerable-sized key, thus ensuring the confidentiality, data privacy and authentication while deeming the decomposition of the algorithm through key search attacks to be difficult. The success of other known attacks will also be eliminated as each device will effectively secure the collected data before transmission.
- High speed in key generation and encryption/decryption processes for quick transmissions and the management of the significant IoT network load. Therefore, the constant availability of the architecture will be ensured because the communication system will be able to handle multiple messages that are rapidly transmitted.
- Flexible design with multiple options for key size and performance speed for various circumstances depending on the user preferences. The scalability of the IoT can thus be confronted by constantly adapting to the current needs of the system.
- Minimal resource utilization of the employed security scheme for efficient implementation in hardware-constrained IoT devices. This way, all devices in the healthcare structure can employ a level of security and hence ensure their integrity. Proper bal-

ance between the resources and the performance throughput must also be ensured. The performance flow of the security scheme must be considered when minimizing the resources of the design. The throughput must remain sufficiently high in order to correspond to high-speed transmissions.

A solution to these concerns is proposed and demonstrated. The proposed lightweight LEAIoT-based security scheme ensures the user authentication and the privacy of the collected data that are shared via IoT networks between resource-constrained devices and via the Internet to cloud-based services. Its resource-efficient implementation in IoT devices quickly encrypts the collected data with flexible and high-speed key generation options while adapting to various communication and computational needs. Therefore, the discussed requirements are fulfilled without adding pressure to the employed hardware-constrained components. The exact methods that lead to the fulfilment of the mentioned requirements in IoT-based healthcare architectures and thus the innovations and contributions of the proposed scheme are listed as follows:

- First, the employed cryptographic primitive, as presented in [17], ensures the maintenance of the main security concepts while being lightweight and having the ability to utilize various sized keys, from 32-bit to 256-bit length. The width of the key is a general parameter of the security level for symmetric encryption, deeming the 256-bit key highly effective for the proposed environment. The LEAIoT algorithm also utilizes asymmetric encryption, which further enhances the security and confidentiality of the system.
- Second, the high computation speed achieved by the hardware-based implementation ensures the quick network management of the IoT's heavy communication load. The performance results of the proposed FPGA-based architecture are compared to the original LEAIoT design [17], which was implemented in a central processing unit (CPU), with a significant improvement of the key generation speed up to 99.9% and the encryption/decryption speed up to 96.2%.
- Third, in the same architecture, four different key sizes are integrated, providing four diverse performance speeds and security efficiencies that the applications can instantly choose from depending on the current network and the computational needs of the application. Therefore, the system offers flexibility and can easily adapt to various circumstances.
- Finally, the cryptographic scheme is properly implemented through innovative design optimizations and achieves novel implementational results that indicate its resource efficiency without an excessive reduction in throughput. Thus, the proposed scheme effectively fulfills the performance requirements for smart health and it can be easily employed to the various devices of the IoT network, providing global security to the healthcare structure while maintaining performance balance between resources, throughput and security. For further proving the architecture's efficiency, the proposed hardware-based scheme is compared to other relative FPGA-based research. Specifically, it was concluded that the proposed implementation utilizes up to 87.9% and 76.9% fewer resources than the lightweight versions of Advance Encryption Standard (AES). It also decreases the resource consumption up to 65.7% and 12.2% compared to SNOW 3G and ZUC ciphers, respectively. Lastly, it achieves an almost double throughput compared to RC4 cipher and overall similar performance results with PRESENT and CLEFIA, even with a significantly lower frequency.

2. IoT-Based Multi-Sensor Healthcare Applications

2.1. Health 4.0 Objectives

Many smart health applications can be created by applying IoT technology, thus achieving the objectives of Health 4.0 [4]. A general framework for Health 4.0 depicting the integrated technologies and the main components of the healthcare architecture is presented in Figure 1. One of the most important objectives is the creation of high-quality services that facilitate people with various healthcare needs. This requires the improvement of the system performance, the efficient utilization of resources and the optimization of utilized tools. Automated procedures and intelligence can enhance the results' accuracy and accelerate basic and repetitive tasks. Moreover, remote access and real-time responses can assist in immediate medical attention and constant monitoring. Finally, the design of appropriate databases with complete and easily accessible medical records can create a more personalized healthcare and a better diagnosis.

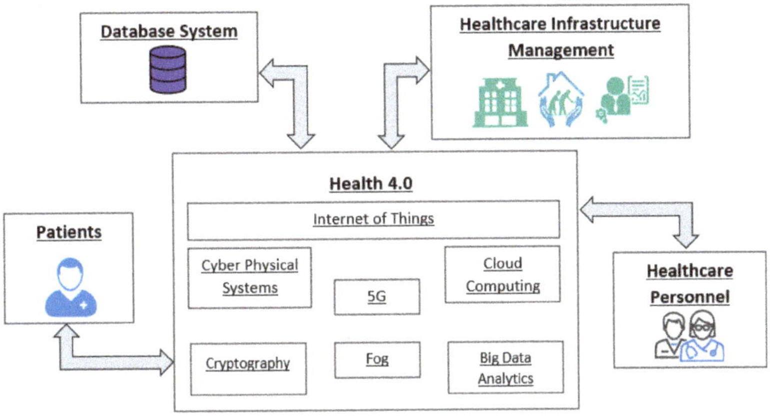

Figure 1. Health 4.0 framework.

Another equally vital objective is the improvement of the operations' effectiveness while simultaneously maintaining minimal costs as well as low resource utilization and energy consumption. Therefore, healthcare applications can be implemented in IoT devices that are mainly resource-constrained with energy limitations. However, the balance between the throughput and resource consumption is critical and depends on the requirements of the system. The ideal approach is the achievement of the highest performance throughput with the utilization of minimum resources. The employment of IoT, independently of its final optimizations, can assist in health monitoring, diagnosis and disease prediction through the enormous number of data that are collected by health sensors. These data are instantly transmitted, analyzed and saved to cloud services, making the diagnosis easier and more accurate. The pressure on healthcare personnel and materials will be reduced because the costly methods of health examination are replaced by cost-efficient, easily accessible, user-friendly and immediately responsive alternatives. Finally, the collaboration and sharing of information between multiple healthcare facilities and providers will be straightforward and timesaving.

2.2. IoT Architecture

First, the IoT architecture is described by a three-layer design [11]. These three layers are the application layer, the network layer and the physical/perception layer. Figure 2 displays the IoT layers in a healthcare architecture with more details. The application layer is the highest layer in the architecture and connects the objects with the IoT network [11]. It consists of various healthcare applications that offer e-health services and functionalities to the user. The network layer enables the IoT components to connect with each other and share data which are collected by the devices of the physical layer through established communication protocols. Some commonly utilized networks are Bluetooth, ZigBee, 5G, Wi-Fi, Radio Frequency Identification (RFID), 6LoWPAN and LoRaWAN. Another established network of nodes that is integrated into the IoT is the Wireless Sensor Network (WSN) [18]. The physical/perception layer is the last layer of the architecture. It consists of all the physical objects employed in IoT systems such as sensors, wearables, actuators, smartphones, antennas, and processors, etc. The purpose of this layer is to collect health signals and convert them to readable data that the network layer can transmit.

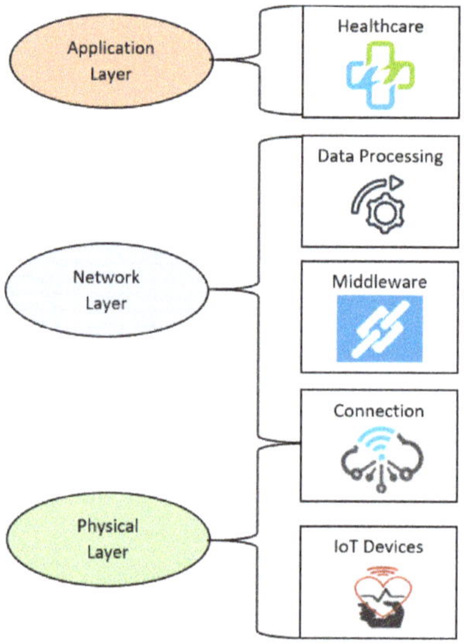

Figure 2. Architecture of IoT layers.

2.3. *Smart Health Multi-Sensors Designs from Literature*

Healthcare services can now be provided outside the hospital facility with the employment of IoT technology. Remote health monitoring, telemedicine, the ambient-assisted living (ALL) of elderly or disabled people and supervision of chronic diseases are some crucial applications that can benefit healthcare [3,19,20]. Specifically, they can improve the effectiveness and accessibility of health services and help alleviate the pressure on hospital resources. Ref. [20] analyzed recent proposals for Internet of Health Things, ambient assisted living and remote healthcare monitoring systems. Furthermore, it provided illustrations of relative architectures. In [21], a decentralized architecture for a smart health system based on Internet of Medical Things was proposed. This design consists of three layers, namely the data producer layer, the hybrid computing layer and the data consumer layer. The first layer includes the IoT sensors producing health data which are later collected and transmitted over to a hybrid computing system benefitting from both edge and cloud paradigms. The decentralized data processing is achieved by utilizing Blockchain technology and Distributed Data Storage System (DDSS) methodology. Moreover, system privacy and security are established by introducing three cryptography algorithms. Another decentralized healthcare architecture was demonstrated in [22]. It aims to monetize the collected health data while offering guidelines for data security and privacy. The technologies employed, namely IoT, AI, Big Data and Blockchain, and all the architecture's elements, were thoroughly described. Ref. [23] displayed the architecture of an IoT-based health application that manages Big Data. Each of its layers and key technologies were analyzed in depth. This design has been employed in smart clothing-based monitoring, telemedicine and emotion interaction services. In [24], a hardware-based implementation with multiple sensors connected to a main processor was designed. The system was employed for the constant monitoring of health variables.

The depiction and development of a Smart Hospital architecture is also a major concern for researchers. Ref. [25] proposed an architecture of a smart hospital that utilizes the low power wide area wireless protocol NarrowBand-IoT (NB-IoT). This protocol renders possible the communication between all IoT devices and sensors that are simultaneously employed in various hospital application scenarios. The paper thoroughly demonstrates the smart hospital's infrastructure and applications. In [26], the implementation of an IoT-based multi-sensor monitoring system, which resembles previously mentioned remote architectures, is demonstrated. The design consists of three layers, namely the node layer that includes the Wireless Sensor Network (WSN) and sensors; the local management layer that is composed by the hospital's computers; and the cloud-based layer, which collects and stores the data. A complete IoT-based tracking system applied in a hospital facility was implemented in [27]. The architecture was composed of multiple smart devices which recorded and transmitted the hand-washing data of the recognized subjects, smart ID cards for easy authentication of hospital personnel, hybrid routers and an IoT gateway for receiving and forwarding data to the last element, namely the cloud server, and sending commands in real-time. The purpose is the prevention of infections inside the hospital by tracking the hand-washing activities of hospital employees.

2.4. IoT-Based Multi-Sensor Healthcare Infrastructure

A complete IoT-based Health architecture is demonstrated in Figure 3. This general design depicts the relations between IoT components and healthcare elements. Smart hospitals, near-patient/personalized smart health systems as well as their connection are presented in detail. The personalized smart health architecture consists of multiple heterogeneous IoT devices, a wireless interface and a connection with a cloud-based database, as described in [2]. First, the devices that are utilized are mostly implantable or wearable medical devices and sensors, such as cardiac pacemakers or defibrillators, blood sugar/blood pressure/heart rate/electroencephalography (EEG) signal monitors, insulin pumps, etc. They are mostly battery dependent, making their implementation's energy efficiency critical. Additionally, they must be able to connect to the wireless network of the system to transmit the "sensed" data. Numerous devices and sensors are simultaneously utilized and thus their interconnection is deemed necessary. Moreover, wearable devices must be portable, relatively small and comfortable while they are worn. Some of the implantable and wearable devices can be re-programmable through wireless communication or they can wirelessly receive various commands such as the adjustment of the drug dosage the device administrates or the configuration of the device settings. Furthermore, the wireless interface must connect to the Internet with the intention of transmitting the collected health data to healthcare personnel such as doctors and nurses. The processing of these data can be handled by the hospital's or private clinic's primary computing system, or by the near-patient IoT system. Some wearable devices may be able to process their collected data and afterwards wirelessly send them to the Internet. Alternatively, a more stationary or mobile device may be used to compensate for the small, wearable and implantable devices that do not possess processing capabilities. These intermediate devices first collect the data from multiple sensors which create the IoT system through the various established IoT networks, before completely or partially processing them and finally transmitting them to back-end systems and databases. Alternatively, they can receive authorized commands or data from back-end systems, process them and correspond through interaction with the sensors. Intelligence and real-time functionalities can be added by giving these devices the capability to make decisions and act accordingly without waiting for the back-end systems' response. Moreover, all IoT devices have limited or no storage capacity. Thus, the data are transmitted and stored in databases to be easily accessible and create a complete medical history.

Nevertheless, IoT can also enhance hospital interactions and its inner functionality [25]. All the previously mentioned implantable and wearable IoT devices can also be utilized inside hospital facilities. Similarly, these multiple sensors and devices must be interconnected through wired and wireless networks and be able to receive from authorized elements or send to appropriate components commands and sensitive data. The main additional features of this hospital architecture are two. First, the clinical beds, which monitor the patient's health through sensors, and all medical equipment, which collect the sensitive health data required for diagnosis, are given communication capabilities and added to the

IoT network. Second, the IoT network is directly connected with the medical records systems and healthcare database located in the hospital facility. Hence, the hospital personnel can receive data in real-time and immediately respond to emergencies. Furthermore, the stored data can be easily accessed by all hospital equipment and promptly provide the patient's medical history. Finally, the hospital's IoT-based network must be connected to the Internet and cloud-based platforms in order to communicate with near-patient systems and other healthcare facilities. The result will be the creation of more personalized and intelligent health services that assist healthcare providers while reducing the stress on hospital resources.

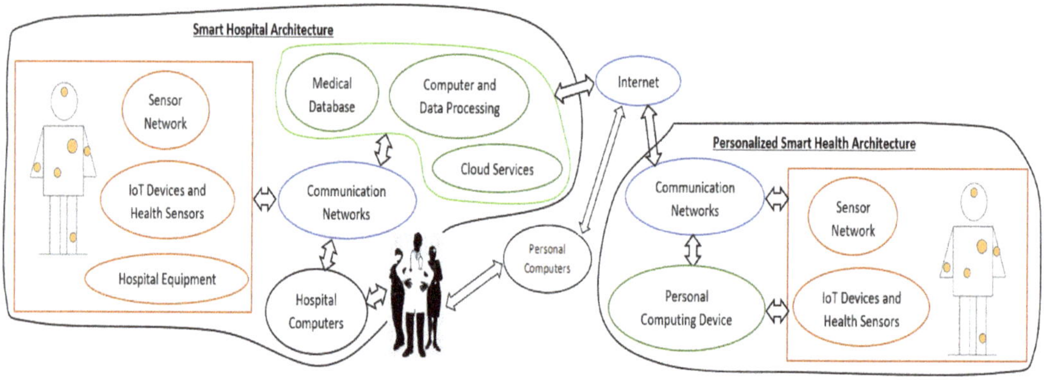

Figure 3. A general Smart Health architecture.

3. Security Requirements

As previously discussed, security is an exceptionally important aspect in the implementation of IoT-based healthcare applications. However, IoT networks that are utilized in the general smart health infrastructure, Figure 3, can be subject to a variety of attacks, with the IoT device being its most susceptible component [2]. An attacker can easily gather the personal information shared by the devices via the IoT network. Eavesdropping and data transmission/traffic monitoring are two main attacks that breach data privacy [11]. Furthermore, without proper data protection, the user authentication is also disrupted. Unauthorized devices can gain access to these data that can then be processed or altered for malicious purposes [28]. They also can fabricate erroneous health data and transmit them to the other components of the IoT network. This leads to inaccurate health diagnosis and unreliable communication between the patient and health provider.

The development of secure communication networks is a major concern for researchers [29,30]. Cryptography is a commonly used practice for ensuring data privacy and user authentication [31]. It utilizes cryptographic algorithms, namely ciphers, for "hiding" the content of various texts through encryption and afterwards "revealing" the original data through decryption. Nevertheless, the known cryptographic primitives cannot be applied to the IoT system because of the devices' resource constraints. The employed cipher must not deprive the valuable resources from other important functionalities that provide healthcare services. Therefore, a more lightweight version must be developed for properly corresponding to the hardware limitations of IoT. Furthermore, the delay of real-time applications because of the algorithms' low speed implementation can be devastating in critical circumstances. Hence, the encryption/decryption speed and the systems' immediate response must be taken into consideration. Finally, the system must provide multiple options with each functionality, satisfying different network and security needs depending on the application's current requirements. Thus, flexibility and scalability mechanisms must be added to the system.

Overall, for properly securing the health data of a smart health application, a lightweight cryptographic primitive must be implemented and a security scheme must be applied. First, the encryption algorithm must encrypt the collected data before they are shared between the utilized IoT devices. Thus, the patient's personal information is protected from malicious attacks. Similarly, the decryption algorithm must decrypt these received data for their further employment depending on the healthcare application. The result is the total content protection of the transmitted data in the employed communication networks, specifically throughout the IoT networks and the Internet that is connected to cloud-based services. Figure 4 depicts all the possible components of a general smart health infrastructure that can implement a lightweight-based security scheme and fulfill all these mentioned requirements.

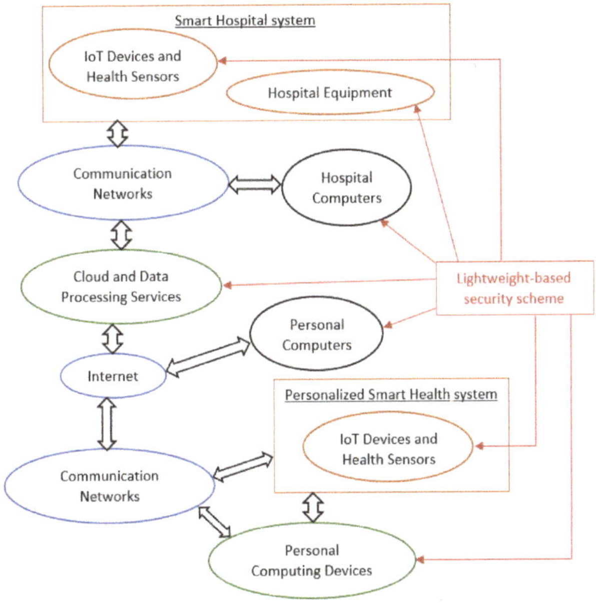

Figure 4. Lightweight-based security scheme.

4. Lightweight-Based Security Scheme

The proposed lightweight-based security scheme employs the LEAIoT cryptographic primitive which efficiently encrypts and decrypts the collected data while providing flexible options regarding the key size and speed implementation. This mechanism is implemented into every IoT device that is utilized in a healthcare system, including both smart hospital and near-patient architectures, and provides data privacy to the communication networks and the Internet. Overall, the proposed lightweight-based security scheme can be applied as depicted in Figure 4.

4.1. LEAIoT: Lightweight Encryption/Decryption Algorithm

LEAIoT is a lightweight cryptographic primitive whose key generation and encryption/decryption speed are higher than other common encryption primitives [17]. The advantages of a lightweight design are fitting in the IoT-based healthcare environment, which has complex communication requirements. Specifically, it provides speed-efficient end-to-end communication with minimum hardware resource utilization. LEAIoT combines a symmetric encryption algorithm and an asymmetric encryption algorithm while aiming to preserve both of their advantages. The symmetric cryptography provides further operational speed with a minimum number of computing resources. The asymmetric primitive has better key distribution and scalability schemes and ensures the confidentiality and authentication of the system.

The LEAIoT encryption process of a plaintext—which is given a synthetic value—first employs the symmetric key encryption with a private key n. Both sender and receiver know the context of this key. Afterwards, the asymmetric linear block cipher (NLBC) is executed with the previously produced ciphertext and two keys, namely a shared key n_1 and one private key k. The result is a completely secure encrypted text. The decryption sequence utilizes the modular inverse of the three encryption keys, via SSK (secure symmetric key), ${n_1}^{(-1)}$ and k'. First, the asymmetric—namely NLBC—decryption is performed with the transmitted ciphertext and the two keys, ${n_1}^{(-1)}$ and k'. These keys are produced by the modular inverse with modulo 37 of keys n_1 and k, respectively. The original plaintext is generated by applying the symmetric decryption with the SSK key, which is the modular inverse of n. For this algorithm, the modular inverse operation is always executed with modulo 37. A detailed description of the encryption and decryption processes is followed.

Encryption sequence:

1. The synthetic values of the plaintext are multiplied with the private key n. Then, modulo 37 is used;
2. A 3×3 matrix is created as the private key k. Additionally, the length of the key n is calculated and utilized as the key n_1;
3. The produced text from step 1 is segregated into blocks b_i matching the key k. Specifically, it is divided to multiple arrays of length equal to 3;
4. Each block b_i is multiplied with the keys k and n_1. Modulo 37 is subsequently applied;
5. The result is the secure ciphertext of the original plaintext;

Decryption sequence:

1. The modular inverse of the keys n_1 and k with modulo 37 is utilized;
2. The received ciphertext is divided to blocks b_i, as with step 3 of the encryption sequence;
3. Each block is multiplied with the two keys k' and ${n_1}^{(-1)}$. Modulo 37 is then applied;
4. The produced text is afterwards multiplied with the key SSK and modulo 37 is once again used;
5. The result is the original plaintext.

A simplified example of the encryption and decryption calculation processes is presented for a better comprehension of these two procedures. The private key n is 12,345, the inputted synthetic values are $\begin{bmatrix} 13 \\ 19 \\ 15 \end{bmatrix}$, and the private key k is $\begin{bmatrix} 1 & 2 & 1 \\ 4 & 5 & 6 \\ 7 & 8 & 9 \end{bmatrix}$.

Encryption sequence:

1. First, the inputted values are multiplied with key n:

$$\left(\begin{bmatrix} 13 \\ 19 \\ 15 \end{bmatrix} * 12345 \right) mod\ 37 = \begin{bmatrix} 16 \\ 12 \\ 27 \end{bmatrix}$$

2. Finally, the result of the previous calculation is multiplied with key k and the length of key n to produce the encrypted message:

$$\left(\begin{bmatrix} 1 & 2 & 1 \\ 4 & 5 & 6 \\ 7 & 8 & 9 \end{bmatrix} * \begin{bmatrix} 16 \\ 12 \\ 27 \end{bmatrix} * 5 \right) mod\ 37 = \begin{bmatrix} 2 \\ 24 \\ 35 \end{bmatrix}$$

Decryption sequence:

1. First, the SSK, k' and ${n_1}^{(-1)}$ are calculated:

$$SSK = \left(n^{(-1)}\right) mod\ 37 = 12345^{(-1)}\ mod\ 37 = 17$$

$$n_1{}^{(-1)} = \left(5^{(-1)}\right) mod\ 37 = 15$$

$$k' = k^{(-1)} = \begin{bmatrix} 18 & 23 & 32 \\ 1 & 25 & 12 \\ 18 & 1 & 18 \end{bmatrix}$$

2. Then, the encrypted message is multiplied with k' and $n_1{}^{(-1)}$:

$$\left(\begin{bmatrix} 18 & 23 & 32 \\ 1 & 25 & 12 \\ 18 & 1 & 18 \end{bmatrix} * \begin{bmatrix} 2 \\ 24 \\ 35 \end{bmatrix} * 15\right) mod\ 37 = \begin{bmatrix} 16 \\ 12 \\ 27 \end{bmatrix}$$

3. Finally, the original message is revealed by multiplying the previous result with the SSK key:

$$\left(\begin{bmatrix} 16 \\ 12 \\ 27 \end{bmatrix} * 17\right) mod\ 37 = \begin{bmatrix} 13 \\ 19 \\ 15 \end{bmatrix}$$

4.2. Implementation of the Proposed Lightweight-Based Security Scheme

In this section, the proposed lightweight-based security scheme is presented. The LEAIoT-based lightweight encryption/decryption implementation has a symmetric and asymmetric encryption and decryption stages. The user can choose the length of the symmetric key between 32, 64, 128 or 256 bits. For the NLBC or asymmetric phase, the size of the private key k is a 3×3 matrix with each element having a value within the range of $[1, 36]$. The overall design of the proposed cryptographic approach is demonstrated in Figure 5. The architecture has additional input and output signals that are not depicted in this figure. The extra input signals concern the start of a new procedure, the selection of the procedure that will be executed and the selection of the key size. The supplementary output signals indicate the completion of each key generation procedure, the encryption process and the decryption process.

The main units of the architecture are analyzed below.

- ENCR_DECR_ROUND unit contains the necessary resources for the encryption and decryption of the input text. It also executes the additions and multiplications of finite elements and sends the results into the appropriate units.
- The MOD37_CALC unit calculates modulo 37 of symmetric key n.
- DET_MOD37_CALC unit's input is a 3×3 matrix, namely key k, and its outputs are modulo 37 of the matrix's determinant and table I, which is an intermediate step for computing the modular inverse of the matrix. Each element of table I, namely i_{ab}, is calculated by the following equation:

$$i_{ab} = det\ 2_{ab} * \left((-1)^{(a+b)}\right) mod\ 37, \tag{1}$$

  where $det\ 2_{ab}$ is the determinant of a 2×2 matrix created after erasing row a and column b of the input matrix.
- MULT_INV_MOD37 unit computes the modular inverse of the input. The input can either be module 37 of the symmetric key n or the determinant's module 37 of key k, which was calculated by DET_MOD37_CALC. The input is selected by a multiplexer.
- MATRIX_INV_CALC unit receives the determinant's modular inverse of key k as input, as computed by MULT_INV_MOD37, and receives the intermediate table I as input, as calculated by DET_MOD37_CALC. The output is the modular inverse of key k.
- LENGTH_CALC_MOD37 unit calculates the length of symmetric key n, namely generating key n_1. The modular inverse of this key can be produced by the MULT_INV_MOD37

unit. The procedure is the same as with the modular inverse computation of symmetric key n.

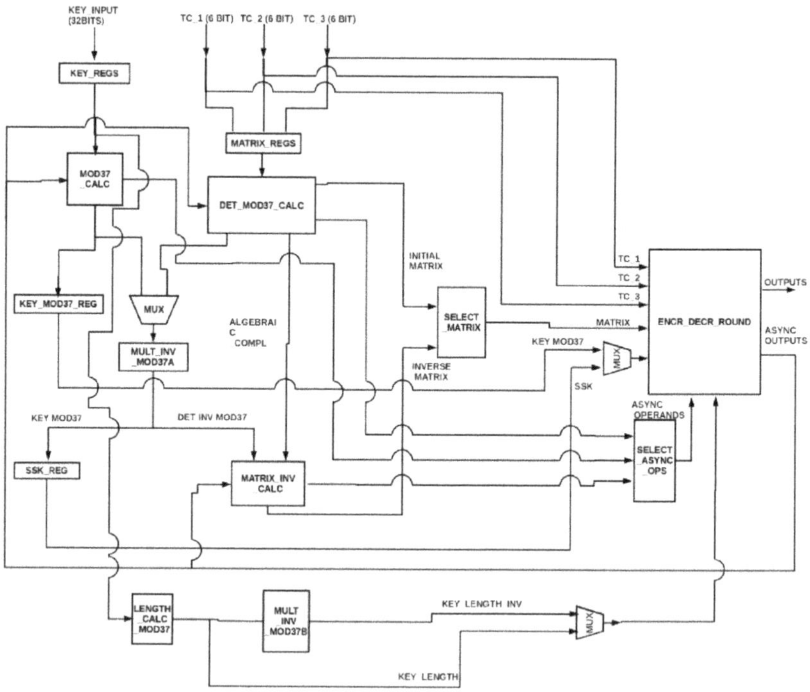

Figure 5. Proposed architecture of cryptographic primitive.

The four main functions of the proposed lightweight architecture are the insertion of a new symmetric key n; insertion of a new asymmetric key k; encryption; and decryption. The first function changes the key n of the symmetric encryption/decryption stage and calculates the new SSK. The user selects the length of the new key, namely 32, 64, 128 or 256 bits. The key is inserted in blocks of 32-bit and is stored in the registers KEY_REGS. The contents of registers KEY_REGS are then given as inputs to the MOD37_CALC, which calculates the $n \bmod 37$. The result is stored in the register KEY_MOD37_REGS, which is the input of the MULT_INV_MOD37A unit. This unit calculates the $n^{-1} \bmod 37$, namely the SSK key. The produced SSK key is stored in the register SSK_REG. Finally, the units LENGTH_CALC_MOD37 and MULT_INV_MOD37B calculate the $n_1 \bmod 37$ and $n_1^{(-1)} \bmod 37$, respectively.

The second function, namely the insertion of a new asymmetric key k, changes the key k and calculates its modular inverse, namely k'. The insertion of the key, which is a 3×3 matrix, is executed in three stages. In each stage, the three elements of a row are inserted via the three inputs, tc_1, tc_2 and tc_3. The elements are stored in the registers MATRIX_REGS. The completely inserted key is then given as input in the DET_MOD37_CALC unit. This unit produces two outputs, namely the determinant's module 37 of the key and table I. The first output is inserted to the MULT_INV_MOD37A unit, which calculates its modular inverse. The second output is given as one of the two inputs of the MATRIX_INV_CALC unit. The other input is produced by the MULT_INV_MOD37A unit modular inverse of the determinant. The final output is the modular inverse of the new asymmetric key k, namely k'.

The encryption function begins by inserting three characters of the plaintext via the inputs tc_1, tc_2 and tc_3 to the ENCR_DECR_ROUND unit. The other inputs of this unit are modulo 37 of key n; modulo 37 of the n key's length; and key k. The result is the ciphertext which is computed by the correct processing of the data. The final function, namely decryption, starts by inserting three characters of the ciphertext via the three inputs tc_1, tc_2 and tc_3 to the ENCR_DECR_ROUND. The other inputs of this unit are the modular inverse of the key n (SSK), the modular inverse of the key n length (${n_1}^{(-1)}$) and the modular inverse of the key k (k').

4.2.1. Modulo 37 Calculation Module

The basic element of this module is an asynchronous unit that calculates module 37 of a 32-bit number. Its operation is based on the algorithm for modulo reduction that was proposed in [32]. Moreover, the multiplications are implemented based on the Wallace method for multiplying integers using only combinational logic [33]. There are three steps for the module's process. First, the key is divided into 32-bit sections and modulo 37 of each section is calculated. The results are stored in registers. For 32-bit keys, the process is completed with this step. The calculation and data storage of each section is completed in one clock cycle.

Afterwards, modulo 37 of each section—excluding the 32 least important bits—is multiplied with modulo 37 of a power of 2, which depends on its position. Specifically, for module 37 of the $[n, n + 31]$ bits, the operation 2^n mod 37 is used. Then, the results are stored in registers. For utilized hardware reduction purposes, the multiplications of module 37 are only implemented in the encryption/decryption unit and not in this unit. The encryption/decryption unit can execute three modulo 37 multiplications anytime. Hence, the process is completed after a clock cycle for 64-bit or 128-bit keys, because one and three multiplications are needed, respectively. For a 256-bit key, seven multiplications are required, thus the process is completed after three clock cycles.

Finally, the sum of the previous results' modulo 37 is calculated. Similar with the multiplications, the additions are only implemented in the encryption/decryption unit, with three of them being ready anytime. For a 64-bit key, only one addition is executed and its completion requires one clock cycle. For a 128-bit key, two clock cycles are needed. Finally, for the 256-bit key, four clock cycles are required.

4.2.2. Three-Dimensional Matrix's Determinant Calculation Module

The input of this module is a 3×3 matrix M and its outputs are the determinant's module 37 and table I. The calculation process also follows three steps. First, the data of the input matrix M are inserted into the module that produces table I. After the computation of element I_{11}, this same element and M_{11} are sent to the encryption/decryption unit. The unit calculates modulo 37 of their multiplication and stores the result in a register. In a similar way, the operations $M_{12} * D_{12}$ mod 37 and $M_{13} * I_{13}$ mod 37 are computed and stored. It is important to clarify that the two elements I_{12} and D_{12} are generated simultaneously. After the complete computation of table I, the results from the two operations $M_{11} * I_{11}$ mod 37 and $M_{13} * I_{13}$ mod 37 are sent to the encryption/decryption unit. Modulo 37 of their sum (S) is calculated. Finally, the difference $Diff = S - (M_{12} * D_{12}\ mod\ 37)$ is computed and the process is completed.

The module that produces table I executes the following procedure. First, the matrix M elements are stored in registers. Afterwards, the cells of table I are generated sequentially. The i_{ab} elements are produced by the det2 $\times$ 2 module. This module receives the data of the 2×2 matrix, which is generated from the input matrix M and calculates modulo 37 of the input's determinant. If $(-1)^{(a+b)} = 1$, then the requested element of table I is the output of det2 $\times$ 2, which is stored in a register. If $(-1)^{(a+b)} = -1$, then the output of det2 $\times$ 2 is subtracted from the value 37. This difference is the requested element. It must be clarified that, during the computation of i_{12}, both the output of the det2 $\times$ 2 module, which is utilized in the calculation of the M matrix's determinant, and the result of this output subtracting with the value 37, which is the element i_{12}, are stored in registers.

The two-dimensional matrix's determinant calculation module receives a 2×2 E matrix as input and generates modulo 37 of the input's determinant. First, the E matrix's elements are stored in registers. These elements are sent to the encryption/decryption unit, which calculates and stores the values $p_1 = E_{11} * E_{22}\ mod\ 37$ and $p_2 = E_{12} * E_{21}\ mod\ 37$ in registers. Afterwards, the value $d_1 = p_1 - p_2$ is computed. If $d_1 \geq 0$, then the E matrix's determinant is equal the value d_1, which is stored to the output register, and the calculation is completed. If $d_1 < 0$, then value $d_2 = p_2 - p_1$ is calculated and stored in the output register. Lastly, the value $d' = 37 - d_2$ is computed and stored in the output register, thus completing the operation.

4.2.3. Modular Inverse Calculation Module and Three-Dimensional Matrix's Modular Inverse Calculation Module

The modular inverse calculation module, which exclusively uses combinational logic, accepts as input a 6-bit value (a) in the range of [0, 36] and returns the value b where $(a * b)\ mod\ 37 = 1$. A system of logic gates assists the module in identifying the value and then selecting the appropriate output b with the usage of multiplexers.

The three-dimensional matrix's modular inverse calculation module receives as input table I and the modular inverse of the M matrix's determinant. The output is the modular inverse of the M matrix (M′). The M′ matrix's elements of each row are calculated by multiplying the elements in table I's corresponding column with the modular inverse of the determinant. Furthermore, the multiplications are executed in the encryption/decryption unit. Therefore, the process is completed after three clock cycles since three multiplications can be executed anytime table I is a 3×3 matrix.

4.2.4. Encryption and Decryption Unit

The proposed architecture of the encryption and decryption unit is depicted in Figure 6. The inputs are the following:

- A 3×3 matrix with 6-bit elements (m11, m12, … , m33);
- Three 6-bit characters (tc_1, tc_2, tc_3) from the text which will be encrypted or decrypted;
- A 6-bit key (sk_o);
- A 6-bit signal (kl_o), which is modulo 37 of the key length or its modular inverse;
- Three single bit signals that indicate the start of a new operation or the applied process, namely encryption or decryption;
- Twelve different data for the generation of the keys, six of them (async_add_opa1, async_add_opa2, async_add_opa3, async_add_opb1, async_add_opb2, async_add_opb3) are added together while the other six (async_mul_opa1, async_mul_opa2, async_mul_opa3, async_mul_opb1, async_mul_opb2, async_mul_opb3) are multiplied with each other;
- A signal (async_op_is_add) which selects whether the operation for the keys' generation is addition or multiplication.

Moreover, the basic modules of this unit are the v_add, v_mul and v_mod. The first one consists of three parallel additions. The second one consists of three parallel multiplications, which are implemented based on the Wallace algorithm. The last one consists of three smaller versions of the modulo computation, which is based on the Barrett algorithm.

In the first clock cycle of the encryption process, the inputs tc_1, tc_2 and tc_3 are multiplied with the sk_o in the v_mult module. The results are then sent to the v_mod module, which calculates modulo 37. The v_mod module's outputs are stored in the output registers. For the next three cycles, each row of the input matrix is multiplied with the array that contains the values of the output registers element by element. Modulo 37 of the produced values is computed and then stored in the registers mr11, … , mr33. In the next two cycles, the columns of the matrix, which contain the values of registers mr11, … , mr33, are being added together in the v_add module, their modulo is computed in v_mod and the result is stored in the output registers. Finally, the values of the output registers are multiplied with the kl_o, their modulo 37 is calculated, the results are stored in the output registers afresh and the signal, which indicates the process completion, is generated.

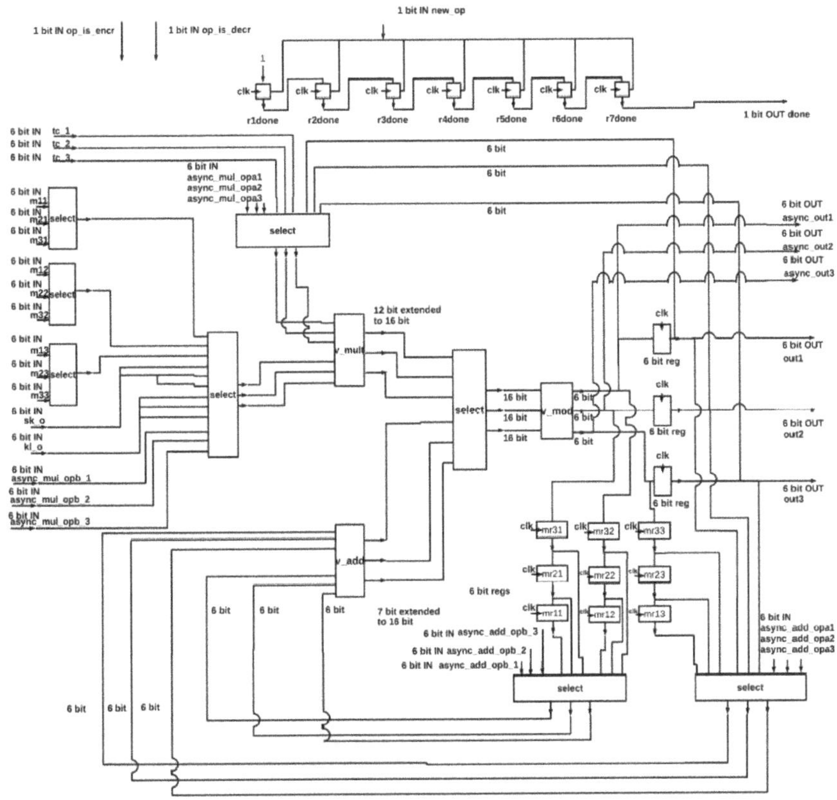

Figure 6. The architecture of the encryption and decryption unit.

For the first three clock cycles of the decryption process, each row of the input matrix is multiplied element by element with the array [tc_1, tc_2, tc_3], modulo 37 of the produced values is computed, and the results are stored in register mr11, ... , mr33. In the next two cycles, the columns of the matrix, which contain the values of registers mr11, ... , mr33, are being added together, their modulo 37 is calculated and the results are stored in the output registers. The values of the output registers are multiplied with the kl_o and modulo 37 of the produced result is stored in the output registers. Finally, the new values of the output registers are multiplied with the ks_o, modulo 37 of the results is stored in the output registers afresh and the signal indicating the process completion is generated.

Additions and multiplications with modulo 37 are required for both the generation of the keys and the encryption and decryption procedure. For efficient hardware utilization, the modules of the encryption/decryption unit are also employed for the key generation functions. Therefore, the unit executes the appropriate additions and multiplications with the twelve given inputs (async_add_opa1, async_add_opa2, async_add_opa3, async_add_opb1, async_add_opb2, async_add_opb3, async_mul_opa1, async_mul_opa2, async_mul_opa3, async_mul_opb1, async_mul_opb2, async_mul_opb3) and returns the results to the key generation modules via the outputs async_out1, async_out2 and async_out3.

5. Synthesis Implementation Results

The proposed architecture is simulated and implemented in FPGA 7z007s-clg400 [34] using VHDL. This FPGA was selected because it had the fewest hardware resources in the

Vivado web pack 2018.3 [35]. For the simulation, the ghdl tool [36] was utilized, and for the synthesis and implementation, the Vivado web pack 2018.3 tool was used.

5.1. Simulations

The insertion and the modular inverse calculation of the symmetric key and the asymmetric key are simulated. For a better comprehension of the two operations, Figures 7 and 8 present some examples for various key sizes.

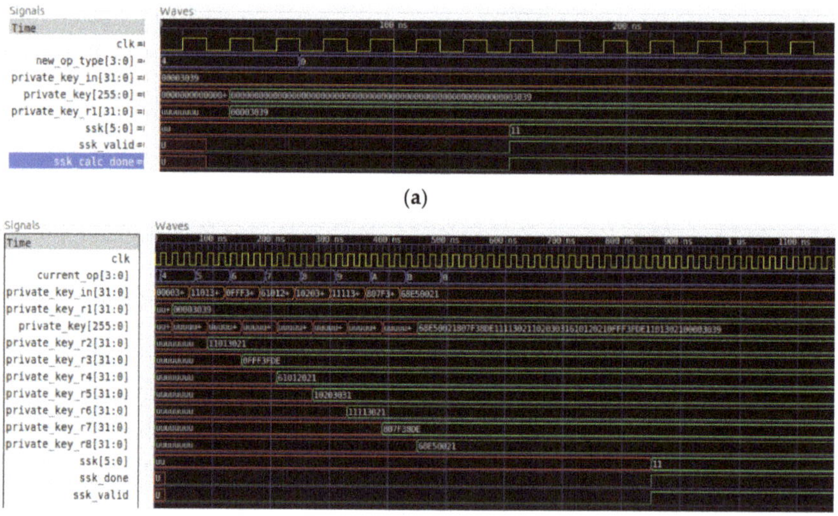

(a)

(b)

Figure 7. Insertion and modular inverse calculation of the symmetric key with: (**a**) 32-bit; and (**b**) 256-bit size.

The operations for inserting the new symmetric keys are denoted by the values 4_{hex} for a 32-bit key and [4_{hex}, B_{hex}] for a 256-bit key. The complete insertion of the matrix, namely the asymmetric key, is denoted by the values C_{hex}, D_{hex} and E_{hex}. Moreover, the 0_{hex} and 1_{hex} indicate the start of the modular inverse operation.

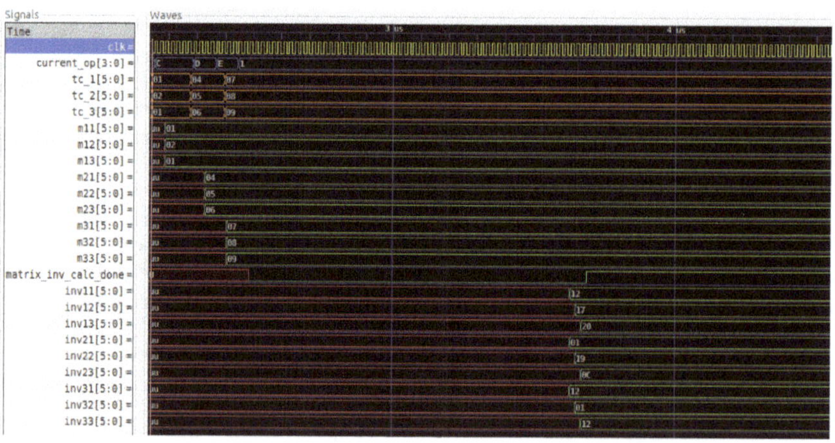

Figure 8. The modular inverse calculation of the asymmetric key.

The encryption, which is denoted with the 2_{hex} value, and the decryption, which is denoted with the 3_{hex} value, are also simulated, as demonstrated in Figure 9.

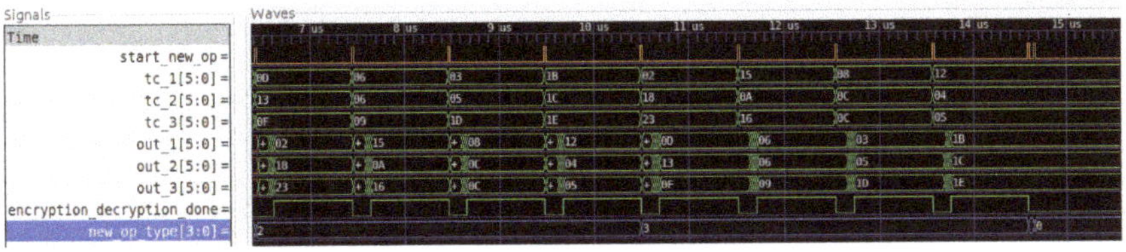

Figure 9. Encryption and decryption operations.

5.2. Implementation

The proposed architecture of the encryption and decryption processes three 6-bit characters in each iteration. Furthermore, each function requires a total of nine clock cycles. The first cycle is needed to properly start the encryption or decryption operation and in the next eight cycles, the operation is executed. The results of the proposed implementations are presented in Table 1. The clock cycles required for the generation of both keys are displayed in Table 2. In these clock cycles, the cycles needed for the start signal generation are included.

Table 1. The characteristics of the proposed lightweight-based security scheme.

LUTs	Flip-Flops	F7-Muxes	F8-Muxes	Execution Time	Frequency	Throughput	Throughput/Slices
2700	815	67	24	261 ns	34.483 MHz	68.965 Mbps	0.0191

Table 2. Clock cycles for the generation of keys.

S.K. * Size (bit)	Insertion of S.K. *	Modular Inverse of S.K. *	Insertion of A.K. **	Modular Inverse of A.K. **	Total
32	2	5	6	61	74
64	4	8	6	61	79
128	8	12	6	61	87
256	16	19	6	61	102

* symmetric key. ** asymmetric key.

6. Performance Results

The proposed architecture of the LEAIoT-based encryption and decryption process is resource and speed efficient. Therefore, it can be employed in IoT-based healthcare applications for lightweight security in the communication network. Moreover, the generation speed of the symmetric and asymmetric keys is exceptionally high. The implementational results, as presented in Table 3, indicate the architecture's superiority over the original CPU-based design in [17]. Specifically, the proposed design has 99.9% higher key generation speed than the original for all key sizes and it achieves 99.3%, 98.4%, 97.5% and 96.2% higher encryption/decryption speed for 100 kilobits, 300 kilobits, 500 kilobits and 1000 kilobits, respectively. There are two reasons for these great differences in key generation speed. The first is the speed acceleration ability of FPGAs over CPUs. A CPU is a chip that executes the instructions of a program sequentially, whereas an FPGA is a configurable chip consisting of logic blocks which is ideal for the parallelization and simultaneous execution of instructions. Thus, the FPGA can be properly modified to execute the specific operations of the program better and faster in a parallel manner in contrast to the general-purpose CPU. The last reason is the design efficiency of the proposed methodology which parallelizes the sequential calculations of the algorithm and exploits the optimized nature of the FPGA.

Table 3. Performance comparisons with the original design in [17].

Operation	Original CPU-Based LEAIoT [17] (ms)	The Proposed FPGA-Based Architecture (ms)
32-bit key generation	4	0.002146
64-bit key generation	10	0.002291
128-bit key generation	16	0.002523
256-bit key generation	22	0.002958
Encr/Decr * of 100 kilobits	210	1.45
Encr/Decr * of 300 kilobits	275	4.35
Encr/Decr * of 500 kilobits	290	7.25
Encr/Decr * of 1000 kilobits	386	14.5

* encryption/decryption.

To the best of the authors' knowledge, there are no other implementations of the lightweight encryption algorithm LEAIoT. Thus, this architecture is the most lightweight and time-efficient implementation of LEAIoT. It also has four different key selection choices that provide flexibility in the system. Hence, depending on the application's requirements, the latency and security of the architecture can be configured.

Furthermore, the proposed architecture is compared to various cryptography implementations from the literature, as presented in Table 4. The comparison parameters are the key size which can be a general representative of the provided security level, the number of implemented Look-Up Tables (LUTs), which are custom truth tables that determine the output value for given inputs and can be compared to a small RAM, and the number of utilized Flip-Flops (FFs), which are binary shift registers that store logical states between the system's clock cycles. Moreover, the frequency and throughput of all these designs, which represent the system's performance efficiency, are also compared for better examining the proposed architecture's execution and implementation advantages.

All the compared cryptography designs are developed in different FPGAs with higher frequencies, contrasting the LEAIoT implementation. Therefore, a proper comparison of the architectures cannot be achieved. Nevertheless, for further proving the efficiency of the proposed system, a typical comparison is performed. First, even though the proposed design has the lowest frequency, it achieves better hardware consumption than most of the other ciphers. It utilizes 87.9% and 76.9% fewer resources than the lightweight versions of AES. It also achieves a 65.7% and 12.2% decrease in resource consumption compared to SNOW 3G and ZUC ciphers. Moreover, it achieves an almost double throughput than RC4 cipher, even with its comparatively lower frequency. Lastly, the proposed architecture has relatively similar performance results with PRESENT and CLEFIA. One significant difference is the lower frequency of the LEAIoT design. Thus, it can be concluded that the hardware and throughput efficiency of the proposed implementation can surpass the PRESENT and CLEFIA approaches by simply increasing the frequency.

Table 4. Performance comparisons from the literature.

Algorithm	Key Size	LUTs	FFs	Frequency (MHz)	Throughput (Mbps)
SNOW 3G [37]	128	7881	2391	28.84	922.88
AES-128 [38]	128	20402/14798	8704/1345	332.34/272.33	4342/3485
AES-MPPRM [39]	128	8129	7119	81.328	42.92
PRESENT [40]	80/128	215/264	153/201	213.81/194.63	102.89/91.59
ZUC [41]	256	2494	1512	209.346	6540
CLEFIA [42]	128/192/256	1725	663	147	990/818/696
RC4 [43]	-	277	197	123.64	41.21
Proposed design	32/64/128/256	2700	815	34.483	68.965

Finally, the proposed design will be evaluated depending on the security and performance requirements that were previously clarified. The employed cryptographic primitive was tested and the provided security was verified through simulation. It has also been implemented with four different key sizes that the user can choose from in order to provide configuration abilities and flexibility for constantly corresponding to varying network speeds and security needs. For example, when the network is overloaded, a smaller key size can be selected to accelerate the encryption process even more. Likewise, for crucially important messages, a higher key size can be utilized to efficiently protect the contents of the data. Furthermore, the key generation and encryption/decryption speed have been accelerated offering great transmission speed to the system and facilitating the maintenance of the system's constant availability. Hence, it can respond to the high-speed demands of the IoT network. Lastly, the synthesis's results together with the comparisons with other related hardware-based works proved the lightweight capability and performance efficiency of the proposed architecture required for IoT-based healthcare systems. The design achieves an effective balance between resources, throughput and security, presenting a novel security implementation for the analyzed IoT-based healthcare architecture.

7. Conclusions

In this paper, the IoT-based multi-sensor architecture was analyzed. This was first achieved by demonstrating the Health 4.0 design framework and investigating the smart health infrastructure. This extensive presentation of the particular environment offers a proper guidance to efficiently employing IoT in the healthcare domain, either for smart hospitals or for personalized smart health systems. Furthermore, a deep comprehension of the domain's current state was achieved with the display of representative research. Security, as one of the main concerns in IoT, was also discussed. The maintenance of data privacy and user authentication is the most important requirement of general smart health infrastructure. This has led to the proposition of a new hardware-based security implementation suitable for IoT devices. The proposed lightweight-based security scheme, which utilizes the LEAIoT encryption/decryption algorithm, provides more key selection possibilities than other compared architectures. Hence, it can offer extra speed when the network is overloaded. Furthermore, its hardware-based design has 99.9% higher key generation speed and 96.2% higher encryption/decryption speed for 1000 kilobits compared to the original CPU-based LEAIoT implementation presented in [17]. Moreover, it is resource-efficient because it employs the fewest number of FFs and a comparative small number of LUTs. Specifically, it achieves a 87.9%, 65.7% and 12.2% decrease in hardware resources compared to the lightweight ciphers, AES, SNOW 3G and ZUC, respectively, which were implemented in similar hardware devices. This trait can be efficiently utilized in an IoT-based multi-sensor environment because even the smallest devices can execute this design and be secured. Lastly, even though it has low throughput, it also has one of the lowest frequencies which better represents the resource limitations of IoT devices. Overall, it is best suited for applications that prioritize resource-efficiency and the high-speed generation of multiple keys. It has also proven capable of efficiently satisfying the main security and performance requirements of the introduced IoT-based smart health structure, providing innovative results and optimizations.

Future works will aim at both design improvements and further application employment. Specifically, the increase in the frequency and the matrix's size are the next two steps that will enhance the efficiency of the implementation. Finally, another future objective will be the simulation of this security architecture to IoT-based environments that collect and transmit healthcare data in real time.

Author Contributions: V.A., K.T. and N.S. have contributed to methodology, research, software, hardware, testbed, validation, results analysis, investigation, resources, writing—review and editing. All authors have read and agreed to the published version of the manuscript.

Funding: This research received no external funding.

Institutional Review Board Statement: Not applicable.

Informed Consent Statement: Not applicable.

Data Availability Statement: All of the reported data are included in the manuscript.

Conflicts of Interest: The authors declare no conflict of interest.

References

1. Fei, H. *Security and Privacy in Internet of Things (IoTs): Models, Algorithms, and Implemenations;* CRC Press: Boca Raton, FL, USA, 2016; ISBN 9781498723183.
2. Stellios, I.; Kotzanikolaou, P.; Psarakis, M.; Alcaraz, C.; Lopez, J. A survey of IoT-Enabled cyberattacks: Assessing attack paths to critical infrastructures and services. *IEEE Commun. Surv. Tutor.* **2018**, *20*, 3453–3495. [CrossRef]
3. Baker, S.B.; Xiang, W.; Atkinson, I. Internet of things for smart healthcare: Technologies, challenges, and opportunities. *IEEE Access* **2017**, *5*, 26521–26544. [CrossRef]
4. Al-Jaroodi, J.; Mohamed, N.; Abukhousa, E. Health 4.0: On the way to realizing the healthcare of the future. *IEEE Access* **2020**, *8*, 211189–211210. [CrossRef] [PubMed]
5. Mohamed, N.; Al-Jaroodi, J. The Impact of Industry 4.0 on healthcare system engineering. In Proceedings of the 2019 IEEE International Systems Conference (SysCon), Orlando, FL, USA, 8–11 April 2019; pp. 1–7.
6. Tsantikidou, K.; Sklavos, N. Vulnerabilities of Internet of Things, for Healthcare Devices and Applications. In Proceedings of the 2021 8th NAFOSTED Conference on Information and Computer Science (NICS), Hanoi, Vietnam, 21–22 December 2021.
7. Simplicio, M.A., Jr.; Silva, M.V.M.; Alves, R.C.A.; Shibata, T.K.C. Lightweight and escrow-less authenticated key agreement for the internet of things. *Comput. Commun.* **2017**, *98*, 43–51. [CrossRef]
8. Makhdoom, I.; Abolhasan, M.; Lipman, J.; Liu, R.P.; Ni, W. Anatomy of Threats to the Internet of Things. *IEEE Commun. Surv. Tutor.* **2019**, *21*, 1636–1675. [CrossRef]
9. Lounis, K.; Zulkernine, M. Attacks and defenses in short-range wireless technologies for IoT. *IEEE Access* **2020**, *8*, 88892–88932. [CrossRef]
10. Butun, I.; Österberg, P.; Song, H. Security of the Internet of Things: Vulnerabilities, attacks, and countermeasures. *IEEE Commun. Surv. Tutor.* **2020**, *22*, 616–644. [CrossRef]
11. Khanam, S.; Ahmedy, I.B.; Idna Idris, M.Y.; Jaward, M.H.; Bin Md Sabri, A.Q. A survey of security challenges, attacks taxonomy and advanced countermeasures in the Internet of Things. *IEEE Access* **2020**, *8*, 219709–219743. [CrossRef]
12. Burg, A.; Chattopadhyay, A.; Lam, K.Y. Wireless communication and security issues for cyber–Physical systems and the internet-of-things. *Proc. IEEE* **2018**, *106*, 38–60. [CrossRef]
13. Tsantikidou, K.; Sklavos, N. Hardware security for IoT-based, healthcare applications. In Proceedings of the New England Hardware Security Day 2021 Workshop, Virtual, 9 April 2021.
14. Bikos, A.N.; Sklavos, N. The future of privacy and trust on the internet of Things (IoT) for healthcare: Concepts, Challenges, and Security Threat Mitigations. In *Recent Advances in Security, Privacy, and Trust for Internet of Things (IoT) and Cyber-Physical Systems (CPS);* CRC Press: Boca Raton, FL, USA, 2020.
15. Panagiotou, P.; Sklavos, N.; Zaharakis, I.D. Design and implementation of a privacy framework for the internet of things (IoT). In Proceedings of the 2018 21st Euromicro Conference on Digital System Design (DSD), Prague, Czech Republic, 29–31 August 2018.
16. Tsavos, M.; Sklavos, N.; Alexiou, G.P. Lightweight security data streaming, based on reconfigurable logic, for FPGA platform. In Proceedings of the 2020 23rd Euromicro Conference on Digital System Design (DSD), Kranj, Slovenia,, 26–28 August 2020.
17. Habib, M.A.; Ahmad, M.; Jabbar, S.; Ahmed, S.H.; Rodrigues, J.J.P.C. Speeding up the Internet of Things: LEAIoT: A lightweight encryption algorithm toward low-latency communication for the Internet of Things. *IEEE Consum. Electron. Mag.* **2018**, *7*, 31–37. [CrossRef]
18. Kocakulak, M.; Butun, I. An overview of wireless sensor networks towards Internet of Things. In Proceedings of the 2017 IEEE 7th Annual Computing and Communication Workshop and Conference (CCWC), Las Vegas, NV, USA, 9–11 January 2017.
19. Al-Garadi, M.A.; Mohamed, A.; Al-Ali, A.K.; Du, X.; Ali, I.; Guizani, M. A Survey of Machine and Deep Learning Methods for Internet of Things (IoT) Security. *IEEE Commun. Surv. Tutor.* **2020**, *22*, 1646–1685. [CrossRef]
20. Rodrigues, J.J.; Segundo DB, D.R.; Junqueira, H.A.; Sabino, M.H.; Prince, R.M.; Al-Muhtadi, J.; De Albuquerque, V.H.C. Enabling Technologies for the Internet of Health Things. *IEEE Access* **2018**, *6*, 13129–13141. [CrossRef]
21. Egala, B.S.; Pradhan, A.K.; Badarla, V.; Mohanty, S.P. Fortified-Chain: A blockchain-based framework for security and privacy-assured Internet of medical things with effective access control. *IEEE Internet Things J.* **2021**, *8*, 11717–11731. [CrossRef]
22. Firouzi, F.; Farahani, B.; Barzegari, M.; Daneshmand, M. AI-Driven Data Monetization: The other Face of Data in IoT-based Smart and Connected Health. *IEEE Internet Things J.* **2020**, *9*, 5581–5599. [CrossRef]
23. Ma, Y.; Wang, Y.; Yang, J.; Miao, Y.; Li, W. Big health application system based on health Internet of Things and big data. *IEEE Access* **2017**, *5*, 7885–7897. [CrossRef]
24. Kamal, N.; Ghosal, P. Three tier architecture for IoT driven health monitoring system using raspberry Pi. In Proceedings of the 2018 IEEE International Symposium on Smart Electronic Systems (iSES) (Formerly iNiS), Hyderabad, India, 17–19 December 2018.

25. Zhang, H.; Li, J.; Wen, B.; Xun, Y.; Liu, J. Connecting intelligent things in smart hospitals using NB-IoT. *IEEE Internet Things J.* **2018**, *5*, 1550–1560. [CrossRef]
26. Cabra, J.; Castro, D.; Colorado, J.; Mendez, D.; Trujillo, L. An IoT approach for wireless sensor networks applied to e-health environmental monitoring. In Proceedings of the 2017 IEEE International Conference on Internet of Things (iThings) and IEEE Green Computing and Communications (GreenCom) and IEEE Cyber, Physical and Social Computing (CPSCom) and IEEE Smart Data (SmartData), Exeter, UK, 21–23 June 2017.
27. Wu, F.; Wu, T.; Zarate, D.C.; Morfuni, R.; Kerley, B.; Hinds, J.; Taniar, D.; Armstrong, M.; Yuce, M.R. An autonomous hand hygiene tracking sensor system for prevention of hospital associated infections. *IEEE Sens. J.* **2021**, *21*, 14308–14319. [CrossRef]
28. Neshenko, N.; Bou-Harb, E.; Crichigno, J.; Kaddoum, G.; Ghani, N. Demystifying IoT security: An exhaustive survey on IoT vulnerabilities and a first empirical look on internet-scale IoT exploitations. *IEEE Commun. Surv. Tutor.* **2019**, *21*, 2702–2733. [CrossRef]
29. Sklavos, N.; Zaharakis, I.D.; Kameas, A.; Kalapodi, A. Security & trusted devices in the context of Internet of Things (IoT). In Proceedings of the 2017 Euromicro Conference on Digital System Design (DSD), Vienna, Austria, 30 August–1 September 2017.
30. Ioannidou, I.; Sklavos, N. On general data protection regulation vulnerabilities and privacy issues, for wearable devices and fitness tracking applications. *Cryptography* **2021**, *5*, 29. [CrossRef]
31. Stallings, W. *Cryptography and Network Security*, 6th ed.; Pearson: Upper Saddle River, NJ, USA, 2014.
32. Barrett, P. Implementing the rivest shamir and adleman public key encryption algorithm on a standard digital signal processor. In *Advances in Cryptology—CRYPTO' 86*; Springer: Berlin/Heidelberg, Germany, 1987; pp. 311–323.
33. Wallace, C.S. A Suggestion for a Fast Multiplier. *IEEE Trans. Electron. Comput.* **1964**, *EC-13*, 14–17. [CrossRef]
34. Xilinx. *Zynq-7000 SoC Data Sheet: Overview*; DS190 datasheet; Xilinx: San Jose, CA, USA, 2018.
35. Xilinx. *Vivado Design Suite User Guide: Release Notes, Installation, and Licensing*; UG973 datasheet; Xilinx: San Jose, CA, USA, 2018.
36. GHDL Documentation. Release 1.0-dev Datasheet. Available online: https://ghdl-rad.readthedocs.io/_/downloads/en/latest/pdf/ (accessed on 29 April 2022).
37. Madani, M.; Benkhaddra, I.; Tanougast, C.; Chitroub, S.; Sieler, L. FPGA implementation of an enhanced SNOW-3G stream cipher based on a hyperchaotic system. In Proceedings of the 2017 4th International Conference on Control, Decision and Information Technologies (CoDIT), Barcelona, Spain, 5–7 April 2017.
38. Lata, K.; Saini, S. Hardware Software Co-Simulation of an AES-128 based data encryption in image processing systems for the internet of things environment. In Proceedings of the 2020 IEEE International Symposium on Smart Electronic Systems (iSES) (Formerly iNiS), Chennai, India, 14–16 December 2020.
39. Kumar, T.; Reddy, K.; Rinaldi, S.; Parameshachari, B.; Arunachalam, K. A Low area high speed FPGA implementation of AES architecture for cryptography application. *Electronics* **2021**, *10*, 2023. [CrossRef]
40. Lara-Nino, C.A.; Diaz-Perez, A.; Morales-Sandoval, M. Lightweight Hardware Architectures for the Present Cipher in FPGA. *IEEE Trans. Circuits Syst. I Regul. Pap.* **2017**, *64*, 2544–2555. [CrossRef]
41. Yang, Y.; Zhao, W.; Xiong, L.; Wang, N.; Ma, Y. Optimized implementations for ZUC-256 on FPGA. *Wirel. Pers. Commun.* **2021**, *116*, 2615–2632. [CrossRef]
42. Cheng, X.; Zhu, H.; Xu, Y.; Zhang, Y.; Xiao, H.; Zhang, Z. A reconfigurable and compact hardware architecture of CLEFIA block cipher with multi-configuration. *Microelectron. J.* **2021**, *114*, 105144. [CrossRef]
43. Taqieddin, E.; Abu-Rjei, O.; Mhaidat, K.; Bani-Hani, R. Efficient FPGA Implementation of the RC4 Stream Cipher using Block RAM and Pipelining. *Procedia Comput. Sci.* **2015**, *63*, 8–15. [CrossRef]

sensors

MDPI

Article

Green Communication in Internet of Things: A Hybrid Bio-Inspired Intelligent Approach

Manoj Kumar [1], Sushil Kumar [1], Pankaj Kumar Kashyap [1], Geetika Aggarwal [2], Rajkumar Singh Rathore [3], Omprakash Kaiwartya [2,*] and Jaime Lloret [4,5]

1 School of Computer and Systems Sciences, Jawaharlal Nehru University, New Delhi 110067, India; manoj26_scs@jnu.ac.in (M.K.); skdohare@mail.jnu.ac.in (S.K.); pankaj76_scs@jnu.ac.in (P.K.K.)
2 School of Science and Technology, Nottingham Trent University, Nottingham NG11 8NS, UK; geetika.aggarwal@ntu.ac.uk
3 Department of Computer Science, Cardiff School of Technologies, Cardiff Metropolitan University, Cardiff CF5 2YB, UK; rsrathore@cardiffmet.ac.uk
4 Department of Communications, Universitat Politècnica de València, 46022 Valencia, Spain; jlloret@dcom.upv.es
5 School of Computing and Digital Technologies, Staffordshire University, Stoke ST4 2DE, UK
* Correspondence: omprakash.kaiwartya@ntu.ac.uk

Abstract: Clustering is a promising technique for optimizing energy consumption in sensor-enabled Internet of Things (IoT) networks. Uneven distribution of cluster heads (CHs) across the network, repeatedly choosing the same IoT nodes as CHs and identifying cluster heads in the communication range of other CHs are the major problems leading to higher energy consumption in IoT networks. In this paper, using fuzzy logic, bio-inspired chicken swarm optimization (CSO) and a genetic algorithm, an optimal cluster formation is presented as a Hybrid Intelligent Optimization Algorithm (HIOA) to minimize overall energy consumption in an IoT network. In HIOA, the key idea for formation of IoT nodes as clusters depends on finding chromosomes having a minimum value fitness function with relevant network parameters. The fitness function includes minimization of inter- and intra-cluster distance to reduce the interface and minimum energy consumption over communication per round. The hierarchical order classification of CSO utilizes the crossover and mutation operation of the genetic approach to increase the population diversity that ultimately solves the uneven distribution of CHs and turnout to be balanced network load. The proposed HIOA algorithm is simulated over MATLAB2019A and its performance over CSO parameters is analyzed, and it is found that the best fitness value of the proposed algorithm HIOA is obtained though setting up the parameters $pop_{size} = 60$, number of rooster $N_r = 0.3$, number of hen's $N_h = 0.6$ and swarm updating frequency $\theta = 10$. Further, comparative results proved that HIOA is more effective than traditional bio-inspired algorithms in terms of node death percentage, average residual energy and network lifetime by 12%, 19% and 23%.

Keywords: Internet of Things; chicken swarm optimization; genetic algorithm; energy optimization

Citation: Kumar, M.; Kumar, S.; Kashyap, P.K.; Aggarwal, G.; Rathore, R.S.; Kaiwartya, O.; Lloret, J. Green Communication in Internet of Things: A Hybrid Bio-Inspired Intelligent Approach. *Sensors* **2022**, *22*, 3910. https://doi.org/10.3390/s22103910

Academic Editors: Alexandru Lavric, Liliana Anchidin and Adrian I. Petrariu

Received: 16 April 2022
Accepted: 18 May 2022
Published: 21 May 2022

1. Introduction

Over the years, revolutionary development in IoT devices has opened the paradigm for dynamic sensing technology that provides seamless communication over the Internet [1]. Wireless sensor networks (WSNs) are prominently used for the collection of data and communication over the fifth generation (5G) and beyond 5G IoT network envision as sixth generation (6G) technology [2]. Moreover, the combination of IoT networks with WSNs has many potentials in various applications, such as precision agriculture, intelligent transport systems, health care, smart cities, military, environment and habitat monitoring, environment anomalies and human intrusion detection [3,4]. However, with all these remarkable properties, the fallout in the unbalanced energy consumption and lower lifetime of battery-enabled IoT devices limits the seamless communication of intelligent devices over the IoT network. Therefore, energy-efficient communication over 5G and beyond 5G (6G) enabled IoT devices is the utmost concern in IoT network use cases.

Clustering is a robust and scalable approach to lower energy consumption with better network throughput [4,5]. It is widely studied as probability-based [6–9], weight-based [10–12] and heuristic-based approaches [13–18] for conservation of network energy. Moreover, local decision and uncertainties of the network dynamics have a huge impact on energy consumption in the optimal cluster head (CH) selection as designated data forwarder. Further, the problem of CH selection is non-deterministic polynomial hard (NP) since optimal data aggregation cannot be efficiently solved in polynomial time to ensure balanced energy consumption in each round using a probability- and weight-based approach [19]. Recent studies have shown that meta-heuristics approaches are more suitable for approximately solving NP problems for CH selection [20]. Consequently, proper optimization methods such as fuzzy logic inference [13,14], bat algorithm [15], particle swarm optimization [16–18], differential evolutionary and harmony search [19,20], genetic algorithm (GA) [21], and bio-inspired chicken swarm optimization (CSO) have the potential to be effectively used for finding the optimal number of CHs [22–24].

A critical investigation of CSO techniques concluded that it has better ability for IoT network-centric feature selection with faster convergence rate over fuzzy logic and genetic algorithm due to the effective balance between network uncertainty and finding the parameter optima [17]. In fuzzy logic, the output depends upon only the knowledge base rule, which is robust in nature, but fuzzy logic may not be applicable in frequent network environment change scenarios, whereas GA has the ability to adapt the environment precisely and CSO has better hierarchal classification and speed reduction design in size for optimization problems with maximum accuracy [25]. Therefore, in this paper, integrating the above three features of the mentioned technique, we propose a novel Hybrid Intelligent Optimization Algorithm (HIOA) to optimize the overall energy consumption in the network by rotating the role of CHs. The presented HIOA integrates Fuzzy logic (FL) and chicken-swarm genetic optimization (CSGO) algorithms that inherently address the problem of repeatedly choosing the same IoT nodes as CH during transmission rounds. CSGO simulates foraging activity by dividing the chicken into smaller groups. In each group, every chicken moves toward the optimal one simultaneously, which motivates the idea of rotating CH in each time slot. The major contributions of the proposed model are as follows:

- The system model includes cluster-based IoT architecture with aid the feature of cloud network, where energy consumption of the cluster network is evaluated.
- An energy optimization problem is formulated in terms of a fitness function that minimizes the intra- and inter-cluster distance in the IoT network.
- We present HIOA to generate an optimal set of CHs to minimize the overall energy consumption. This employs FL for the creation of the initial population. Further, CSGO divided the IoT node into a hierarchal structure to increase the population diversity that optimizes the formulated fitness function using crossover and mutation.
- Finally, extensive simulation over different CSGO parameters and comparison with the state-of-the-art algorithms has been performed for critical performance evaluation.

The rest of the paper is divided into the following sections. Section 2 shows the related literature. Section 3 describes the presented system model, problem formulation and proposed algorithm in detail. In Section 4, the simulation results and analysis are discussed. The conclusion of the work and future perspective are presented in Section 5.

2. Related Works

As mentioned above, recent studies [13–17,21–24] through simulation have observed that artificial intelligence and precisely bio-inspired techniques are preferable to traditional probabilistic and deterministic approaches subject to optimizing the energy consumption in IoT networks. In [17,21], the authors proved the proposed algorithm based on CSO obtains excellent performance over traditional approaches such as PSO, FL and GA for robust beam-forming approach, whereas authors in papers [24,25] CSO-based clustering routing protocol plays a significant role in reducing energy consumption over integration with bio-inspired approaches.

In paper [26], a Fuzzy-based routing protocol (FRP-LEACH) is proposed to enhance the traditional Low Energy Adaptive Clustering Hierarchy (LEACH) protocol that forms the clusters in an energy efficient manner. The proposed FRP-LEACH works over cross-layers, and the authors claim that cloud-based services such as the proposed algorithm protect medical staff and patients from the ongoing COVID-19 pandemic. However, the proposed algorithm is restricted to a lower dataset as the COVID-19 dataset increases, then fuzzy logic fails to adapt the changes delay in transferring the packet and node death ratio increases according to data growth rate. The authors in paper [27] enhanced the performance of LEACH using a fuzzy logic inference system and renamed LEACH-FL. However, both approaches carry forward the clustering process efficiently by considering the fuzzy logic inputs as closeness to the base station and residual energy but left out the node density parameter. This cause's optimum number of CHs generated through approaches to cover the entire network was not ensured.

In [28], authors have proposed EC-PSO models based on a standard PSO approach to optimize the energy consumption of the network to overcome the hotspot problem. When some of the sensor nodes may be left out from the coverage area of CHs and live spontaneously, they are called isolated nodes. These nodes continuously search for CHs and forces to communicate base station directly to exhaust more energy. However, the authors completely ignore the node distance from the base station, which causes it to affect the fitness function tremendously and ultimately consume uneven energy in each time slot. In [29], the authors proposed an enhanced version of a cluster-based genetic algorithm referred to as CRCGA by coding the fitness function to minimize the energy consumption and load balancing of the network. This proposed algorithm mainly considers three factors: formation of clusters, finding the best route and then maintaining the clusters appropriately in each time slot. Further, adaptive round-trip time is considered over the traditional TDMA schedule to further improve network performance. However, CRCGA outperforms the traditional algorithm but fails to consider the scenarios of different sink node positions. In the paper [30], authors have reduced the localization error in WSN using the CSO technique, and further statistical analysis compared to benchmarks optimization technique PSO and GA reveals that CSO provides an upper hand to robustness, precision and performance in terms of convergence speed over the IoT network. To boost the performance of CSO, a cuckoo search is integrated called CSCSO [31] to find the optimal route for data transfer between node and base station. An enhanced version of LEACH using CSO is presented as LEACH-PSO [32] to form optimal clusters and routing paths. Whereas in the paper [33], authors have addressed the problem of balance between power supply and demand using the CSO technique in the residential area and tertiary industry. Further simulation proved that the improved CSO algorithm outperforms in interruptible load scheduling over peak demand in real time over the GA and PSO algorithms. The above-mentioned algorithms missed the multi-hop routing scenario that extends to inter-clustering routing and intra-clustering routing exchange a greater number of packets that maximize the overall energy consumption of the network.

From the above comparative study of Table 1, it is clearly demanding that bio-inspired CSO and genetic candidate are the best suited approach to handle the uneven clusters formation and efficiently optimize the overall energy consumption. In this paper, clustering is divided into two phases. (i) Fuzzy logic is used to produce tentative CHs. (ii) Thereafter, CSO with GA is used to produce the final optimal number of CHs at each round considering the minimization of total energy consumption per round focuses on enhancing the IoT network lifetime.

Table 1. Comparative study of recent research works.

Characteristics / Protocols	Issues	Techniques	Contributions	Metrics	Limitations	Publication Year
GAOC [21]	Selection of the optimum number of CHs	Genetic Algorithm	Multiple data sinks to overcome hotspot problem	residual energy, node density and node distance	Parameter intra-cluster distance have been not taken	2019
LEACH-FL [27]	Clustering process and routing	Fuzzy Logic Inference System	Improved clustering process	residual energy	Left out Node density	2020
EC-PSO [28]	Hotspot problem	Particle swarm intelligence	Improved fitness function	residual energy,	Left out node distance to base station causes exhaust more energy	2019
CRCGA [29]	Load-balance clustering process with routing	Genetic Algorithm	Select optimal CHs and best route	Encode them into single chromosome	Inter/Intra clustering distance ignored	2020
ICSO-LA [30]	Localization error	CSO	Prevent from IoT nodes to falling into local optimum	Update the distance from real node to base station	Node delay and node density is not covered that arises the problem of hotspot	2021
SEOANS [31]	Optimize beam pattern in WSNs	Cuckoo Search and CSO	Calculation method for node location	Adopt chaos theory and grade scheme to improve CSO	Node density is left out	2018
Interruptible load scheduling protocol [33]	Power balance between supply and demand	CSO	Solve the interruptible load scheduling on peak demand	Alleviate the peak load by reducing cost	Green energy resource and delay constraint is neglected	2021

3. Green Communication in the Internet of Things: A Hybrid Bio-Inspired Intelligent Approach

3.1. Network Model

In the energy-constraint IoT network, all the smart IoT nodes are deployed in the environment randomly. These sensor nodes form clusters as a Fuzzy logic-based chicken swarm genetic optimization (HIOA) algorithm. Only CH is able to send aggregated data to nearby edge nodes. Finally, edge node transferred data to the cloud network for data storage that can be used for data analytics with the following assumptions about IoT nodes below. All the smart IoT sensor nodes are homogenous in nature in terms of energy and computation capabilities (such as data collection, transmission and data aggregation). All these wireless IoT nodes periodically observe the environment and send data to their nearby CH. Initially, all the IoT nodes start with an equal quantity of energy stored in the battery. Using the received signal strength indicator (RSSI), IoT nodes calculate their distance from edge nodes by finding their co-ordinate (X_I, Y_I). An equal amount of energy is consumed by the symmetrical channel in packet transfer in either direction from A to B or B to A. In addition, IoT nodes can adjust their transmission power for packet transfer according to distance either from the edge node or CH. The operational period is divided into two phases: the setup phase and the steady state phase. A crystal-clear observation of a single transmission period is further explained in Section 3.3.4 Transmission Procedure of HIOA Algorithm.

3.2. Energy Model

The IoT nodes set their states between transmitter and receiver before the start of the transmission period. The total energy ($E_S(y,z)$) consumed to send y bit of packet is twofold (a) energy (E_T) exhaust in transceiver circuit to process the packet (b) energy ($E_A(y,z)$) consumption takes place to amplify the y bit of packet over z distance either in free-space energy (∂_{fs}) or multipath energy (∂_{mp}) corresponds to threshold distance z_0 similar to the first order radio model [3] as follows:

$$E_S(y,z) = E_T + E_A\ (y,z) = \begin{cases} y\ E_T + y\ \partial_{fs}\ z^2,\ if\ z < z_0 \\ y\ E_T + y\ \partial_{mp}\ z^4,\ if\ z \geq z_0 \end{cases} \tag{1}$$

Additionally, energy (E_r) consumed by IoT node to receive y bit of packet as follows:

$$E_R(y,z) = yE_T \tag{2}$$

We consider N IoT nodes are evenly distributed into W clusters such that N/W nodes are allocated to each cluster. Further, CHs use the TDMA slotted period to get data from their member IoT nodes, where IoT nodes are able to transmit the sensed data to their CH only in wakeup mode after that switch to sleep mode. The total energy (E_P) exhaust by each CH in a single transmission period is the energy payoff in (a) (E_g) gathering the data from its member's node and (b) (E_e) transmitted the received data to the edge node as aggregated data following the multipass model can be written as:

$$E_P = E_g + E_e \tag{3}$$

where, $E_g = yE_T\left(\frac{N}{w} - 1\right)$, and $E_e = yE_{ay}\frac{N}{W} + yE_T + y\partial_{mp}\ z_{pe}^4$ is consumed energy in the multipass model. E_{ay} represents the data aggregation energy and y_{ze} denote the distance between CH and edge node. The energy consumed (E_c) by child IoT nodes to transfer gathered data to their parent CH uses a free space model over average distance ($z_{cp} = (1/2\pi)(R^2/w)$, R is diameter of network) between parent and child node in one round as follows:

$$E_c = yE_T + y\partial_{fs}z_{cp}^2 \tag{4}$$

Overall, each cluster exhausts the total energy is the summation of energy consumed by parent CH and its child IoT nodes during one round as follows:

$$E_{cluster} = E_P + E_c = y\left(2\ N\ E_T + NE_{ay} + w\ \partial_{mp}\ z_{pe}^4 + w\ \partial_{fs}z_{cp}^2\right) \tag{5}$$

$$= y\left(2\ N\ E_T + NE_{ay} + w\ \partial_{mp}\ z_{pe}^4 + w\ \partial_{fs}\ \frac{1}{2\pi}\ \frac{R^2}{w}\right) \tag{6}$$

3.3. Hybrid Intelligent Energy Optimization

The primary objective of the proposed HIOA approach is to generate an optimal set of CHs to minimize the overall energy consumption and prolong IoT network lifetime, as shown in Figure 1. It consists of two optimization phases. First, a tentative set of CHs is selected through fuzzy logic inference system (FLIS) optimization. In the second optimization phase, the chicken swarm genetic optimization (CSGO) algorithm uses the output of the FLIS labeled as input and treated as the first initial population. Further, the

CSGO algorithm optimizes the election process by choosing the appropriate IoT node as CH to successfully bring out optimum clustering operation.

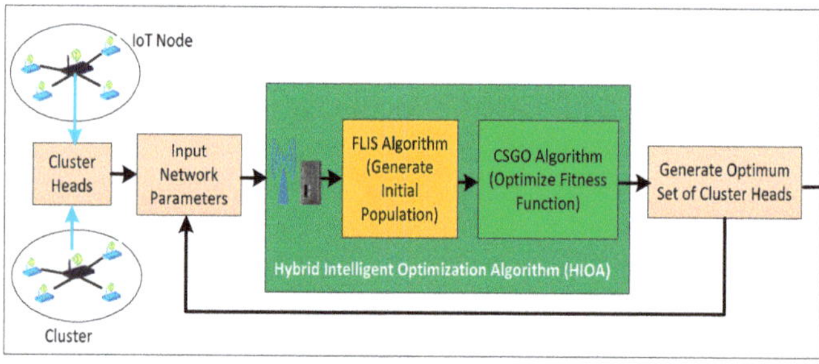

Figure 1. Block diagram of HIOA.

3.3.1. Adapted Fuzzy Logic Interference System for Green Communication

In the first optimization phase, a tentative number of CHs is selected based on three parameters in an IoT network: residual energy (I_{re}), node density (I_{nd}), distance to edge node (I_{de}). The reason behind the selection of these parameters as follow: I_{re} relate to IoT node becoming a CH having enough energy to successfully execute the operation of clustering. I_{nd} relates to the number of neighboring IoT nodes corresponding to the selected node as CH, if I_{nd} is in sparse of nature to the CH, then communication cost increases. Moreover, the worst-case cluster member's nodes exhaust all energy in the transmission of observed information because it is far from CH. I_{de} defining the CH should be at an optimal position so that it manages the communication cost both from cluster member's nodes and edge nodes.

The linguistic variable for mamdani type FLIS as input: I_{re} = {below (BW), upright (UT), great (GT)}, I_{nd} = {sparse (SE), average (AE), compact (CT)} and I_{de} = {nearby (NB), midway (MD), far (FR)}. These parameters are fed into FLIS and in turn output probability as a chance calculated for an IoT node to become CH or not. The linguistic variable of the output probability chance as follow: I_{ch} ={beginner (BR), naive (NE), amateur (AR), strong (SG), hard (HD), harder (HR), hardest (HT)}. The extreme linguistic variables are denoted by trapezoidal membership functions (MFs), and the middle ones are from triangular MFs, as shown in Figure 2a–d. The If-Then rules (input $3^3 = 27$) for evaluation of output probability are presented in Table 2.

Table 2. Fuzzy logic rules.

Rule	If			Then	Rule	If			Then
	I_{re}	I_{nd}	I_{de}	I_{ch}		I_{re}	I_{nd}	I_{de}	I_{ch}
1.	BW	SE	NB	BR	15.	UT	AE	FR	AR
2.	BW	SE	MD	BR	16.	UT	CT	NB	SG
3.	BW	SE	FR	NB	17.	UT	CT	MD	SG
4.	BW	AE	NB	NE	18.	UT	CT	FR	AR
5.	BW	AE	MD	BR	19.	GT	SE	NB	HD
6.	BW	AE	FR	BR	20.	GT	SE	MD	HD
7.	BW	CT	NB	NE	21.	GT	SE	FR	SG
8.	BW	CT	MD	NE	22.	GT	AE	NB	HT
9.	BW	CT	FR	BR	23.	GT	AE	MD	HR
10.	UT	SE	NB	NE	24.	GT	AE	FR	HD
11.	UT	SE	MD	AR	25.	GT	CT	NB	HT
12.	UT	SE	FR	NE	26.	GT	CT	MD	HR
13.	UT	AE	NB	SG	27.	GT	CT	FR	HD
14.	UT	AE	MD	AR					

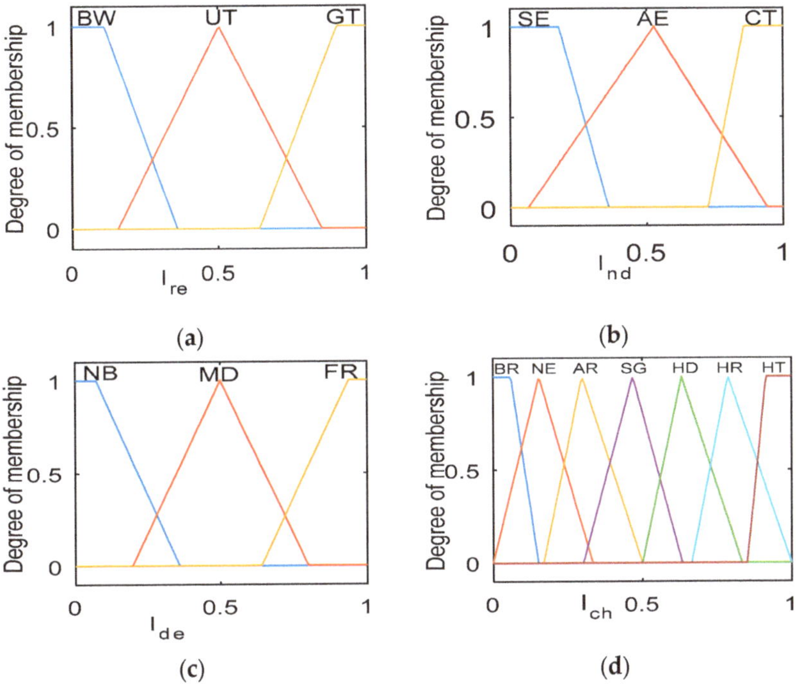

Figure 2. Membership function: (**a**) Residual energy ($\boldsymbol{I_{re}}$), (**b**) Node density ($\boldsymbol{I_{nd}}$), (**c**) Distance to edge node ($\boldsymbol{I_{de}}$), and (**d**) Probability chance ($\boldsymbol{I_{ch}}$).

The FLIS evaluates the output probability in four steps as follows: (i) Fuzzification—this step creates MFs of the crisp input variables according to intersection point. (ii) Fuzzy rule-base—all 27 If-Then rules executed parallel on the given three input variables to generate single output. This can be done by using a fuzzy minimum AND operator from the selection of three input MF parameters. (iii) Aggregation—as there is multiple output value generated corresponding to 27 rules, to aggregate as one single output, maximum union operator OR is used. (iv) Defuzzification—the center of area is used for defuzzification of aggregated value into a crisp output. After the selection of CHs process, the remaining nodes join the nearest CHs through the join message.

3.3.2. Adapted Chicken Swarm Genetic Optimization for Green Communication

In the second optimization phase, the set of tentative CHs resulting from FLIS optimization serves as the initial population of the CSGO to generate a better set of CH compared to the first FLIS optimization phase. In the proposed CSGO algorithm, binary representation is used to represent the IoT nodes as CH (1) or a normal IoT node (0). The binary index (b_{IN}) value of an IoT node can be calculated using the sigmoid function as follows:

$$b_{IN} = \begin{cases} 1, \, if \; sigm \; f(I_{ch}) > 0.5 \\ \quad 0, \, otherwise \end{cases} \tag{7}$$

where, $sigm\ f(I_{ch}) = 1/1 + e^{-I_{ch}}$ represent the sigmoid function. The CSGO includes adapting GA crossover and mutation processes into traditional chicken swarm optimization to enhance the diversity in the population. In the presented algorithm, each cluster is represented as pack classified into Rooster, hens and chicks. Chickens have the best fitness value represented as Rooster (CHs) and chicks are the chickens with the worst fitness value. Most chickens are hens labeled as normal IoT nodes. The hens and chicks form mother–

child relationships arbitrarily. Moreover, the mother–child relationship and dominance relation are unaltered in a pack and updated every distinct (θ) swarm updating frequency as time steps. The position of the rooster is updated using $rand(0, \sigma^2)$ Gaussian distribution with mean zero and standard deviation (σ^2) as follow:

$$Y_{m,D}^{t+1} = Y_{m,D}^{t} * \left(1 + rand\left(0, \sigma^2\right)\right) \tag{8}$$

$$\sigma^2 = \begin{cases} 1\, if\ \{(F_m) \leq \{(F_r) \\ \exp\left(\frac{\{(F_r) - \{(F_m)}{|\{(F_m)| + £}\right),\ otherwise \end{cases},\ r, m \in [1, N],\ r \neq m \tag{9}$$

where, $Y_{m,D}^{t+1}$ represent the position of mth rooster in the D-dimension space, $\{(F_r)$ define the fitness function of the randomly selected rth rooster and £ is the smallest constant to avoid zero-error division. The position of the mth hens' is updated as follows:

$$Y_{m,D}^{t+1} = Y_{m,D}^{t} + \rho_1 * rand(Y_{\omega 1,D}^{t} - Y_{m,D}^{t}) + \rho_2 * rand(Y_{\omega 2,D}^{t} - Y_{m,D}^{t}) \tag{10}$$

$$\rho_1 = exp\left(\frac{\{(F_m) - \{(F_{\omega 1})}{|\{(F_m)| + £}\right) \text{ and } \rho_1 = \exp(\{(F_{\omega 2}) - \{(F_m)) \tag{11}$$

where, $\rho_1, \rho_2 \in [1, 2 \ldots .N]$ are the index of randomly chosen rooster and chickens (hens' or chicks), respectively, such that $\rho_1 \neq \rho_2$. The $rand(\ )$ function generates a uniform random number between 0 and 1. Similarly, the position of mth chicks follows their mother position (FL) as follows:

$$Y_{m,D}^{t+1} = Y_{m,D}^{t} + FL(Y_{\mathfrak{M},D}^{t} - Y_{m,D}^{t}) \tag{12}$$

where, $Y_{\mathfrak{M},D}^{t}$ stands for the position of mth chick's mother, such that $\mathfrak{M} \in [1, n]$. $FL(\ )$ function allows chick's to choose any random value between zero and two. The proposed HIOA algorithm (refer to Algorithm 1) for the generation of the best set of CH works in two phases shown in Figure 3. The first phase generates the tentative list of CHs using FLIS. The second phase (CSGO) generates the optimum set of (rooster) CHs and genetic operators such as crossover and mutation are applied over hens' and chicks to enhance their fitness value. In addition, CSGO includes three steps: (i) initialization step—set the K number of CHs obtained from FLIS optimization, number of rooster (N_r), hens (N_h) and chicks (N_c), population size (pop_{size}), maximum number of swarm updating frequency (θ), single point crossover (ϑ) and mutation rate (τ) with maximum iteration. (ii) Selection phase—it updates the position of rooster, hens' and chicks using Equations (7)–(12). Moreover, crossover and mutation operators were applied over hens' and chicks to enhance their fitness value. (iii) Output step—individuals having the best fitness value generated as the optimum set of CHs.

Algorithm 1: HIOA

1. **Begin**
2. Read network configuration
3. ***First phase:*** **FLIS optimization**
4. ***Input:*** I_{re}, I_{nd}, I_{de}
5. Execute the FLIS engine based on rules define in Table 2
6. Return the chance of an IoT node selected as CH
7. ***Output:*** Tentative set W number of CH
8. ***Second phase:*** **CSGO algorithm**
9. ***Initialization:***
    a. assign pop_{size}, t_{max}= maximum generation, τ, ϑ, τ, θ
    b. Initialize the population matrix U by random values from 0 and 1.
    c. Include the output of FLIS (set of k number of CHs) as one feasible solution.
10. ***Selection:***
11. Reconstruction of infeasible solution takes places those have CHs less than k.
12. Calculate the fitness value of each row of U.
13. Optimum set (O_{CH}) = row of U having best fitness value.
14. **For** t = 1 to t_{max} **do**
15. **If** (t mod $\theta == 0 \,||\, t == 1$) **then**
    a. Sort the individuals into ascending order according to their fitness value.
    b. Divide into three category rooster (N_r), hens'(N_h) and chicks (N_c).
    c. Determine the relationship between mother-child in a pack.
16. **End If**
17. **for** each mth individuals in each Y row of U **do** // pop_{size}
18. **If m== rooster then** update its position using Equation (8).
19. **Else if m== hen then** update its position using Equation (10).
20. **Else if m==chick then** update its position using Equation (12).
21. **end if**
22. Convert Y into binary form using Equation (7)
23. // Single-point Crossover
24. **for** $D = N_r + 1$ to pop_{size} **do**
25. generate two child offspring y_{o1} and y_{o1} from two parent chromosome Y_D and Y_{D+1}
26. set $Y_D = y_{o1}$ and $Y_{D+1} = y_{o2}$
27. **end for** // end of for loop in line no. 22
28. // Mutation
29. **for** $D = N_r + 1$ to pop_{size} **do**
30. $r =$ generate $rand(0, 1)$
31. **if** $r < \tau$ **then**
32. **integer** $\rho_1 =$ generate $rand(0, n)$
33. **if** $Y(\rho_1) == 1$ **then**
34. set $Y(\rho_1) = 0$
35. **else**
36. set $Y(\rho_1) = 1$
37. **end if**
38. **end for** // end of for loop in line no. 27
39. Reconstruction of infeasible solution takes places those have CHs less than k.
40. Updating the fitness value of each Y row of U.
41. If fitness value of new solution is better than previous one, then update optimum CH set with new solution.
42. **End for** // end of for loop in line no. 17.
43. **End for** // end of for loop in line no. 14
44. **Output:** Return the optimum set (O_{CH}) of CHs.

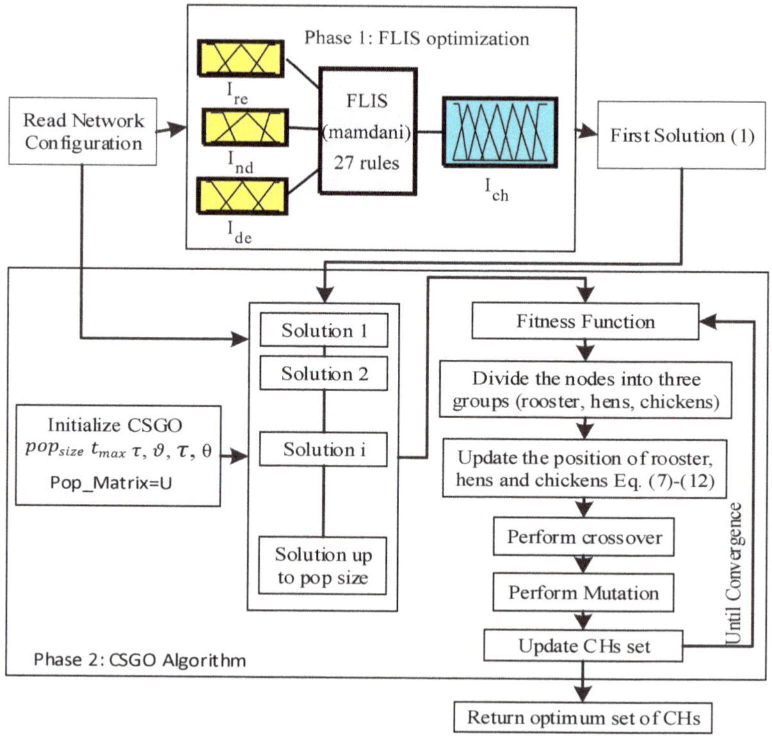

Figure 3. Workflow of the HIOA Algorithm.

3.3.3. Complexity Analysis of HIOA

The presented HIOA algorithm works in two phases. The time complexity of FLIS optimization consists of maximum comparison for each elected CH is $(N^2 - N)$, where N is the number of IoT nodes. Thus, the time complexity of the first phase is $O(N^2)$. In the CSGO phase, the initialization step takes a constant time of $O(1)$. In addition, line no. 11 to line no. 13 are executed in $O(1)$ time. Further, the computational complexity of the second phase mainly depends upon line no. 14 up to line 43, which consists of four nested loops. Line no. 14, 17, runs up to maximum generation t_{max} and possible solution pop_{size}, respectively; further Line no. 24 and 29 both executed up to N number of nodes. Therefore, the time complexity of the second phase can be evaluated as $O(1) + O(1) + O(t_{max} \times pop_{size}(N + N))$ $\approx O(N)$, where t_{max} and pop_{size} are very small compared to the number of IoT nodes and can be neglected. Thus, the overall time complexity of the presented HIOA algorithm is the summation of first phase and second phase $\left[O(N^2) + O(N)\right]$.

3.3.4. Operational Procedure for HIOA

An operational period for the presented HIOA algorithm operates in a setup and steady-state period that is repeated in each round. Thus, each IoT node has the chance to play the role of CHs in blanching the energy consumption in the IoT network. Firstly, in the setup period, CH election and binding of IoT nodes subject to cluster formation takes place. To get knowledge about IoT nodes, edge nodes advertise beacon messages over the network. In turn, IoT nodes respond with their ID, co-ordinate and residual energy. Using the RSSI model, IoT nodes calculate their distance from the edge node by finding their co-ordinate (X_l, Y_l) as follow:

$$\text{RSSI} = -(10\varphi \, log_{10} \, z_{ie} + \text{A}) \tag{13}$$

where, φ represents the coefficient of signal propagation parameter alias exponent, z_{ie} is the usual distance between edge node to an IoT node and A denotes the obtained signal strength in one meter distance without obstacle. Further, the edge node uses the presented HIOA for the election of K number of CHs. Furthermore, edge nodes send messages to each IoT node in the K pack to inform nodes as CHs. Each IoT node in the pack advertises its role as cluster member CM or CHs containing its ID. Those remaining nodes that do not belong to any pack choose the nearby CH based on minimum communication cost through the join message. Second, in the steady state period, the major goal is to avoid data collisions occurring during the gathering of data by parent CH from its child CM nodes. Then, CHs schedule the data gathering using the TDMA technique. Now, CM sends their observed data to their respective CH that include ID and residual energy. This local information is useful in deciding on the role of CH for the next round in a pack. CH applies a data aggregation algorithm to remove redundancy in the gathered data. Finally, CH sends a packet of fixed size to the edge node with the local information of CM's.

4. Results and Discussion

The performance of the HIOA algorithm is compared with a state-of-the-art algorithm using MATLAB 2019A. This section is divided into two parts: (i) CSGO parameter analysis and (ii) network parameter analysis. In addition, 200 IoT nodes are randomly distributed over the IoT network size 150 × 150 m^2 with edge node set at the corner position of the network. The initial power of the nodes is 1 J, similar to the first order radio model used in paper [20]. CSO is initializing with a random solution in the search space and applied to minimize the fitness function that yields a single solution with all features selected as the initial solution. For CSO, the maximum generation is set to be 60, the number of roosters is between 0.1 to 0.4 percentages randomly, and the number of hens is between 0.2 to 0.8 randomly. The list of other parameters and their values is listed in Table 3, which is similar to paper [29,31].

Table 3. Simulation parameters.

Parameter	Value
Number of nodes (N)	200
Network size	150 × 150
Percentage of CH	5
Packet size	4000 bit with 100 bit header
Initial Energy	1 J
E_{ay}	5 nJ/bit/message
E_T	50 nJ/bit
∂_{fs}	10 pJ/bit/m^2
∂_{mp}	0.0013 pJ/bit/m^4
ϑ	0.3
τ	0.006
Cycle time	60 µs
Crossover rate	0.7
Mutation rate	0.1
Population size	60

4.1. CSGO Parameter Analysis

In this subsection, the consequences of four different parameters on the fitness performance of the presented algorithm HIOA are analyzed. These parameters are (i) population size (pop_{size}) (ii) number of rooster (N_r) (iii) number of hens' (N_h) and (iv) swarm updating frequency θ. In addition, the number of chicks can be evaluated as $N_c = N - N_r - N_h$.

From the above analysis of the result from Figure 4a–d, we observed that the presented algorithm achieves a better fitness value of about 0.625 within 60 iterations and after that more iteration, the variation in fitness value is negligible. This can be attributed to the reason the presented algorithm evaluates the fitness value on the basis of three optimization functions: inter-cluster, intra-cluster distance and energy consumption. This reveals that the presented algorithm is able to balance the energy consumption and average distance properly. Thus, the best fitness value of the proposed algorithm HIOA is obtained through setting up the parameters $pop_{size} = 60$, $N_r = 0.3$, $N_h = 0.6$ and $\theta = 10$. In addition, the number of chicks can be evaluated as $N_c = 1 - 0.3 - 0.6 = 0.1$. In the next upcoming simulation, the above parameters are set with their values.

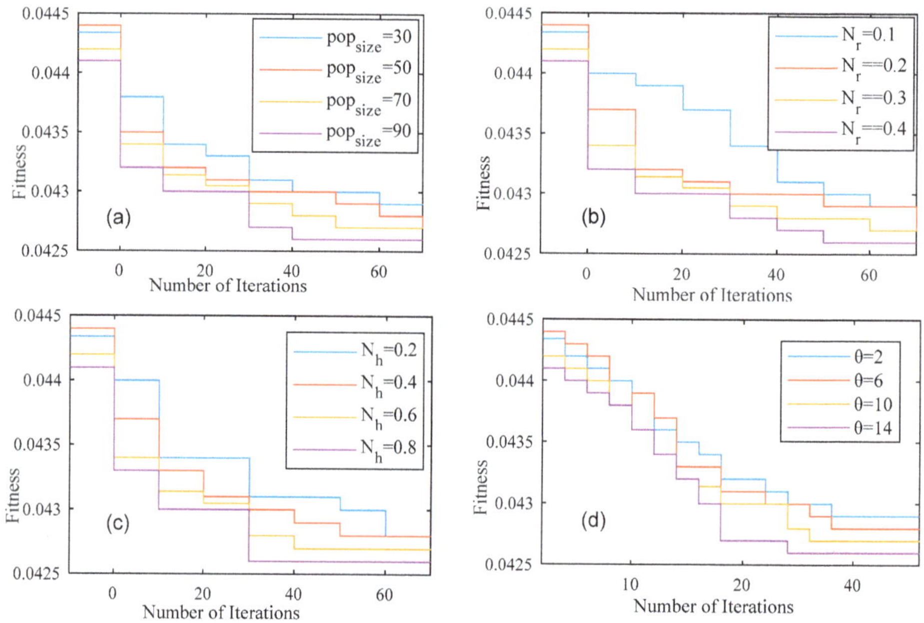

Figure 4. Optimization performance of HIOA over iterations (**a**) pop_{size} (**b**) N_r (**c**) N_h (**d**) θ.

4.2. Network Parameter Analysis

In this section, a comparison of the HIOA with LEACH-FL [27], CRGA [29] and EC-PSO [28] algorithms has been analyzed to show the effectiveness in terms of network lifetime, average energy consumed per node at each round and the total average energy consumed per round.

4.2.1. Comparison of Active Nodes over Rounds

In Figure 5, the number of active IoT nodes is plotted per round to show the effectiveness of the proposed algorithms compared to state-of-the-art algorithms. It is evident from the result for the starting phase of all the algorithms, up to 300 rounds around 98% IoT nodes are alive/active except in LEACH-FL only 87%. Further, on enhancement in rounds, the proposed algorithm shows the best performance compared to EC-PSO, CRGA and LEACH-FL. This is due to the fact that each chicken in the proposed algorithm acts as an agent, which updates its position in the network areas to become CH or not. In addition, this agent (chicken) has information about the whole group (cluster) and fairly balances the energy consumption load using GA crossover and mutation operator. Further, the EC-PSO algorithm does not have the ability to multi-swarm optimization and lacks adaptation to changes in the network. Whereas, CRGA do not has agent so that keep information's of other chromosomes in a population. It is also worth noting that the number of active nodes in LEACH-FL is only 20 after crossing 800 rounds and further increments in the number of rounds in turn all the nodes die. This result confirms that LEACH-FL is not able to achieve fairness in network load balance and residual energy of IoT nodes. Thus, overall, the proposed algorithm HIOA outperforms EC-PSO, CRGA and LEACH-FL by 12%, 19% and 29%, respectively, in terms of the number of active nodes remaining in the 900 rounds.

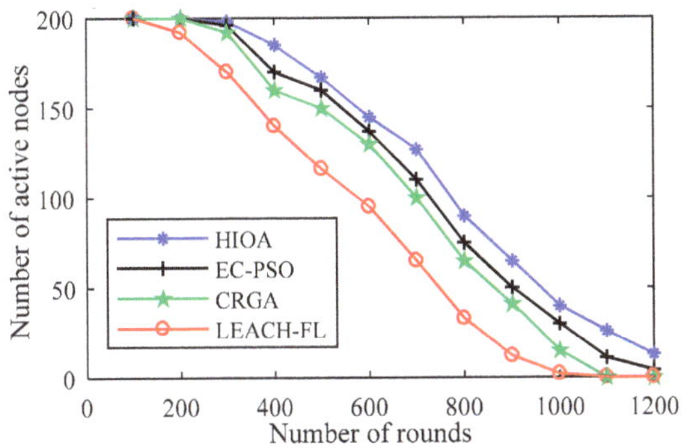

Figure 5. Number of active nodes over round.

4.2.2. Comparison of Network Lifetime over Rounds

A comparison of the network lifetime of the proposed algorithm and state-of-the-art algorithms is shown in Figure 6. The network lifetime is the time at which the first node dies (FND) or half of the nodes die (HND) and the last node dies (LND). It can be clearly observed from the result the proposed algorithm HIOA achieves better lifetime in terms of FND is 11%, 18% and 48% from EC-PSO, CRGA and LEACH-FL, respectively. Further, HND is 27%, 35% and 58% compared with EC-PSO, CRGA and LEACH-FL, respectively. Furthermore, in terms of LND is 35%, 46% and 81% compared to EC-PSO, CRGA and LEACH-FL, respectively. This can be attributed to the reason that the proposed algorithm uses both fuzzy logic optimization with a chicken swarm genetic algorithm that helps in building a more feasible solution rather than only using particle swarm optimization in EC-PSO or a genetic algorithm in CRGCA. This is also observed from the result that LEACH-FL shows the worst performance in terms of lifetime, where FND, HND and LND occur at 100, 336 and 982 rounds. It is due to the reason LEACH-FL selects the CH based only on residual energy and node distance, where node density is left out. There is no optimization, such as PSO, or GA is used to enhance performance.

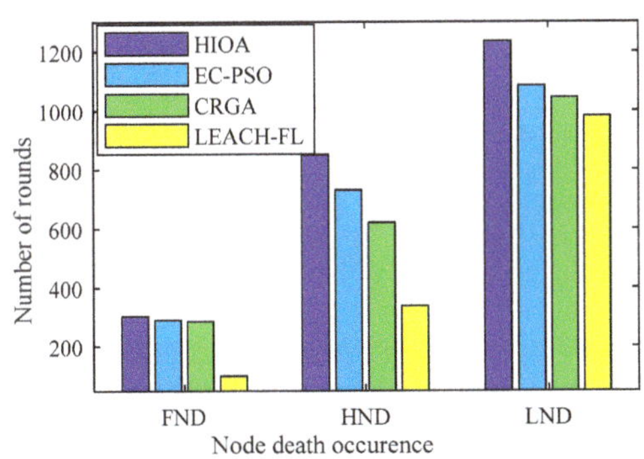

Figure 6. Network lifetime over round.

4.2.3. Comparison of Average Energy Consumed over Rounds

A comparison of average consumed energy over rounds of the proposed algorithm HIOA and state-of-the-art algorithms is shown in Figure 7. Initially, all the state-of-the-art algorithms except LEACH-FL [27], up to 300 rounds, consume almost similar amounts of energy. Thereafter, on the enhancement of the number of rounds, the proposed algorithm

HIOA consumes much less energy than other state-of-the-art algorithms and goes up to above 1200 rounds with a consumption of 0.75 Joule. In addition, it can be also observed that CRGA [29] and EC-PSO [28] consume almost 0.95 Joule and 0.85 Joule of energy within 972 rounds and 1030 rounds, respectively. This is due to the fact presented HIOA algorithm inherit CSGO that designated to inmate multi-swarm optimization group with crossover and mutation technique of genetic algorithm, which enhances the efficient use of energy in the IoT network. EC-PSO is a single group swarm optimization technique that is lacking in the generation of feasible solutions of CHs throughout the IoT network. In addition, the fitness function of CRGA only includes residual energy parameter and left the inter-cluster distance and intra-cluster distance parameter, so that selection of CHs is not optimal. Thus, overall, the proposed algorithm HIOA outperforms EC-PSO, CRGA and LEACH-FL by 17%, 26% and 38%, respectively.

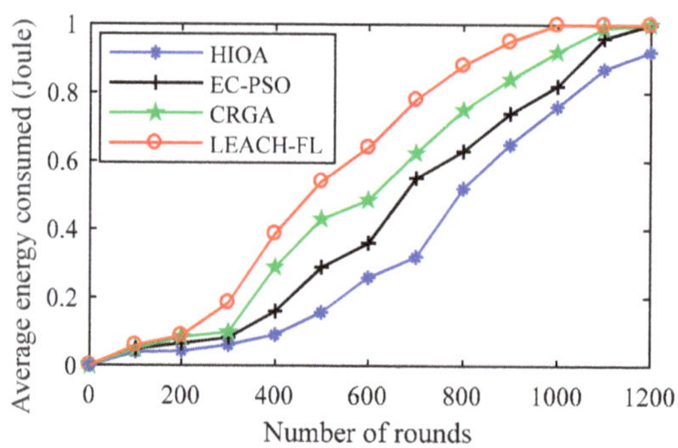

Figure 7. Average energy consumed over round.

4.2.4. Comparison of Average Residual Energy over Rounds

A comparison of the average residual energy per IoT node with respect to the round between the proposed algorithms and state-of-the-art algorithms is shown in Figure 8. The average residual energy is calculated as dividing the overall residual energy per round by the total number of IoT nodes. It is obvious that the residual energy of IoT nodes that uses the presented HIOA has more residual energy (0.12 Joule) after completion of 1200 rounds, whereas CRGA [29] and EC-PSO [28] algorithms consume more energy in data communication and IoT nodes have only 0.05 Joule and 0.02 Joule of energy art 1000 rounds. This is due to the fact that the proposed algorithm HIOA optimally selects the CHs at each round and balances the energy consumption load among all the nodes within a cluster. This can be attributed to the proper division of IoT nodes in rooster, hens and chicks, where the mother–child (hens & chicks) relationship is also balanced. In addition, crossover and mutation operators also help in enhancing the selection of optimum set of CHs. It is also worth noting that initially the LEACH-FL [27] algorithm consumes less energy and IoT nodes have a handsome amount of residual energy left out. However, in the increment of rounds, the residual energy of IoT nodes declined sharply and almost zero within 892 rounds. This is because LEACH-FL ignores the inter-cluster and intra-cluster communication costs in the selection of CHs that consume extra energy. Thus, overall, the proposed algorithm HIOA has more residual energy compared to EC-PSO, CRGA and LEACH-FL by 11%, 16% and 21%, respectively.

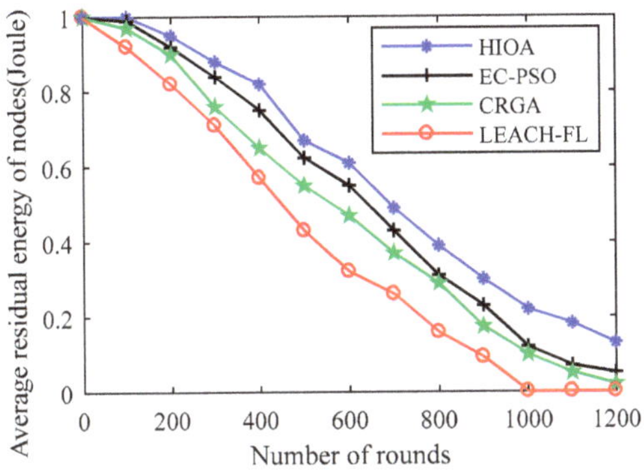

Figure 8. Average residual energy over rounds.

4.2.5. Comparison of Standard Deviation over Rounds

A comparison of the standard deviation of residual energy and energy load balance by the CHs between the proposed algorithm and state-of-the-art algorithms is shown in Figure 9a and b, respectively. The standard deviation of residual energy measures the variability and consistency of the residual energy of the IoT nodes in the population. This is evident from the result that the standard deviation of residual energy and energy load of the selected CHs for the proposed HIOA algorithms is more balanced compared to EC-PSO [28], CRGA [29] and LEACH-FL [27]. This can be attributed to the reason that it's minimizing the energy consumption during the selection of the fitness function of CHs by considering inter-cluster and intra-cluster distance. In addition, GA operators (crossover and mutation) help in reducing the convergence time in selecting the global optimum set of CHs, which increases population diversity and not to select local optimum CH sets. EC-PSO forms the clusters according to swarm particle optimization, and normal nodes join the cluster according to the distance from the edge node and left behind the number of neighboring IoT nodes. This can be attributed to unbalanced energy consumption in the cluster.

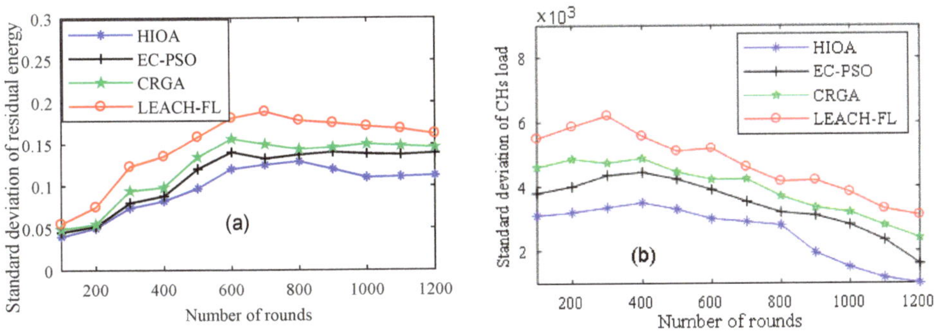

Figure 9. Standard deviation over rounds: (**a**) residual energy and (**b**) CH load.

It can also be noticed that CRGA and LEACH-FL show variations in the standard deviation of the IoT node's residual energy and load balancing by CHs. It is due to the fact both CRGA and LEACH-FL algorithms choose the CHs randomly, in turn, unevenly distributing the load on the CHs and reducing residual energy. The worst performance is shown by LEACH-FL because probabilistic selection generates some isolated CHs that consume much more energy in the transmission of data to edge nodes. Thus, the overall result proved that the standard deviation of residual energy of the proposed algorithm

HIOA was lower than the EC-PSO, CRGA and LEACH-FL by 18%, 26% and 42%, respectively. The load balancing of the proposed algorithm is also 23%, 32% and 48% lower than state-of-the-art algorithms.

5. Conclusions and Future Scope

In this paper, we proposed a new hybrid algorithm for clustering in the IoT network based on FL and CSO employing a GA to minimize energy dissipation. To this end, tentative CHs are selected using FL by considering essential parameters. Further, enhanced CSO with GA utilizes the concept of hierarchal order (rooster →hen →chicks) to divide the population and their fitness values are evaluated in order to find the best of nodes as CHs using crossover and mutation operation of GA. Simulations conclude that our proposed algorithm HIOA is robust to different CSO parameters and finds a near-optimal solution with low time complexity and higher convergence speed over state-of-the-art algorithms. The particle implementation of the simulate model can be used for load balancing or self-organizing structure in IoT nodes (electric vehicles) for the energy buying market.

The presented HIOA model effectively captures the rapid and frequently changing traffic patterns due to the inherent feature of GA and the dynamic classification property of CSO with a lower rate of latency for convergence. As the proposed model has no prediction ability for a sustainable network where electric vehicles are used as nodes, to add these features in the future, we included a neural network or deep learning approach for the formation of clusters based on past experience [34]. The nearby models include training and testing and deployment as well in different network scenarios.

Author Contributions: Conceptualization, M.K.; Formal analysis, M.K.; Investigation, M.K.; Methodology, P.K.K.; Resources, M.K., G.A. and R.S.R.; Supervision, S.K., O.K. and J.L.; Validation, R.S.R.; Writing, M.K.; Review & Editing, P.K.K., O.K. and R.S.R. All authors have read and agreed to the published version of the manuscript.

Funding: This research was funded by Jawaharlal Nehru University, New Delhi, India, and partially supported by Cardiff Metropolitan University, UK.

Acknowledgments: The research work is scientifically supported by Nottingham Trent University, UK.

Conflicts of Interest: The authors declare no conflict of interest.

References

1. Aanchal, A.; Kumar, S.; Kaiwartya, O.; Abdullah, A.H. Green computing for wireless sensor networks: Optimization and Huffman coding approach. *Peer-to-Peer Netw. Appl.* **2016**, *10*, 592–609. [CrossRef]
2. Kumar, V.; Kumar, S.; AlShboul, R.; Aggarwal, G.; Kaiwartya, O.; Khasawneh, A.; Lloret, J.; Al-Khasawneh, M. Grouping and Sponsoring Centric Green Coverage Model for Internet of Things. *Sensors* **2021**, *21*, 3948. [CrossRef] [PubMed]
3. Rehman, A.; Haseeb, K.; Saba, T.; Lloret, J.; Sendra, S. An Optimization Model with Network Edges for Multimedia Sensors Using Artificial Intelligence of Things. *Sensors* **2021**, *21*, 7103. [CrossRef] [PubMed]
4. Rani, R.; Kumar, S.; Kaiwartya, O.; Khasawneh, A.; Lloret, J.; Al-Khasawneh, M.; Mahmoud, M.; Alarood, A. Towards Green Computing Oriented Security: A Lightweight Postquantum Signature for IoE. *Sensors* **2021**, *21*, 1883. [CrossRef] [PubMed]
5. Kashyap, P.K.; Kumar, S.; Jaiswal, A.; Kaiwartya, O.; Kumar, M.; Dohare, U.; Gandomi, A.H. DECENT: Deep Learning Enabled Green Computation for Edge centric 6G Networks. *IEEE Trans. Netw. Serv. Manag.* **2022**, *22*, 1–15. [CrossRef]
6. Al-Sodairi, S.; Ouni, R. Reliable and energy-efficient multi-hop LEACH-based clustering protocol for wireless sensor networks. *Sustain. Comput. Inform. Syst.* **2018**, *20*, 1–13. [CrossRef]
7. Vijayalakshmi, V.; Senthilkumar, A. USCDRP: Unequal secure cluster-based distributed routing protocol for wireless sensor networks. *J. Supercomput.* **2019**, *76*, 989–1004. [CrossRef]
8. Al-Shalabi, M.; Anbar, M.; Wan, T.-C.; Alqattan, Z. Energy efficient multi-hop path in wireless sensor networks using an enhanced genetic algorithm. *Inf. Sci.* **2019**, *500*, 259–273. [CrossRef]
9. Heinzelman, W.R.; Chandrakasan, A.P.; Balakrishnan, H. Energyefficient communication protocol for wireless sensor networks. In Proceedings of the 33rd Annual Hawaii International Conference on System Sciences, Honolulu, HI, USA, 4–7 January 2000.
10. Zhu, F.; Wei, J. An energy-efficient unequal clustering routing protocol for wireless sensor networks. *Int. J. Distrib. Sens. Netw.* **2019**, *15*, 1550147719879384. [CrossRef]

11. Genta, A.; Lobiyal, D.K.; Abawajy, J.H. Energy Efficient Multipath Routing Algorithm for Wireless Multimedia Sensor Network. *Sensors* **2019**, *19*, 3642. [CrossRef]
12. Kaiwartya, O.; Kumar, S. Cache agent-based geocasting in VANETs. *Int. J. Inf. Commun. Technol.* **2015**, *7*, 562–584. [CrossRef]
13. Sert, S.A.; Alchihabi, A.; Yazici, A. A two-tier distributed fuzzy logic based protocol for efficient data aggregation in multihop wireless sensor networks. *IEEE Trans. Fuzzy Syst.* **2018**, *26*, 3615–3629. [CrossRef]
14. Prasad, M.; Liu, Y.T.; Li, D.L.; Lin, C.T.; Shah, R.R.; Kaiwartya, O.P. A new mechanism for data visualization with TSK-type preprocessed collaborative fuzzy rule based system. *J. Artif. Intell. Soft Comput. Res.* **2017**, *7*, 2017. [CrossRef]
15. Cai, X.; Sun, Y.; Cui, Z.; Cui, Z.; Zhang, W.; Chen, J. Optimal LEACH protocol with improved bat algorithm in wireless sensor networks. *KSII Trans. Internet Inf. Syst.* **2019**, *13*, 2469–2490.
16. Sahoo, B.M.; Amgoth, T.; Pandey, H.M. Particle swarm optimization based energy efficient clustering and sink mobility in heterogeneous wireless sensor network. *Ad Hoc Netw.* **2020**, *106*, 102237. [CrossRef]
17. Deb, S.; Gao, X.-Z.; Tammi, K.; Kalita, K.; Mahanta, P. Recent studies on chicken swarm optimization algorithm: A review (2014–2018). *Artif. Intell. Rev.* **2020**, *53*, 1737–1765. [CrossRef]
18. Sambo, D.W.; Yenke, B.; Förster, A.; Dayang, P. Optimized clustering algorithms for large wireless sensor networks: A review. *Sensors* **2019**, *19*, 322. [CrossRef]
19. Gong, D.; Yang, Y.; Pan, Z. Energy-efficient clustering in lossy wireless sensor networks. *J. Parallel Distrib. Comput.* **2013**, *73*, 1323–1336. [CrossRef]
20. Sohrabi, M.K.; Alimirzaee, S. Improving performance of node clustering in wireless sensor networks using meta-heuristic algorithms and a novel validity index. *J. Supercomput.* **2019**, *75*, 7550–7572. [CrossRef]
21. Verma, S.; Sood, N.; Sharma, A.K. Genetic algorithm-based optimized CH selection for single and multiple data sinks in heterogeneous wireless sensor network. *Appl. Soft Comput.* **2019**, *85*, 105788. [CrossRef]
22. Cui, L.; Zhang, Y.; Jiao, Y. obust Array Beamforming via an Improved Chicken Swarm Optimization Approach. *IEEE Access* **2021**, *9*, 73182–73193. [CrossRef]
23. Pitchaimanickam, B.; Murugaboopathi, G. A hybrid firefly algorithm with particle swarm optimization for energy efficient optimal CH selection in wireless sensor networks. *Neural Comput. Appl.* **2020**, *32*, 7709–7723. [CrossRef]
24. Devassy, D.; Immanuel Johnraja, J.; Paulraj, G.J.L. NBA: Novel bio-inspired algorithm for energy optimization in WSN for IoT applications. *J. Supercomput.* **2022**, *22*, 1–18. [CrossRef]
25. Shi, W.; Wang, W.; Yu, Y.; Zhang, S.; Cao, Y.; Yan, S.; Gao, J. Optimal Deployment of Phased Array Antennas for RFID Network Planning Based on an Improved Chicken Swarm Optimization. *IEEE Internet Things J.* **2021**, *8*, 14572–14588. [CrossRef]
26. Nasri, M.; Helali, A.; Maaref, H. Energy-efficient fuzzy logic-based cross-layer hierarchical routing protocol for wireless Internet-of-Things sensor networks. *Int. J. Commun. Syst.* **2021**, *34*, e4808. [CrossRef]
27. el Alami, H.; Najid, A. Fuzzy logic based clustering algorithm for wireless sensor networks. In *Sensor Technology: Concepts, Methodologies, Tools, and Applications*; IGI Global: Hershey, PA, USA, 2020; pp. 351–371.
28. Wang, J.; Gao, Y.; Liu, W.; Sangaiah, A.; Kim, H.-J. An improved routing schema with special clustering using PSO algorithm for heterogeneous wireless sensor network. *Sensors* **2019**, *19*, 671. [CrossRef]
29. Wang, C.; Liu, X.; Hu, H.; Han, Y.; Yao, M. Energy-Efficient and Load-Balanced Clustering Routing Protocol for Wireless Sensor Networks Using a Chaotic Genetic Algorithm. *IEEE Access* **2020**, *8*, 158082–158096. [CrossRef]
30. Sandeli, M.; Bouanaka, M.A.; Kitouni, I. An Efficient Localization Approach in Wireless Sensor Networks Using Chicken Swarm Optimization. In Proceedings of the 2021 International Conference on Information Systems and Advanced Technologies (ICISAT), Tebessa, Algeria, 27–28 December 2021; pp. 1–6. [CrossRef]
31. Sun, G.; Liu, Y.H.; Liang, S.; Chen, Z.Y.; Wang, A.M.; Ju, Q.A.; Zhang, Y. A sidelobe and energy optimization array node selection algorithm for collaborative beamforming in wireless sensor networks. *IEEE Access* **2018**, *6*, 2515–2530. [CrossRef]
32. Wang, Q.; Zhu, L. Optimization of wireless sensor networks based on chicken swarm optimization algorithm. *AIP Conf.* **2017**, *1839*, 020197.
33. Wang, J.; Zhang, F.; Liu, H.; Ding, J.; Gao, C. Interruptible load scheduling model based on an improved chicken swarm optimization algorithm. *CSEE J. Power Energy Syst.* **2021**, *7*, 232–240. [CrossRef]
34. Kashyap, P.K.; Kumar, S.; Jaiswal, A.; Prasad, M.; Gandomi, A.H. Towards Precision Agriculture: IoT-Enabled Intelligent Irrigation Systems Using Deep Learning Neural Network. *IEEE Sens. J.* **2021**, *21*, 17479–17491. [CrossRef]

sensors

Article

Cable Monitoring Using Broadband Power Line Communication

Lukas Benesl [1,†], Petr Mlynek [1,*,†], Michal Ptacek [2], Vaclav Vycital [2], Jiri Misurec [1], Jan Slacik [1], Martin Rusz [1] and Petr Musil [1]

[1] Department of Telecommunications, Brno University of Technology, Technicka 12, 61600 Brno, Czech Republic; xbenes44@vut.cz (L.B.); misurec@vut.cz (J.M.); xslaci00@vut.cz (J.S.); xruszm00@vut.cz (M.R.); xmusil56@vut.cz (P.M.)

[2] Department of Electrical Power Engineering, Brno University of Technology, Technicka 12, 61600 Brno, Czech Republic; ptacekm@vut.cz (M.P.); vycital@vut.cz (V.V.)

* Correspondence: mlynek@vut.cz

† These authors contributed equally to this work.

Abstract: Power line communication (PLC) is considered one of the possible communication technologies for applications in the field of smart metering, smart substations, smart homes, and recently for the management of renewable resources or micro grid control. This article deals with the use of PLC technology to determine the technical condition of the cable. This coefficient can help distribution system operators (DSO) to assess the condition of their cable routes. In this way, possible cable breakdowns and subsequent power outages can be prevented. The resulting methodology for calculating the coefficient is presented in two specific examples of routes, in which a significant benefit for DSO's can be found.

Keywords: cable diagnostics; smart grid monitoring; health cable monitoring; technical cable coefficient

Citation: Benesl, L.; Mlynek, P.; Ptacek, M.; Vycital, V.; Misurec, J.; Slacik, J.; Rusz, M.; Musil, P. Cable Monitoring Using Broadband Power Line Communication. *Sensors* **2022**, 22, 3019. https://doi.org/10.3390/s22083019

Academic Editors: Alexandru Lavric, Liliana Anchidin and Adrian I. Petrariu

Received: 21 March 2022
Accepted: 11 April 2022
Published: 14 April 2022

1. Introduction

Broadband power line communication (BPL) technology is today well known in the commercial sector as power line adapters for home networking [1]. In the industry, BPL was considered in particular for smart meters, but today, thanks to selective roll-out, radio and mobile technologies for point-to-point connection are predominantly considered [2,3].

Nowadays, possible usage of BPL in smart grids is considered mainly for smart secondary substation applications [4,5]. The need to build a smart secondary substation arises from the requirements to more accurately control and monitor the energy system, especially for:

- Power Quality: The distributed power system has been increased with solar and wind power local facilities, which make the grid more heterogeneous and difficult to control.
- Electric Vehicles: Charging and discharging of electric vehicles will be based on available production from solar power generation or other renewable energy sources. Therefore, communication between electric vehicle management, renewable energy sources and data collection supervisory control and data acquisition (SCADA) systems will be necessary.
- Distribution Generation: Implementation of distribution generation requires advanced tools, standards, and guidelines for the secure, reliable and resilient operation of smart grid systems to ensure grid stability, power quality, and cost-effectiveness on the entire value chain of the power network.

BPL is therefore considered for implementation in underground secondary substations, where the signal of mobile networks is not sufficient [4,6].

There are three main advantages of BPL:

- With the growing number of connections of new elements, growing demands on broadband communication and high requirements for cyber security, BPL appears to be a suitable technology that will meet the expected requirements with its parameters. It is also an independent communication network under the administration of utilities (e.g., dependencies on telecommunications operators).
- BPL technology can be deployed directly to existing transformer stations (existing medium voltage (MV) lines), without excavations, and without major intervention. It can be considered for a transitional period and in terms of investment and operating costs, until there is an optical network available everywhere, as the BPL technology is completely sufficient in terms of communication parameters [7].
- Compared to GSM (Groupe Spécial Mobile) and mobile technologies, BPL is a suitable technology for underground transformer stations, where there is no signal and no possibility to pull out an antenna. The most suitable technology would be an optical fiber, but building optical networks is not so simple, especially in city centers. PLC/BPL can show better performance in terms of network-latency, while LTE is proven to be less susceptible to short-term interruptions, resulting in a higher overall reliability [8].

Technology also has disadvantages that many DSOs can discourage from acquiring:

- Communication distance may not be satisfactory. For communication over longer distances, the signal must be amplified or repeated. This brings delays into the whole system and another element that may or may not be necessary for a given route. See also Section 4.1.
- The transmitted signal on the route can attenuate, with a multipath effect [9] where the branches are located, various types of noise [10,11] that can be caused by the power line itself or equipment connected to the network.

Besides the primary use of power line communication (PLC) technology for substation automation and smart metering, the secondary use of PLC could be for cable health monitoring and control in distribution networks.

Cable health monitoring is not a new field of research [12–14], but these referenced solutions require dedicated test equipment or manual waveform analysis for data interpretation. Cable health monitoring is an important method in grid monitoring and prevention of grid faults for utilities [15]. Grid faults result in power outages and financial losses, and could also lead to potentially hazardous situations and loss of lives. According to [16–24], cable health monitoring and cable fault diagnosis is still an open and challenging research issue in terms of the market and social impact.

The idea used in PLC/BPL modems designated for smart grids communication can also serve as a tool for cable health monitoring, which is a challenging task.

In this paper, we propose a method that exploits the topological and communication parameters to determine the health of distribution cables. The cable health monitoring method based on PLC parameters and cable topological parameters could be used as a utilities diagnostics method for cable renovation planning that can be integrated into SCADA or as a smart grids concept. Monitoring power cables enables utilities to prevent cable failures, which can potentially lead to improving operation reliability and thus improved reliability indexes like the System Average Interruption Duration Index (SAIDI), and the System Average Interruption Frequency Index (SAIFI).

The contribution of this article is threefold. Firstly, the communication testbed for testing and evaluation of communication technologies for the new generation smart substation secondary system was proposed. Secondly, the methodology for verifying the communication parameters of broadband communications considered for the smart substation concept was investigated. Thirdly, the BPL technology was evaluated in a testbed thanks to a universal load generator (traffic generator) which enables the emulation and simulation of the various data flows of smart substations based on IEC 60870-5-104, IEC 61850, and DLMS protocols. The main contribution of the research is the repeatable methodology for evaluation of different communications, standards, or vendors for the smart substation.

Based on the evaluation, the recommendation for carrying out BPL technology for the smart substation is proposed.

Besides the existing literature listed in Table 1, our cable diagnostics method uses network topological parameters, cable physical properties, and measured communication parameters of BPL networks.

Table 1. State of the art.

No.	Authors	Year	Method	Purpose
[16]	Y. Huo et al.	2018	S [1]	diagnostic tool for degradation of cables
[17]	G. Prasad et al.	2019	S [1]	diagnostic tool based on existing PLCs that can measure SNR
[18]	Y. Huo et al.	2018	S [1]	diagnostic tool for degradation of cables
[19]	A. Poluektov et al.	2018	S [1]/L [5]	diagnostic tool for degradation of cables based on BIS, only for LV
[20]	Y. Huo et al.	2019	S [1]	monitoring cable health conditions based on machine learning framework
[21]	A. Pinomaa et al.	2015	L [5]	diagnostic tool for degradation of cables based on BIS, only for LV
[22]	Y. Huo et al.	2019	S [1]	neural networks for cable diagnostics using power line modems
[23]	Y. Huo et al.	2019	S [1]	automated machine learning based cable diagnostics design
[24]	L. Förstel et al.	2017	S [1]	PLC as a diagnostic tool for cable aging
[25]	Y. Ohtomo et al.	2010	C [2]	node detection in topology
[26]	M. Solaz et al.	2014	W [3]	field and laboratory tests have been run successfully
[27]	C. Freitag et al.	2013	M [4]	mathematical description of cable degradation without using PLC/BPL
[28]	S. Abeysinghe et al.	2021	A [6]/S [1]	modeling of electrical networks in rural, suburban, and urban areas
[29]	A. Siswoyo et al.	2021	S [1]/E [7]	simulation and verification by experimental measurements based on BIS
[30]	Y. Kakimoto et al.	2020	S [1]/E [7]	partial discharge monitoring system based on HD-PLC communication
[31]	N. Hopfer et al.	2019	L [5]/F [8]	analysis of the technical condition of the cable line using BPL
[32]	S. Hu et al.	2018	L [5]	cable fault diagnosis by SSTDR

[1] Simulation, [2] Concept, [3] Working, [4] Mathematical, [5] Laboratory tests, [6] Analysis, [7] Experiment, [8] Field tests.

Section 2 summarizes the current state of PLC/BPL technology as a diagnostic tool. Section 3 presents work motivation and goals. The theoretical description of individual parameters is contained in Section 4. The proposed methodology and description are part of Section 5. Section 6 contains measurements according to the proposed methodology.

2. Related Works—PLC/BPL as a Diagnostic Tool

According to analysis of related works in this Section 2, the method for cable health monitoring is based on topological parameters of the network, physical properties of the cable and measured Quality of Services (QoS) parameters. A similar method of diagnostics is summarized in Table 1, but the works are only based on a mathematical model, which is no longer applied to a specific use case in a real field. Other articles are based on simulations, where a hypothesis is made, and in the end, it is proved by the result of the simulation. Similar works do not consider all three parts of diagnostics (topological parameters of the network, physical properties of the cable, and measured QoS parameters). Thus, this work is different and offers a new perspective on diagnostic solutions.

Publication [16] mentions the use of PLC/BPL technology as a diagnostic tool for detecting cable degradation due to age. The authors examine the theoretical assumption by using simulation, where they create a mathematical model of the cable, including the effects of thermal degradation, and then create a PLC channel. With the help of machine learning and a larger number of samples, they obtain successful results. The authors of article [25] present a method of searching new nodes in already installed PLC/BPL topologies. They use the ALOHA access method for detection, in which they use ACK/NACK messages. It detects a new node by transmission or, conversely, removes it when the node is unavailable, thus determining the status of the line. Article [26] presents several different scenarios that can be used to detect the following types of network errors using BPL technology: backbone fault, BPL device fault, BPL device link fault, coupler fault and connection fault. The authors conducted laboratory and field follow-up tests which proved their assumptions to be successfully verified. Another researcher, Freitag [27], deals with the degradation of paper insulated lead covered (PILC) and cross-linked polyethylene (XLPE) cables depending on the dissipation factor. The author points out that more diagnostic parameters need to be included to increase the accuracy of the diagnostics. Work [17] presents that even older PLC modems can be used to diagnose cable life to prevent malfunctions. The proposed solution is the analysis of available data that the PLC produces and, with the help of machine learning the created algorithm and with a certain probability, can prevent route faults. Simulations in [18] show that water tree (WT) degradation can be detected with an accuracy of more than 90%, assuming a plausible design of channel frequency response (CFR) generation for a specific topology. Simulation design works only for cables with XLPE insulation. Article [19] deals with the use of Broadband Impedance Spectroscopy (BIS) as another possibility of using BPL technology. The cable fault diagnosis method uses the measured channel response and channel gain data, which are further processed. The authors performed a test at the low voltage (LV) level on an AXMK cable with an aluminum core. Fault detection estimation worked relatively accurately. The disadvantage is that the algorithm can only be implemented in BPL modems that use orthogonal frequency division multiplexing (OFDM) modulation. The authors also state that the diagnostic tool can only be used up to 500 m. The model described in article [28] divides the distribution network into two parts, suburban and urban. It takes into account parameters such as area, population density, total number of nodes, total number of branches, number of outgoing feeders from the primary substation, installed capacity of the primary substation, number of secondary substations in the network, total installed capacity of the secondary substation, and maximum feeder length. The resulting model should be used for generation of network statistics. Article [29] deals with the analysis of the localization and detection of degraded parts of a cable using the method BIS. It can be used with both the inverse transformation method and the inverse fast Fourier transform (IFFT) method. Research [30] has focused on partial discharge (PD) monitoring using PLC communication and physical layer (PHY) speed. Temporary changes in PHY speed may indicate PDs. Testing was performed only up to 3 kV voltage. A model based on the use of reflectometry joint time frequency domain reflectometry (JTFDR) using PLC modems to determine the condition of the cable, classification of the cable aging profile, assessment of the severity of cable degradation, and the exact location of the degradation in case of localized degradation, is described in the publication [20]. The authors of [21] performed laboratory testing for a proposed algorithm for cable fault diagnosis, which is designed to detect and locate a location with an accuracy of less than one meter. Testing was performed only on LV cables. The neural network described in article [22] should be used for cable diagnostics. The online monitoring solution uses PLC modems to intelligently diagnose the status of underground power cables through inherently estimated information on the status of the power line communication channel. In [23], the authors propose cable management diagnostics based on machine learning. The performed simulations suggest that the location of cable degradation can be detected with good accuracy only up to 300 m. A diagnostic tool [24] that primarily detects cable aging is involving the WT. The problematic part of the proposed solution was variations caused by cable aging and those due to load changes. Publication [31] shows that attenuation decreases during laboratory testing with cabling degradation. This statement was also confirmed by field testing. Signal to noise ratio (SNR) values in a 15-min interval were used to determine the current state of the cable. Time variation, which may occur due to interference, has also been identified. Article [32] describes the possible detection of cable fault diagnosis by spread spectrum time domain reflectometry (SSTDR), which is a non-intrusive method. The authors test their idea using a simulation, and then experiment on a short cable that is approximately 20 m long. The cable fault was detected within approximately 30 cm of the true fault location.

Dozens of works deal with the use of diagnostics in order to determine the technical conditions of cables. Based on this analysis, the authors of these articles focused on the research of topological properties of underground power line cables with a correlation of QoS parameters on the same route. The list of considered parameters is given in Section 4.

3. Motivation and Goals

Prevention and estimation of grid faults is essential for utilities. Grid faults result in power outages and financial losses, and could also lead to potentially hazardous situations and loss of lives.

According to analysis of related works in Section 2, the autonomous method for cable health monitoring based on topological parameters of the network, physical properties of the cable, and measured communication parameters of BPL networks, is missing.

As a part of testing the BPL in a pilot deployment at the distribution company E.ON, problematic power lines were revealed for the BPL communication itself, where communication was not possible at all, but the power line was electrically connected. An electrically interconnected route where BPL communication was not possible at all is described in Section 6.1.

Moreover, other power lines with significantly worse BPL performance and reliability parameters were measured in comparison with similar power lines (BPL network sections). In Section 6.2, a very low transmission throughput for short distances between BPL modems (ideal conditions without branches) is described. The route achieves significantly lower throughput in comparison with other similar routes in a given locality and under the same conditions.

These results led us to further consider what causes inaccessible communication in an electrically connected line, lower throughput or unstable communication (connection failures) for particular lines. Thanks to inaccessible communication and subsequent analyses, problematic sections of lines were discovered, such as wet couplings or poor-quality line couplings, and led and motivated us to create a methodology for using BPL communication outside data transfer as an auxiliary monitoring and diagnostic tool.

The main goals of the article are the following:

- Introduce possible physical properties of underground power line cables, topological parameters of BPL networks and measured communication parameters of BPL networks for a power line cable monitoring and diagnostic method.
- Provide measurements of power line physical parameters and measurements of their influence on BPL performance and power cable life.
- Provide autonomous methodology for cable health monitoring, which could be used by utilities for cable recovery planning.

4. Topological Properties of Underground Power Line Cables

The cable's QoS, or its reliability, can be deduced from various indications. For example, the cable age, its type, and the type of cable joints might be the main driving factors suggesting the cable's remaining life expectancy. However, as was discussed in Section 2, the PLC/BPL communication technology might also be an early indicator of cable deteriorating conditions. In the following subsections, the power line (and especially the power cable) parameters that might be used for the detection of line deteriorating conditions will be discussed (i.e., short remaining life expectancy and decreased line reliability with higher risk of fault occurrence). The discussed parameters are:

- Distance between BPL modems/cable length,
- cable type,
- cable age,
- cable cross section,
- number of joints on the route,
- joint type installation,
- joint age,
- power loading of the cable,
- partial discharge measurement,
- cable sheath bonding.

These parameters are the main factors that can affect the monitoring of cable health. A detailed explanation of individual parameters is given in Sections 4.1–4.11.

4.1. Distance between BPL Modems/Cable Length

The distance that the BPL signal must travel is one of the most important parameters that will affect the resulting communication. According to simulation and real measurements, BPL communication can be expected in tens of Mbps up to a distance of 600 m [26,33,34] for underground MV power lines. The distances between distribution transformer stations (DTS) are mostly in the range of tens of meters to the higher hundreds of meters for power grids in the Czech Republic. As shown in Figure 1, the field measurements of BPL throughput for point-to-point connection without repeaters has an exponential declining trend for underground MV power lines without branches. In the picture, there are blue dots that are near the curve which, with their predispositions, correspond with the approximate communication throughput concerning the given distance. However, the dots marked in red are too far from the curve, so it is necessary to investigate why. This could be, for example, due to a bad cable, a faulty cable cross section, or a cable joint with pervaded moisture.

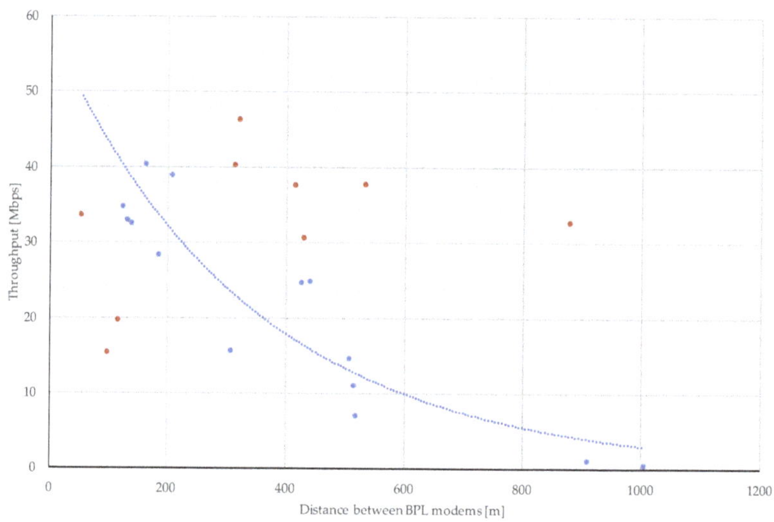

Figure 1. The trend in transmission control protocol (TCP) throughput depending on communication distances between BPL modems on MV lines.

The cable length also correlates with the cable failure probability. For example, if the failure probability per unit length of the cable is X, and the cable is n times longer, the likelihood that there will be a failure will be higher due to the unconditional nature of this random process [35]. So, it is very reasonable to include the cable length in the proposed coefficient of the cable condition.

4.2. Cable Type

Cables are basic elements of the electrical network. They ensure the interconnection of individual network elements such as transformer stations, electrical sources, and the connection of end customers. High-voltage cables are most often placed in the ground in built-up areas, in most cases to a depth of 80 cm. This is also due to regulations, but it is still possible to meet with overhead power lines. Overhead lines have a wider protection zone for safety reasons, while they have a longer service life and lower construction costs [36]. High voltage cables can be divided into two basic types in terms of the insulation used. The structurally older type of cable is PILC, with paper insulation impregnated with oil. In contrast, newer types are cables with cross-linked polyethylene insulation, referred to as XLPE. XLPE is gradually replacing older PILC cables. At voltage levels of 3 and 6 kV, it is also possible to meet cables insulated with polyvinyl chloride (PVC) or polyethylene (PE), or with cables that have combined insulation PVC with PE.

- PILC cables—on older cable routes, in some cases, high-voltage PILC cables still occur. A tape of cable paper is wound on the surface of the core of the PILC cable, which can reach a thickness of several millimeters to tens of millimeters. After winding, the layer goes through a drying process and then the insulation is impregnated with cable oil. As a result, this oil provides the cable with electrical strength. Impregnating oil together with cellulose paper is the biggest weakness of these cables. Over time, the oil begins to dry, which reduces the electrical strength of the fabric, thus deteriorating the insulating and transmission properties of the cable. Another disadvantage of the PILC cable is its higher weight due to the sheathing of the cable with a lead layer, which results in a more complicated construction of the network [37].
- XLPE cables—cable with cross-linked polyethylene is a variant of linear polyethylene linked polyethylene (LPE). Compared to LPE cables, XLPE cables excel in better mechanical properties at higher thermal loads, usually at operating temperatures up to 90 °C. LPE cross-linking can be achieved by two technologies, the first technology is electron beam irradiation. The second technology is extrusion, in which a layer of LPE is applied and then heated by pressure with added peroxides. This process then results in the required cross-linking [37].

PLC communication will perform better on XLPE cables than on PILC cables. However, it is also possible to use PLC communication for diagnostic purposes on PILC cables [16]. The utility provider has data on cables entered in the Global Information System (GIS), but not everything is always correct. Some records may not be preserved. If BPL technology were used on a cable route, it would be possible to estimate which type of cable it could be according to the throughput speed. In this way, BPL technology can describe the cable and, according to the methodology, also identify the health condition of the cable.

4.3. Cable Age

The service life of power lines can fluctuate significantly, mainly due to location (overhead lines/underground lines), as well as due to the geographical characteristics of the environment. The cable life will be different in dry or humid environments, as well as in areas with constant temperature compared to areas with high temperature fluctuations. Overhead lines have an expected service life between 40 and 60 years, while in the case of underground lines, this service life is halved, from 25 to 35 years [38]. The results revealed that the fault current level in the case of an overhead line is significantly smaller than the fault current level in an underground cable.

4.4. Cable Cross Section

Power cables for MV are usually produced in a variety of different sizes, i.e., 50, 70, 95, 120, 150, 185, 240, ..., 630 mm^2. Cables with greater cross sections tend to also have a thicker sheath and insulation layers. Therefore, it might take a slightly longer for a water tree defect to evolve into a cable failure [35]. Thus, the cable cross section will be considered for the cable condition coefficient.

4.5. Number of Joints on the Route

The power cable failure happens quite often in the cable transition zones, i.e., at the cable ends, or/and at the location of cable joints. Besides both cable ends, there might be a number of cable joints on the cable due to past cable failures, or at installation convenient points (junctions, obstacles etc.). Thus the growing number of junctions on the cable route might increase the cable likelihood to failure. Furthermore, a higher number of cable joints will increase the probability that the BPL communication will not be operational.

Based on cooperation with utility providers, there are theoretical assumptions that cable joints have a big influence on the life of the cable line and also on the communication speed. According to real measurements, the BPL communication route was operational with 16 connectors on one cable route. The total length of the cable was 910 m. If it were a single solid cable, the throughput would be higher than the current lower units of Mbps.

4.6. Joint Type Installation

There is a large number of cable joint types, but they are divided into two main groups. The first type is the plastic junction and the second type is the heat-shrinkable junction. The process of replacing or installing a junction is a time-consuming operation in which a mistake can be made very easily. If moisture gets into the insulation, the cable will start to break very soon. The problem can be revealed only by a voltage test. As can be seen in Section 6.2, the junction installation type parameter has a significant effect on the resulting communication.

4.7. Junction Age

If there is a problem with the cable route, the utility provider must dig up the cable and replace the problematic section with a new cable, connecting the new cable to the old cable using a cable junction. These cable joints are subject to the effects of aging in the same way as cables. Cable joints are prone to moisture ingress, which can shorten the overall life of the cable route. From the point of view of PLC communication, wet cable joints can function as high-frequency filters, which can prevent communication partially or completely. This issue is described in Section 6.1.

4.8. Power Loading of The Cable

PLC communication is unstable over time, as different appliances are used at different times, both in households and industrial companies. In cases where there is a large load on the cable route, the capacity of the channel decreases, and thus the quality of the condition of the transmission path deteriorates [39]. A further influence on the load can be caused by the increasing number of branches that can be located on a given route [40].

The different cable loading can also impact the remaining life of the cable because of the varying electrothermal stresses imposed on the cable insulation layers [41]. During the cable nominal operation the cable temperature can increase by several degrees. The generated heat due to resistive and dielectric losses have to be dissipated into the neighboring environment. However, it must be noted that most cables are operated well below their rated loadings and thus they tend to live longer than their designed life.

4.9. Correlation of Topological Properties with BPL Communication Parameters

According to analysis of related works in Section 2, the existing methods for cable monitoring or diagnostics are not considering measured communication parameters of BPL networks for correlation with the topological parameters of underground power cables. The BPL throughput and other communication parameters can only be considered where BPL communication is already fully operational. If the communication is operational and some devices are communicating, the throughput, latency, jitter and reliability on the route can be detected or measured in real-time. Thanks to these measurements, these parameters could be used for online autonomous diagnostics. After the installation of BPL, a decrease in average throughput or communication outages can be noted for further analysis. The network administrator can then evaluate that there is something wrong with the route and propose countermeasures. The route can be measured using another diagnostic tool, for example, the very low frequency (VLF) method to prevent possible failure of the cable route. A similar method of evaluation is described in Section 6.

4.10. Partial Discharge Measurement

Besides methods taking advantage of new PLC/modem communication monitoring methods, there are still a lot of classical methods for investigating possible QoS of selected lines/cables. Widely adopted methods by the utility companies are the partial discharge and tangent delta measurement. These methods have been proven as a source of information indicating technical condition of power cables.

One of the problems [42] with using these methods is that there is still an ongoing investigation into how to properly read the measurement and extract maximum information about possible improper cable conditions. Thus, the field of correct measurement data interpretation retains a lot of attention and so new assessment frameworks are being proposed [43]. Another problem of the partial discharge measurement is that the measured cable needs to be put out of service and the measurement is usually conducted by the distribution system company employees (i.e., measuring truck, etc.), making this measurement by manpower expensive.

This measurement is therefore conducted quite scarcely, mainly for cables with indices of problems, or as part of routine preventive maintenance. It is quite usual that not every cable condition is assessed by this measurement very often and, for example, in the Czech Republic DSO they are supposed to conduct this measurement only once every four years. Thus, the partial discharge measurement as an early and online indicator of cables deteriorating conditions is not yet very favourable. A possible solution of using this measurement as an early online cable condition indicator might be the adoption/installation of measurement systems such as Smart Cable Guard [44]. The results of partial discharge and tangent delta measurement will be not included in the condition coefficient in the current proposal, however, in case of abundant data availability (like with Smart Cable Guard), it might make the method much more reliable.

4.11. Cable Sheath Bonding

High voltage cables are made with sheaths that should increase the protection level in case of failure as well as in case of normal operation. Although most of the concerns with cable sheaths are focused on reducing the level of induced voltage, or the voltage levels during faults, one of the problems with cable sheaths is the presence of circulating currents (CC). The CC usually develop due to the presence of electromagnetic fields (e.g., from other parallel lines) or as potential difference at earthing nodes. The presence of CC is in fact a non-desirable effect because they create additional thermal stress on the cable due to increased power losses, and thus might have a slight effect on cable life expectancy. On the other hand, making the cable sheath grounded on both ends, or even at multiple points, might be beneficial from a PLC/BPL communication point of view, as the cable gets closer to the ideal telegraph cable. From the possible cable bonding/grounding options there are basically four options—no grounding at either end, single end grounding, both ends grounding or multiple points grounding [45] (i.e., respectively no-bonding, single bonding, both end bonding, cross bonding). The optimum case from both a thermal stress and communication interference point of view can be expected for the cross bonding variant. The optimum case for thermal stress alone would be the single end bonding, where no bonding would be impractical and also hazardous, and both end bonding would lead to increased thermal stress. The thermal stress can be further influenced by the fact that some MV cables are still three core cables, or by another factor that is the effect of different cable laying methods.

5. Methodology

BPL technology can be considered as an active on-line method of determining the technical health condition of the cable route. It is also possible to use this technology as a diagnostic tool. For this reason, it was possible to design a methodology and create a coefficient for the MV network, which can clearly assess the condition of the cable route.

Each parameter has already been described in Section 4 together with explanation on why it is considered for the methodology. Table 2 indicates what values an individual parameter can take, what the interval of values used in the calculation is (as will be discussed later), and also what the total share of the parameter in the final value of the coefficient is. The parameter of partial discharge measurement is not considered here because the Czech distribution system operator does not have sufficient data that can be used in calculated examples.

Table 2. Rating of individual parameters of the cable health on the MV network.

Parameter	Range of Values	Interval	Part of Coeff. Val.
Distance between BPL modems	1–1200 [m]	0–0.05	5%
Cable type	0 or 1	0 or 0.05	5%
Cable age	0–40 [year(s)]	0 or 0.15	15%
Number of cable joints	0–20	0–0.15	15%
Cable joint type installation	0 or 1	0 or 0.02	2%
Cable joint age	0–40 [year(s)]	0 or 0.05	5%
Load	0–100 [%]	0–0.035	3.5%
Bonding	0 or 1	0 or 0.01	1%
Cross section	50–630 [mm^2]	0–0.035	3.5%
Average TCP throughput	0–50 [Mbps]	0–0.45	45%

Part of the coefficient value is set according to the needs of Czech utility providers. These values can be adjusted as needed by the DSO. The concept of the coefficient is designed so that it can be modified and freely used by others.

The total coefficient starts with number 1 and, by successively subtracting all the 10 subcoefficients, we get a value in the range 0.000–1.000. Some of the parameters (distance between BPL modems/cable length, number of joints on the route, power loading of the cable, and average TCP throughput) can take values within their intervals. To achieve this, a generic linearization of the actual values to the range interval was used as

$$\text{Coefficient} = \frac{\text{Actual Value}}{\text{Max Range}} \cdot \text{Max Coeff Interval} \quad (1)$$

As the dependency in case of cable cross section and TCP throughput are reversed (as lower the actual value as worse the conditions), the formula also needs to be reversed as

$$\text{Coefficient} = \frac{\text{Max Range} - \text{Actual Value}}{\text{Max Range} - \text{Min Range}} \cdot \text{Max Coeff Interval} \quad (2)$$

For further understanding of using these formulas, see the calculation example in the following Section 5.1. Other parameters are evaluated according to the logical condition: type of cable, whether the junction is heat-shrunk, whether it was installed before 2000, and whether the cable is cross bonded. If the condition is true then the parameter value takes either the upper or lower bound of the interval.

The higher the resulting value of the coefficient, the greater the presumption that the given cable route is in good condition and its early replacement is not expected. If the resulting value of the coefficient is too low, the utility provider should be alerted and first measure the section using VLF methods (tan δ, and partial discharge). This should lead to more frequent voltage tests. If the service life of the cable routes is 50 years, and if we multiply the service life by a coefficient, the result should reduce the service life. If we arrive to a value lower than 5 years, the utility provider should decide, based on both measurements, whether to operate a high-risk cable route or to allocate funds for the renewal of the cable route.

The parameters share on the total coefficient have been chosen so that there is high emphasis on the measured TCP throughput, with this being the leading indicator. The rest of the parameters are more likely connected to the topological properties of the cable, and so the percentage representation of topological and communication influence on the resulting coefficient is set almost identical (45:50). However, it needs to be kept in mind that the PLC/BPL communication is also dependent on some of the topological parameters. The setting of the parameters share has been proposed empirically while taking into account the experience with measurement in the Czech DSOs network and also experience from cited articles of Sections 2 and 4. It ought to be noted that a more detailed statistical investigation would be beneficial and might be used in future studies to adjust the parameters share of the total coefficient. The current setting of parameters share should avoid the method of falling into deadlock where only long and aging cables would be visited regardless of bad communication performance.

5.1. Example of Coefficient Calculation

Thanks to cooperation with the Czech Distribution System Operator (DSO) E.ON, it was possible to perform measurements on a circular topology in the Brno center area. The topology is composed of a total of 26 stations, each station containing a headend and a repeater. The measurement took place between two stations (point-to-point connection). The physical and topological parameters of all measured routes were also provided from the GIS system.

The BPL solution was based on IEEE 1901 OFDM (FFT Access) with a 2–30 MHz frequency band. An example of how to calculate the coefficient for a particular route is given below. Description of the communication route with all available parameters:

Route A: Substation—DTS1

Distance between BPL modems: 515.2 m
Cable type: AXEKCY, AXEKCEY, AXEKCEY, AXEKCY, AXEKCEY
Cable age: 1995, 1995, **1979**, **1979**, 1995
Number of cable joints: 4
Cable joint type installation: plastic
Cable joint age:
Load: unknown
Bonding: unknown
Cross section: 240 mm^2
Average TCP throughput: 8.23 Mbps

$$\text{Route A: Coefficient}_{\text{Distance}} = \frac{515.2}{1200} \times 0.05 = 0.0215 \quad (3)$$

$$\text{Route A: Coefficient}_{\text{Cable type}} = (\text{cable not risky}) = 0 \quad (4)$$

$$\text{Route A: Coefficient}_{\text{Cable age}} = \frac{(\text{actual year} - 1979)}{40} \times 0.15 = (\text{older than 40 years}) = 0.1500 \quad (5)$$

$$\text{Route A: Coefficient}_{\text{Number of cable joints}} = \frac{4}{20} \times 0.15 = 0.0300 \quad (6)$$

$$\text{Route A: Coefficient}_{\text{Cable joint type}} = (\text{plastic not risky, but unknown}) = 0.0200 \quad (7)$$

$$\text{Route A: Coefficient}_{\text{Cable joint age}} = \frac{(\text{actual year} - 2000)}{20} \times 0.15 = (\text{unknown}) = 0.1500 \quad (8)$$

$$\text{Route A: Coefficient}_{\text{Load}} = \frac{\text{load}}{100} \times 0.035 = (\text{unknown}) = 0.0350 \quad (9)$$

$$\text{Route A: Coefficient}_{\text{Bonding}} = (\text{unknown}) = 0.0100 \quad (10)$$

$$\text{Route A: Coefficient}_{\text{Cross section}} = \frac{240 - 630}{50 - 630} \times 0.05 = 0.0336 \quad (11)$$

$$\text{Route A: Coefficient}_{\text{TCP throughput}} = 0.45 - \frac{8.23 \times 0.45}{50} = 0.3750 \quad (12)$$

$$\text{Route A: Coefficient}_{\text{TOTAL}} = 1 - 0.826 = \mathbf{0.174} \quad (13)$$

Route A shows medium performance, which is mainly influenced by the length of the route itself, which is operated on an old section of cable with a medium number of cable joints. These results correlate with the maximum achieved TCP throughput of slightly over 8 Mbps. The utility provider should regularly monitor this route to see if the maximum throughput is decreasing. Reducing throughput could lead to the detection of a problematic route and prevent cable breakage and subsequent power outages.

Route B: DTS2—DTS3

Distance between BPL modems: 118.8 m
Cable type: AXEKCY
Cable age: 1990
Number of cable joints: 0
Junction type installation: without junction
Junction age: without junction
Load: unknown
Bonding: unknown
Cross section: 240 mm^2
Average TCP throughput: 36.7 Mbps

$$\text{Route B: Coefficient}_{\text{Distance}} = \frac{118.8}{1200} \times 0.05 = 0.0050 \quad (14)$$

$$\text{Route B: Coefficient}_{\text{Cable type}} = (\text{cable not risky}) = 0 \quad (15)$$

$$\text{Route B: Coefficient}_{\text{Cable age}} = \frac{(\text{actual year} - 1979)}{40} \times 0.15 = 0.1163 \quad (16)$$

$$\text{Route B: Coefficient}_{\text{Number of cable joints}} = \frac{0}{20} \times 0.15 = 0 \quad (17)$$

$$\text{Route B: Coefficient}_{\text{Cable joint type}} = (\text{no cable joint}) = 0 \quad (18)$$

$$\text{Route B: Coefficient}_{\text{Cable joint age}} = (\text{no cable joint}) = 0 \quad (19)$$

$$\text{Route B: Coefficient}_{\text{Load}} = \frac{\text{load}}{100} \times 0.035 = (\text{unknown}) = 0.0350 \quad (20)$$

$$\text{Route B: Coefficient}_{\text{Bonding}} = (\text{unknown}) = 0.0100 \quad (21)$$

$$\text{Route B: Coefficient}_{\text{Cross section}} = \frac{240 - 630}{50 - 630} \times 0.05 = 0.0336 \quad (22)$$

$$\text{Route B: Coefficient}_{\text{TCP throughput}} = 0.45 - \frac{36.7 \times 0.45}{50} = 0.1197 \quad (23)$$

$$\text{Route B: Coefficient}_{\text{TOTAL}} = 1 - 0.340 = \mathbf{0.660} \quad (24)$$

Route B shows almost no wear down. This route is relatively shorter and contains only one undivided cable, so there is no junction on the route. Additionally, the maximum TCP throughput is more than 28 Mbps. The utility provider does not have to supervise this route regularly. The route should be able to operate without any restrictions.

If we compare routes A and B, the coefficient tells us that we need to pay more attention to route A. The cable health is almost half, so the utility provider should monitor the TCP throughput. If the throughput starts to decrease, the route should be measured using the VLF method.

Interpretation of results:

Cable life varies with each DSO. Some may be 30 years, others up to 50 years. In our case, we will state, for example, the service life of a cable line at 30 years. If the coefficient for Route A was 0.174, then we multiply this value by the service life of the cables, so it will be

5.22 years. During this time, we should make control measurements using the VLF method at the latest. Route B has a final coefficient of 0.66, if we perform a similar calculation, the final time of further measurement by the VLF method is approximately 20 years. These status messages can help the DSO better alert you to early problem detection before a fault occurs.

6. Experimental Measurements of BPL Parameters for Cable Health Monitoring

The VLF method belongs to experimental measurements that are used to determine the service life of a given line. The VLF method uses loss factor measurements tan δ (TD) and PD measurements, which are able to determine the state of the cable insulation.

The loss factor TD determines the degree of real power loss in the dielectric material, and thus its losses. The loss factor is more suitable for determining the general condition of the cable insulation, as a whole route. In modeling, the cable insulation system is represented by a simple equivalent circuit consisting of two elements: a resistor and a capacitor. After applying a voltage to the cable, the total current I will consist of the capacitor current I_C and the resistance current I_R. The loss factor is the ratio of the current of the resistor and the current of the capacitor. the angle TD is the angle between the total current and the charging current when displayed as phasors.

The measurement of partial discharges is closely related to the formation of WT. By measuring PD with source location, it is possible to directly determine the activity of PD in cable sections, connections, and terminals. The passage of the partial discharge pulses depends on the damping in the cable. The measured level, therefore, depends on the distance from the end of the source of PD. Only the time delay between the first and the reflected pulse is important for locating the source of PD. PD inside the cable causes a short-term breakdown of the cable insulation. The pulse-shaped charging current thus generated is detected by means of a coupling capacitor above the measuring device and converted into an equivalent voltage signal. Subsequently, this voltage signal is recorded by the PD detection system and displayed as a pulse on the monitor.

6.1. Measurements of Partial Discharges

Experimental measurement with BPL technology was carried out on the distribution territory of a Czech Republic utility provider. A BPL solution was based on IEEE 1901 OFDM (FTT Access) with a 2–30 MHz frequency band. The BPL infrastructure is made up of one substation and 25 stations that are in circular topology. After installing the technology and initial testing, it was found that communication cannot be established between two particular stations. Furthermore, after a closer examination of this route, it was found that the route is electrically connected and operable, although it cannot be communicated using BPL modems. For this reason, a cable measuring car was called that can diagnose the cable. The test includes a voltage test and measurement of partial discharges.

Over time, cables age and their insulation degrades. Thermal overload, humidity ingress, or poorly processed cable joints and cable terminations also contribute to losses. These processes take place over a long period of time. Measuring the loss factor tan δ at a frequency of 0.1 Hz makes it possible, as an integral measuring method, to reliably distinguish between new, weak, and heavily aged cables. The measurement of the tan δ is an established method for integral diagnostics of the insulation condition on MV cable systems. During the measurement, the absolute limit values are compared with the reference value tan δ. The measurement allows the definition of individual evaluation criteria and creation of a reference database.

Figure 2 shows the analysis of TD trends, where the measurement takes place at three voltage levels (12.7, 19.0 and 25.4 kV) and there is a gradual increase in the loss factor on all three phases by almost more than twice the original value. A slight standard deviation is evident between the individual phases. Low PD activity with a mild presence of WT is also indicated.

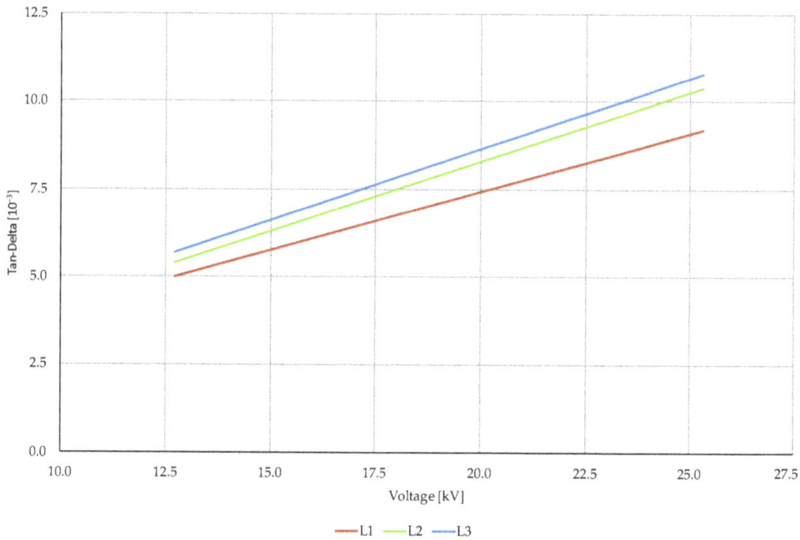

Figure 2. Measurement of tan δ using a measuring car.

Thanks to cooperation with the local DSO, data from GIS were analyzed. Based on the analysis, routes with older cable sections or cable joints were identified, which were then measured using a cable measuring car. Figure 3 shows the entire measured route. The measurement took place from station DTS1 to DTS2. The figure shows a clear description of the topological parameters of the route (cable type, cable age, junction type, junction installation method). In the picture, PDs are visualized using a yellow flash, including the distance from DTS1. These discharges indicate that the cable outer jacket is damaged and the humidity has entered the cable core through the outer jacket. There are also several cable joints on this route. The distance of the discharges from the beginning of the measurement indicates that cable joints 1 and 2 are also affected by humidity. The wet cable joint began to function as a high-frequency filter, making it impossible to communicate using BPL technology. Because of this, part of the cable and the two problematic cable joints are replaced.

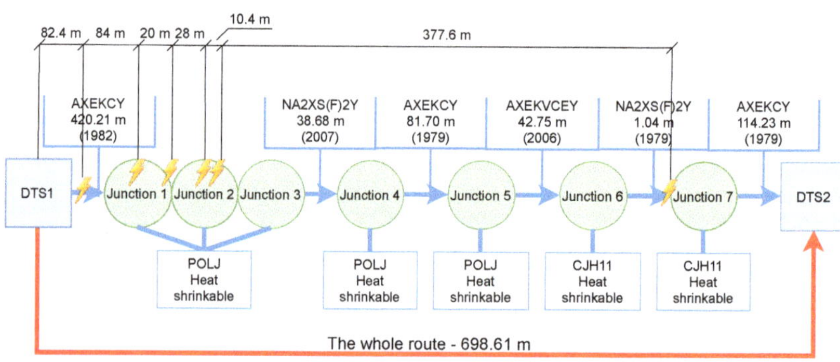

Figure 3. Measurement of the whole route showing cable types, junction types and measurements indicating partial discharges.

After replacement of the problematic cable joints, the subsequent measurements of BPL technology connectivity were made. The results showed the possibility of setting up a connection (communication was enabled), but the connection was not stable. A common type of BPL modems were replaced by the Longhaul type, which operates in the frequency range of 0.75–5 MHz. The disadvantage of Longhaul type BPL modems is low throughput (the measurements show throughput of 7–8 Mbit/s). This set of events revealed another use case of PLC technology. Thanks to the installation of this technology, a problematic section was discovered, which could become inoperable at any time and the customer would be without electricity. A situation where a customer is without electricity is often sanctioned in some countries. Thanks to the early discovery, no customer was disconnected, as the utility provider could prepare in advance for the planned outage and connect the customer using another route.

6.2. Measurements of Communication Parameters

Based on the methodology described in Section 5, individual sections of the entire topology of one substation and 25 transformer stations were measured (point-to-point measurements). The goal was to find the bottleneck bandwidth—the lowest value of the bandwidth of the entire measured route [4].

BPL communication can have very unexpected results. The distance between DTS stations is not the only parameter that affects communication. Based on theoretical assumptions [46], throughput decreases with distance. Based on the measurement of the whole topology, two sections were selected for comparison. These sections, named as the longest and the worst, did not correspond to the theoretical assumptions, and the results of the value of the measured throughput also stood out from the remaining sections.

Two routes were compared. Let us call the first route the worst. The route measures 519 m, contains four cable joints and five cables. Let us call the second route the longest. This route measures 880 m, contains five cable joints and six cables. It can therefore be assumed that the worst route will have higher throughput. However, the result is quite the opposite. The communication throughput was approximately six times higher than the worst route in the case of the longest route. The average throughput of the worst route was measured at 5.35 Mbit/s and the longest route reached an average TCP throughput of 32.21 Mbit/s. A comparison of the resulting worst and the longest route communication can be seen in Table 3. The measurement was performed using the EXFO FTB–Pro testers, where the RFC 6349 test methodology was used. The main advantage of the Internet Engineering Task Force (IETF) method RFC 6349 is the fact that it uses the TCP protocol for the measurement itself, which is now predominantly used for non-real-time communication on the internet. This method is standardized for measuring network devices. The size of the TCP window was used by the tester itself, as the most optimal size.

Table 3. TCP throughput the worst vs. the longest.

TCP Throughput [Mbps] (TCP Window 43.8 Kbyte)	Worst (519 m)	Longest (880 m)
Average	5.35	32.21
Median	5.29	32.00
Standard deviation	1.35	1.18
Minimum	2.72	28.30
Maximum	8.89	35.70

Figures 4 and 5 show the entire measured routes using VLF methods. In both pictures you can see the total length, number and type of cable joints, as well as the lengths and types of individual cable sections. Not all data are always included in GIS data. According to GIS, we know that the junction is located there, but no record of the junction type has been preserved. For this reason, there will be an unknown. The places where the PD's were detected are marked with a yellow flash, the distance from the starting position is also recorded.

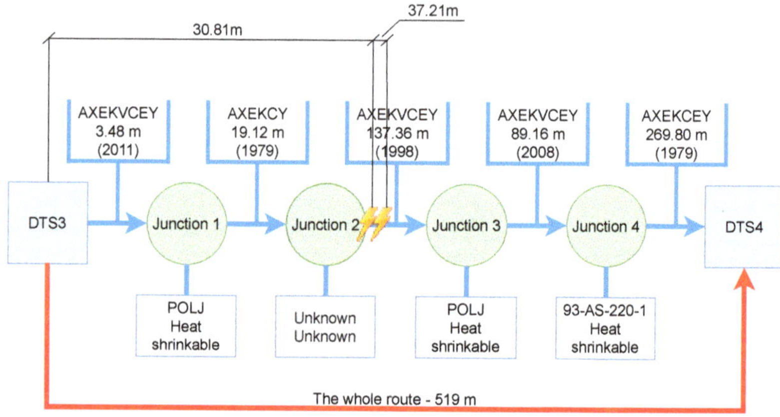

Figure 4. Measurement of the whole **worst route** showing cable types, junction types and measurements indicating partial discharges.

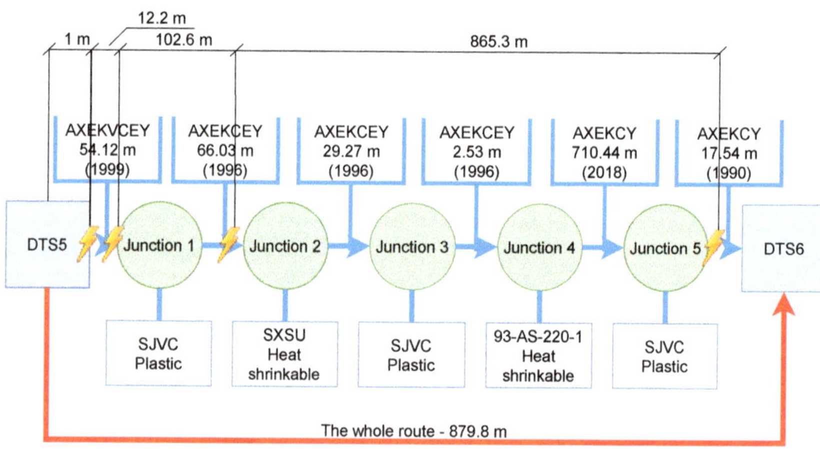

Figure 5. Measurement of the whole **longest route** showing cable types, junction types and measurements indicating partial discharges.

In Table 4, it is possible to clearly compare both measured routes. The table contains an exact description of the topological parameters of the route, which can be found with the GIS tool owned by each utility provider. Data from GIS systems may not always correlate with reality. According to the data, it is not always possible to join two cables with a specific junction. It is therefore necessary to examine the data to see whether they are valid.

Table 4. Comparison of the worst and the longest measured route.

Worst (519 m)					
	Cables			Cable joints	
No.	**Year**	**Length [m]**	**Type**	**Model**	**Year**
1	2011	3.5			
2	**1979**	19.1	**Heat shrink.**	93-AS-220-1	2011
3	1998	137.4	Heat shrink.	POLJ	2008
4	2008	89.2	Unk.	Unk.	Unk.
5	**1979**	269.8	**Heat shrink.**	POLJ	2008
Longest (880 m)					
	Cables			Cable joints	
No.	**Year**	**Length [m]**	**Type**	**Model**	**Year**
1	1999	54.1			
2	1996	66	Heat shrink.	SXSU	1999
3	1996	29.3	Plastic	SJVC	1996
4	1996	2.5	Plastic	SJVC	1996
5	2018	710.4	Plastic	SJVC	1996
6	1990	17.5	Plastic	SJVC	1990

A closer look at the table shows that a substantial part of the cable section of the longest route was replaced in 2018. This change of cable had a positive effect on the possible throughput. Although there are several cable joints on the route, they are mostly the plastic junction type that is not as prone to failure and humidity penetration as the heat-shrinkable one. The worst route contains a substantial part of the cable, which was laid in 1979. This cable is then connected to other parts of the cables using heat-shrinkable cable joints.

According to this experience, it can be said very clearly that throughput and QoS parameters also determine the technical condition of the cable.

7. Discussion

The use of PLC/BPL modems designed for communication in smart grids as a tool for monitoring the status of the cable is not a completely new idea. Even so, many researchers do not devote themselves to this idea. Our proposed coefficient of the technical condition of the cable route considers a total of 10 parameters that can significantly affect the communication, and thus detect the health of the route itself. The coefficient has a set weighting that corresponds to a certain setting in the Czech distribution network. For the sake of accuracy, it is possible that this weighting will change with an increasing amount of available data, thus increasing the possibility of fault detection.

The data we have so far are only from the MV voltage level. Therefore, this coefficient is focused only for medium voltage networks. Our goal is to expand the coefficient for LV voltage levels and thus increase the possibility of fault detection on more types of lines and make more appropriate use of the already installed BPL technology.

If we compare the current method of underground line diagnostics, it is necessary to find out the main disadvantage. The main disadvantage is that the existing diagnostic tools are mainly offline. This means that for the DSO, it is necessary to plan a shutdown, to secure the supply of energy from other sources, if it is not possible to secure the supply of energy, the affected customers must be informed. In the case of using online diagnostics, it is not necessary to disconnect the measured route, but the measurement/detection of faults takes place constantly. For this reason, the method is more appropriate, not expensive, and if the PLC/BPL infrastructure is available, nothing extra is needed. The method proposed by [21] is based on measuring the broadband impedance response of the power cable. The goal was to create a faster and easier way to monitor cable status. The authors managed to create a method and verify it on AMCMK and AXMK cable types. The cable was broken during testing, so it was an offline method. The proposed algorithm was able to locate the fault within a distance of one meter, but the detection could only be between phases. Our proposed method is independent of the cable type and uses the permeability of the cable joint as one of the main parameters affecting the life of the cable. The proposed method aims to monitor the life of cables, predict when a breakthrough may occur, and prevent this breakdown by the timely and optimal dispatch of a cable measuring car, which used the VLF method for fault localization, loss factor measurement, and partial discharge.

8. Conclusions and Future Work

The PLC/BPL technology is in decline due to the roll-out of wireless communications, but where the technology is already installed, it would be a great pity not to use its full potential. The article introduced the secondary use of PLC/BPL communication, especially for cable health monitoring. The article proposed a method for evaluating the condition of cables based on 10 parameters, where each parameter is weighted by its significance. The weighting of the coefficient can be freely adjusted according to the needs of the particular DSO's. Furthermore, some parameters can be omitted or added according to the needs of the DSO. Thanks to the cooperation with the Czech DSO, the method has been validated in real conditions and the technical condition coefficients of the individual routes were determined. The use of the coefficients can help to optimize the allocation of funds in case of renewal of the power line infrastructure.

The future work will be focused on further measurements in real conditions. This includes working with DSOs who have a PLC/BPL infrastructure. Data collection will be used to refine the coefficient. Obtaining data from cable measurements could help to find a closer connection with the emergence of WT and PD in correlation with the transmission rate. The next step should include verification of the coefficient or adjustment of the weighting of the coefficient. Data should also be collected on transmission parameters of high current cables, such as SNR value, noise, and CFR.

Author Contributions: Conceptualization, L.B. and P.M. (Petr Mlynek); methodology, P.M. (Petr Mlynek) and L.B.; validation, P.M. (Petr Mlynek), J.S. and P.M.; formal analysis, V.V., M.P., J.M.; investigation, P.M. (Petr Mlynek) and L.B.; resources, J.S., M.R. and P.M. (Petr Musil); writing—original draft preparation, L.B., P.M. (Petr Mlynek), V.V., M.P.; writing–review and editing, P.M. (Petr Musil), J.S., J.M. and M.R.; supervision, P.M. (Petr Mlynek), M.P. All authors have read and agreed to the published version of the manuscript.

Funding: This research received no external funding.

Institutional Review Board Statement: Not applicable.

Informed Consent Statement: Not applicable.

Data Availability Statement: Author do not will to publish data online.

Acknowledgments: This research work has been carried out in the Centre for Research and Utilization of Renewable Energy (CVVOZE). Authors gratefully acknowledge financial support from the Ministry of Education, Youth and Sports of the Czech Republic under BUT specific research programme (project No. FEKT-S-20-6449).

Conflicts of Interest: The authors declare no conflict of interest.

References

1. Devolo. Magic 2 WiFi Next. 2020. Available online: https://www.devolo.global/magic-2-wifi-next (accessed on 15 September 2021).
2. Hidayati, A.; Reza, M.; Adriansyah, N.M.; Nashiruddin, M.I. Techno-Economic Analysis of Narrowband IoT (NB-IoT) Deployment for Smart Metering. In Proceedings of the 2019 Asia Pacific Conference on Research in Industrial and Systems Engineering (APCoRISE), Depok, Indonesia, 18–19 April 2019; pp. 1–6. [CrossRef]
3. Tretinjak, R.; Pehrsson, T. E.ON Sweden's 2nd Smart Meter Roll-Out. 2021. Available online: https://www.mpo.cz/assets/cz/energetika/strategicke-a-koncepcni-dokumenty/narodni-akcni-plan-pro-chytre-site/2021/1/Prezentace-EON.pdf (accessed on 15 September 2021).

4. Slacik, J.; Mlynek, P.; Musil, P.; Benesl, L.; Hlavnicka, J. Smart Substation Emulation for BPL Evaluation. In Proceedings of the 2020 21st International Scientific Conference on Electric Power Engineering (EPE), Prague, Czech Republic, 19–21 October 2020; pp. 1–4. [CrossRef]
5. Kelm, P.; Wasiak, I.; Mieński, R.; Wędzik, A.; Szypowski, M.; Pawełek, R.; Szaniawski, K. Hardware-in-the-Loop Validation of an Energy Management System for LV Distribution Networks with Renewable Energy Sources. *Energies* **2022**, *15*, 2561. [CrossRef]
6. Havel, J.; Michal, H. BPL komunikace na vrchním vedení VN. In Proceedings of the Conference CK CIRED 2017, Tabor, Czech Republic, 7–8 November 2017; pp. 1–5.
7. Segatto, M.E.V.; de Oliveira Rocha, H.R.; Silva, J.A.L.; Paiva, M.H.M.; do Rosário Santos Cruz, M.A. 14—Telecommunication Technologies for Smart Grids: Total Cost Optimization. In *Advances in Renewable Energies and Power Technologies*; Yahyaoui, I., Ed.; Elsevier: Amsterdam, The Netherlands, 2018; pp. 451–478. [CrossRef]
8. Früh, H.; Rudion, K.; von Haken, A.; Wasowicz, B.; Gerber, M. Field-test based comparison of LTE and PLC communication technologies for smart grid applications. In Proceedings of the CIRED 2020 Berlin Workshop (CIRED 2020), Online, 22–23 September 2020; Volume 2020, pp. 378–381. [CrossRef]
9. Zimmermann, M.; Dostert, K. A multipath model for the powerline channel. *IEEE Trans. Commun.* **2002**, *50*, 553–559. [CrossRef]
10. Zhao, Y.; Zhou, X.; Lu, C. A new channel emulator for low voltage broadband power line communication. In Proceedings of the 2013 IEEE 10th International Conference on ASIC, Shenzhen, China, 28–31 October 2013; pp. 1–4. [CrossRef]
11. Zimmermann, M.; Dostert, K. Analysis and modeling of impulsive noise in broad-band powerline communications. *IEEE Trans. Electromagn. Compat.* **2002**, *44*, 249–258. [CrossRef]
12. Wang, J.; Crapse, P.; Shin, Y.J.; Dougal, R. Diagnostics and Prognostics of Electric Cables in Ship Power Systems via Joint Time-Frequency Domain Reflectometry. In Proceedings of the 2008 IEEE Instrumentation and Measurement Technology Conference, Victoria, BC, Canada, 12–15 May 2008; pp. 917–921. [CrossRef]
13. Coats, D.; Alam, M.N.; Deng, Q.; Ali, M.; Shin, Y.J. Joint time-frequency optimized reference for surface wave reflectometry-based insulation health assessment. In Proceedings of the 2012 11th International Conference on Information Science, Signal Processing and their Applications (ISSPA), Montreal, QC, Canada, 2–5 July 2012; pp. 1135–1140. [CrossRef]
14. Lars T. Berger, K.I. *Smart Grid Applications, Communications, and Security*; Wiley: Hoboken, NJ, USA, 2012.
15. Hartlein, R.A.; Hampton, R.N. *Cable Diagnostic Focused Initiative*; Georgia Institute of Technology: Atlanta, GA, USA, 2010. [CrossRef]
16. Huo, Y.; Prasad, G.; Lampe, L.; Leung, V.C.M. Cable Health Monitoring in Distribution Networks using Power Line Communications. In Proceedings of the 2018 IEEE International Conference on Communications, Control, and Computing Technologies for Smart Grids (SmartGridComm), Aalborg, Denmark, 29–31 October 2018; pp. 1–6. [CrossRef]
17. Prasad, G.; Huo, Y.; Lampe, L.; Mengi, A.; Leung, V.C.M. Fault Diagnostics with Legacy Power Line Modems. In Proceedings of the 2019 IEEE International Symposium on Power Line Communications and its Applications (ISPLC), Aalborg, Denmark, 29–31 October 2019; pp. 1–6. [CrossRef]
18. Huo, Y.; Prasad, G.; Atanackovic, L.; Lampe, L.; Leung, V.C.M. Grid surveillance and diagnostics using power line communications. In Proceedings of the 2018 IEEE International Symposium on Power Line Communications and its Applications (ISPLC), Manchester, UK, 8–11 April 2018; pp. 1–6. [CrossRef]
19. Poluektov, A.; Pinomaa, A.; Romanenko, A.; Kosonen, A.; Ahola, J. Diagnostics of Low-Voltage Power Cables by Frequency-Response Analysis of Power-Line Communication Channel. In Proceedings of the 2018 20th European Conference on Power Electronics and Applications (EPE'18 ECCE Europe), Riga, Latvia, 17–21 September 2018; pp. P.1–P.10.
20. Huo, Y.; Prasad, G.; Atanackovic, L.; Lampe, L.; Leung, V.C.M. Cable Diagnostics With Power Line Modems for Smart Grid Monitoring. *IEEE Access* **2019**, *7*, 60206–60220. [CrossRef]
21. Pinomaa, A.; Ahola, J.; Kosonen, A.; Ahonen, T. Diagnostics of low-voltage power cables by using broadband impedance spectroscopy. In Proceedings of the 2015 17th European Conference on Power Electronics and Applications (EPE'15 ECCE-Europe), Geneva, Switzerland, 8–10 September 2015; pp. 1–10. [CrossRef]
22. Huo, Y.; Prasad, G.; Lampe, L.; Leung, V.C.M. Advanced Smart Grid Monitoring: Intelligent Cable Diagnostics using Neural Networks. In Proceedings of the 2020 IEEE International Symposium on Power Line Communications and its Applications (ISPLC), Malaga, Spain, 11–13 May 2020; pp. 1–6. [CrossRef]
23. Huo, Y.; Prasad, G.; Lampe, L.; Victor Leung, C.M. Smart-Grid Monitoring: Enhanced Machine Learning for Cable Diagnostics. In Proceedings of the 2019 IEEE International Symposium on Power Line Communications and its Applications (ISPLC), Prague, Czech Republic, 3–5 April 2019; pp. 1–6. [CrossRef]
24. Förstel, L.; Lampe, L. Grid diagnostics: Monitoring cable aging using power line transmission. In Proceedings of the 2017 IEEE International Symposium on Power Line Communications and its Applications (ISPLC), Madrid, Spain, 3–5 April 2017; pp. 1–6. [CrossRef]
25. Ohtomo, Y.; Yamazato, T.; Katayama, M. An access control method for multipoint cyclic data gathering over a PLC network. In Proceedings of the ISPLC2010, io de Janeiro, Brazil, 28–31 March 2010; pp. 285–290. [CrossRef]
26. Solaz, M.; Simon, J.; Sendin, A.; Andersson, L.; Maurer, M. High Availability solution for medium voltage BPL communication networks. In Proceedings of the 18th IEEE International Symposium on Power Line Communications and Its Applications, Glasgow, UK, 30 March–2 April 2014; pp. 162–167. [CrossRef]

27. Freitag, C.; Mladenovic, I.; Weindl, C. An interpretation approach for in field measured dissipation factor values of MV cable lines. In Proceedings of the 2013 IEEE International Conference on Solid Dielectrics (ICSD), Bologna, Italy, 30 June–4 July 2013; pp. 96–99. [CrossRef]
28. Abeysinghe, S.; Abeysekera, M.; Wu, J.; Sooriyabandara, M. Electrical properties of medium voltage electricity distribution networks. *CSEE J. Power Energy Syst.* **2021**, *7*, 497–509. [CrossRef]
29. Siswoyo, A.; Zhang, G.J.; Suwarno, H. Broadband Impedance Spectroscopy For Locating and Detecting Various Type of Degraded Portion Along with Polymeric Cable. In Proceedings of the 2021 11th International Conference on Power, Energy and Electrical Engineering (CPEEE), Shiga, Japan, 26–28 February 2021; pp. 1–7. [CrossRef]
30. Kakimoto, Y.; Yoshikawa, H.; Jogo, T.; Wakisaka, T.; Kozako, M.; Hikita, M.; Sato, H.; Tagashira, H.; Soeda, M. Construction and Experimental Verification of Novel Online Partial Discharge Monitoring System Using Power Line Communication. *IEEE Trans. Dielectr. Electr. Insul.* **2020**, *27*, 2165–2171. [CrossRef]
31. Hopfer, N.; Rezaei, H.; Zdrallek, M.; Krampf, M.; Karl, F.; Dietzler, U. Analysis of Broadband PLC Characteristics as a Second Use Case for Distribution System Operators. In Proceedings of the 2019 IEEE International Symposium on Power Line Communications and its Applications (ISPLC), Prague, Czech Republic, 3–5 April 2019; pp. 1–6. [CrossRef]
32. Hu, S.; Wang, L.; Gao, C.; Zhang, B.; Liu, Z.; Yang, S. Non-Intrusive Cable Fault Diagnosis Based on Inductive Directional Coupling. *Sensors* **2018**, *18*, 3724. [CrossRef] [PubMed]
33. Olsen, R. Technical considerations for Broadband Powerline (BPL) communication. In Proceedings of the 2005 Zurich Symposium on Electromagnetic Compatibility, Zurich, Switzerland, 13–18 February 2005; p. 6.
34. Slacik, J.; Mlynek, P.; Fujdiak, R.; Musil, P.; Voznak, M.; Orgon, M.; Hlavnicka, J. Capabilities and Visions of Broadband Power-Line in Smart Grids Applications. In Proceedings of the 2019 20th International Scientific Conference on Electric Power Engineering (EPE), Kouty nad Desnou, Czech Republic, 15–17 May 2019; pp. 1–5. [CrossRef]
35. Takeda, T.; Okamoto, T.; Suzuki, H.; Morikawa, M. A Study on the Actual Failure Situation of XLPE Cable and the Order of Priority of Degradation Diagnosis. *Electr. Eng. Jpn.* **2004**, *148*, 50–58. [CrossRef]
36. Mueller, C.; Keil, S.; Bauer, C. Underground Cables vs. Overhead Lines: Quasi-Experimental Evidence for the Effects on Public Perceptions and Opposition. 2018. Available online: https://www.researchgate.net/publication/326804018_Underground_cables_vs_overhead_lines_quasi-experimental_evidence_for_the_effects_on_public_perceptions_and_opposition?channel=doi&linkId=5b64269e458515298ce15039&showFulltext=true (accessed on 15 September 2021).
37. Wagenaars, P.; Wouters, P.; Van der Wielen, P.; Steennis, E. Measurement of transmission line parameters of three-core power cables with common earth screen. *Sci. Meas. Technol. IET* **2010**, *4*, 146–155. [CrossRef]
38. Kim, J.H.; Kim, J.Y.; Cho, J.T.; Song, I.K.; Kweon, B.M.; Chung, I.Y.; Choi, J.H. Comparison between Underground Cable and Overhead Line for a Low-Voltage Direct Current Distribution Network Serving Communication Repeater. *Energies* **2014**, *7*, 1656–1672. [CrossRef]
39. Anatory, J.; Theethayi, N.; Thottappillil, R.; Kissaka, M.M.; Mvungi, N.H. The Influence of Load Impedance, Line Length, and Branches on Underground Cable Power-Line Communications (PLC) Systems. *IEEE Trans. Power Deliv.* **2008**, *23*, 180–187. [CrossRef]
40. Liu, T. The influence of the branch loads on the Chinese low-voltage power line communication channel. In Proceedings of the 2017 IEEE 17th International Conference on Communication Technology (ICCT), Chengdu, China, 27–30 October 2017; pp. 199–202. [CrossRef]
41. Mazzanti, G. The Effects of Seasonal Factors on Life and Reliability of High Voltage AC Cables Subjected to Load Cycles. *IEEE Trans. Power Deliv.* **2020**, *35*, 2080–2088. [CrossRef]
42. Ghaffarian, M.; Wang, X.; Clemence Kiiza, R. Review of Partial Discharge Activity Considering Very-Low Frequency and Damped Applied Voltage. *Energies* **2021**, *14*, 440. [CrossRef]
43. Siodla, K.; Rakowska, A.; Noske, S. The Proposal of a New Tool for Condition Assessment of Medium Voltage Power Cable Lines. *Energies* **2021**, *14*, 4116. [CrossRef]
44. Denny, H.; Stefan, L.; Van Minnen, F.; Paul, W. Accurate on-line fault location and PD activity location results obtained with SCG—A long-term utility experience. *CIRED-Open Access Proc. J.* **2017**, *2017*, 1–5.
45. Czapp, S.; Dobrzynski, K. Safety Issues Referred to Induced Sheath Voltages in High-Voltage Power Cables—Case Study. *Appl. Sci.* **2020**, *10*, 6706. [CrossRef]
46. Sláčik, J.; Mlynek, P.; Rusz, M.; Musil, P.; Benešl, L.; Ptacek, M. Broadband Power Line Communication for Integration of Energy Sensors within a Smart City Ecosystem. *Sensors* **2021**, *21*, 3402. [CrossRef] [PubMed]

sensors

MDPI

Article

Fast Constant-Time Modular Inversion over $\mathbb{F}_p$ Resistant to Simple Power Analysis Attacks for IoT Applications

Anissa Sghaier [1], Medien Zeghid [1,2], Chiraz Massoud [1], Hassan Yousif Ahmed [2], Abdellah Chehri [3,*] and Mohsen Machhout [1]

[1] Electronics and Micro-Electronics Laboratory, Faculty of Sciences, University of Monastir, Monastir 5000, Tunisia; sghaier.anissa@gmail.com (A.S.); medien.zeghid@fsm.rnu.tn (M.Z.); massoud.chiraz@hotmail.fr (C.M.); machhout@yahoo.fr (M.M.)

[2] Electrical Engineering Department, College of Engineering at Wadi Aldawaser, Prince Sattam Bin Abdulaziz University, Wadi Aldawaser 11991, Saudi Arabia; hassanuofg@gmail.com

[3] Department of Applied Sciences, University of Quebec in Chicoutimi (UQAC), Chicoutimi, QC G7H 2B1, Canada

* Correspondence: achehri@uqac.ca

Abstract: The advent of the Internet of Things (IoT) has enabled millions of potential new uses for consumers and businesses. However, with these new uses emerge some of the more pronounced risks in the connected object domain. Finite fields play a crucial role in many public-key cryptographic algorithms (PKCs), which are used extensively for the security and privacy of IoT devices, consumer electronic equipment, and software systems. Given that inversion is the most sensitive and costly finite field arithmetic operation in PKCs, this paper proposes a new, fast, constant-time inverter over prime fields $\mathbb{F}_p$ based on the traditional Binary Extended Euclidean (BEE) algorithm. A modified BEE algorithm (MBEEA) resistant to simple power analysis attacks (SPA) is presented, and the design performance area-delay over $\mathbb{F}_p$ is explored. Furthermore, the BEE algorithm, modular addition, and subtraction are revisited to optimize and balance the MBEEA signal flow and resource utilization efficiency. The proposed MBEEA architecture was implemented and tested on Xilinx FPGA Virtex #5, #6, and #7 devices. Our implementation over $\mathbb{F}_p$ (length of p = 256 bits) with 2035 slices achieved one modular inversion in only 1.12 µs on Virtex-7. Finally, we conducted a thorough comparison and performance analysis to demonstrate that the proposed design outperforms the competing designs, i.e., has a lower area-delay product (ADP) than the reported inverters.

Keywords: IoT; PKCs; prime field; modular inversion; BEEA; modular addition and subtraction; SPA; ADP; FPGA

Citation: Sghaier, A.; Zeghid, M.; Massoud, C.; Ahmed, H.Y.; Chehri, A.; Machhout, M. Fast Constant-Time Modular Inversion over $\mathbb{F}_p$ Resistant to Simple Power Analysis Attacks for IoT Applications. *Sensors* **2022**, *22*, 2535. https://doi.org/10.3390/s22072535

Academic Editors: Alexandru Lavric, Liliana Anchidin and Adrian I. Petrariu

Received: 26 January 2022
Accepted: 22 March 2022
Published: 25 March 2022

Publisher's Note: MDPI stays neutral with regard to jurisdictional claims in published maps and institutional affiliations.

1. Introduction

The IoT encompasses the idea that everyday objects can be connected to the Internet and interfere with each other. Thus, these objects are capable of exchanging and storing data. The concept of IoT is to create a link between the real world and the digital world. However, several high-profile incidents have highlighted the vulnerability of IoT security because a common device was used to penetrate a larger network.

Connected objects are being used more and more, especially with the new connected cities projects. This increases the security risks of the IoT, especially when there is no monitoring or management of the devices. For example, any security breach in medical devices in healthcare applications (wearable or implantable ones) needs increasing attention to protect patient privacy.

In an IoT environment, there are several challenges to securing devices and ensuring end-to-end security. Additionally, since IoT is still an emerging market, many manufacturers and designers are more concerned about launching their products to market than about designing and building security at the beginning.

One of the most frequently cited security concerns with IoT is the use of hard-coded or default passwords.

Furthermore, since a large number of IoT devices are designed to be "set and forget"—placed on a machine or the field and left at the end of their lifespans—they rarely receive security or patch updates. Adding security upfront can be costly, slow development, and prevent the device from working properly.

Another security challenge is connecting legacy assets that are not designed for IoT connectivity. It would be prohibitively expensive to replace legacy infrastructure with connected technologies. In spite of this, there are still some objects that probably have never been updated or protected against modern threats. As a result, the potential attack surface has grown.

There are also a limited number of industry-recognized standards for IoT security. Despite the existence of several IoT security frameworks, there is no single framework that has been agreed upon.

Security and privacy must be prioritized by product manufacturers and service providers. For example, default encryption and authorization should be included.

Smart homes, connected cars, and manufacturing plants are all examples of environments that may experience IoT security breaches. For example, an attack that disables the brakes on a connected car or on a connected healthcare device can have life-threatening effects. Likewise, an attack on critical infrastructures—such as an oil well, energy grid, or water supply—could have dire consequences. Therefore, relevant industries are adopting procedures and implementing measures to ensure their safety. An analysis method that processes the data generated from all security equipment, and a measure based on previous attacks against control systems, have been developed [1].

The rise of IoT has sparked worries about the security of data transmitted between IoT devices and the edge. Indeed, Kim et al. in 2019 [2] conducted a study to address the security flaws of existing IoT devices such as sensor multi-platforms, and they proposed a model that addressed their security vulnerability.

The Elliptic/Hyperelliptic Curve (ECC/HECC) has gained increased attention in recent years because it provides shorter private key lengths with the same level of security as other PKCs such as RSA [3]. At present, cryptosystems based on asymmetric algorithms such as ECC/ECDSA are employed by many IoT devices to safeguard their data and connections [4–6].

In the ECC/ECDSA algorithms, protecting the private and ephemeral keys (d, k) is essential because, if an adversary obtains these keys, they could modify messages and signatures, thus making the algorithms useless. Several physical attacks aim to retrieve private and ephemeral keys (d, k) [7,8]. A side-channel attack (SCA) in cryptography is used to extract cryptographic keys and other secret information from a device such as IoT sensors, a smart card, and an integrated circuit.

The power consumption of an electronic device while processing secret data can reveal some of those details. SPA, which uses the existence of visually recognizable power consumption patterns that may expose the sequence of operations conducted by an algorithm, is one of the approaches that can be used to recover hidden information [9,10]. Hence, leakage power consumption is determined by tracking voltages of the device and then dividing by the transition count leakage and Hamming weight, with the first reflecting the number of 1-bit bits treated in time and the second reflecting the number of state variables loaded at a time [11]. Thus, if a secret value is required for this sequence of operations, the implementation is SPA-vulnerable.

Many state-of-the-art studies have examined the protection of the IoT environment against power analysis attacks [12]. Through a bit-checking mechanism, Moon et al. [13] proposed a side-channel attacks countermeasure for IoT systems. In order to eliminate branching in modulus operations, bit checking was introduced. In 2021, a statistical experimental design was proposed to optimize the power attack parameters of an IoT transducer with a minimum cost [14].

In the same year, against differential power analysis (DPA) attacks, two novel countermeasures were proposed. The two methods are based on the back-gate bias technique of fully depleted silicon on insulator (FD-SOI) technology [15]. Using the proposed countermeasures, the required number of traces to recover the secret was increased, as demonstrated by the experimental results.

Finally, SPA has been successfully implemented against a variety of cryptographic algorithms. The ECC/ECDSA key generation technique in the IoT environment is one of the algorithms that has been targeted by this type of attack.

Jérémy et al. [16] published a review on passive attacks on ECC scalar multiplication algorithms in 2016, including leakage sources and frequent errors exploited to attack the ECDSA system. This work described the link between lattice attacks and partial leakage to show how tiny leaks affect ECDSA security. In the same year, Genkin et al. [17] investigated the susceptibility of mobile devices' ECC implementations to side-channel key extraction, finding that these implementations are vulnerable to electromagnetic and power side-channel assaults.

The authors demonstrated partial key leakage from OpenSSL running on Android and from iOS' CommonCrypto, and complete key extraction from OpenSSL running on iOS devices. In 2018, to recover the ECDSA secret key to SM2 Digital Signature Algorithm (SM2-DSA), Zhang et al. [18] extended the new lattice-based attack introduced by Later Nguyen and Shparlinski. SM2-DSA is a Chinese version of ECDSA. They tested the security of the SM2-DSA on the Atmega128 microcontroller using a lattice attack.

In 2019, Wunan et al. [19] published a threat analysis on the broken digital signature of data transactions and an improved SPA against ECDSA. ECDSA's private key can be obtained by using the described attack method combined with a power trace. Moreover, Wunan et al. proposed a side-channel attacks countermeasure for blockchain devices by inserting empty operations into ECC-(point doubling/point addition) operations. In 2021, Thiebault et al. [20] presented a high-quality hardware ECDSA core and provided a complete open-source ECDSA attack artifact. They demonstrated an effective PAA against its FPGA implementation.

In ECC/ECDSA cryptosystems, the SPA leakage-based side-channel research concentrates on the modular-inversion process necessary to generate an ECC/ECDSA private key. The conversion of coordinates in an ECC implementation from projective to affine based on conventional modular inversion, for example, can lead to the disclosure of information, and an adversary could obtain the secret key. Additionally, ECDSA inverts the per-message random secret after scalar multiplication to generate a digital signature [21].

A challenge for cryptographic implementations is a modular inversion because it is one of the most time-consuming field operations in ECC computations. Computing modular inverses can be performed in a variety of ways. Using Fermat's little (FLT) theorem, the Extended Euclidean algorithm (EEA), or any binary variant based on Montgomery's Modular Inverse (MMI) algorithm, it is possible to compute the inverse function. Figure 1 shows the two popular methods, which are the FLT (its variation is the Itoh–Tsujii technique) and the EEA (its variations are the Binary Inversion Algorithm, Left Shift Binary Algorithm, Right Shift Binary Algorithm, and the Montgomery Inversion Algorithm) [22].

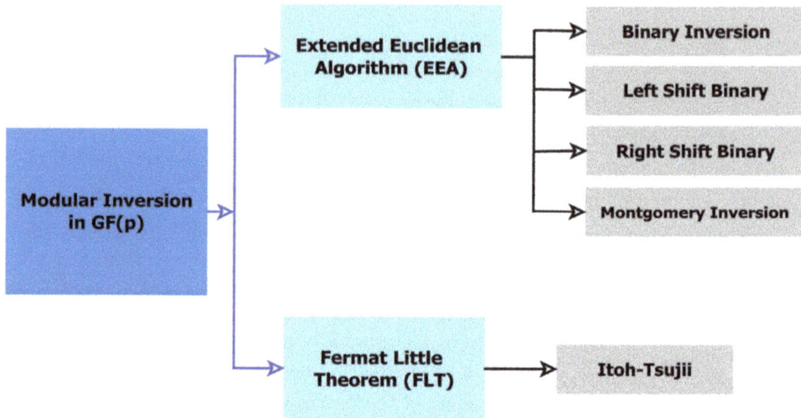

Figure 1. Modular inversion methods in $\mathbb{F}_p$.

A comparison between the FLT, EEA, and BEEA methods is presented in Table 1. It shows that the BEEA binary version has high performance and efficiency, which meets the requirements of electronic devices (smart cards, RFID tags, mobile phones, IoT devices, etc.).

Table 1. Comparison between FLT, EEA, and BEEA methods.

FLT	EEA	BEEA
• More complicated than EEA • Higher complexity O (log^3 n) • Slower (consumes a lot of time compared to EEA) • Uses a large number of repetitive multiplications • Secure against SPA and timing attack	• More efficient than FLT • Less complexity O (log^2 n) • Very fast and commonly used for large operands	• Suitable for hardware implementation because it replaced expensive divisions with shift-right operations • Faster than EEA • Less complicated (uses only ordinary additions and subtractions) • No need for multiplications or divisions

The BEEA method is commonly favored to eliminate the divisions needed in the EEA algorithm since it uses right-shift operations to replace multi-precision divisions. This improvement results in high-performance software-hardware implementations. The BEEA's execution flow is heavily reliant on its inputs. This attribute was investigated in an attempt to find side-channel vulnerabilities that could seriously affect the privacy of a secret variable processed by BEEA.

It is possible to leak information as a result of a typical BEEA implementation. This is because BEEA's execution flow is heavily reliant on its inputs. Hence, the time required to calculate the result is dependent on the inputs. As a countermeasure, one of the most important elements is to ensure that the implementation has constant run-time (worst-case), meaning that the implementation's execution time is unaffected by the input. This is usually accomplished by removing any data-dependent BEEA code branches. Hence, the main idea of this work is to compute an inverse using the same number of constant-time iterations each time. Figure 2 shows the class diagram of this work.

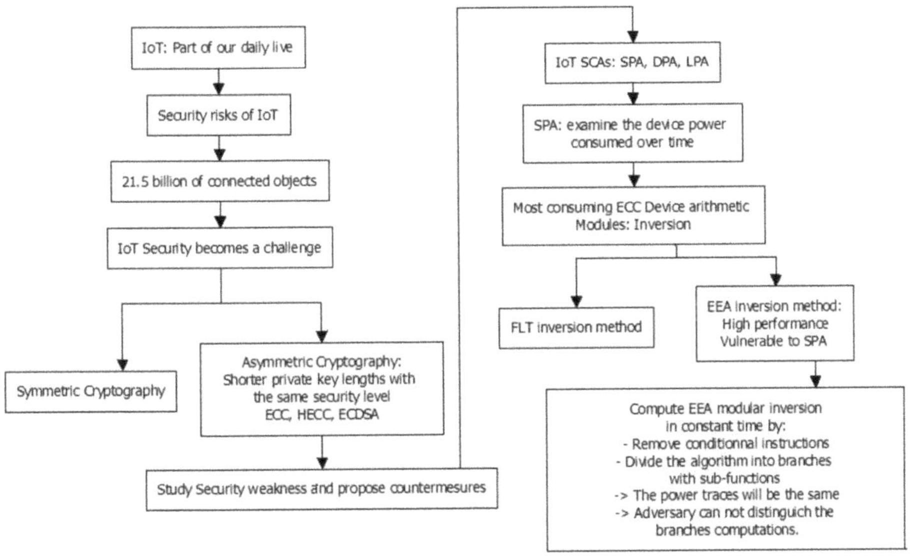

Figure 2. UML class diagram of IoT/modular inversion/SPA.

In addition, a high-performance modular inverter is required to speed up the calculation of a PKC system. However, key sizes grow as security levels increase, and this becomes a limiting factor in inverter design over prime field $\mathbb{F}_p$ due to the carry propagation problem. As a result, adding two prime numbers (A and B) in $\mathbb{F}_p$, where p is a prime of length > 128 bits has a direct impact on the efficiency of calculations in hardware/software (HW/SW) inverter implementations.

Over $\mathbb{F}_p$, carry skip (CSK), carry select (CSL), carry-save (CSA), carry-lookahead (CLA), and parallel prefix (PPF) adders are some of the fast binary adders proposed in the literature [23]. The Kogge–Stone adder (KSA) is the fastest adder in the literature because it has a lower fan-out at each stage, which then increases its performance, and is thus widely considered as a standard adder in the industry for high-performance arithmetic circuits [24]. However, it takes more area to be implemented because it computes the carries in parallel. Therefore, KSA performs parallel additions in microprocessors, DSPs, mobile devices, and other high-speed applications.

Based on the above discussions, this paper proposes an alternative method for computing modular inversion in constant time in order to prevent SPA attacks. We modify the (non-constant-time) BEEA approach by removing the conditional instructions and dividing the algorithm into four branches with sub-functions. Since the power traces will be the same regardless of input changes, the adversary cannot distinguish the branch's computation. Therefore, a novel BEEA-based inverter over prime fields is proposed and implemented (MBEEA).

The BEE algorithm, modular addition, and subtraction are revisited to perform MBEEA concurrently, resulting in competitive time and area complexities. Since addition is an essential operation in MBEEA, this paper introduces the generic G-KSA/S adder/subtractor to be reused for any prime number length.

The remainder of the paper is structured as follows: The (modular inversion-SPA) related works are introduced in Section 2. A brief description of BEEA is given in Section 3. Section 4 develops the constant-time BEEA-based inverter. The MBEEA architecture design is presented in Section 5. Section 6 contains the results and analysis of performance. The conclusion is given in Section 7.

2. Related Works

In the literature, several algorithms have been proposed for computing modular inversion. Some works were interested in analyzing SPA leakages of modular inversion implementation and their countermeasures. In 2014, to prevent combinational attacks (SPA and lattice techniques), Joppe W. Bos [25] modified the non-constant-time approach form of the Montgomery Inversion based on the EE greatest common divisor algorithm to compute both the classical and the MMI in constant time. He demonstrated that when the modulus has a special form, such as the Curve25519 (Elliptic curve with 256-bit key size offering 128-bit security), the FLT performance is comparable to the BEEA. However, the software implementation is done on ARM 32-bit, and the results showed that the constant-time almost inversion is much slower than the classic one.

The famous and easiest method to resist the leakage of information is the blinding technique using simple multiplicative masking [26]. When Z is the z-coordinate in projective coordinates of a given point and u is a random element in $\mathbb{F}_p$ (unknown to the attacker), the inversion will be performed for the product Zu instead of directly inverting Z using the BEEA; finally, the result $(Zu)^{-1}$ is multiplied by u to obtain Z^{-1}. However, this method requires two additional modular multiplication operations. Since the modular inversion based on FLT is based on modular exponentiation, Xu et al. constructed, in 2017, a secure and efficient modular exponentiation using Montgomery friend primes described by NIST to obtain an efficient and constant-time modular inversion over $\mathbb{F}_p$ [27]. Their improvement allows reducing the number of modular multiplications (a saving of 90%), which are computed using only subtractions and shifts.

In [28], the power consumption traces of two different BEEA implementations are investigated in detail in order to extract the SPA leakages that the binary EEA implementation may exhibit during the computation of the RSA key generation algorithm.

In 2018, Savaş et al. proposed two efficient constant-time Montgomery Inversion Algorithms that can be used to counter SCA [29].

The first algorithm is based on the BEEA method, and the second uses Kaliski's method. The two algorithms have comparable performance. The number of iterations for the Montgomery Inversion Algorithm based on BEEA was $2n$ (where n is the bit length of modulo p) which is less than for the Montgomery Inversion Algorithm using Kaliski's method ($2n + 1$). Furthermore, to speed up (upper-bounded by 2) computation in the software implementation of a multi-core processor, the author proposed a simple parallel algorithm of Montgomery Inversion.

In 2019, Bernstein et al. introduced streamlined constant-time variants of Euclid's algorithm for polynomial and integer inputs [30]. They presented simple, fast, constant-time "division steps" that work the same for integer and polynomial inputs. After applying the main algorithm to GCD, they studied the software speed of modular inversion of polynomials as part of a key generation in the NTRU cryptosystem.

In 2021, Awaludin et al. presented a high-speed ECC processor for arbitrary Weierstrass curves over GF (p) [31]. The proposed processor works in constant time and is suitable for applications that require high speed and SCA resistance. To preserve the SCA-resistance, they used constant-time Fermat's little theorem to perform field inversion operations.

Similarly, Sarna et al. presented a constant-time modular inversion algorithm development process capable of achieving a high level of security against timing and SPA Attacks [32].

In their work, Aldaya et al. present a novel SPA of the BEEA algorithm that reveals some exploitable power consumption-related leakages [33]. Using traces, they used the ECDSA protocol to apply SPA to reveal standardized private key sizes. An investigation was conducted into three countermeasures for eliminating SPA leakages from BEEA implementations.

3. BEE Algorithm

Several cryptographic algorithms use modular inverses. The BEEA is one of the most common ways to carry out the inversion operation. Given two integer numbers '*a*' and '*p*', BEEA computes x and y such that $ax + py = gcd(a, p)$. When $gcd(a,\ p) = 1$, the calculated value for x refers to the multiplicative inverse of (a mod p).

The BEEA has been reported in a few different forms in the literature. The original BEEA can be shortened when it is known in advance that the modulus p is a prime number, as in ECDSA. Algorithm 1 illustrates the updated version of the BEEA.

According to Algorithm 1, 'v-loop' refers to the loop that divides 'v' by 2, while 'u-loop' refers to the loop that divides 'u' by 2.

Additionally, the term sub-step is associated with subtraction operations executed in step 5.5 to update u and x and v and y, respectively.

According to Algorithm 1, in accordance with the least significant bit (LSB) of 'a', only the u-loop during the first iteration can be executed since 'v' is equal to 'p' and 'p' is a prime number. Additionally, for the rest of the iterations, due to the subtraction at step 5.5, only one loop is executed per iteration. Hence, 'u' and 'v' are odd integers just before the execution of the sub-step stage.

A BEEA side-channel analysis was presented in [33]. It concluded that a full recovery of the algorithm's inputs can be achieved when the adversary can collect Z_i and SUBS[i] for all iterations, where Z_i is the number of times that an x-loop (x = u or v) is executed at iteration 'i' and SUBS[i] is the output of the sub-step function of Algorithm 1 at iteration. When $u \geq v$, SUBS[i] = u, else SUBS[i] = v.

Algorithm 1: Pseudo-code of the BEE algorithm.

```
1.  Input: Integers a and p such that gcd(a, p) = 1
2.  Output: b= a^-1 mod p
3.  Define; u = a; v = p
4.  Define; x1 = 1; x2 = 0
5.  While (u ≠ 1 and v≠ 1){
5.1. While u is even {                      ┐
   5.1.1. u←u/2;                            │
   5.1.2. If x1 is even {                   │
         5.1.2.1. x1 ← x1/2;                ├-- u-loop
         5.1.2.2. Else                      │
         5.1.2.3. x1 ← (x1 + p)/2;          │
   5.1.3. }                                 │
5.2. }                                      ┘
5.3. While v is even {                      ┐
   5.3.1. v←v/2;                            │
   5.3.2. If x2 is even {                   │
         5.3.2.1. x2 ← x2/2;                ├ v-loop
         5.3.2.2. Else                      │
         5.3.2.3. x2 ← (x2 + p)/2;          │
   5.3.3. }                                 │
5.4. }                                      ┘
5.5. If u ≥ v {                             ┐
   5.5.1. u←u-v;                            │
   5.5.2. x1 ← x1- x2;                      │
   5.5.3. Else                              ├ sub-step
   5.5.4. v←v-u;                            │
   5.5.5. x2 ← x2- x1;                      │
5.6. }                                      ┘
5.7. }
6.  If(u = 1){
6.1. Return x1 mod p;
6.2. Else
6.3. Return x2 mod p;
6.4. }
```

By knowing only Z_i, the authors were able to recover some SUBS[i] and to express a certain number of bits from one of the BEEA inputs as a function of the other.

It is very interesting to consider this approach in the context of power-based side-channel analysis since it appears very difficult to extract the SUBS[i] directly from power traces, while it is easier to extract the Z_i when the sub-step can be differentiated.

We can segment BEEA iterations into x-loops followed by sub-steps. Consequently, the Z_i is based on the duration between two consecutive sub-steps. Therefore, if an adversary can distinguish between the sub-steps of the BEEA in the power trace, then he can exploit the Z_i.

Remarkably, Algorithm 1 is based on conditional instructions (loops while and if statements). Therefore, the execution time of Algorithm 1 depends on 'a' and 'p' values. Furthermore, the input 'a' presents the scalar multiplication algorithm results in affine coordinates, and a modular inversion will convert it to projective coordinates. Hence, the inversion should be secure against SPA attacks to prevent information leakage. For this reason, our contribution is to ensure a countermeasure to prevent Algorithm 1 from a specific SPA attack.

4. Constant-Time Modular Inversion

Input-dependent execution flows of the BEEA have prompted the development of anti-side-channel measures. A modified version of BEEA that operates in constant time and is resistant to SPAs is proposed in this section. As a countermeasure, we propose eliminating the conditional branches in Algorithm 1 in order to reduce the data dependency.

Algorithm 1's execution time depends on both 'a' and 'p'. In such an algorithm, when different inputs are used, the number of iterations in the while loop that have to be removed at the end of the algorithm may vary significantly.

Algorithm 1 may be converted into a constant-time algorithm by meeting the following conditions:

- The same amount of time is always taken to compute an iteration. In constant time, this requires computing all four branches of Algorithm 1 and selecting the appropriate values. As a result, the computed time of each iteration is independent of the branch taken; however, it may be increased to (at most) that of the computation of the sum of all the branches.
- The algorithm always has the same number of iterations. As a result, the worst-case number of 'K' iterations should always be calculated. It can be achieved by determining when Algorithm 1 terminates (when we reach $v = 1$ or $u = 1$).

The constant-time version of Algorithm 1 is described in Algorithm 2. The first two branches are computed in every cycle (constant run-time) regardless of whether the values of u and v are even or odd. Thus, the sequence of the parity computation deduced from the power trace does not reveal any information about the inputs (the bits of a).

In Algorithm 2, the comments showing which branches from Algorithm 1 are being computed are displayed after an '#'. As shown in Algorithm 2, the "while loops" of Algorithm 1 are replaced by "for loops" whose higher bound is 'K'. 'K' presents the cycle number of the biggest number inverse in the finite field $\mathbb{F}_p$, such that ($a < p$). Therefore, K presents the worst case of iterations.

It is not difficult to determine the number of iterations ('K') in the worst-case scenario of Algorithm 1. In every iteration, either 'u' or 'v' are reduced by at least a factor of two, so the maximum number of iterations is $2log_2(p)$, where p is a prime number of n-bit length. It follows that the minimum number of iterations is $log_2(p)$. This reveals the bounds on the exponent 'k' when the algorithm terminates. The functions presented in Algorithm 2 are:

- The shift-by-one function denoted by $lshift_1(z, x)$: this function shifts x by one position to the left and stores the result in z.

- The subtraction and addition functions are denoted respectively by sub (z, x, y) and add $(z, x, y))$ computing $z \leftarrow x - y$ and $z \leftarrow x + y$. Those two functions are computed by the adder/subtractor G-KSA/S which is discussed in the next section.

For the function add $(x_1, x_1, (p' \wedge x_1\ (0))$, we create a bitmask $x_1(0)$ since the LSB of x_1 determines the parity of x_1 (1 indicates odd, 0 indicates even) in order to calculate the iteration 5.1.2 in Algorithm 1.

$$\text{add}\ (x_1,\ x_1, (p' \wedge x_1\ (0)) = \begin{cases} if\ \ x_1 \text{ is even then } x_1\ (0)\ = 0 \rightarrow x_1 = \frac{x_1}{2} \\ if\ \ x_1 \ \text{ is odd then } x_1\ (0)\ = 1 \rightarrow x_1 = \frac{x_1}{2} + p' \end{cases} \quad (1)$$

- The function mem_u *and* mem_v are used respectively to save the values of u and x_1 if u is even (mem_u = 0, RAM active 0), and the values of v and x_2 if v is even (mem_v = 0, RAM active 0). As the LUT area is $2n$, we need four look-up tables to memorize u, x_1, v, and x_2.
- The comparison function is denoted by comp (u, v). This function is used to compare 'u' and 'v' values.

$$\begin{cases} if\ u \geq v \text{ and } \ u \neq 1 \ \text{ then } \ sel_1 = 1,\ else\ sel_1 = 0 \\ if\ v \geq u \text{ and } u \neq 1 \text{ then } sel_2 = 1,\ else\ sel_2 = 0 \end{cases} \quad (2)$$

- The function denoted $mem_u(sel_1)$

$$\begin{cases} \ if\ sel_1 = 1 \text{ then } u \text{ and } x_1 \text{ alookuped in the look up tables} \\ else\ no\ values\ are\ stored \end{cases} \quad (3)$$

- The occurrence function is denoted by occur (RAM, u, 1, i), which means the first occurrence of 1 in RAM of u indicates the corresponding i, then:

$$\begin{cases} If\ i < j \quad then \quad b = x_1 \\ If\ j > i \quad then\ b = x_2 \end{cases} \quad (4)$$

Algorithm 2: Pseudo-code of the proposed modified BEE algorithm

1. **Input:** p and $a \in \mathbb{F}_p$
2. ***Output:*** $b = a^{-1}$ *mod p*
3. ***Define;*** **u = a; v = p**
4. ***Define;*** ***x1 = 1; x2 = 0***
5. ***Define;*** p' = *Rshift*(p,1)

Inversion *steps*

6. ***for*** $(i = 1;\ i \leq k;\ i{+}{+})\ \{$

6.1.	$lshift_1(u,u)$	# $u \leftarrow SHIFT(u,1)$
6.2.	$lshift_1(x_1,x_1)$	# $x_1 \leftarrow SHIFT(x_1,1)$
6.3.	$add(x_1, x_1,(p' \wedge x_1\ (0))$	# $x_1 \leftarrow SHIFT(ADD(x_1,p),1)$
6.4.	$mem_u\ (u(0))$	
6.5.	$lshift_1(v,v)$	# $v \leftarrow SHIFT(v,1)$
6.6.	$lshift_1(x_2,x_2)$	# $x_2 \leftarrow SHIFT(x_2,1)$
6.7.	$add(x_2, x_2,(p' \wedge x_2\ (0))$	# $x_2 \leftarrow SHIFT(ADD(x_2,p),1)$
6.8.	$mem_v\ (v(0))$	
6.9.	$comp(u,v)$	# If $u \geq v$
6.10.	$sub(u,u,v) \wedge sel_1$	# $u \leftarrow SUB\ (u,v)$
6.11.	$sub(x_1,x_1,x_2) \wedge sel_1$	# $x_1 \leftarrow SUB\ (x_1,\ x_2)$
6.12.	$mem_u\ (sel_1)$	# $v \leftarrow SUB\ (v,u)$
6.13.	$sub(v,v,u) \wedge sel_2$	# $x_2 \leftarrow SUB\ (x_2,\ x_1)$
6.14.	$sub(x_2,x_2,x_1) \wedge sel_2$	
6.15.	$mem_v\ (sel_2)$	
6.16.	*occur(RAM, u, 1,i)*	# Return x_1 mod p if u equal 1
6.17.	*occur(RAM, v, 1,j)*	# Return x_2 mod p if v equal 1
6.18.	*return (i,j,x)*	

7. }

Overall, Algorithm 2 in $\mathbb{F}_p$ has two specified inputs: an element 'a' and a prime number 'p'. The output of the algorithm is the multiplicative inverse of 'a', i.e., $ba = 1 \ mod \ p$. Algorithm 2 employs three steps in total to produce the multiplicative inverse result p:

- The initial step: 'a' and 'p' are assigned to 'u' and 'v', respectively. Moreover, in this step, '1' and '0' are assigned to $x1$ and $x2$, respectively.
- Updating step: u, v, $x1$, and $x2$ values are updated by three different operations: add, sub, left shift ($lshift$), mem_u, and mem_v. The updating process is executed throughout the entire loop ((1 < i < k) execution time) and managed by two signals: sel1 and sel2. The new (u, v, $x1$, and $x2$) values are assigned according to sel1 and sel2, and these control signals are updated in step 12.9 during each iteration of the "for loop".
- Return 'b' after k iterations.

5. MBEEA Implementation

As shown in Algorithm 2, inversion is realized mainly by subtracting and shifting operations. Additionally, shift operations are easily implemented in hardware at no cost. We can therefore consider in our design a case that allows performing as many shifts as possible in one clock cycle to reduce the number of clock cycles.

As seen in Algorithm 2, 'Reg_u', 'Reg_v', 'Reg x1', and 'Regx2' n-bit registers are essential for implementing the hardware architecture of MBEEA over a prime field. By applying this algorithm, the calculation of division, such as '$u/2$', '$v/2$', '$x1/2$', and '$y1/2$', is based on comparisons of parity and magnitude. There are two multiplexers for selecting 'u', 'v', '$x1$', and '$x2$'. However, exact comparisons can only be made through full n-bit subtractions, and this has the effect of delaying decisions about the next calculation. To perform the additions or subtractions, we used n-bit KSA. In the implemented design, all possible 'u', 'v', '$x1$', and '$x2$' updated values are computed at once with multiplexers, and the new values for 'u' and 'v' are selected.

Figure 3 depicts the proposed MBEEA inverter design, which is based on Algorithm 2 and includes the following units:

- *Controller unit:* This generates control signals for all the MBEEA architecture units and the data flow in the inverter design, and the movement of data between the adder-subtractor units, the memory units, the comparator unit, and the occurrence units.
- *Adder/Subtractor (G-KSA/S):* This performs the addition or the subtraction of two values according to the controller decision.
- *Memory unit (MU):* The main purpose of this unit is to store different parameters such as u, v, $x1$, and $x2$, and their intermediate results. It constitutes of four blocks of RAM (RAMu, RAMv, RAMx1, and RAMx2) and multiplexers that are used to read operands (u, v, $x1$, and $x2$) from the MU using the corresponding control signals.
- *Comp unit:* This is a comparator between u and v. It has a 3-bit output to indicate whether $u > v$ or $u < v$ or $u = v$.
- *Occur unit:* This searches for $u = 1$ occurrence. Return x 1 mod p if u equal 1. Otherwise, if v is 1, return $x2$ mod p.

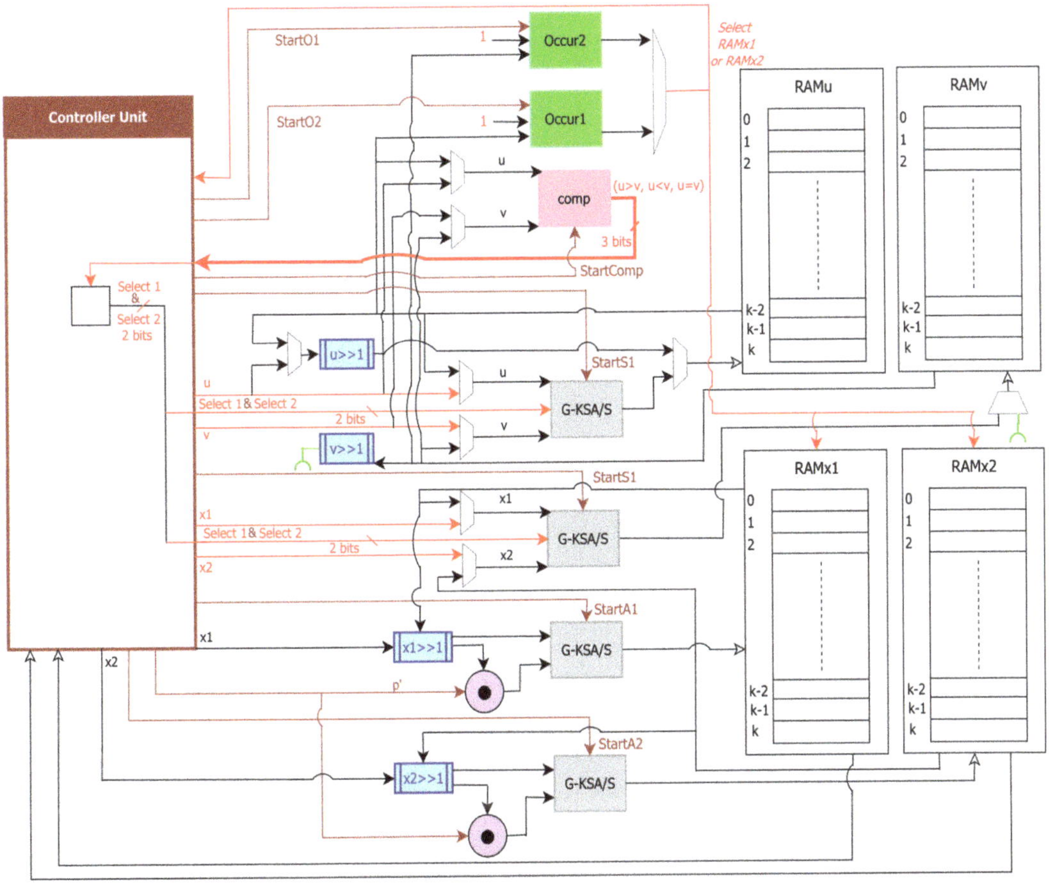

Figure 3. Proposed MBEEA architecture.

5.1. Kogge–Stone Adder/Subtractor Unit: Design and Implementation

Adders are widely recognized as the fundamental building blocks for more complex arithmetic operators such as multipliers and inverters. Usually, subtraction and addition are implemented using a single circuit. The addition of x, $\overline{y}$, and 1 can be used to compute the subtraction of $x - y$, where y is the bitwise complement of y. Therefore, hardware addition can be used to support the subtraction function. To perform the addition and subtraction in Algorithm 2, four adders/subtractors are needed. The data path of the adder directly affects the delay of the arithmetic inversion data path. KSA is the fastest adder and shows good hardware performance. Thus, KSA was selected for use in the MBEEA architecture for performing addition and subtraction. A parallel prefix adder such as KSA is suitable for additions with longer word lengths. The tree network of KSA allows reducing the latency to $O(log_2 n)$ where 'n' represents the number of bits.

The KSA algorithm is composed of three stages: preprocessing, parallel prefix network, and post-processing. In the preprocessing stage, two signals are calculated simultaneously for each input: Propagate (P) and Generate (G). The intermediate carries are then generated using these signals in the parallel prefix network. Finally, in the post-processing, we calculate the sum by XORing the intermediate carries and the propagate signals.

In the MBEEA design, a generic adder/subtractor based on the G-KSA/S algorithm was developed. The bit-level version of the G-KSA/S algorithm is shown in Algorithm 3, which employs n-bit registers to compute three intermediate results P, G, and C. Overall, Algorithm 3 has two specified inputs: A and B in $\mathbb{F}_p$ and a pre-computed vector S. The pre-computed vector S is represented by a set of elements 's_i'= 2^i where $i \in [0,\ m-1]$, and m represents the stage number which depends on the length of the prime number p (n), such that $m = log_2(n)$. Table 2 presents the 'm' and 'S' values for a prime number of lengths n = 4, 8, 16, 32, 64, 128, and 256. Hence for m = 3, $S = \{s_0, s_1, s_2\} = \{2^0, 2^1, 2^2\}$.

Table 2. Values of m and S for n = 4, 8, 16, 32, 64, 128, and 256 bits.

n	m	S
4	2	1,2
8	3	1,2,4
16	4	1,2,4,8
32	5	1,2,4,8,16
64	6	1,2,4,3,16,32
128	7	1,2,4,3,16,32,64
256	8	1,2,4,3,16,32,64,128

A and B in $\mathbb{F}_p$, two n-bit inputs, are processed in the first stage bit-by-bit, to compute the generation signals G_0 and propagation signals P_0 which are presented in steps 2.1 and 2.2 of Algorithm 3. To perform the subtraction, the *carry_in* and the input B are just XORed, such as presented in step 1 in Algorithm 3.

Algorithm 3: G-KSA/S algorithm

Input: $(A,B) \in \mathbb{F}_p$
Define: n; n = *bit length of p*
Pre-computed:S = [1,2,4,8,16,32,64,128], $j \leftarrow 1$, $k \leftarrow 1$
Output: $(Sum = (A+B) mod\ p;\ c_{out})$
Preprocessing
Step 1: For (i = 0; $i \leq$ to $n-1$; i++) {
C(i) := $c_{in} \oplus B(i)$;
};
Step 2: For (i = 0; $i \leq n-1$; i++) {
2.1:$P_0(i) := A(i) \oplus C(i)$;
2.2:$G_0(i) := A(i) \otimes C(i)$;
};
Parallel Prefix Network
Step 3: For (i = 0; $i \leq j$-1; i++) {
3.1: $G_k(i) := G_{k-1}(i)$,
3.2: $P_k(i) := P_{k-1}(i)$
};
Step 4: For (i = 0; $i \leq n-j$-1){
4.1:$G_k(i+j) := (P_{k-1}(i+j) \otimes\ G_{k-1}(i))) \vee G_{k-1}(i+j)$;
4.2: $P_k(i+j) := (P_{k-1}(i+j) \otimes; P_{k-1}(i))$;
};
$k++$;j := $S[k\text{-}1]$, T := $m-1$;
IF($T > 0$) go to step 3 ELSE go to step 5
Step 5: For (i = 0; $i \leq n$-1; i++) {
5.1: $C(i) := G_{NP-1}(i) \vee (c_{in} \otimes P_{NP-1}(i))$;
};
Post-processing
Step 6: 6.1:sum(0) = $P_0(0) \oplus c_{in}$;
6.2: For (i = 1; $i \leq n$-1; i++){
6.2.1: sum(i) = $P_0(i) \oplus C(i-1)$;
};
Step 7: c_{out} = $C(N-1)$;
Return Sum, c_{out}

Figure 4 presents the KSA process for 8 bits. As shown in Figure 3, the parallel prefix network for G-KSA/S is composed of three levels. The carry propagation network is in charge of transmitting the carry signal from the preceding bit lines. P_k and G_k are generated via 'carry propagation' steps, steps 4.1.1 and 4.1.2, as shown in Algorithm 3. Finally, the carry signal obtained in step 5 is used to calculate the sum in the last stage (step 6). The output step performs the XOR operation between the previous bits of the 'carry' signal (c_{i-1}) and the current bits of the propagation signal (Pi), as shown in steps 6 and 7.

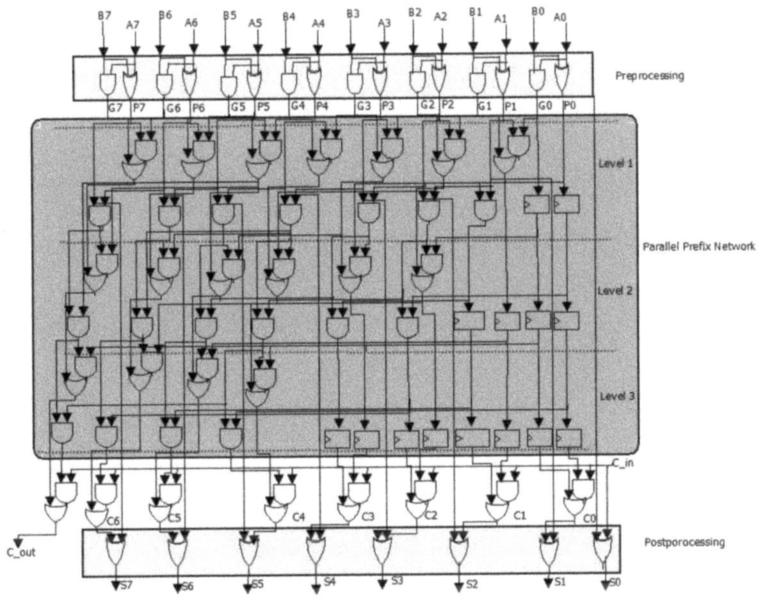

Figure 4. Eight-bit G-KSA/S data-path.

5.2. The Modular Reduction

The modular reduction in large values is a fundamental operation in the majority of common PKs that involve intensive computations in prime fields. At the end of Algorithm 2, the result will be reduced modulo the prime number 'p'. The prime number should be chosen carefully to ensure efficient reduction and high performance. The modular reduction can be calculated in three ways:

- General module forms (i.e., Barrett and Montgomery algorithms), which are slower and present an expansive part of the arithmetic operation.
- Based on the LUT method (which is based on pre-computed values); however, this requires a large amount of memory.
- Modulo form of prime numbers, such as pseudo-Mersenne numbers. Their special form makes them appropriate for modular reduction. The pseudo-Mersenne prime number p is presented with the special form: $p = 2^{\alpha} - c$, where $\alpha = n$ (n represents the security level) and 'c' is a positive integer that is relatively small compared to the modulus. An integer $0 \leq z < (2^{\alpha} - c)^2$ can be represented in radix-2^{α} by spilling z up into a lower part Z_L and a higher part Z_H : $Z_H 2^{\alpha} + Z_L$. Then, using the fact that $2^{\alpha} \equiv c \bmod p$, we have $Z_H 2^{\alpha} + Z_L \equiv Z_H c + Z_L \bmod p$ where $0 \leq Z_H c + Z_L < (c + 1)2^{\alpha}$. Hence, a multiplication of Z_H by the constant c is needed and the final result is obtained after the addition of Z_L. For the three values of n (128, 192, and 256), the resulting primes satisfy $p \equiv 3 \pmod 4$.

5.3. Controller Unit

The main component of the proposed inverter design is the control unit (CU), which is in charge of all communications between all MBEEA units.

At the beginning of the algorithm, the start signal is equal to '1', and initial values 'a', '*p*', '1', '0', and '*p*/2' are assigned to '*u*', '*v*', '*x*1', '*x*2', and '*p*' registers, respectively, by input multiplexers. The computed intermediate results are stored in the memory units. The control unit activates the addition/subtraction chains after receiving the input data (*p* and *a*) and the signals "CLK", "Reset" = '1', and "Start" = '1'. Hence KSA-start takes '1' and Cf becomes '0'. If KSA completes the addition function ("KSA-done" = '1'), the control unit sends signals of writing to the memory unit to update the '*u*', '*v*', '*x*1', and '*x*2' values. The update process is controlled by five functions belonging to three blocks: addition and subtraction functions (KSA block), left shifting function (shift registers), writing, and reading functions (block memory). These new values are determined by two distinct control signals: sel1 and sel2. These control signals are updated after the KSA block performs the '*u*-*v*' function as shown in step 12.9 in Algorithm 2. Hence, the CU generates the signals for the G-KSA/S components, and the read and write addresses for the memory units. The MBEEA state machine describing the design data flow is presented in Figure 5.

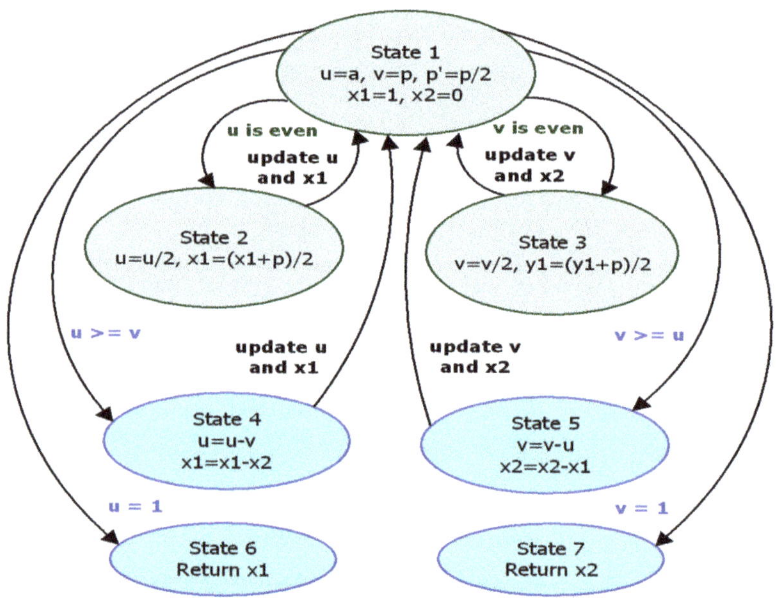

Figure 5. MBEEA state machine.

To implement the MBEEA, FSM is made up of seven states: St#1 is a state of inactivity, whereas during St#2 to St#7, necessary signals for updating '*u*', '*v*', '*x*1', and '*x*2' are generated. Of these seven states, two states (S#2, S#3) are executed in parallel. Hence two additions, four shifting operations, and four reading/writing operations are required. Similarly, in states 4 and 5, two subtractions and two reading/writing operations are needed. Finally, based on the output of the comparator unit, '*u*' and '*x*1' or '*v*' and '*x*2' are updated.

6. Results and Performance Analysis

The proposed MBEEA inverter and G-KSA/S designs were then developed for a different prime field of lengths 8, 16, 32, 64, 128, and 256 bits. Modelsim was used to validate the proposed designs, which were coded in VHDL. The G-KSA/S and MBEEA designs were then implemented on various devices (Spartan 3E, Virtex-E, Virtex-II, Virtex 5, Virtex 7) using Xilinx 14.7. Tables 3 and 4 show the obtained results after place and route, which include the maximum frequency (Fmax, MHz), area usage (slices/CLB), latency (µs), and ADP (ADP = #slices × Latency).

6.1. G-KSA/S Implementation Results

The G-KSA/S algorithm was successfully implemented for $\mathbb{F}_p$ (p is a prime number of length 8, 16, 32 64, 128, 256 bits). Figure 6 presents the required logic gates (AND, OR, and XOR) for the GKSA design (for G-KSA/S, we have to add one XOR gate for subtraction). We can see that the number of required logic gates grows linearly.

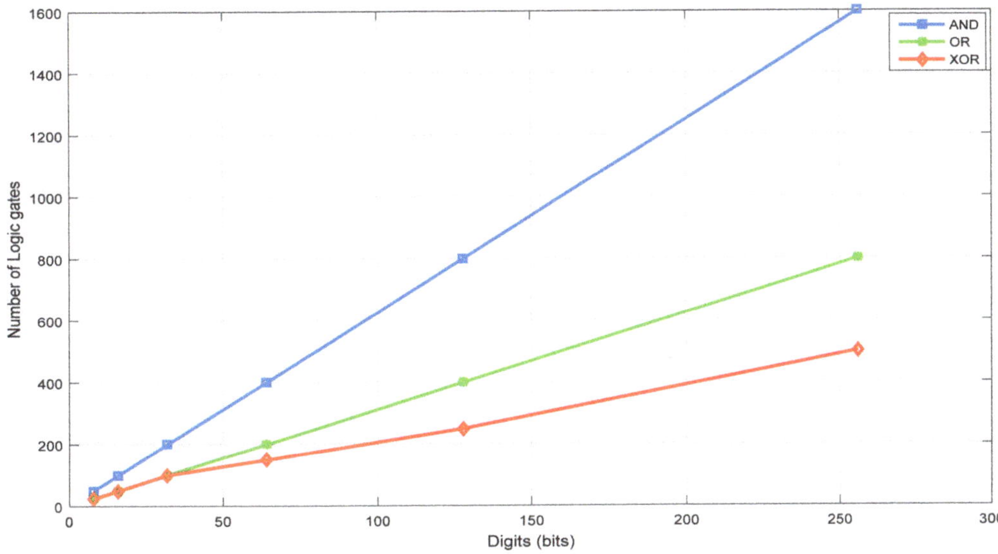

Figure 6. Logic gates versus $\mathbb{F}_p$.

For a fair comparison with related works, the G-KSA/S algorithm was implemented on Spartan3E and Virtex-5, as shown in Table 3. As shown in Table 3, the proposed design clearly outperforms the existing different prime field-length KSA implementations. We can see from Table 3 that our proposed design is much smaller (almost 66% for $\mathbb{F}_p$ (p is a prime of length 8, 16, and 32), and almost 41% for $\mathbb{F}_p$ (p is a prime of length 64, 128, and 256)) and considerably faster than those of [34,35].

Furthermore, the G-KSA/S proposed design always achieves the best ADP value in all the prime fields compared to the existing designs. For example, over $\mathbb{F}_p$ with a length of p = 256 bits, compared with [35] on the Virtex-5 device, the ADP of the proposed design has a 28.62% smaller ADP.

Table 3. Comparison of the proposed and existing G-KSA/S designs for different prime field lengths.

Designs	Platform	n = Bit Length of p	Area (Slices)	Delay (ns)	ADP (10^{-9})	Gain %
[34]	Spartan-3E	8	83	5.776	479.408	
G-KSA/S	Spartan-3E		47	3.6	169.2	74.71%
[34]	Spartan-3E	16	166	10.85	1801.1	
G-KSA/S	Spartan-3E		98	7.3	715.4	61.28%
[34]	Spartan-3E	32	332	20.56	6825.92	
G-KSA/S	Spartan-3E		174	12.3	2140.2	68.65%
[35]	Virtex-5	64	449	30.5	13,694.5	
G-KSA/S	Virtex-5		289	27.9	8063.1	41.13%
[35]	Virtex-5	128	1111	57.3	63,660.3	
G-KSA/S	Virtex-5		641	64.4	41,280.4	35.16%
[35]	Virtex-5	256	1345	106.7	143,511.5	
G-KSA/S	Virtex-5		737	139	102,443	28.62%

6.2. MBEEA Implementation Results

Table 4 summarizes the MBEEA hardware implementation results on a contemporary Xilinx Virtex-7 (x7vx330t-2.g1157) FPGA. The proposed MBEEA generic architecture designs were implemented over different prime field lengths (n (bit length of p) = 8, 16, 32, 64, 128, and 256 bits). In practice, 128- and 256-bit lengths for ECC/HECC and pairings systems, over prime fields, are very useful for modern security applications. The post place and route-static timing report was used to calculate the minimum clock period.

Table 4. FPGA implementation performance for the proposed MBEEA design in Virtex-7.

Design	n = Bit Length of p	Freq. (MHz)	Area (Slices)	Latency (µs)
MBEEA	8	530	545	0.179
	16	480	770	0.27
	32	420	1060	0.346
	64	380	1237	0.428
	128	310	1532	0.851
	256	250	2035	1.24

In Table 5, we compare the corresponding FPGA implementation results with those of reports available in the literature to additionally assess the actual performance of the proposed inverter design over $\mathbb{F}_{256}$. $\mathbb{F}_p$ for a prime of length 256 bits is considered to be the large prime field size. Table 5 lists all related performance metrics: area, frequency (MHZ), latency (µs), and ADP. As an overall performance metric, we used ADP for all related designs to ensure a fair comparison.

Table 5. Performance analysis of the proposed and the existing modular inversion designs over F_{256}.

Ref.	FPGA Device	Freq (MHz)	Time (µs)	Area (Slices)	ADP ($\times10^{-9}$)
[36]	Virtex-7	146.23	2.329	1480	3.44
[37]	Kintex 7	142.38	2.33	1480	3.45
[38]	Virtex-6	151	3.39	1190	4.04
[39]	Virtex-6	146	3.52	1340	4.72
[40]	Virtex-5	129	7.937	592	4.7
[41]	Virtex-7	138.3	2.45	1577	3.87
[42]	Virtex-II	55.70	6.2	5863	36.35
[43]	Virtex-II	37	4.98	9213	45.88
[44]	Virtex-II	68.17	11.60	2085	24.19
[10]	Virtex-II	34	14.6	9146	133.53
[45]	Virtex-II	40.68	15.22	14,844	225.26
[46]	Virtex-II	50	6.4	5477	35
MBEEA	Virtex-E	106	2.92	2830	8.26
MBEEA	Virtex-II	175	1.77	2530	4.47
MBEEA	Virtex-5	208	1.49	2318	3.45
MBEEA	Virtex-6	240	1.29	2140	2.76
MBEEA	Virtex-7	276	1.12	2035	2.28

Table 5 clearly shows that the proposed design outperforms the existing large prime field-size modular inversion implementations.

In terms of latency, on the Virtex-7 device, the proposed design is 51.2% and 54.29% faster than those of [36,41], respectively. At the same time, on the Virtex-6 device, the proposed design is 61% and 63.35% faster than the competing designs of [38,39], respectively. On the Virtex-II, the proposed design remarkably outperforms all the existing designs of [10,42–46] in terms of latency and area. For instance, the works presented in [43,46] utilize 72.54% and 53.81%, respectively, more FPGA slices than this work.

At the same time, Table 5 shows that the proposed design ensures significantly lower ADP than the existing designs. Compared with [36,37,41] on the Virtex-7 device, the ADP of the proposed design has 23.81%, 23%, and 41.18% smaller ADP, respectively.

7. Conclusions

Securing the IoT is one of the major, if not the major, challenges for IT systems today. If the IoT is not sufficiently protected, critical events can occur, such as water or power outages, or worse, manipulated processes resulting in bacteria in water and faulty products such as cars, etc., that pose security risks. This urgent need for security, combined with the lack of security of IoT devices that are simultaneously connected to the Internet, which has a high risk of threat, illustrates the importance of the topic. The data and connections of many IoT devices are secured by cryptographic algorithms such as ECC and ECDSA. ECC/ECDSA is one of the algorithms targeted by the SPA attack. The modular BEEA inversion process used for generating an ECC/ECDSA private key is the focus of the leakage-based side-channel study. During the execution of the cryptographic algorithm, SPA can easily distinguish the conditional branches outcomes since a device consumes power differently and the execution time is not constant.

A new version of the BEE inversion algorithm was proposed that works in constant time by avoiding the data dependency of the classical BEEA. To ensure the traces are the same, thus preventing an attacker from distinguishing the computation across branches, the proposed method removes conditional instructions and divides the algorithm into separate branches with sub-functions. The classical BEEA is secured against SPA attacks due to the new countermeasure.

The new algorithm was designed so that the calculation of the modular inversion is easy and efficient for hardware implementation. The algorithm complexity is suitable even for a device having limited resources such as an IoT device. To achieve high performance on an FPGA, a variety of optimization techniques, including algorithmic reformulation and architectural optimization, are used. On a Xilinx Virtex-7 FPGA, our design can achieve a maximum clock rate of 276 MHz and it takes only 1.12 μs to perform the 256-bit prime modular inversion.

The proposed design for $\mathbb{F}_{256}$ provides an ADP figure of 2.28 on a Virtex-7, which is less than the relevant state-of-the-art solutions. Furthermore, our architecture outperforms others in terms of FPGA area (slices) and delay. Therefore, the proposed design is suitable for constrained implementations of cryptographic primitives, such as IoT, wireless sensor nodes, and RFID devices.

Author Contributions: A.S., M.Z., C.M. and H.Y.A. designed the research. A.S., M.Z., H.Y.A. and A.C. performed the code programming. C.M. carried out simulation analysis. A.S., M.Z., C.M., H.Y.A. and M.M. performed the analysis of the results and the interpretation of the results. A.S., M.Z. and A.C. wrote the first draft together, supported by H.Y.A. and M.M. commented on the draft and all authors finalized it. A.C. is the corresponding author. All authors have read and agreed to the published version of the manuscript.

Funding: This research received no external funding.

Institutional Review Board Statement: Not applicable.

Informed Consent Statement: Not applicable.

Conflicts of Interest: The authors declare no conflict of interest.

References

1. Lee, S.; Huh, J.-H. An effective security measures for nuclear power plant using big data analysis approach. *J. Supercomput.* **2019**, *75*, 4267–4294. [CrossRef]
2. Kim, S.-K.; Kim, U.-M.; Huh, J.-H. A Study on Improvement of Blockchain Application to Overcome Vulnerability of IoT Multiplatform Security. *Energies* **2019**, *12*, 402. [CrossRef]
3. Miller, V.-S. Use of elliptic curves in cryptography. In *Advances in Cryptology—CRYPTO '85 Proceedings. CRYPTO 1985*; Lecture Notes in Computer Science; Springer, Inc.: New York, NY, USA, 1986; pp. 417–426.
4. Islam, T.; Youki, R.-A.; Chowdhury, B.-R.; Hasan, A.-S.M.T. An ECC Based Secure Communication Protocol for Resource Constraints IoT Devices in Smart Home. In *Proceedings of the International Conference on Big Data, IoT, and Machine Learning*; Lecture Notes on Data Engineering and Communications Technologies; Arefin, M.S., Kaiser, M.S., Bandyopadhyay, A., Ahad, M.A.R., Ray, K., Eds.; Springer: Singapore, 2022; Volume 95.
5. Lohachab, K.A. ECC based inter-device authentication and authorization scheme using MQTT for IoT networks. *J. Inf. Secur. Appl.* **2019**, *46*, 1–12. [CrossRef]
6. Marin, L.; Pawlowski, M.-P.; Jara, A. Optimized ECC Implementation for Secure Communication between Heterogeneous IoT Devices. *Sensors* **2015**, *15*, 21478–21499. [CrossRef]
7. Varchola, M.; Drutarovsky, M.; Repka, M.; Zajac, P. Side-channel attack on multi-precision multiplier used in protected ecdsa implementation. In Proceedings of the 2015 International Conference on ReConFigurable Computing and FPGAs (ReConFig), Riviera Maya, Mexico, 7–9 December 2015; pp. 1–6.
8. Fan, J.; Verbauwhede, I. An updated survey on secure ECC implementations: Attacks, countermeasures and cost. In *Cryptography and Security: From Theory to Applications*; Springer: Berlin/Heidelberg, Germany, 2012; pp. 265–282.
9. Rafik, M.-B.-O.; Mohammed, F. The impact of ecc's scalar multiplication on wireless sensor networks. In Proceedings of the 2013 11th International Symposium on Programming and Systems (ISPS), Algiers, Algeria, 22–24 April 2013; pp. 17–23.
10. Ghosh, S.; Mukhopadhyay, D.; Roychowdhury, D. Petrel: Power and timing attack resistant elliptic curve scalar multiplier based on programmable GF(p) arithmetic unit. *IEEE Trans. Circuits Syst. I Regul. Pap.* **2011**, *58*, 1798–1812. [CrossRef]
11. Kocher, P.C. Timing Attacks on Implementations of Diffie-Hellman, RSA, DSS, and Other Systems. In *Advances in Cryptology—CRYPTO'96*; Lecture Notes in Computer Science; Koblitz, N., Ed.; Springer: Berlin/Heidelberg, Germany, 1996; Volume 1109.
12. Shanmugham, S.-R.; Paramasivam, S. Survey on power analysis attacks and its impact on intelligent sensor networks. *IET Wirel. Sens. Syst.* **2018**, *8*, 295–304. [CrossRef]
13. Moon, J.; Junga, I.-Y.; Park, J.-H. IoT application protection against power analysis attack. *Comput. Electr. Eng.* **2018**, *67*, 566–578. [CrossRef]
14. Arpaia, P.; Bonavolontà, F.; Cioffi, A.; Moccaldi, N. Reproducibility Enhancement by Optimized Power Analysis Attacks in Vulnerability Assessment of IoT Transducers. *IEEE Trans. Instrum. Meas.* **2021**, *70*, 1–8. [CrossRef]
15. Dao, B.A.; Hoang, T.T.; Le, A.T.; Tsukamoto, A.; Suzaki, K.; Pham, C.K. Exploiting the Back-Gate Biasing Technique as a Countermeasure Against Power Analysis Attacks. *IEEE Access* **2021**, *9*, 24768–24786. [CrossRef]
16. Dubeuf, J.; Hely, D.; Beroulle, V. ECDSA Passive Attacks, Leakage Sources, and Common Design Mistakes. *ACM Trans. Des. Autom. Electron. Syst.* **2016**, *21*, 1–24. [CrossRef]
17. Genkin, D.; Pachmanov, L.; Pipman, I.; Tromer, E.; Yarom, Y. ECDSA Key Extraction from Mobile Devices via Nonintrusive Physical Side Channels. In Proceedings of the ACM SIGSAC Conference on Computer and Communications Security, Vienna, Austria, 24–28 October 2016; pp. 1626–1638.
18. Zhang, K.; Xu, S.; Gu, D.; Gu, H.; Liu, J.; Guo, Z.; Liu, R.; Liu, L.; Hu, X. Practical Partial-Nonce-Exposure Attack on ECC Algorithm. In Proceedings of the 13th International Conference on Computational Intelligence and Security (CIS), Hong Kong, China, 15–18 December 2017.
19. Wunan, W.; Hao, C.; Jun, C. The Attack Case of ECDSA on Blockchain Based on Improved Simple Power Analysis. In *Artificial Intelligence and Security. ICAIS 2019*; Lecture Notes in Computer Science; Springer: Cham, Switzerland, 2019; Volume 11635.
20. Thibault, J.-P.; O'Flynn, C.; Dewar, A. Ark of the ECC: An Open-Source ECDSA Power Analysis Attack on an FPGA Based Curve P-256 Implementation. Cryptology ePrint Archive: Report 2021/1520. 2021. Available online: https://eprint.iacr.org/2021/1520 (accessed on 25 January 2022).
21. Sghaier, A.; Zeghid, M.; Massoud, C.; Mahchout, M. Design and Implementation of Low Area/Power Elliptic Curve Digital Signature Hardware Core. *Electronics* **2017**, *6*, 46. [CrossRef]
22. Choi, P.; Lee, M.K.; Kong, J.T.; Kim, D.K. Efficient Design and Performance Analysis of a Hardware Right-shift Binary Modular Inversion Algorithm in GF(p). *J. Semicond. Technol. Sci.* **2017**, *17*, 425–437.
23. Alhazmi, B.; Gebali, F. Fast Large Integer Modular Addition in GF(p) Using Novel Attribute-Based Representation. *IEEE Access* **2019**, *7*, 58704–58719. [CrossRef]
24. Janwadkar, S.; Dhavse, R. Qualitative and Quantitative Analysis of Parallel-Prefix Adders. In *Advances in VLSI and Embedded Systems*; Lecture Notes in Electrical Engineering; Patel, Z., Gupta, S., Kumar, Y.B.N., Eds.; Springer: Singapore, 2020; Volume 676.
25. Bos, J.-W. Constant time modular inversion. *J. Cryptogr. Eng.* **2014**, *4*, 275–281. [CrossRef]
26. Liu, Z.; Großschädl, J.; Li, L.; Xu, Q. Energy-efficient elliptic curve cryptography for MSP430-based wireless sensor nodes. In Proceedings of the 21st Australasian Conference on Information Security and Privacy, Melbourne, Australia, 4–6 July 2016; Springer: Cham, Switzerland, 2016; Volume 9722, pp. 94–112.

27. Xu, S.; Gu, H.; Wang, L.; Guo, Z.; Liu, J.; Lu, X.; Gu, D. Efficient and Constant Time Modular Inversions Over Prime Fields. In Proceedings of the 13th International Conference on Computational Intelligence and Security (CIS), Hong Kong, China, 15–18 December 2017; pp. 524–528.
28. Aldaya, A.C.; Márquez, R.C.; Sarmiento, A.-J.C.; Sánchez-Solano, S. Side-channel analysis of the modular inversion step in the RSA key generation algorithm. *Int. J. Circuit Theory Appl.* **2017**, *45*, 199–213. [CrossRef]
29. Savaş, E.; Koç, Ç.K. Montgomery inversion. *J. Cryptogr. Eng.* **2018**, *8*, 201–210. [CrossRef]
30. Bernstein, D.-J.; Yang, B.-Y. Fast constant-time gcd computation and modular inversion. *IACR Trans. Cryptogr. Hardw. Embed. Syst.* **2019**, 340–398. [CrossRef]
31. Awaludin, A.-M.; Larasati, H.-T.; Kim, H. High-Speed and Unified ECC Processor for Generic Weierstrass Curves over GF(p) on FPGA. *Sensors* **2021**, *21*, 1451. [CrossRef]
32. Sarna, S.; Czerwinski, R. RSA and ECC universal, constant time modular inversion. In Proceedings of the AIP Conference Proceedings, Crete, Greece, 29 April–3 May 2020; Volume 2343. [CrossRef]
33. Aldaya, A.-C.; Sarmiento, A.-J.C.; Sánchez-Solano, S. Spa vulnerabilities of the binary extended Euclidean algorithm. *J. Cryptogr. Eng.* **2017**, *7*, 273–285. [CrossRef]
34. Rajnish, D.; Jitendra, J. An Efficient Processing by using Kogge-Stone High-Speed Addition Technique. *Int. J. Comput. Appl.* **2015**, *131*, 21–23.
35. Vitoroulis, K.; Al-Khalili, A.-J. Performance of Parallel Prefix Adders implemented with FPGA technology. In Proceedings of the 2007 IEEE Northeast Workshop on Circuits and Systems, Montreal, QC, Canada, 5–8 August 2007. [CrossRef]
36. Hossain, M.-S.; Kong, Y. High-Performance FPGA Implementation of Modular Inversion over F256 for Elliptic Curve Cryptography. In Proceedings of the 2015 IEEE international conference on Data Science and Data Intensive Systems, Sydney, NSW, Australia, 11–13 December 2015; pp. 169–174. [CrossRef]
37. Hossain, M.-S.; Kong, Y.; Saeedi, E.; Vayalil, N.C. High-performance elliptic curve cryptography processor over NIST prime fields. *IET Comput. Digit. Tech.* **2016**, *11*, 33–42. [CrossRef]
38. Javeed, K.; Wang, X. Low latency flexible FPGA implementation of point multiplication on elliptic curves over GF (p). *Int. J. Circuit Theory Appl.* **2017**, *45*, 214–228. [CrossRef]
39. Javeed, K.; Wang, X.; Scott, M. High-performance hardware support for elliptic curve cryptography over general prime field. *Microprocess. Microsyst.* **2017**, *51*, 331–342. [CrossRef]
40. Mrabet, A.; El-Mrabet, N.; Bouallegue, B.; Mesnager, S.; Machhout, M. An efficient and scalable modular inversion/division for public-key cryptosystems. In Proceedings of the 2017 International Conference on Engineering & MIS (ICEMIS), Monastir, Tunisia, 8–10 May 2017; pp. 1–6. [CrossRef]
41. Kudithi, T.; Sakthivel, R. High-performance ECC processor architecture design for IoT security applications. *J. Supercomput.* **2019**, *75*, 447–474. [CrossRef]
42. Liu, Z.; Liu, D.; Zou, X. An efficient and flexible hardware implementation of the dual field elliptic curve cryptographic processor. *IEEE Trans. Ind. Electron.* **2017**, *64*, 2353–2362. [CrossRef]
43. Lee, J.W.; Chung, S.C.; Chang, H.-C.; Lee, C.Y. Efficient power analysis resistant dual field elliptic curve cryptographic processor using heterogeneous dual processing element architecture. *IEEE Trans. Very Large Scale Integr. (VLSI)* **2014**, *22*, 49–61. [CrossRef]
44. Vliegen, J.; Mentens, N.; Genoe, J.; Braeken, A.; Kubera, S.; Touha, A.; Verbauwhede, I. A compact FPGA based architecture for elliptic curve cryptography over prime fields. In Proceedings of the ASAP 2010—21st IEEE International Conference on Application-Specific Systems, Architectures and Processors, Rennes, France, 7–9 July 2010; pp. 313–316. [CrossRef]
45. McIvor, C.J.; McLoone, M.; McCanny, J.V. Hardware elliptic curve cryptographic processor over GF(p). *IEEE Trans. Circuits Syst. I Regul. Pap.* **2006**, *53*, 1946–1957. [CrossRef]
46. Daly, A.; Marnane, W.; Kerins, T.; Popovici, E. An FPGA Implementation of a GF(p) ALU for encryption processors. *Microprocess. Microsyst.* **2004**, *28*, 253–260. [CrossRef]

sensors MDPI

Article

An Ultra-Low-Cost RCL-Meter

Pedro M. C. Inácio [1], Rui Guerra [1,2,*] and Peter Stallinga [1,3]

1 CEOT—Center for Electronics, Optoelectronics and Telecommunications, University of Algarve, Campus Gambelas, 8005-139 Faro, Portugal; pminacio@ualg.pt (P.M.C.I.); pjotr@ualg.pt (P.S.)
2 Department of Physics, University of Algarve, Campus Gambelas, 8005-139 Faro, Portugal
3 Department of Electronics and Computer Engineering, University of Algarve, Campus Gambelas, 8005-139 Faro, Portugal
* Correspondence: rguerra@ualg.pt

Abstract: An ultra-low-cost RCL meter, aimed at IoT applications, was developed, and was used to measure electrical components based on standard techniques without the need of additional electronics beyond the AVR® micro-controller hardware itself and high-level routines. The models and pseudo-routines required to measure admittance parameters are described, and a benchmark between the ATmega328P and ATmega32U4 AVR® micro-controllers was performed to validate the resistance and capacitance measurements. Both ATmega328P and ATmega32U4 micro-controllers could measure isolated resistances from 0.5 Ω to 80 MΩ and capacitances from 100 fF to 4.7 mF. Inductance measurements are estimated at between 0.2 mH to 1.5 H. The accuracy and range of the measurements of series and parallel RC networks are demonstrated. The relative accuracy (a_r) and relative precision (p_r) of the measurements were quantified. For the resistance measurements, typically a_r, p_r < 10% in the interval 100 Ω–100 MΩ. For the capacitance, measured in one of the modes (fast mode), a_r < 20% and p_r < 5% in the range 100 fF–10 nF, while for the other mode (transient mode), typically a_r < 20% in the range 10 nF–10 mF and p_r < 5% for 100 pF–10 mF. a_r falls below 5% in some sub-ranges. The combination of the two capacitance modes allows for measurements in the range 100 fF–10 mF (11 orders of magnitude) with a_r < 20%. Possible applications include the sensing of impedimetric sensor arrays targeted for wearable and in-body bioelectronics, smart agriculture, and smart cities, while complying with small form factor and low cost.

Keywords: impedance meter; RCL-bridges; portable instrument; AVR® micro-controller; low-cost; internet of things

Citation: Inácio, P.M.C.; Guerra, R.; Stallinga, P. An Ultra-Low-Cost RCL-Meter. *Sensors* **2022**, *22*, 2227. https://doi.org/10.3390/s22062227

Academic Editors: Alexandru Lavric, Liliana Anchidin and Adrian I. Petrariu

Received: 18 January 2022
Accepted: 8 March 2022
Published: 14 March 2022

1. Introduction

The Internet of Things (IoT) entails a network of physical objects—'things'—that are embedded with sensors, electronics, software, etc. for the purpose of communicating with other devices over the Internet. Caused by the sheer number of things connected in this way, these data-acquisition devices obviously need to be of low-cost and fulfill certain tasks: sensing, electronic processing and connecting to the Internet. Impedimetric sensor arrays are an emerging field of study that is concerned with sense, processing and casting the measured data to the Internet [1–3]. Typical applications are wearable and in-body electronics [4–8] and plants and smart agriculture [9–13]. A common aspect shared between different devices is that the electronic interfaces for detection, processing and connection to the Internet are mainly carried out by separate external systems specifically optimized for each of these functions. While designing a system with specific units may often be advantageous to enhance the overall performance of the device, it also leads to higher manufacturing cost. For example, most biosensors feature a transduction mechanism to couple the physical and/or chemical changes in the system under measurement to the electronic circuitry of the measuring device. The latter is specifically optimized for the sensing interface and is followed by an analog-to-digital conversion unit (ADC). In the most usual scenario, the end user has access to a plug-and-play instrument box. But even in this case, the transduction mechanism may require finding signal conditioning strategies in order to maximize the linear range of the analog signal and the corresponding measurement accuracy [14–17].

Furthermore, the measurement unit, which often replaces a benchtop instrument, must be designed for small form factor devices and a low power profile, while maintaining performance close to gold standard instruments. Application-specific integrated circuit (ASIC) based devices fulfill these requirements, such as the ones developed to perform online electrochemical impedance spectroscopy (EIS) for characterization of lithium-ion battery packs [18–20]. This technology can be extended to other applications in areas where size and power consumption are crucial. For instance, in our research we have been developing technology for using admittance spectroscopy to determine the physical state of plants [21]. However, this study relied on bulky and expensive lock-in detectors, not appropriate for an IoT implementation. A cheaper solution developed by S. Grassini is to use an Arduino-based electrochemical impedance spectroscopy (EIS) system [22] for in situ corrosion monitoring of metallic works of art [23]. The Arduino-based EIS is already a huge improvement over conventionally used RCL-bridges (resistance, capacitance, inductance) and lock-in detectors. Similarly, the ASIC-based miniaturized system for Online-EIS proposed by Manfredini [19] shows how versatile the ASIC device is, being capable of measuring not only the impedance of commercial batteries, but also capacitive and resistive sensors. The device is based on the SENSIPLUS, which is a System on a Chip (SoC) solution that uses minimal external hardware, and shows performance on a par with gold standard instruments. The Arduino has been also used as a platform to measure capacitances, as explained, for example, by Campbell [24]. It has been used to deploy a digital LCR meter [25] to measure single parameters (not combinations), although this depends on the known nominal values of external components.

The current report describes a very-low-cost solution for an admittance meter, resulting from an effort to lower the cost and the size of these instruments. Specifically, it uses a micro-controller for directly measuring admittance without the need for any additional electronic circuitry beyond the micro-controller hardware itself. The proposed implementation differs from the instruments developed by [22,24] in two main features: (i) it is based on a time-domain approach, since the measurements are performed with signal transients rather than with sinusoidal signals; (ii) it avoids the need for external circuits, since all the signal generation and detection are performed by the micro-controller. Moreover, when used in popular platforms such as Arduino, full functionality is available, from sensing to communicating, for a price that lies in the order of mere euros per unit, thus fulfilling the requirement of the Internet of Things. Additionally, it paves the way to cheaper wearable and in-body bioelectronic sensors, allowing a live feed of collected data from impedimetric sensor arrays to the IoT, constituting an all-in-one solution: Sensing, Processing and Connection to the Internet. The technique that we will describe here relies on determining the behavior after applying voltage steps, a standard technique used in circuit analysis, which can readily measure resistances and capacitances, either isolated or in series and parallel. It is based on the low-cost AVR® micro-controller series commonly used in the Arduino® platform. The accuracy and precision of the measurements are discussed based on the relative parameters of measurement uncertainty (u_r), accuracy (a_r) and precision (p_r). All the C++ codes for Arduino and the MATLAB scripts used in this manuscript are available in the Supplementary Materials.

2. Materials and Methods

2.1. *AVR® Micro-Controllers Based RCL-Meter*

Two development boards manufactured by Arduino®, namely the Uno and Leonardo boards, were used to develop the ultra-low-cost RCL-meter. The Arduino® Uno board makes use of an ATmega328P micro-controller, and the Arduino® Leonardo uses an ATmega32U4. Figure 1e shows the block diagram of the proposed measurement system based exclusively on the internal circuitry of the AVR® micro-controller. Both ATmega328P [26] and ATmega32U4 [27] are low-power AVR® 8-bit micro-controllers that share similar features, namely equal clock frequency up to 16 MHz, operational voltage range between 2.7 to 5.5 V, 32 kB of flash memory (enough to store the firmware and processing code), 2 kB (ATmega328P)/2.5 kB (ATmega32U4) of static random-access memory (SRAM) (to store the main variables and acquisition samples) and an analog-to-digital unit (ADC) with 10-bit resolution, the core of the system. Both the Uno and Leonardo boards were supplied through the local-PC connection via USB interface, and the default settings were used, meaning the reference voltage (V_{REF}) is the internal voltage source of 5 V.

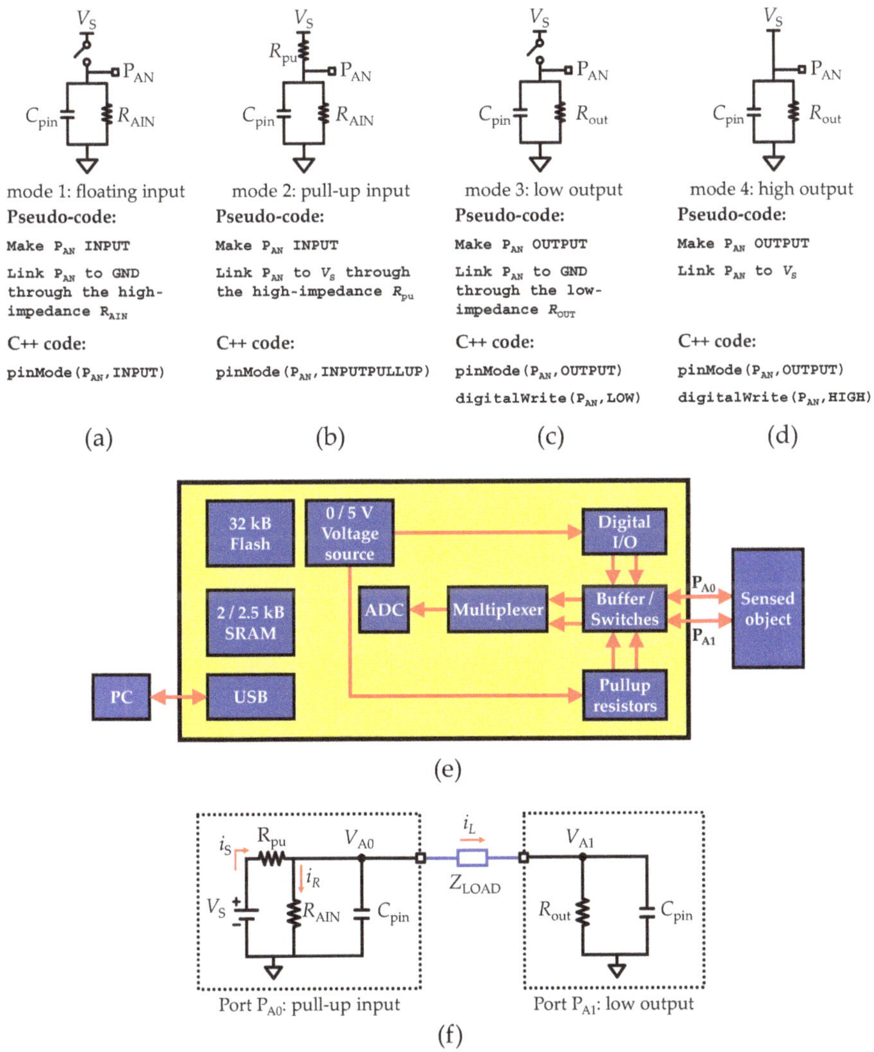

Figure 1. Equivalent circuits for the four operation modes available to configure each analog input/output (I/O) port. (**a**) mode 1, floating input. (**b**) mode 2, pull-up input. (**c**) mode 3, low output. (**d**) mode 4, high output. Each equivalent circuit is adapted from the schematics provided in the datasheets. (**e**) Block diagram of the proposed measurement system, including the serial communication and voltage source through USB interface to local PC, flash and SRAM memory, internal voltage source, and internal circuitry to program I/O ports to digital or alternate functionalities. (**f**) Equivalent circuit of two analog I/O ports bridged with a load impedance (Z_{LOAD}).

2.2. Analog I/O Operation Modes

The AVR® micro-controller family provides access to several digital and analog ports with input and output (I/O) functionalities. Both digital and analog I/O ports share a common control unit design, often called "General Digital I/O" [26,27]. In this work, the digital I/O ports (PDN) are labeled by an alpha-numerical code, denoted by the letter "D" proceeded by the port bit number "N", while the analog I/O ports (PAN) are represented by the letter "A". The analog ports have exclusive access to the alternate functions, featuring an I/O source-measuring unit (SMU) and an analog-to-digital converter (ADC) unit. By default, P_{AN} ports can be configured into four distinct operation modes: (mode 1) floating input; (mode 2) pull-up input; (mode 3) low voltage output; (mode 4) high voltage output. Each operation mode shares the same basic electrical structure, consisting of the supply voltage source (V_S), common ground (GND), a stray capacitance (C_{pin}) and location of the I/O port (P_{An}) for the SMU and ADC unit. By default, vs. is equal to the reference voltage (V_{REF}). Figure 1 shows the equivalent circuit that describes each analog I/O port operation mode and was adapted from references [26,27]. Operation modes 1 and 2, shown in Figure 1a,b, respectively, share the same internal circuitry components to form a high-impedance input port due to the presence of the analog input resistance (R_{AIN}). C_{pin} is also included in parallel with R_{AIN} to account for the stray capacitance. The unique difference between operation mode 1 and 2 is the state of the internal pull-up resistance (R_{pu}), which is connected to an internal bias voltage. In the case of operating mode 1, the R_{pu} is inactive, leading to a floating input configuration, while in the case of operation mode 2, the R_{pu} sets a SMU configuration. This aspect will be further explored to implement the RCL-meter. Operation modes 3 and 4, shown in Figure 1c,d, relate to the "General Digital I/O" functions. Operation mode 3 consists of a floating and impedance low output configuration, and the operation mode 4 consists of a voltage source (V_S), thus forcing a high output configuration. In the absence of load impedance connected to P_{AN}, the configuration of the operation mode 4 sinks current through the parallel RC network formed by the output resistance (R_{out}) and C_{pin}. Table 1 shows the typical values of the internal components of the I/O ports [26,27].

Table 1. Typical values of the operational voltage of the circuits (V_S), the ADC unit, the internal circuitry to each I/O port and the TTL unit to each AVR® micro-controller [26,27].

Voltage Source	ADC Unit		Internal Circuitry Parameters				TTL Unit			
							ATmega328P		ATmega32U4	
V_S (V)	n	N_{max} ($2^n - 1$)	R_{AIN} (MΩ)	R_{out} (Ω)	R_{pu} (kΩ)	C_{pin} (pF)	V_{IL} (V)	V_{IH} (V)	V_{IL} (V)	V_{IH} (V)
5	10	1023	100	600	32	24	−0.5–1.5	3.0–5.5	−0.5–0.9	1.9–5.5

The R_{pu} and C_{pin} values listed in Table 1 are representative values. In case of the R_{pu}, both micro-controller datasheets [26,27] characterize the range of R_{pu} between 20 kΩ to 50 kΩ, and solely for the purpose of the ADC unit it refers a typical reference input resistance (R_{REF}) value of 32 kΩ. As for the C_{pin}, the micro-controller datasheets [26,27] do not provide a typical value; only a conceptual component is shown in the equivalent circuits of the analog input circuitry representing the overall stray capacitance (also named C_{pin}). Therefore, the typical stray capacitance listed in Table 1 considers all capacitive sources in the internal analog input circuitry, such as: (i) the sampling and holder capacitance ($C_{S/H}$), approximately equal to 14 pF and (ii) the input capacitance (C_i) of each I/O pin, approximately equal to 10 pF. Both $C_{S/H}$ and C_i are grounded; therefore, it is assumed that the parallel of both $C_i \parallel C_{S/H}$ is the minimum value of C_{pin} and is approximately equal to $C_{pin} \cong C_i \parallel C_{S/H} \cong 24$ pF.

In practical terms, the operation mode of each port available on an AVR® micro-controller must be configured according to the pseudo-code described in Figure 1 through a low-level instruction set, which is not a user-friendly environment. However, Arduino®

has developed an integrated development environment (IDE) platform with support of high-level C++ functions such as pinMode() and digitalWrite(), an easy and user-friendly method to configure the ports. Figure 1 describes the C++ code required to properly configure each operation mode. In the following sections, it will be shown how the four different configurations for the ports may be used to build an RCL bridge. First, it is shown how to measure isolated resistors, capacitors and inductors, and then how to measure combinations of resistors and capacitors in series or parallel.

2.3. Recording Circuit

The recording circuit to measure resistances, capacitances and inductances, either isolated or in series and parallel is composed of the same internal components by setting two I/O ports with alternate functions, namely P_{A0} and P_{A1} ports configured for pull-up input (mode 1) and low output (mode 3), respectively, and as described in Section 2.2. Figure 1f shows the equivalent circuit resulting from the combination of the internal circuitry of each I/O port bridged by a load impedance (Z_{LOAD}). In the following Sections 2.4–2.8, the equivalent circuit shown in Figure 1f is detailed by replacing Z_{LOAD} with resistance, capacitance and inductance, either isolated or in series and parallel.

2.4. Measurements of an Isolated Load Resistance (R–Meter)

Considering that in the equivalent circuit shown in Figure 1f, Z_{LOAD} is composed of an isolated load resistance (R_{LOAD}), the pull-up resistor (R_{pu}) at port P_{A0} drives a current source that will sink through the parallel resistance path (R_P) formed by the analog input resistance (R_{AIN}) and the sum of the load and output resistances (R_{LOAD} + R_{out}). Thus, for circuit analysis purposes, in Figure 2a, a reduction of the entire equivalent circuit to a voltage divider circuit is shown. In Figure 2a, the variable N_{A0} is introduced as the digital counterpart of the analog variable V_{A0}, where $N_{A0} = \lfloor N_{max} \cdot (V_{A0}/V_S) \rfloor$. The stray capacitances (C_{pin}) were neglected from the equivalent circuit, since on DC measurements, the capacitances behave as open circuit. In addition, to avoid interference of any stray capacitance, the procedure is to implement a reasonable delay time (say 1 ms) after setting the port P_{A0} to operation mode 2 (pullup input). This procedure assures that the stray capacitances are open circuit ($t \gg \tau = R_T C_{pin}$). Using C_{pin} = 25 pF and R_{pu} = 32 kΩ, one obtains $\tau < 1$ µs.

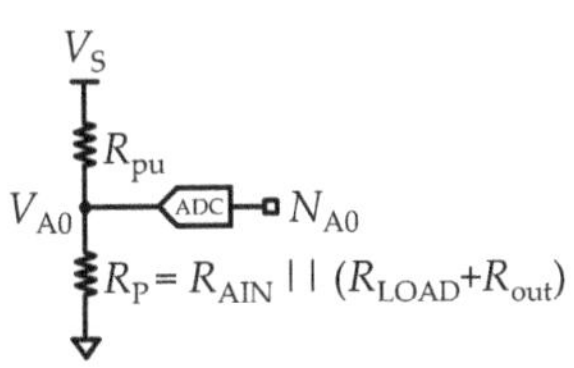

(a)

Pseudo-code:

```
Set port PA0 to INPUT PULLUP
Set port PA1 to LOW OUTPUT
Sample port PA0:
   NA0 measures VA0
RLOAD is given by Eq.(2)
```

(b)

Figure 2. Set-up of a pure load resistance meter (R-meter). (**a**) Reduction of the equivalent circuit to a voltage divider circuit. (**b**) Pseudo-code used to implement the resistance meter mode.

In Figure 2a, V_{A0} is the measured voltage at port P_{A0} and is given by:

$$V_{A0} = V_S \frac{R_p}{R_{pu} + R_p} \tag{1}$$

where R_P is given by R_{AIN} || (R_{LOAD} + R_{out}), and vs. is equal to the reference voltage source (V_{REF}). Solving the voltage divider Equation (1), R_{LOAD} is found as:

$$R_{LOAD} = \frac{R_{AIN}R_{out} - K(R_{AIN} + R_{out})}{K - R_{AIN}}, K = R_{pu}\frac{V_{A0}}{V_S - V_{A0}} \tag{2}$$

where K is an auxiliary variable used to shorten the length of Equation (2). All the parameters described in Equation (2) are known values, and are available on the datasheet of the micro-controller. The pseudo-code for a routine to measure R_{LOAD} is described in Figure 2b.

2.5. Measurements of an Isolated Load Capacitance (C–Meter)

Two methodologies are reported in this section: (i) fast acquisition mode, which is based on the immediate response of the C_{LOAD} charging cycle, and (ii) transient acquisition mode, based on measuring the charging time until a threshold voltage is reached. It is essential that prior to initiating measurements of C_{LOAD}, all capacitive components are fully discharged. A typical procedure to discharge all capacitances is achieved by configuring the operation mode of the two I/O ports in use to low output (mode 3) and waiting, for example, a period of $\Delta t \geq 1$ ms.

2.5.1. Fast Acquisition Mode of an Isolated Load Capacitance

As an exception to all other methods presented in this manuscript, the fast acquisition mode to measure an isolated load capacitance (C_{LOAD}) uses a different internal circuitry. Therefore, Figure 3a shows the equivalent circuit resulting from the combination of the internal circuitry of two I/O ports, P_{A0} and P_{A1}, bridged by a pure load capacitance (C_{LOAD}). Essentially, the port P_{A1} is set in high output (mode 4) and links the voltage source (V_{S}) to the C_{LOAD} terminal behaving as an input current source (i_{S}), and the port P_{A0} is set to floating input (mode 1), consisting of a measuring unit that links the input current to the ground through a high impedance resistor (R_{AIN}). Both the P_{A1} and P_{A0} ports consider the leakage current path to the ground through a stray capacitance (C_{pin}). However, the C_{pin} located at P_{A1} can be neglected, since the characteristic time constant (τ_{A1}) of P_{A1} is $\tau_{\mathrm{A1}} = R_{\mathrm{out}}C_{\mathrm{pin}} \cong 15$ ns, and therefore, because the fastest clock cycle (τ_{clk}) of the micro-controllers in use are approximately equal to $\tau_{\mathrm{clk}} \approx 5\tau_{\mathrm{A1}}$, after a single clock cycle the voltage in port P_{A1} (V_{A1}) is stationary.

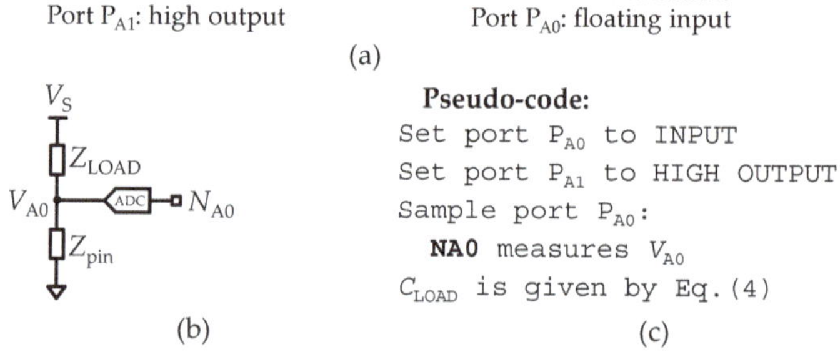

Figure 3. Set-up of a pure load capacitance meter (C-meter). (**a**) Equivalent circuit of two analog I/O ports bridged with a load capacitance (C_{LOAD}). (**b**) Reduction of the equivalent circuit to an impedance divider. (**c**) Pseudo-code used to implement the fast acquisition mode.

For circuit analysis purposes, assuming all capacitances are fully discharged at the instant $t = 0$, all capacitances are shorted and all currents flow through the capacitive path. In these circumstances, the circuit analysis is simpler considering that at $t = 0$ the current path flow exclusively through the capacitive path (C_T) formed by the load (C_{LOAD}) and stray (C_{pin}) capacitors in series, given by $C_T = C_{LOAD}C_{pin}/(C_{LOAD} + C_{pin})$. This allows the reduction of the entire equivalent circuit to a voltage divider, as shown in Figure 3b, and is only valid for measuring the impulse response. Thus, V_{A0} must be read immediately after defining P_{A0} to floating input and P_{A1} to high output.

One approach to find the load capacitance is through the analysis of the charge level (Q) of the circuit. Since both C_{LOAD} and C_{pin} are in series, the charge level must be equal in both capacitances ($Q_{LOAD} = Q_{pin}$), thus allowing the expression of V_{A0} as:

$$V_{A0} = V_S \frac{C_{pin}}{C_{pin} + C_{LOAD}} \tag{3}$$

where the solution given by Equation (3) assumes $Q_{LOAD} = C_{LOAD}\,(V_S - V_{A0})$ and $Q_{pin} = C_{pin}V_{A0}$. Rearranging (3), C_{LOAD} is found as:

$$C_{LOAD} = C_{pin} \frac{V_{A0}}{V_S - V_{A0}} \tag{4}$$

The size of C_{pin} is made available through the datasheet of the micro-controller, and is typically 25 pF, while vs. is equal to the reference voltage source (V_{REF}), and V_{A0} is the measured voltage at port P_{A0}. Thus, all parameters are known, and C_{LOAD} can be determined. Lastly, the fast acquisition mode methodology is limited by the size of the C_{pin}, such that, if $V_{A0} \rightarrow N_{max}$, the maximum range of (4) is an asymptote with $C_{LOAD} \rightarrow \infty$. Therefore, an approximation to determine the range of measurable C_{LOAD} is:

$$\frac{C_{pin}}{N_{max} - 1} \leq C_{LOAD} \leq C_{pin}(N_{max} - 1) \tag{5}$$

where the total range is obtained for $1 \leq N_{A0} < N_{max} - 1$. The pseudo-code for a routine to measure C_{LOAD} through the fast acquisition mode is described in Figure 3c.

2.5.2. Transient Acquisition Mode of an Isolated Load Capacitance

As pointed out before, the fast-acquisition mode methodology has limitations for large capacitances, thus requiring a different measurement strategy, namely monitoring the transient during the charging cycle of a C_{LOAD} bridging the ports P_{A0} and P_{A1} in the equivalent circuit shown in Figure 1f. The stray capacitances (C_{pin}) are not larger than some tens of pico-Farads (pF) and can therefore be neglected.

The remaining circuit analysis is simpler and is preferably done using the impedance analysis. Figure 4a shows the impedance representation of the equivalent circuit, where Z_{in} represents the impedance due to the pull-up resistance (R_{pu}) at the input port P_{A0}, and Z_{out} is the output impedance formed by the parallel RC network between the analog input resistance (R_{AIN}) and the series RC network ($Z_{LOAD} + R_{out}$). Z_{LOAD} is the impedance representation of the load capacitance (C_{LOAD}). Therefore, an estimation of C_{LOAD} is obtained using the step response of the impedance divider circuit shown in Figure 4b, which is given by:

$$V_{A0}(t) = V_S\left(1 - e^{\frac{-t}{\tau}}\right),\ \tau = R_T C_{LOAD} \tag{6}$$

where R_T is the total resistance path contributing for the potential difference on the capacitance terminals given by $R_T = R_{AIN} \,||\, R_{pu} + R_{out}$. Thus, C_{LOAD} can be found at any time t by rearranging Equation (6) to:

$$C_{LOAD} = \frac{t}{R_T \cdot \ln\left(\frac{V_S}{V_S - V_{A0}(t)}\right)} \tag{7}$$

where t is the time since starting the charging of the capacitance. All parameters in Equation (7) are known except for the time (t). Thus, to find the C_{LOAD} value, the elapsed time (Δt) must be determined, since $t = 0$ until a threshold voltage (V_{th}) is reached. For instance, solving Equation (6) with $t = \tau$ allows the definition of $V_{th} = (1 - e^{-1})V_S$ and therefore solves Equation (7). The elegant procedure for testing V_{th} makes use of the built-in transistor-transistor logic (TTL) unit to monitor V_{A0} by setting the TTL unit high voltage threshold (V_{IH}) to the V_{th} ($V_{th} = V_{IH}$), allowing not only the avoidance of the computation of the V_{th}, but also the enhancement of the accuracy of the measurement. In practical terms, the elapsed time (Δt) since C_{LOAD} initiates the charging cycle ($t = 0$) until the TTL unit changes to the logical state high '1' ($V_{A1} \geq V_{IH}$) must be measured, as well as proceeding to the reading of V_{A0} using the ADC unit (N_{A0}). Then, C_{LOAD} is found by replacing V_{A0} and t by the measured values of N_{A0} and Δt in Equation (7). Figure 4c depicts the voltage step response of the impedance divider circuit. The highlighted region delimited between the voltage level V_{IH} and vs. represents the operation region where the TTL logic unit changes of logic state low '0' to high '1'. For reference, Table 1 includes the typical values of V_{IH} for the ATmega328P and ATmega32U4 micro-controllers.

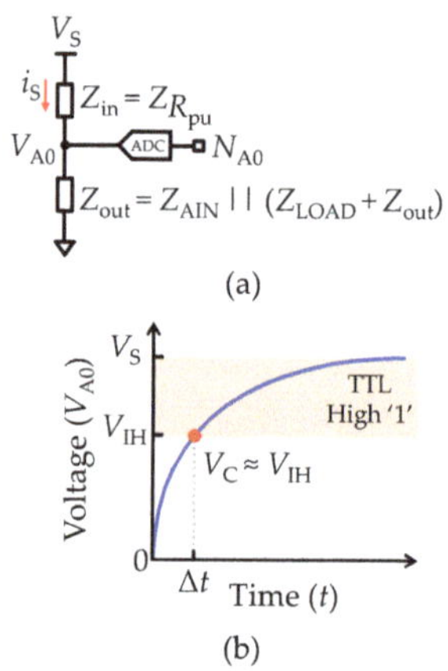

Pseudo-code:

```
Set port PA0 and PA1 to LOW OUTPUT
Wait 1 millisecond: delay (1)
Set port PA0 to INPUT PULLUP
Read timer: TS measures microseconds
do
  Sample digital port PA0: NA0 measures VA0
while (NA0 == 0)
Read timer: TE measures microseconds
Sample port PA0: NA0 measures VA0
Determine elapsed time (Δt): T = TE - TS
CLOAD is given by Eq.(7): Replace VA0 by NA0
                                 t by T
```

(c)

Figure 4. Set-up for recording a pure load capacitance (C_{LOAD}) through the transient acquisition mode. (**a**) Impedance representation of the reduced equivalent circuit. (**b**) Illustration of the step response in voltage measured at the input terminal of C_{LOAD}. V_{IH} is defined in Table 1. (**c**) Pseudo-code used to implement the transient acquisition mode.

Additionally, the transient acquisition mode methodology is limited by the characteristic time constant (τ) of the circuit, such that if $\tau \to 0$, the charging velocity (dv/dt) of the C_{LOAD} increases to proportions where high temporal resolution is required to measure the N_{A0} accurately, since $N_{A0} \to N_{max}$ just before the first reading is taken, causing the ADC unit to overflow. An approximation of the ranging limits of the transient acquisition mode is given by:

$$\frac{\Delta t}{R_T \cdot \ln(N_{max})} \leq C_{LOAD}(t) \leq \frac{\Delta t}{R_T \cdot \ln\left(\frac{N_{max}}{N_{max}-1}\right)} \quad (8)$$

where in Equation (8) the lower and upper resolution of the ADC unit (N_{A0}) are solved assuming $1 \leq N_{A0} \leq N_{max} - 1$. Using the typical values provided in Table 1, the detection limit of the interface assuming the fastest reading of the ADC unit could be achieved after one clock-cycle ($\Delta t = 62.5$ ns). Then, the minimum range of the transient acquisition method varies between 214 fF $\leq C_{min} \leq$ 392 fF with 22.4 k$\Omega \leq R_{pu} \leq$ 41.6 kΩ. Nevertheless, the impact of the stray capacitances (C_{pin}) was neglected, and therefore such a range is merely indicative, since $C_{pin} \gg C_{min}$. As for the maximum range of the transient acquisition method, the maximum Δt is limited by the size of the unsigned long variables given by $\Delta t = (2^{32} - 1) \times 1$ µs $\cong$ 4295 s. Thus, the maximum range corresponds to a battery-like storage unit rather than a capacitor, as it is some hundreds of farads. The pseudo-code of a routine to perform the measurement of C_{LOAD} through the transient acquisition mode is described in Figure 4c.

2.6. Measurements of a Serial RC Network (RC–Meter Mode)

Considering that in the equivalent circuit shown in Figure 1f, Z_{LOAD} is composed by a series RC network, the strategy to extract C_{LOAD} and R_{LOAD} resembles the previously described techniques to measure a pure resistor in Section 2.4 and a pure capacitor through the transient acquisition mode in Section 2.5.2. However, the TTL-based technique used to determine C_{LOAD} cannot guarantee that a TTL transition will occur in the transient response of the circuit, since the voltage V_{A0} might start at a value above the TTL threshold (V_{IH}). Instead, the monitoring process of the transient response must be carried out exclusively using the analog functions available on the AVR® micro-controller ports. First, the C_{LOAD} must be completely discharged, which requires defining both ports P_{A0} and P_{A1} to low output configuration for a reasonable time, and then defining P_{A0} to a high-impedance SMU, while maintaining P_{A1} as low output. For the sake of simplicity, the stray capacitances (C_{pin}) are neglected, which is allowed if they are not larger than some tens of pico-farads (pF).

With C_{LOAD} fully discharged, after setting the port P_{A0} to a high-impedance SMU ($t = 0$), a voltage step is applied on the serial RC circuit. Since C_{LOAD} is empty, it behaves as a short-circuit, which results in an equivalent circuit, as shown in Figure 2a, thus allowing the extraction of the R_{LOAD} value through use of the technique described in Section 2.4 to measure a pure resistor (note that the port configuration is the same). The measured voltage (V_{A0}) at the instant $t = 0$ defines an offset voltage (V_R). For any instant $t > 0$, C_{LOAD} starts to accumulate charge, and the same circuit analysis described in Section 2.5.2 applies to determine the transient dynamics of the voltage step response. For $t \to \infty$, C_{LOAD} is fully charged and behaves as an open circuit, thus forcing the current to flow exclusively through the analog input resistance (R_{AIN}), as represented in Figure 1f by the current path 'i_R'. Thus, considering the influence of the charging current (i_L) due to the series R_{LOAD}, an estimation of C_{LOAD} is obtained using the step response of the equivalent circuit shown in Figure 1f that is given by:

$$V_{A0}(t) = V_R + V_{REF}\left(1 - e^{\frac{-t}{\tau}}\right),\ V_R = \frac{V_S}{R_{pu}}\frac{(R_{out} + R_{LOAD})(R_{AIN} \parallel R_{pu})}{R_{out} + R_{LOAD} + R_{AIN} \parallel R_{pu}}, \tag{9}$$

where V_R is the offset voltage due to the series R_{LOAD}, V_{REF} is the new reference voltage of the circuit given by V_{REF} = vs. $- V_R$, τ is the characteristic time constant of the equivalent circuit given by the $\tau = R_T C_{LOAD}$ and R_T is the total resistance path contributing to the potential difference on the capacitance terminals given by $R_T = R_{AIN} \mid\mid R_{pu} + R_{LOAD} + R_{out}$. Thus, C_{LOAD} can be found at any time t by rearranging Equation (9) to:

$$C_{LOAD} = \frac{t}{R_T \ln\left(\frac{V_{REF}}{V_{REF} - V_{A0}(t)}\right)} \tag{10}$$

where t is the time that takes charges to accumulate in the capacitance. Figure 5a illustrates the transient response to a voltage step expressed by Equation (10), where the highlighted region consists of the offset voltage (V_R) due to the bridge of ports P_{A0} and P_{A1} with the serial RC network. As for the reading capacitance voltage level (V_C) it is determined when $t = \tau = R_T C_{LOAD}$, and is given by $V_C = V_R + (1 - e^{-1})(V_S - V_R)$. Therefore, all parameters in Equation (10) are known except for the time (t). To extract C_{LOAD}, a routine to determine the elapsed time (Δt) until the measured voltage (V_{A1}) is greater than or equal to V_C ($V_{A0} \geq V_C$) must be implemented. Figure 5b depicts the pseudo-code of a routine to implement the measurement of a serial RC network. In practical terms, C_{LOAD} is found by replacing in Equation (10) V_R with the measured value of N_{A0} at $t = 0$ (N_R), vs. with the maximum resolution of the ADC unit (N_{max}) and V_{A0} and t with the measured value of N_{A0} and Δt after $V_{A0} \geq V_C$, respectively.

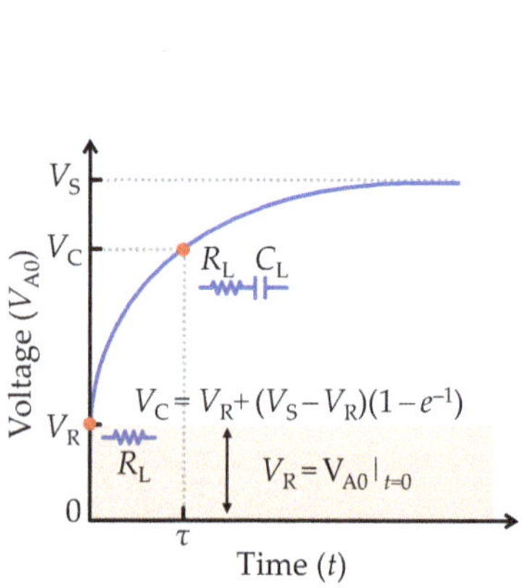

(a)

Pseudo-code:

```
Set port PA0 and PA1 to LOW OUTPUT
Wait 1 millisecond: delay (1)
Set port PA0 to INPUT PULLUP
Sample port PA0: VR measures VA0
Read timer: TS measures microseconds
Define stop condition:
  VTAU = VR + 0.6322 × (VS - VR)
do
  Sample port PA0: VC measures VA0
while (VC <= VTAU)
Read timer: TE measures microseconds
Determine elapsed time (Δt): T = TE - TS
RLOAD is given by Eq.(2)  : Replace VA0 by VR
CLOAD is given by Eq.(10): Replace VA0 by VC
                                   t by T
```

(b)

Figure 5. Set-up for recording a load impedance (Z_{LOAD}) formed by a serial RC network. (**a**) Illustration of the transient response to a voltage step at the input terminal (P_{A0}). (**b**) Pseudo-code used to implement the measurement of a serial RC network.

2.7. Measurements of a Parallel RC Network (RC–Meter Mode)

Considering that in the equivalent circuit shown in Figure 1f, Z_{LOAD} is composed of a parallel RC network ($C_{LOAD}||R_{LOAD}$), the strategy to extract C_{LOAD} and R_{LOAD} resembles the previously described technique in Section 2.6, although it must be noted that the change of the load capacitance (C_{LOAD}) from serial mode to parallel mode causes R_{LOAD} to saturate at the reference voltage ($V_F = V_R$) level when $t \to \infty$ and the voltage offset is null at $t = 0$. Thus, the step response of the equivalent circuit shown in Figure 1f is given by:

$$V_{A0}(t) = V_F\left(1 - e^{\frac{-t}{\tau}}\right), \tag{11}$$

where V_F is given by the expression of V_R that is defined in Equation (9), τ is the characteristic time constant of the equivalent circuit given by $\tau = R_T C_{LOAD}$ and R_T is the total resistance path given by $R_T = (R_{AIN} \mid\mid R_{pu}) \mid\mid (R_{LOAD} + R_{out})$. Figure 6a illustrates the transient response to a voltage step expressed by Equation (11). For $t \to \infty$, C_{LOAD} is fully charged and behaves as an open circuit, forcing the current (i_L) represented in the equivalent circuit shown in Figure 1f to flow exclusively through the R_{LOAD} to the ground. Therefore, the extraction of R_{LOAD} is analytically indeterminate, unless an approximation is made, such as, considering for any period (Δt) larger than five times the characteristic time constant (τ) ($\Delta t > 5\tau$), Equation (11) is approximately equal to $V_{A0} = V_F(1 - e^{-5}) \approx V_F$, where the term $(1 - e^{-5}) \approx 0.993$ shows that the approximation has a maximum error of 0.7% when determining the R_{LOAD} value. Thereby, assuming for any $\Delta t > 5\tau$ the plateau $V_{A0} = V_F$ is reached, the R_{LOAD} value is obtained using the technique described in Section 2.4 to measure a pure resistor, replacing in Equation (2) V_{A0} by V_F, where V_F is the measured voltage for any instant $\Delta t > 5\tau$. The next step is to determine the C_{LOAD} value through the arrangement of Equation (11) with $\tau = R_T C_{LOAD}$ by:

$$C_{LOAD} = \frac{t}{R_T \cdot \ln\left(\frac{V_F}{V_F - V_{A0}(t)}\right)} \tag{12}$$

where t is any instant of the step-response and V_{A0} the correspondent measured voltage. The determination of V_F is critical to find both C_{LOAD} and R_{LOAD} values. The simplest approach to find V_F using the transient response to a voltage step is to define the maximum characteristic time constant (τ_{max}) to measure. In these circumstances all parameters are known, and both C_{LOAD} and R_{LOAD} values can be measured using Equations (2) and (10). Figure 6b depicts the pseudo-code of a routine to implement the measurement of a parallel RC network.

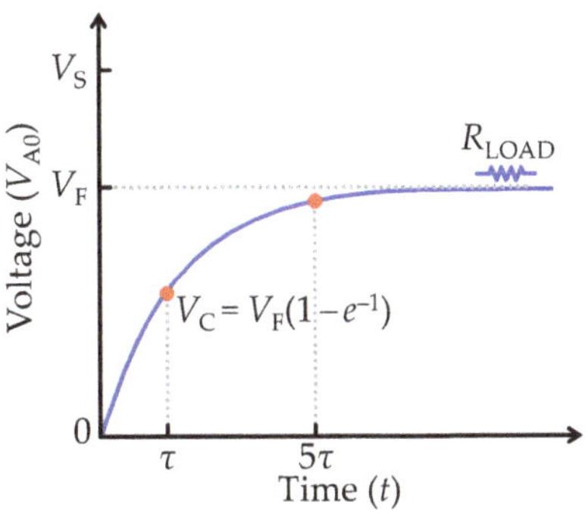

(a)

Pseudo-code:

```
Define highest time constant (τ) to be
measured: tau
Determine time (Δt) until threshold
voltage (V_F): T = 5 × tau
Set port P_A0 and P_A1 to LOW OUTPUT
Wait 1 millisecond: delay (1)
Set port P_A1 to INPUT PULLUP
Wait until Δt = τ : delay (tau)
Sample port P_A1: VC measures V_A1
Wait until Δt = 5τ : delay (T - tau)
Sample port P_A1: VF measures V_A1
R_LOAD is given by Eq.(2)  : Replace V_A0 by VF
C_LOAD is given by Eq.(12): Replace V_A1 by VC
                                        t by T
```

(b)

Figure 6. Set-up for recording a load impedance (Z_{LOAD}) formed by a parallel RC network. (**a**) Illustration of the transient response to a voltage step at the input terminal (P_{A0}). (**b**) Pseudo-code used to implement the measurement of a parallel RC network.

2.8. Measurements of an Isolated Load Inductance (L–Meter Mode)

Considering that in the equivalent circuit shown in Figure 1f, Z_{LOAD} is composed of an isolated load inductance (L_{LOAD}), and that for the sake of simplicity, the stray capacitances (C_{pin}) are neglected, which is allowed if they are not larger than some tens of pico-farads (pF), the circuit analysis is simpler than and preferable to using the impedance analysis. Figure 7a shows the impedance representation of the equivalent circuit, where Z_{in} represents the impedance due to the pull-up resistance (R_{pu}) at the input port P_{A1}, and Z_{out} is the output impedance formed by the parallel RC network between the analog input resistance (R_{AIN}) and the series RC network ($Z_{LOAD} + R_{out}$). Z_{LOAD} is the impedance representation of the load inductance (L_{LOAD}), given by $Z_{LOAD} = j\omega L_{LOAD}$. Therefore, an estimation of L_{LOAD} is obtained using the step response of the impedance divider circuit shown in Figure 7a, which is given by:

$$V_{A0}(t) = V_S\left(1 - e^{\frac{-t}{\tau}}\right),\ \tau = L_{LOAD}/R_T, \tag{13}$$

where R_T is the total resistance path contributing to the potential difference on the inductance terminals given by $R_T = R_{AIN} \mid\mid R_{pu} + R_{out}$. Thus, L_{LOAD} can be found at any time t by rearranging Equation (13) to:

$$L_{LOAD} = \frac{tR_T}{\ln\left(\frac{V_S}{V_S - V_{A0}(t)}\right)}, \tag{14}$$

where t is the time that takes charges to accumulate in the inductance. Except for the time (t), all parameters in Equation (14) are known and available in Table 1. Therefore, the strategy to solve the time (t) of Equation (14) requires a similar approach, as previously described for the measurement of an isolated capacitance through the transient response. In this case, it must be monitored when the threshold voltage (V_{th}) over L_{LOAD} is approximately equal to $V_{th} = V_S \cdot (1 - e^{-1})$. As for the isolated capacitance measurement described in Section 2.5.2, the elegant and simpler procedure for testing V_{th}, involves making use of the built-in transistor-transistor logic (TTL) unit to monitor V_{A0} by setting the TTL unit low voltage threshold (V_{IL}) the V_{th} ($V_{th} = V_{IL}$). In practical terms, the elapsed time (Δt) since L_{LOAD} initiates the cycle ($t = 0$) until the TTL unit changes to the logical state low '0' ($V_{A0} \leq V_{IL}$) must be measured, proceeding afterwards to the reading of V_{A0} using the ADC unit (N_{A0}). Then, L_{LOAD} is found by replacing V_{A0} and t with the measured values of N_{A0} and Δt in Equation (14). Figure 7b depicts the voltage step response of the impedance divider circuit. The highlighted region delimited between the voltage level V_{IL} and vs. represents the operation region where the TTL logic unit changes of logic state high '1' to low '0'.

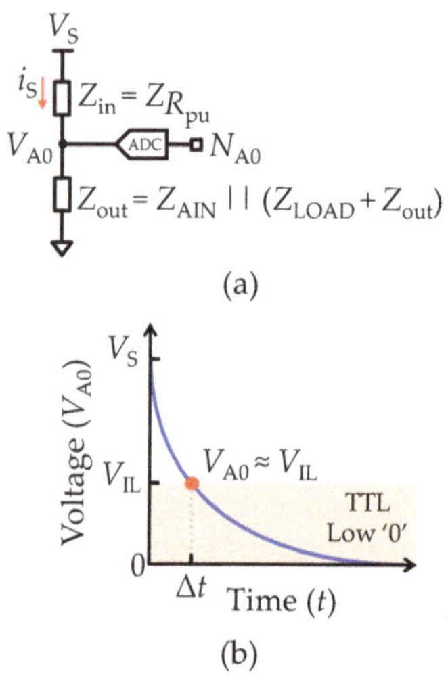

Pseudo-code:

```
Set port PA0 and PA1 to LOW OUTPUT
Wait 1 millisecond: delay (1)
Set port PA0 to INPUT PULLUP
Read timer: TS measures microseconds
do
  Sample digital port PA0: NA1 measures VA0
while (NA0 == 1)
Read timer: TE measures microseconds
Sample port PA0: NA0 measures VA0
Determine elapsed time (Δt): T = TE - TS
LLOAD is given by Eq.(14): Replace VA0 by VC
                                    t by T
```

(c)

Figure 7. Set-up for recording an isolated load inductance (L_{LOAD}) through the transient acquisition mode. (**a**) Impedance representation of the reduced equivalent circuit. (**b**) Illustration of the step response in voltage measured at the input terminal of L_{LOAD}. V_{IL} is defined in Table 1. (**c**) Pseudo-code used to implement the inductance transient acquisition mode.

The measurement range of the inductance is limited due to the ADC unit resolution, where the technique described above to monitor the transient response to a voltage step is limited by the characteristic time constant (τ) of the circuit, such that if $\tau \to 0$, the charging current velocity ($\mathrm{d}i/\mathrm{d}t$) of the L_{LOAD} decreases to such proportions that accurately measuring the N_{A0} requires high temporal resolution, since $N_{A0} \to 0$ just before the first reading is taken, causing the ADC unit to overflow. An approximation of the ranging limits of the transient acquisition mode is given by:

$$\frac{\Delta t \cdot R_T}{\ln(N_{max})} \leq L_{LOAD}(t) \leq \frac{\Delta t \cdot R_T}{\ln\left(\frac{N_{max}}{N_{max}-1}\right)} \tag{15}$$

where in Equation (15) the lower and upper resolution of the ADC unit (N_{A0}) are solved assuming $1 \leq N_{A0} \leq N_{max} - 1$. With the same values and assumptions used for the transient C-method, the range of measurement will vary between 0.2 mH $\leq L_{LOAD} \leq$ 1.5 H with R_{pu} = 22.4 kΩ, and 0.4 mH $\leq L_{LOAD} \leq$ 2.7 H with R_{pu} = 41.6 kΩ. These measurable values of L_{LOAD} are not very useful in real life applications, and for that reason in this manuscript only the methodology is provided as proof of concept to perform measurements with AVR® micro-controllers of an isolated inductance. In spite of this, the pseudo-code is provided in Figure 7c to implement the measurements of an isolated L_{LOAD} through the transient acquisition mode built-in AVR® micro-controllers.

2.9. Data Acquisition and Analysis

The open-source Arduino® IDE software was used to program and upload the scripts on the AVR® micro-controllers. All routines that were programmed to perform the measurements of the RCL-meter are based on the pseudo-codes previously described in each method. The data acquisition was carried out with the serial interface made available on the Arduino® IDE software. The collected data were handled with MATLAB® to perform the data analysis. All measurements were performed at room temperature, about 25 °C.

2.10. Noise and Uncertainty of the Measurements

To achieve the highest measurement accuracy of the RCL parameters, the parasitic capacitances must be minimized by leaving a space of two analog pins between the measured ports, e.g., ports P_{A0} and P_{A3}, and shorten the connection cables to a minimum size.

Additionally, the impact of noise sources in the measurements must be considered. This includes the thermal and $1/f$ contributions, which are not easily modelled, but are included in the global noise measurements. The measured noise of the voltage source (ΔV_n) was 6.7 mV (with the Arduino board powered through the USB interface connected to a local-PC), comparable to the digitalization uncertainty of the ADC (5 mV) described below. This value permits an enhancement of the accuracy of the ADC unit by oversampling techniques. This shows that the digitalization dominates over thermal and $1/f$ noise contributions. Thus, in this analysis we only calculate the effect of digitalization uncertainty. Using the default ADC sampling time of 100 μs, the digitalization uncertainty of a 10-bit ADC unit relative to the reference voltage (V_{REF}) of 5 V, the least significant bit (LSB) voltage is 4.9 mV. To analyze the impact of this digitalization uncertainty, the relative uncertainty (u_r) associated with the analog to digital round off was determined as:

$$u_r = \frac{f(N_{meas}+0.5) - f(N_{meas} - 0.5)}{f(N_{meas})}, \quad (16)$$

where $f(N)$ represents one of the previous Equations (2), (4), (7), (10), (12) and (14) used to determine the R_{LOAD}, C_{LOAD} and L_{LOAD} values. Then, the oversampling technique allow for reducing the digitalization uncertainty (u_r) by a factor of $1/N$ but limited by other sources of uncertainty (noise).

2.11. Relative Accuracy and Precision of the Measurements

The errors associated with the accuracy and precision of the measurements were analyzed through the relative accuracy (a_r) and the relative precision (p_r). The relative precision (p_r) measures the dispersion of the measured impedance values (Z_{meas}) normalized to their average ($\overline{Z}_{meas}$) and was estimated by $p_r = SD(Z_{meas})/\overline{Z}_{meas}$, where SD represents the standard deviation of the measured values. The relative accuracy (a_r) measures the closeness of the measured Z_{meas} to the true or reference value, $Z_{nominal}$, and was estimated by $a_r = |\overline{Z}_{meas} - Z_{nominal}|/Z_{nominal}$.

2.12. Linearization of the ADC Unit

Work was done to improve the linearization of the ADC unit output. By linearization is here meant the maximization of the correlation between measured and nominal values, which depends on the correct knowledge of the micro-controller's parameters. Since all methods presented previously to measure an unknown load impedance (Z_{LOAD}) use the same ports configuration (except for the measurements of an isolated load capacitance using the fast acquisition mode), the simplest method to optimize the linearization of the measurements is to replace Z_{LOAD} with a pure known load resistance (R_{LOAD}) and use different resistor sizes to test and maximize the range and accuracy of the ADC unit. Then, Equation (2) must be used as a fitting function, where the R_{LOAD} values must be replaced by the nominal values provided by the manufacturer of the resistor, and V_{A0} and vs. by the ADC unit values. Figure 8a shows the N_{A0} values collected with through-hole resistors bridging two ports of the micro-controllers. All samples consist of an average of 100 consecutive measurements. The data were distributed in a logarithmic scale along the horizontal axis according to the nominal resistance value provided by the manufacturer. The red triangles represent the samples measured with the ATmega328P and the blue circles represent the ATmega32U4. Ordinary Least Squares (OLS) were used to fit the data shown in Figure 8a, but with a small twist, which consisted of applying the natural logarithm (ln) to the OLS objective fit function and to the measured N_{A0} values. This proved able to cope better with the large range of resistance values, encompassing several orders of magnitude. Then, to extract the optimized values of R_{AIN}, R_{out} and R_{pu}, the user must provide a guess estimation and use a nonlinear programming solver to find the minimum values of the OLS objective function around the estimated values, taken from the manufacturer datasheets. Table 2 shows the extracted values of R_{AIN}, R_{out} and R_{pu} for both ATmega328P and ATmega32U4 AVR® micro-controllers used in the current work.

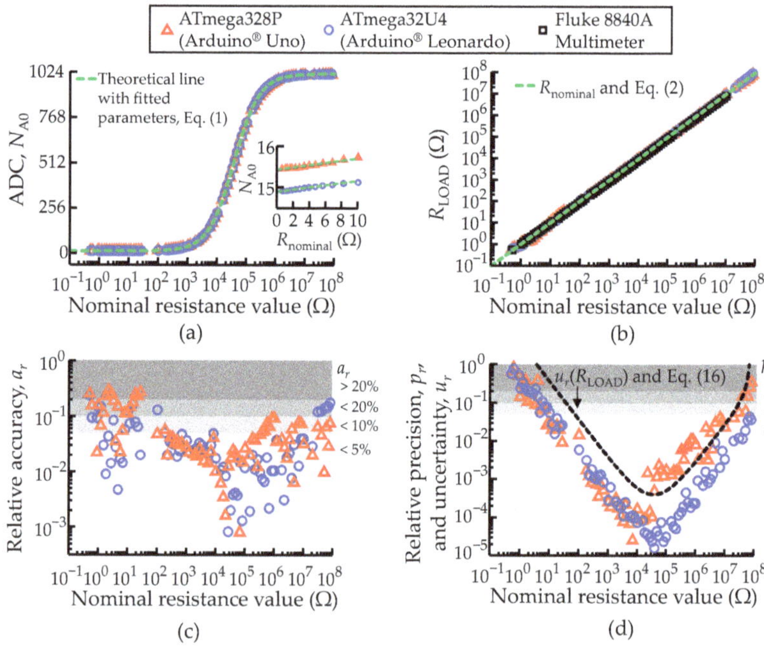

Figure 8. Comparison between the ATmega328P and ATmega32U4 AVR® micro-controllers configured to record an isolated load resistance (R_{LOAD}). (**a**) Measured ADC unit discrete values at port P_{A0} (N_{A0}). Each sample consists of an average of 100 consecutive measurements. (**b**) Measured load resistance (R_{LOAD}) values according to Equation (2). The green dashed lines represent the theoretical lines. (**c**) Relative accuracy (a_r) and (**d**) relative precision (p_r) of the measurements in function of $R_{nominal}$. The black dashed line represents the relative uncertainty (u_r) of the R_{LOAD} measurements according to Equation (16). The white and grey shading areas highlight the levels of u_r, a_r and p_r better than 5%, 10% and 20%. A legend to describe the color scheme used in all plots of Figure 8 was included.

Figure 8a shows the ADC readings as a function of the nominal ADC values. Actual values for both ATmega328P and ATmega32U4 micro-controllers are represented by symbols, and the corresponding fitted lines from Equation (2) (overlapped) in dashed green. The figure inset provides a closer view of the measured samples between 0.5 Ω to 10 Ω. The fitted parameters allow a broad working range of about 8 orders of magnitude for the measured resistance. Otherwise, using typical values provided by the manufacturer would result in a working range of only 2 orders of magnitude. This procedure was performed only once and the values obtained for R_{AIN}, R_{out} and R_{pu} were used in all the subsequent measurements.

Table 2. Optimized values of the R_{AIN}, R_{out}, R_{pu} and C_{pin} for both ATmega328P and ATmega32U4 AVR® micro-controllers.

ATmega328P				ATmega32U4			
R_{AIN} (MΩ)	R_{pu} (kΩ)	R_{out} (Ω)	C_{pin} (pF)	R_{AIN} (MΩ)	R_{pu} (kΩ)	R_{out} (Ω)	C_{pin} (pF)
3.537	36.89	565.8	23.48	5.451	36.66	542.2	25.5

Likewise, the same fitting analysis previously described was used to optimize the value of C_{pin}, which is required to perform measurements of an isolated load capacitance through the fast acquisition mode. In this case, through-hole capacitors of different sizes were used to bridge two I/O ports of the micro-controller, and Equation (4) was used as a fitting function for the measurements of the N_{A0} values shown in the Figure 9a. The C_{pin} values extracted for each micro-controller are shown in Table 2. The MATLAB scripts written to perform the OLS fitting are available in the Supplementary Materials.

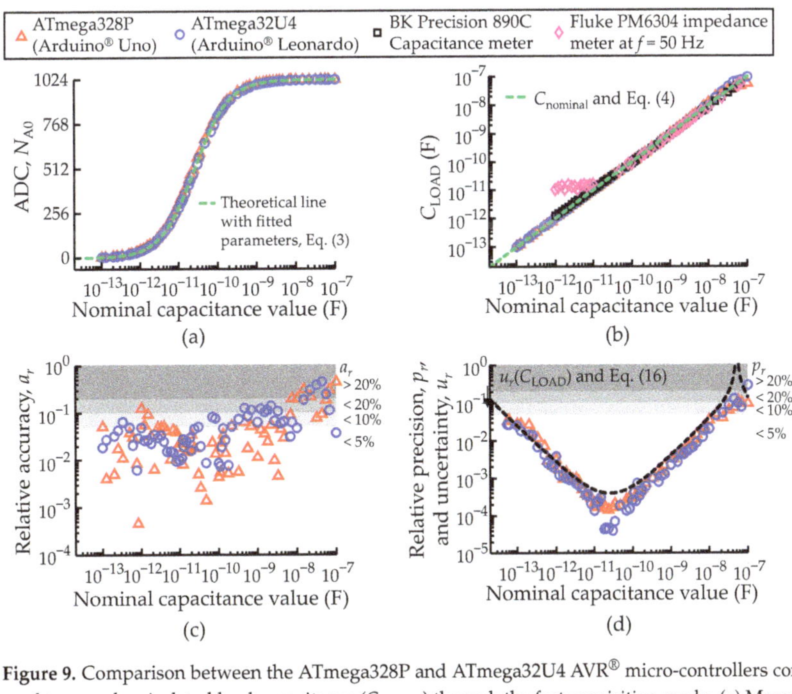

Figure 9. Comparison between the ATmega328P and ATmega32U4 AVR® micro-controllers configured to record an isolated load capacitance (C_{LOAD}) through the fast acquisition mode. (**a**) Measured ADC values at port P_{A0} (N_{A0}). Each ADC sample consists of an average of 100 consecutive measurements. (**b**) Measured load capacitance (C_{LOAD}) given by Equation (4). The green dashed lines represent the theoretical lines. (**c**) Relative accuracy (a_r) and (**d**) relative precision (p_r) of the measurements in function of $C_{nominal}$. The black dashed line represents the relative uncertainty (u_r) of the C_{LOAD} measurements according to Equation (16). The white and grey shading areas highlight the levels of u_r, a_r and p_r better than 5%, 10% and 20%. A legend to describe the color scheme used in all plots of Figure 9 was included.

3. Results

Commercial through-hole resistors and capacitors in the ranges 0.5 Ω–80 MΩ and 100 Ff–4.7 mF (±5%), respectively, were used to perform measurements of either isolated or in series and parallel electrical components with the AVR® micro-controllers. The capacitances in the fF range were of SMD type (Kyocera AVX, Fountain Inn, SC, USA), and all capacitances above 1 µF were aluminum electrolyte capacitor type. The validations were made by comparing the measurements results with the components' nominal values. In all cases, each data point corresponds to the average of 100 measurements performed in a continuous loop.

3.1. *Characterization of Isolated Resistance Measurements*

The resistance measurements were performed with a set of resistors independent of those used to linearize the ADC. Using the methodology described in Section 2.4, the voltages at port P_{A0} (V_{A0}) were recorded with the ADC unit (N_{A0}), and the measurements were inserted in Equation (2), to obtain the R_{LOAD} values shown in Figure 8b. This figure also includes the same resistances measurements recorded with a commercial instrument, the Fluke 8840A multimeter (Fluke Corporation, Everett, WA, USA), for reference and validation. These are excellent results for a low-cost technique. However, it is important to keep in mind that each individual point is the average of 100 consecutive measurements, performed in a loop. Table 3 provides numeric detail on the data shown in Figure 8b.

Table 3. Comparison of the ATmega328P, ATmega32U4 and Fluke 8840A measured load resistance (R_{LOAD}) values ± standard deviation (SD).

$R_{nominal}$	ATmega328P	ATmega32U4	Fluke 8840A		ATmega328P	ATmega32U4	Fluke 8840A
	R_{LOAD} ± SD			$R_{nominal}$	R_{LOAD} ± SD		
0.5 Ω	0.6 ± 0.4	0.62 ± 0.4	0.548 ± 0.001	5.6 kΩ	5.679 ± 0.001	5.7261 ± 0.0006	5.5390 ± 0.0002
1 Ω	1.2 ± 0.2	0.86 ± 0.2	1.044 ± 0.001	8.2 kΩ	8.539 ± 0.009	8.6078 ± 0.0009	8.1837 ± 0.0002
2.2 Ω	1.6 ± 0.3	2.30 ± 0.5	2.247 ± 0.001	10 kΩ	10.101 ± 0.002	10.1772 ± 0.0008	9.9313 ± 0.0002
5.6 Ω	5.7 ± 0.4	5.70 ± 0.6	5.645 ± 0.001	22 kΩ	21.9 ± 0.3	22.1858 ± 0.0008	21.846 ± 0.0000
8.2 Ω	9.1 ± 0.2	7.72 ± 0.6	8.251 ± 0.001	56 kΩ	55.0 ± 0.1	55.789 ± 0.002	55.8529 ± 0.0003
10 Ω	10.5 ± 0.3	8.77 ± 0.5	10.031 ± 0.005	82 kΩ	80.8 ± 0.1	81.891 ± 0.002	81.914 ± 0.002
22 Ω	22.4 ± 0.4	20.1 ± 0.5	21.901 ± 0.002	100 kΩ	97.3 ± 0.2	98.812 ± 0.003	98.622 ± 0.008
56 Ω	59.7 ± 0.4	55.7 ± 0.5	56.023 ± 0.004	220 kΩ	211.9 ± 0.4	217.55 ± 0.01	219.267 ± 0.007
82 Ω	83.3 ± 0.3	83.1 ± 0.4	82.532 ± 0.0014	560 kΩ	535 ± 5	549.7 ± 0.1	561.045 ± 0.009
100 Ω	102.2 ± 0.3	101.9 ± 0.4	99.44 ± 0.02	820 kΩ	777 ± 7	811.4 ± 0.2	824.71 ± 0.01
220 Ω	224.3 ± 0.3	224.2 ± 0.4	220.560 ± 0.004	1 MΩ	0.906 ± 0.007	0.9894 ± 0.0002	1.0140 ± 0.00004
560 Ω	576.9 ± 0.4	573.6 ± 0.8	556.14 ± 0.01	2.2 MΩ	2.14 ± 0.05	2.193 ± 0.004	2.2171 ± 0.0002
820 Ω	835.8 ± 0.6	833.840 ± 0.0001	817.19 ± 0.03	6.8 MΩ	6.3 ± 0.3	6.89 ± 0.01	6.986 ± 0.002
1 kΩ	1.0348 ± 0.0006	1.0331 ± 0.0004	0.9945 ± 0.0001	8.2 MΩ	8.8 ± 0.3	7.87 ± 0.05	8.2066 ± 0.0001
2.2 kΩ	2.242 ± 0.001	2.2596 ± 0.0005	2.1955 ± 0.0001	10 MΩ	10.3 ± 0.2	9.60 ± 0.03	10.129 ± 0.008

Figure 8c shows the relative accuracy (a_r) of the measurements as a function of the nominal resistance, as defined in Section 2.11. Figure 8d shows their relative precision, p_r, also defined in Section 2.11 (data points), and the estimated upper limit for the relative uncertainty caused by the digitalization round-off error, u_r, defined in Equation (16) and represented by the dashed line. Note that the standard deviations are calculated on samples that are already averages of 100 points, which means that u_r should be divided by 10, for a correct comparison with p_r. Both plots include white and grey shading regions to highlight the levels of a_r and p_r better than 5%, 10% and 20%. The two plots show better performance of the method in the intermediate resistance range and its degradation in the regime of very low or very high resistances. This is because the lower and upper limits for R_{LOAD} correspond to N_{A0} tending to the higher (N_{max}) and lower (≈ 0) values of the digital output, respectively, where the roundoff errors introduced by digitalization become more important, affecting both precision and accuracy.

Most of the data points lay below the u_r curve. This is because u_r is merely an upper limit for the digitalization noise, which is actually lower. In any case, the plot indicates that the main uncertainty source in the determination of resistance is the digitalization roundoff.

In terms of performance, the ATmega32U4 delivers better results. The range defined by $p_r < 5\%$ is approximately 10 Ω–10 MΩ for the ATmega328P and 10 Ω–80 MΩ for the ATmega32U4, while the range defined by $a_r < 5\%$ is 100 Ω–100 kΩ for the ATmega328P and 100 Ω–10 MΩ for the ATmega32U4.

The sensitivity of the R-meter (minimum detectable increment in resistance) was assessed at a representative value of 1 kΩ, as well at increments of 0.5, 1 and 4.7 Ω. 50 measurements were acquired at each resistance value. The average of the 50 ADC counts were plotted against the resistance values and a local slope ADC counts/Ω was determined. The standard deviations of the 4 measurements were also calculated and averaged to get a typical value. A conservative estimative of the sensitivity was then performed by calculating the increase in resistance needed to shift the ADC count by two standard deviations, which was about 1.2 Ω or 0.1% of the nominal value. Additionally, a student *t*-test analysis was performed by comparing all the 4 datasets of 50 measurements against each other to conclude that they were all different ($p < 0.05$). This suggested that even an increment of 0.5 Ω is enough to change the output of the R-meter, which is about 2 times less than the previous conservative estimate.

3.2. Characterization of Isolated Capacitance Measurements: Fast Acquisition Method

Using the methodology described in Section 2.5.1 with different commercial capacitors varying from 100 fF to 100 nF, measurements of the voltage at port P_{A0} (V_{A0}) were recorded with the ADC unit (N_{A0}) and shown in Figure 9a. Figure 9a also includes a green dashed line that represents the theoretical lines for the two micro-controllers (overlapped), obtained from Equation (3) with the fitted C_{pin} values shown in Table 2. The measured N_{A0} and fitted C_{pin} values were replaced in Equation (4) to determine the load capacitance (C_{LOAD}), shown in Figure 9b. At the end of Section 3.2, Table 4 was included, providing numeric detail on the data shown in Figure 9b. For reference and validation, this plot also includes the measurements recorded with two commercial instruments, the BK Precision 890C capacitance meter (B&K Precision Corporation, Yorba Linda, CA, USA) and the Fluke PM6304 impedance meter (Fluke Corporation, Everett, WA, USA) at the lowest frequency available (f = 50 Hz). The green dashed line is $C_{LOAD} = C_{nominal}$, showing that the C_{LOAD} values determined by both micro-controllers match the target values within the range 100 Ff–10 nF, thus achieving a range of about 5 orders of magnitude.

Figure 9c,d allow a more detailed view of the quality of this match and display a_r, p_r and u_r (in the same way as in Figure 8c,d). The accuracy a_r drops significantly above 10 nF because the voltage drop at the load capacitance tends towards the circuit voltage source ($N_{A0} \rightarrow N_{max}$), as predicted by Equation (5), inducing large errors in the determination of C_{LOAD}.

This is evidenced in the u_r curve represented (black dashed line in Figure 9d), showing the same V-shaped distribution u_r, as described in Figure 8d, for the resistance measurement. The overlap between u_r and p_r means that the main source of variability is the digitalization noise, to which the fast C-meter adds almost no contribution.

Table 4. Comparison of the ATmega328P, ATmega32U4 and Fluke 8840A measured load capacitance (C_{LOAD}) values ± standard deviation (SD) through the fast acquisition mode.

$C_{nominal}$	ATmega-328P	ATmega-32U4	Fluke PM6304	BK 890C		ATmega-328P	ATmega-32U4	Fluke PM6304	BK 890C
	C_{LOAD} ± SD				$C_{nominal}$	C_{LOAD} ± SD			
1 pF	1.126 ± 0.004	1.050 ± 0.005	12 ± 8	1.3 ± 0.6	560 pF	550.0 ± 0.8	510.89 ± 0.05	526 ± 3	531 ± 3
1.5 pF	1.640 ± 0.002	1.541 ± 0.003	16 ± 18	1.9 ± 0.3	680 pF	644 ± 1	600.30 ± 0.05	678 ± 5	677 ± 7
2.7 pF	2.834 ± 0.003	2.782 ± 0.004	14 ± 8	3.3 ± 0.6	1 nF	1.012 ± 0.003	0.93558 ± 0.00004	0.999 ± 0.009	1.008 ± 0.004
3.9 pF	4.101 ± 0.003	3.998 ± 0.004	12 ± 5	8 ± 1	1.5 nF	1.445 ± 0.007	1.34302 ± 0.00003	1.493 ± 0.005	1.517 ± 0.002
5.8 pF	5.860 ± 0.002	5.709 ± 0.003	17 ± 19	7 ± 3	2.2 nF	2.08 ± 0.01	1.88744 ± 0.00004	2.185 ± 0.005	2.24 ± 0.01
8.2 pF	8.527 ± 0.003	8.278 ± 0.003	17 ± 5	9.3 ± 0.4	3.3 nF	3.29 ± 0.03	2.93701 ± 0.00003	3.314 ± 0.006	3.401 ± 0.003
10 pF	10.231 ± 0.002	9.900 ± 0.002	18 ± 5	11.2 ± 0.3	6.8 nF	7.4 ± 0.2	6.58318 ± 0.00003	7.096 ± 0.02	7.13 ± 0.06
20 pF	20.711 ± 0.003	19.672 ± 0.002	25 ± 4	22.300 ± 0.000	7.5 nF	9.0 ± 0.2	7.98201 ± 0.00004	7.56 ± 0.05	7.7540 ± 0.0004
47 pF	46.934 ± 0.009	46.59 ± 0.04	49 ± 7	49.3 ± 0.5	10 nF	12.0 ± 0.4	9.51585 ± 0.00003	10.15 ± 0.07	9.668 ± 0.001
82 pF	80.88 ± 0.02	81.01 ± 0.02	89 ± 7	83.4 ± 0.5	15 nF	20 ± 1	17.8541 ± 0.00004	15.44 ± 0.06	15.570 ± 0.0000
100 pF	99.56 ± 0.04	100.87 ± 0.03	99 ± 5	103.183 ± 0.04	22 nF	27 ± 1	28.6855 ± 0.00003	21.623 ± 0.007	22 ± 1
180 pF	179.1 ± 0.1	178.6 ± 0.2	187 ± 1	185 ± 1	56 nF	46 ± 5	70.0663 ± 0.00002	57.21 ± 0.04	57.2 ± 0.9
220 pF	210.7 ± 0.2	208.5 ± 0.2	221 ± 12	217.6 ± 0.5	68 nF	47 ± 5	75.7749 ± 0.00002	68.39 ± 0.07	69 ± 2
470 pF	440.0 ± 0.7	409.77 ± 0.05	469 ± 6	449.9 ± 0.8	100 nF	55 ± 5	96.2257 ± 0.00003	101.3 ± 0.02	101.536 ± 0.005

The sensitivity estimates were performed according to the same lines described in Section 3.1, this time for the representative values of 18 pF and 22 pF, with small increments of 1, 1.2 and 1.5 pF. The sensitivity was estimated to be about 10–20 fF in both cases. The student *t*-test analysis also concluded that all the capacitance measurement datasets were different from each other ($p < 0.05$).

3.3. Characterization of Isolated Capacitance Measurements: Transient Acquisition Method

The capacitance meter in the transient mode was tested according to the methodology described in Section 2.5.2 with different commercial capacitors ranging from 100 pF to 4.7 mF. The transient acquisition mode requires waiting until the TTL logic unit returns '1' and the elapsed time (Δt) until that transition. The measurements of port P_{A0} (N_{A0}), taken at the transition and the corresponding elapsed time (Δt) are shown in Figure 10a.

The upper graph shows N_{A0} as symbols. Note that by definition these N_{A0} readings correspond to TTL parameter V_{IH}—High-Level Input Voltage (because they are acquired at the transition). The same graph shows that the lower the capacitance, the higher the V_{IH} of the TTL unit. This aspect is consistent with the expected operation mode of the TTL unit. In fact, the TTL unit uses the output high-level current (i_{HL}) as a test condition, thence, as $C_{LOAD} \rightarrow C_{pin}$, the more leakage current flows through C_{LOAD}, leading to a non-constant V_{IH}. It should be remarked that the inconstancy of the threshold does not represent a problem for the application of Equation (7), since it only requires a given time and the associated reading V_{A0}.

V_{IH} stabilizes above 22 nF because in that range i_{HL} remains essentially undisturbed. This allows the estimation of the "basal" V_{IH} (ΔN_{A0}) averaging the values of N_{A0} at the transition for all samples above 22 nF, specifically $\Delta N_{A0} \cong 536$ for the ATmega328P and $\Delta N_{A0} \cong 331$ for the ATmega32U4, which are represented by the green dashed lines. The measured values of ΔN_{A0} are consistent with the typical values described in Table 1.

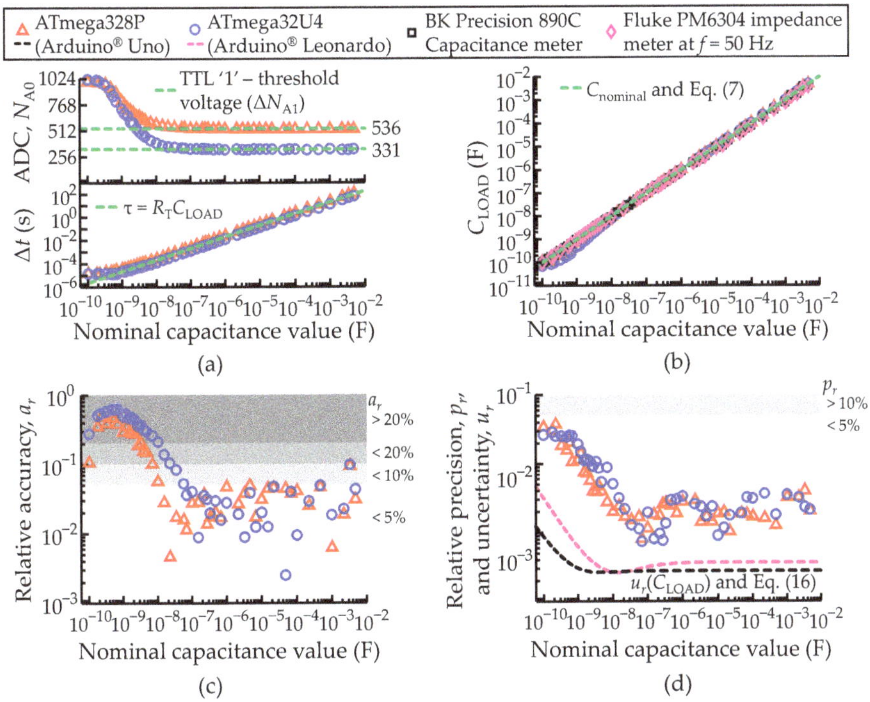

Figure 10. Comparison between the ATmega328P and ATmega32U4 AVR® micro-controllers configured to record an isolated load capacitance (C_{LOAD}) through the transient acquisition mode. (**a**) Measured ADC value and time (Δt) until the TTL unit changes to digital state high, logic '1' at port P_{A0} (N_{A0}). Each ADC and Δt sample consist of an average of 100 consecutive measurements. (**b**) Measured load capacitance (C_{LOAD}) given by Equation (7). The green dashed lines represent the theoretical lines. (**c**) Relative accuracy (a_r) and (**d**) relative precision (p_r) of the measurements in function of $C_{nominal}$. The black dashed line represents the relative uncertainty (u_r) of the C_{LOAD} measurements according to Equation (16). The white and grey shading areas highlight the levels of u_r, a_r and p_r better than 5%, 10% and 20%. A legend to describe the color scheme used in all plots of Figure 10 was included.

The measured elapsed time (Δt) shown in the bottom graph of Figure 10a closely matches the green dashed line, which represents the characteristic time constant $\tau = R_T C_{LOAD}$, where $R_T = R_{AIN} \,||\, R_{pu} + R_{out}$. The linearity still holds in spite of a non-constant V_{IH} due to the role of i_{HL}, since the leakage current is also determined by the RC constant of the circuit. The RT values differ only slightly for the two micro-controllers ($R_T \cong 35$ kΩ for ATmega328P and $R_T \cong 37$ kΩ for ATmega32U4), causing overlap of the corresponding time constant lines.

For the smallest capacitances ($C_{LOAD} < 1$ nF), Δt tends toward a plateau (Δt_{min}) given by 12.5 µs for both micro-controllers, because $C_{LOAD} \rightarrow C_{pin}$. Thus, the C_{LOAD} values determined with Equation (7), and shown in Figure 10b, are affected by larger errors in the low capacitance range. This is not very apparent in this figure, where the data points seem very close to the green dashed line ($C_{LOAD} = C_{nominal}$) because of the logarithmic scale. Still, at the end of Section 3.3, Table 5 provides numeric detail on the data shown in Figure 10b.

Table 5. Comparison of the ATmega328P, ATmega32U4 and Fluke 8840A measured load capacitance (C_{LOAD}) values ± standard deviation (SD) through the transient acquisition mode.

$C_{nominal}$	ATmega-328P	ATmega-32U4	Fluke PM6304	BK 890C		ATmega-328P	ATmega-32U4	Fluke PM6304	BK 890C
	C_{LOAD} ± SD				$C_{nominal}$	C_{LOAD} ± SD			
100 pF	111 ± 4	72 ± 1	99 ± 5	101.41 ± 0.04	2.2 μF	2.259 ± 0.007	2.158 ± 0.006	2.2742 ± 0.0004	2.292 ± 0.005
1 nF	0.703 ± 0.008	0.49 ± 0.01	0.999 ± 0.009	1.008 ± 0.004	4.7 μF	4.933 ± 0.008	4.883 ± 0.009	4.24 ± 0.04	4.5010 ± 0.0006
2.2 nF	1.591 ± 0.009	1.22 ± 0.01	2.185 ± 0.005	2.24 ± 0.01	6.8 μF	6.92 ± 0.01	6.71 ± 0.02	7.1687 ± 0.0001	7.1870 ± 0.0001
4.7 nF	3.99 ± 0.02	3.26 ± 0.02	4.77 ± 0.03	4.90 ± 0.03	10 μF	10.31 ± 0.02	10.204 ± 0.009	9.76 ± 0.01	9.998 ± 0.02
6.8 nF	6.10 ± 0.02	5.01 ± 0.05	7.10 ± 0.02	7.13 ± 0.06	22 μF	23.01 ± 0.02	23.06 ± 0.04	20.98 ± 0.03	21.29 ± 0.03
10 nF	9.43 ± 0.02	7.85 ± 0.04	10.15 ± 0.07	9.667 ± 0.001	47 μF	49.10 ± 0.09	47.12 ± 0.09	42.69 ± 0.04	43.0 ± 0.5
22 nF	22.10 ± 0.04	19.66 ± 0.04	21.623 ± 0.007	22 ± 1	68 μF	70.6 ± 0.1	70.8 ± 0.2	66.39 ± 0.02	68.68 ± 0.06
56 nF	56.89 ± 0.09	53.03 ± 0.06	57.21 ± 0.04	57.2 ± 0.9	100 μF	104.2 ± 0.2	101.0 ± 0.3	98.14 ± 0.04	102.30 ± 0.08
68 nF	67.22 ± 0.06	65.52 ± 0.05	68.39 ± 0.07	69 ± 2	220 μF	225.6 ± 0.4	226.5 ± 0.8	200.29 ± 0.08	206.7 ± 2
100 nF	97.2 ± 0.1	95.9 ± 0.2	101.34 ± 0.02	101.536 ± 0.005	470 μF	491.9 ± 0.6	446.6 ± 0.8	455.7 ± 0.2	472.6 ± 3
220 nF	228.3 ± 0.4	212.2 ± 0.2	221.79 ± 0.03	221.900 ± 0.000	1 mF	1.006 ± 0.003	0.981 ± 0.004	0.9608 ± 0.0006	0.994 ± 0.006
470 nF	480 ± 1	456 ± 1	468.54 ± 0.05	469.800 ± 0.000	2.2 mF	2.240 ± 0.007	2.250 ± 0.007	2.1667 ± 0.0005	2.1987 ± 0.0000
680 nF	692 ± 1	669 ± 2	665.67 ± 0.2	680.000 ± 0.000	3.3 mF	3.62 ± 0.02	3.631 ± 0.009	3.1891 ± 0.0007	3.441 ± 0.002
1 μF	1.047 ± 0.003	0.972 ± 0.003	0.983 ± 0.001	0.989 ± 0.007	4.7 mF	4.85 ± 0.01	4.91 ± 0.01	4.6754 ± 0.0005	4.935 ± 0.003

The plots of a_r in Figure 10c illustrate better the difficulties in the low capacitance ranges. The relative accuracy drops significantly (worse than 5%) below 70 nF for the ATmega32U4 and below 10 nF for the ATmega328P. However, both remain, at least, on a 5% accuracy level above those critical values, representing a linear response of the instrument across 6 decades.

Figure 10d shows p_r (points) and u_r (dashed lines). Contrary to the fast C-meter case, here the u_r and p_r lines are clearly above u_r, which means that the method introduces sources of noise other than digitalization. This is probably because variations of some ADC units in the threshold level impact much more the measurement than one ADC unit only, related to the digitalization error. These variations may be caused by the internal noise of the micro-processors. The performance degradation in the low capacitance regime is also evident from this figure.

The sensitivity estimates were performed according to the same lines described in Section 3.1, for the representative value of 1 μF, with small increments of 1.8, 2.2, 2.7, 3.3, 3.9 and 4.7 nF. The sensitivity was estimated to be about 10–20 fF in both cases. The student *t*-test analysis also concluded that all the capacitance measurement datasets were different from each other ($p < 0.05$).

3.4. Characterization of Measurements for Serial RC Networks

Using different sets of commercial resistors varying in factors of 10, from 10 Ω to 10 MΩ, and capacitors with 2 samples per order of magnitude, from 100 pF to 4.7 mF, seven trials were made to test the methodology described in Section 2.6. Each trial consisted of keeping the R_{LOAD} constant and varying the C_{LOAD}. Figure 11 compiles a total of 14 trials carried out with the two micro-controllers. The white points with red edges represent the data measured with the ATmega328P, and the blue points represent the ATmega32U4. They are mostly overlapped.

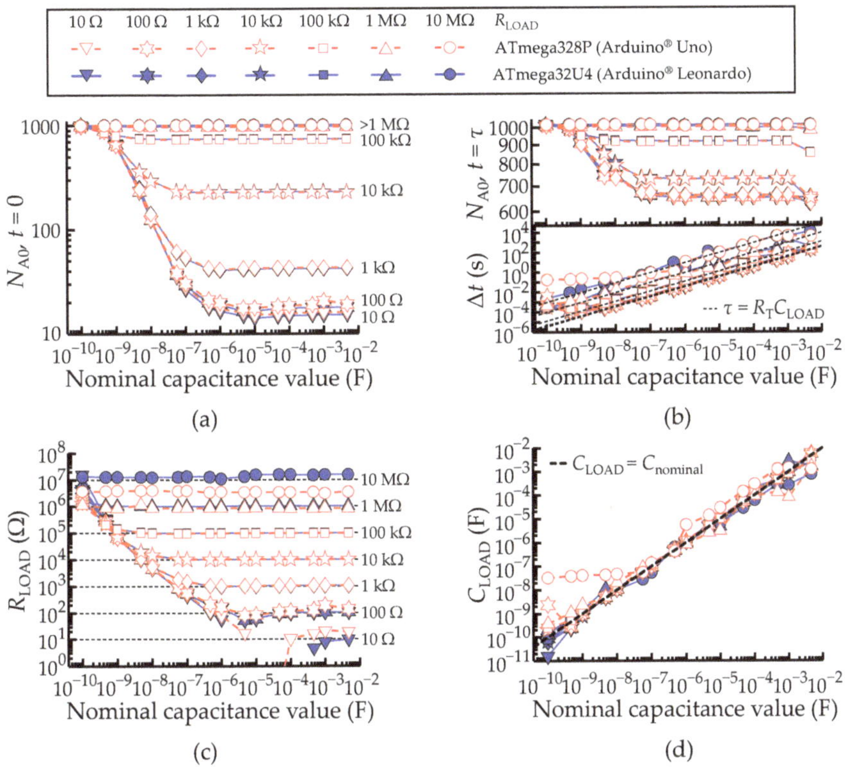

Figure 11. Comparison between the ATmega328P and ATmega32U4 AVR® micro-controllers configured to measure a serial RC network. (**a**) Measured ADC value at $t = 0$. (**b**) Measured ADC value and time (Δt) until $V_{A1} \geq V_C$. (**c**,**d**) Obtained R_{LOAD} and C_{LOAD} values according to Equations (2) and (10), respectively. A legend to describe the color scheme used in all plots of Figure 11 was included. The black dashed lines always represent the theoretical lines.

The method implies measurement of the port P_{A0} voltage level (V_{A0}) at $t = 0$ to find the offset voltage (V_R) and at $t = \Delta t \geq \tau$, to find the voltage level V_C (defined in Section 2.6).

Figure 11a,b show the measured discrete values (N_{A0}) at the instants $t = 0$ and $t \geq \tau$, respectively, with both micro-controllers. Figure 11b also includes in the bottom graph the measured elapsed time (Δt) until $V_{A1} \geq V_C$, together with the theoretical time constants, $\tau = R_T C_{LOAD}$. Note that, from Equation (9), $N_{A0}(t = 0) = V_R$, which is independent of C_{LOAD}. This means that all the curves in Figure 11a should be horizontal lines. However, the minimum available acquisition time (ca. 3 µs) is insufficient to capture the initial curve values for $C_{LOAD} < 1$ µF. Thus, the captured N_{A0} values at $t = 0$ tend to N_{max} in the limit of very small capacitances. This explains why the (ideal) straight lines become distorted in Figure 11a, especially for lower R_{LOAD}. Likewise, the plateau N_{A0} ($t \geq \tau$) = V_C is reached for any $C_{LOAD} > 100$ nF, which hampers the measurements of N_{A0} after $t \geq \tau$ in this range and induces the same type of distortion in Figure 11b, top.

R_{LOAD} and C_{LOAD} were determined through the use of Equations (2) and (10), with the results shown in Figure 11c,d, respectively. The horizontal lines in Figure 11c represent the ideal result, $R_{measured} = R_{LOAD}$. The measured R_{LOAD} values are unreliable within a domain in the R-C plane approximately defined by $R_{nominal}\, C_{nominal} < 10^{-4}$ s.

As discussed above, the system loses accuracy for smaller values of R_{LOAD}. In fact, when $R_{LOAD} \ll (R_{AIN} \mid\mid R_{pu}) + R_{out}$, then $R_T = (R_{AIN} \mid\mid R_{pu}) + R_{LOAD} + R_{out} \approx R_{pu} + R_{out}$, and the information about R_{LOAD} is lost. For this reason, the higher the R_{LOAD}, the better the accuracy. This observation is valid for both micro-controllers, except for the trials performed with a R_{LOAD} of 10 MΩ since $N_{A0} \rightarrow N_{max}$ at $t = 0$.

The measurements of the elapsed time (Δt), after $V_{A0} \geq V_C$, shown in the bottom graph of Figure 11b, exhibit a relative shift in the vertical axis due to the different R_{LOAD} values. Deviations from the straight line are consequence of the excess digitalization uncertainty. This aspect is evidenced after determination of the C_{LOAD} values using Equation (10) and shown in Figure 11d, where the deviations from the black dashed line ($C_{LOAD} = C_{nominal}$) relate to the excess digitalization uncertainty.

Overall, there is a trade-off between the R_{LOAD} and C_{LOAD} ranges for best accuracy. There are three extreme regimes: (case 1) R_{LOAD} is large (>1 MΩ), irrespective of the C_{LOAD} value: the accuracy of the measurements is good for the resistance and poor for the capacitance; (case 2) small R_{LOAD} (<100 kΩ) and large C_{LOAD} (>1 nF): measurements with poor accuracy for the resistance, good for the capacitance; (case 3) small R_{LOAD} (<100 kΩ) and small C_{LOAD} (<0.1 nF): measurements with very poor accuracy for the resistance, poor for the capacitance. Outside these extreme regimes, the accuracy is at least acceptable for R_{LOAD} and C_{LOAD} simultaneously.

3.5. *Characterization of Measurements for Parallel RC Networks*

The experimental procedures to perform the characterization of measurements for parallel RC networks were the same as those of the previous section.

For the parallel RC, the saturation voltage (V_F) lies below V_S, and it is necessary to determine both V_F and τ from the data. There are simple and computationally light algorithms allowing the identification of a stationary plateau, such as that occurring at V_F. These have been tested and verified, but including here the description of such methods would increase the length of this report. Thus, the subsequent analysis assumes that τ is already known. No generality is lost with this assumption.

Therefore, to each combination of a parallel RC network ($R_{LOAD} \mid\mid C_{LOAD}$) that was measured, the time constant τ was directly assumed as $R_T C_{LOAD}$. The total acquisition time was set to 10τ, V_F was read from V_{A0} at $t = 5\tau$ and V_C was read from V_{A0} at $t = \tau$.

Figure 12 aligns the results in two columns. The left column [(a) and (c)] refers to the measurements of R_{LOAD} values and the right one [(b) and (d)] refers to the measurements of C_{LOAD} values. The N_{A0} values at the instants $t = 5\tau$ and $t = \tau$ are shown in Figure 12a,b, respectively. Figure 12c,d show the R_{LOAD} and C_{LOAD} values, which were obtained from Equations (2) and (12), respectively.

The R_{LOAD} values are generally in line with the measurements carried on with isolated resistors shown in Figure 8b, but with few deviations to linearity (at 10 Ω and any C_{LOAD}, for the ATmega328P; at 1 MΩ and $C_{LOAD} > 1$ µF, for both micro-controllers).

As discussed before, in Section 3.4, a too large minimal acquisition time and/or the excess of digitalization noise are the reason the collected samples deviate from the theoretical lines represented by the horizontal black dashed lines. Similarly, the C_{LOAD} values are in line with the measurements carried on with isolated capacitors shown in Figure 9b. However, the accuracy of the measurements is variable, for the same reasons mentioned above.

Overall, there is a trade-off between the R_{LOAD} and C_{LOAD} ranges for best accuracy. There are two extreme regimes: (case 1) C_{LOAD} is small (<1 µF), and R_{LOAD} is large (<1 kΩ): the accuracy of the measurements is good for the resistance and poor for the capacitance; (case 2) large R_{LOAD} (>1 MΩ) and small C_{LOAD} (<1 µF): measurements with very poor accuracy for the capacitance, and poor for the resistance. Outside these extreme regimes, the accuracy is at least acceptable for R_{LOAD} and C_{LOAD} simultaneously.

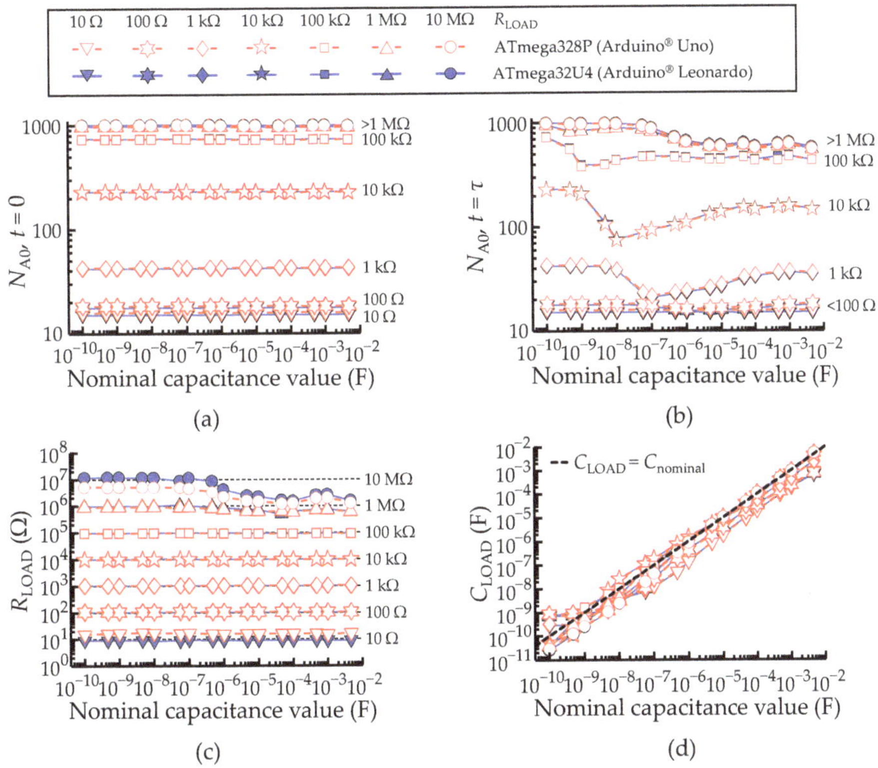

Figure 12. Comparison between the ATmega328P and ATmega32U4 AVR® micro-controllers configured to measure a parallel RC network. (**a**) Measured ADC value at $t = 5\tau$. (**b**) Measured ADC value at $t = \tau$. (**c**,**d**) Obtained R_{LOAD} and C_{LOAD} values according to Equations (2) and (12), respectively. A legend to describe the color scheme used in all plots of Figure 12 was included. The black dashed lines always represent the theoretical lines.

4. Discussion and Conclusions

This work described and characterized methods to accurately measure impedance using Arduino® boards with built-in AVR® micro-controllers. This is highly remarkable considering the ultra-low cost of the hardware. The measurement method allows the extraction of the resistance (R) and capacitance (C) values of either isolated or series and parallel configuration. Furthermore, inductance (L) measurements can be also performed, yet the range of measurable values is not very useful.

To check the cross-platform applicability of our proposed RCL-meter, two different AVR® micro-controllers assembled on Arduino® boards were selected, namely the ATmega328P assembled on an Arduino® Uno and the ATmega32U4 assembled on an Arduino® Leonardo. A benchmark was made to test the performance of micro-processors.

As for the measurements of isolated resistances and capacitances, the ATmega32U4 outperforms the ATmega328P, and some specific differences were identified. For instance, the ATmega32U4 was revealed to be more efficient in terms of acquisition time, providing a significant improvement when recording long-term transients. In the worst-case scenario, when recording an isolated capacitance (C_{LOAD} = 4.7 mF) through the transient acquisition mode, the ATmega32U4 performs two times faster than the ATmega328P. In addition, when measuring both C_{LOAD} and R_{LOAD} values of in-series or parallel RC networks, the ATmega32U4 takes a slight advantage over the ATmega328P for larger values of R_{LOAD}, while, conversely, the ATmega32U4 performs better for larger values of C_{LOAD}.

The noise profiles of the direct measurements (R-meter and fast C-meter) are essentially defined by the digitalization noise, while the transient C-meter brings an important extra noise source from the variability in the determination of instants of time required to perform the calculations.

The fast and transient C-meter methods complement each other, since the fast method is best for low capacitances and the transient method best for high capacitances. Operated together, they are able to deliver a relative accuracy equal to or better than 20% in the range 100 fF–10 mF, that is, across 11 orders of magnitude. Furthermore, the accuracy is equal to or better than 5% in the ranges 100 fF–100 pF and 100 nF–10 mF. The series and parallel RC combinations are also able to deliver good measurements of R and C in specific domains of the R-C plane.

Additional investigations were made to analyze the performance of both Arduino® boards supplied via a large power bank (Litionite Tanker 90 W/50,000 mAh) and using a data logger shield (RobotDyn™, Zhuhai, China) to store the measurements in a local micro-SD card. This work led to the conclusion that no substantial improvements are achieved by using a low-noise and low-uncertainty voltage supply, and therefore all presented data considers the typical noise and uncertainty from a common USB interface voltage supply source. However, the voltage supply unit determines the overall measurement quality in regions close to the ADC unit threshold values (N) for $N < 20$ or $N > 1000$.

Additionally, measurements of the same R_{LOAD} and C_{LOAD} values were made with commercial instruments and presented in the manuscript to provide insight on the overall performance of the ATmega328P and ATmega32U4 micro-controllers as a low-cost alternative to more expensive and sophisticated instruments.

Moreover, the concept proposed in this work, based on AVR® micro-controllers, may possibly be extended to other microcontrollers such as the STM32 family based on ARM architecture. The latter have more or improved integrated hardware features relative to the former, such as more flash memory, higher resolution ADC drive and faster clocks. On the other hand, using more sophisticated microcontrollers brings the disadvantage of idle but power-consuming internal hardware, for example, the digital-to-analog unit (DAC), which is not required in the present work. In any case, the inherent specificities of each architecture imply different implementations, difficult to cover in a single report.

The work carried out in the investigation of an ultra-low-cost RCL meter was mainly targeted towards impedimetric biosensor measurements, in order to facilitate the integration of the sensing and processing layers to the IoT. Its simplicity opens new possibilities for the improvement of ongoing and future projects in the field of smart sensing. The long-term goal of this work is to integrate the control of the sensing and processing layers into the Web of Things (WoT), which is an upper layer of interaction between devices that may be managed by artificial intelligence.

Supplementary Materials: The following supporting information can be downloaded at: https://www.mdpi.com/article/10.3390/s22062227/s1, C++ code A.1: High-level routine to measure isolated resistances; C++ code A.2: High-level routine to measure isolated capacitances through the fast acquisition mode; C++ code A.3: High-level routine to measure isolated capacitances through the transient acquisition mode; C++ code A.4: High-level routine to measure a series RC-network; C++ code A.5: High-level routine to measure a parallel RC-network; C++ code A.6: High-level routine to measure an isolated inductance; MATLAB code B.1: High-level routine for fitting the Ordinary Least Squares (OLS) function to the resistance values; MATLAB code B.2: High-level routine for fitting the Ordinary Least Squares (OLS) function to the capacitance values.

Author Contributions: Conceptualization, P.M.C.I. and P.S.; methodology, P.M.C.I. and P.S.; software, P.M.C.I.; validation, P.M.C.I.; formal analysis, P.M.C.I.; investigation, P.M.C.I. and R.G.; resources, R.G.; data curation, P.M.C.I.; writing—original draft preparation, P.M.C.I.; writing—review and editing, R.G. and P.S.; supervision, P.S.; project administration, R.G.; funding acquisition, R.G. All authors have read and agreed to the published version of the manuscript.

Funding: This research was funded by the FCT—Fundação para a Ciência e a Tecnologia, Portugal, through the CEOT project UIDB/00631/2020 CEOT BASE and UIDP/00631/2020 CEOT PROGRAMATICO. This work was also developed under the FEDER/Portugal2020 project NIBAP (ALG-01-0247-FEDER-037303).

Institutional Review Board Statement: Not applicable.

Informed Consent Statement: Not applicable.

Data Availability Statement: Not applicable.

Conflicts of Interest: The authors declare no conflict of interest.

References

1. Cima, M.J. Next-generation wearable electronics. *Nat. Biotechnol.* **2014**, *32*, 642–643. [CrossRef] [PubMed]
2. Stoppa, M.; Chiolerio, A. Wearable electronics and smart textiles: A critical review. *Sensors* **2014**, *14*, 11957–11992. [CrossRef] [PubMed]
3. Heo, J.S.; Eom, J.; Kim, Y.-H.; Park, S.K. Recent Progress of Textile-Based Wearable Electronics: A Comprehensive Review of Materials, Devices, and Applications. *Small* **2017**, *14*, 1–16. [CrossRef] [PubMed]
4. Joo, H.; Lee, Y.; Kim, J.; Yoo, J.-S.; Yoo, S.; Kim, S.; Arya, A.K.; Kim, S.; Choi, S.H.; Lu, N.; et al. Soft implantable drug delivery device integrated wirelessly with wearable devices to treat fatal seizures. *Sci. Adv.* **2021**, *7*, eabd4639. [CrossRef] [PubMed]
5. Matsukawa, R.; Miyamoto, A.; Yokota, T.; Someya, T. Skin Impedance Measurements with Nanomesh Electrodes for Monitoring Skin Hydration. *Adv. Healthc. Mater.* **2020**, *9*, e2001322. [CrossRef] [PubMed]
6. Chung, M.; Fortunato, G.; Radacsi, N. Wearable flexible sweat sensors for healthcare monitoring: A review. *J. R. Soc. Interface* **2019**, *16*, 20190217. [CrossRef] [PubMed]
7. Boutry, C.M.; Kaizawa, Y.; Schroeder, B.C.; Chortos, A.; Legrand, A.; Wang, Z.; Chang, J.; Fox, P.; Bao, Z. A stretchable and biodegradable strain and pressure sensor for orthopaedic application. *Nat. Electron.* **2018**, *1*, 314–321. [CrossRef]
8. Oren, S.; Ceylan, H.; Schnable, P.S.; Dong, L. High-Resolution Patterning and Transferring of Graphene-Based Nanomaterials onto Tape toward Roll-to-Roll Production of Tape-Based Wearable Sensors. *Adv. Mater. Technol.* **2017**, *2*, 1700223. [CrossRef]
9. Giraldo, J.P.; Wu, H.; Newkirk, G.M.; Kruss, S. Nanobiotechnology approaches for engineering smart plant sensors. *Nat. Nanotechnol.* **2019**, *14*, 541–553. [CrossRef] [PubMed]
10. Catini, A.; Papale, L.; Capuano, R.; Pasqualetti, V.; Di Giuseppe, D.; Brizzolara, S.; Tonutti, P.; Di Natale, C. Development of a sensor node for remote monitoring of plants. *Sensors* **2019**, *19*, 4865. [CrossRef] [PubMed]
11. Mizukami, Y.; Sawai, Y.; Yamaguchi, Y. Moisture content measurement of tea leaves by electrical impedance and capacitance. *Biosyst. Eng.* **2006**, *93*, 293–299. [CrossRef]
12. Li, M.Q.; Li, J.Y.; Wei, X.H.; Zhu, W.J. Early diagnosis and monitoring of nitrogen nutrition stress in tomato leaves using electrical impedance spectroscopy. *Int. J. Agric. Biol. Eng.* **2017**, *10*, 194–205. [CrossRef]
13. Diacci, C.; Abedi, T.; Lee, J.W.; Gabrielsson, E.O.; Berggren, M.; Simon, D.T.; Niittylä, T.; Stavrinidou, E. Diurnal in vivo xylem sap glucose and sucrose monitoring using implantable organic electrochemical transistor sensors. *iScience* **2020**, *24*, 101966. [CrossRef] [PubMed]
14. Lopez-Martin, A.J.; Carlosena, A. Sensor signal linearization techniques: A comparative analysis. In Proceedings of the 2013 IEEE 4th Latin American Symposium on Circuits and Systems, LASCAS 2013-Conference Proceedings, Cusco, Peru, 27 February–1 March 2013; IEEE: Piscataway, NJ, USA, 2013; pp. 1–4.
15. Kraska, M. Digital linearization and display of non-linear analog (sensor) signals. In Proceedings of the IEEE Workshop on Automotive Applications of Electronics, Dearborn, MI, USA, 19 October 1988; IEEE: Piscataway, NJ, USA, 1988; pp. 45–51. [CrossRef]
16. Islam, T.; Mukhopadhyay, S.C. Linearization of the sensors characteristics: A review. *Int. J. Smart Sens. Intell. Syst.* **2019**, *12*, 1–21. [CrossRef]
17. Alegria, F.; Moschitta, A.; Carbone, P.; Serra, A.D.C.; Petri, D. Effective ADC linearity testing using sinewaves. *IEEE Trans. Circuits Syst. I Regul. Pap.* **2005**, *52*, 1267–1275. [CrossRef]
18. Gong, Z.; Liu, Z.; Wang, Y.; Gupta, K.; da Silva, C.; Liu, T.; Zheng, Z.H.; Zhang, W.P.; van Lammeren, J.P.M.; Bergveld, H.J.; et al. IC for online EIS in automotive batteries and hybrid architecture for high-current perturbation in low-impedance cells. In Proceedings of the 2018 IEEE Applied Power Electronics Conference and Exposition (APEC), San Antonio, TX, USA, 4–8 March 2018; pp. 1922–1929. [CrossRef]
19. Manfredini, G.; Ria, A.; Bruschi, P.; Gerevini, L.; Vitelli, M.; Molinara, M.; Piotto, M. An ASIC-Based Miniaturized System for Online Multi-Measurand Monitoring of Lithium-Ion Batteries. *Batteries* **2021**, *7*, 45. [CrossRef]
20. Crescentini, M.; De Angelis, A.; Ramilli, R.; De Angelis, G.; Tartagni, M.; Moschitta, A.; Traverso, P.A.; Carbone, P. Online EIS and Diagnostics on Lithium-Ion Batteries by Means of Low-Power Integrated Sensing and Parametric Modeling. *IEEE Trans. Instrum. Meas.* **2020**, *70*, 1–11. [CrossRef]
21. Monteiro, F.D.R.; Stallinga, P. Using an Off-the-Shelf Lock-In Detector for Admittance Spectroscopy in the Study of Plants. *Agric. Sci.* **2020**, *11*, 390–416. [CrossRef]

22. Grassini, S.; Corbellini, S.; Angelini, E.; Ferraris, F.; Parvis, M. Low-cost impedance spectroscopy system based on a logarithmic amplifier. *IEEE Trans. Instrum. Meas.* **2014**, *64*, 1110–1117. [CrossRef]
23. Grassini, S.; Corbellini, S.; Parvis, M.; Angelini, E.P.M.V.; Zucchi, F. A simple Arduino-based EIS system for in situ corrosion monitoring of metallic works of art. *Measurement* **2018**, *114*, 508–514. [CrossRef]
24. Campbell, S. How to Make and Arduino Capacitance Meter. Available online: https://www.circuitbasics.com/how-to-make-an-arduino-capacitance-meter/ (accessed on 17 March 2021).
25. Viraktamath, S.V.; Kannur, K.; Kinagi, B.; Shinde, V.R. Digital LCR meter using arduino. In Proceedings of the 2017 International Conference on Computing Methodologies and Communication (ICCMC), Erode, India, 18-19 July 2017; pp. 805–808. [CrossRef]
26. MICROCHIP. *8-Bit AVR Microcontroller with 32K Bytes In-System Programmable Flash, ATmega328P DATASHEET, Rev.: 7810D–AVR–01/15*; Microchip Technology Inc.: Chandler, AZ, USA, 2015.
27. MICROCHIP. *8-Bit Microcontroller with 16/32K Bytes of ISP Flash and USB Controller, ATmega16U4/ATmega32U4 DATASHEET, Rev.: Atmel-7766I-USB-ATmega16U4-32U4 07/2015*; Microchip Technology Inc.: Chandler, AZ, USA, 2015.

sensors

MDPI

Article

Blockchain Based Authentication and Cluster Head Selection Using DDR-LEACH in Internet of Sensor Things

Sana Amjad [1], Shahid Abbas [1], Zain Abubaker [1], Mohammed H. Alsharif [2], Abu Jahid [3], Nadeem Javaid [1,4,*]

[1] Department of Computer Science, COMSATS University Islamabad, Islamabad 44000, Pakistan; sanaamjad702@gmail.com (S.A.); shahidabbas1260@gmail.com (S.A.); zainmalik.gcuf@gmail.com (Z.A.)
[2] Department of Electrical Engineering, Sejong University, Seoul 05006, Korea; malsharif@sejong.ac.kr
[3] School of Electrical Engineering and Computer Science, University of Ottawa, Ottawa, ON K1N 6N5, Canada; ajahi011@uottawa.ca
[4] School of Computer Science, University of Technology Sydney, Ultimo, NSW 2007, Australia
* Correspondence: nadeemjavaidqau@gmail.com

Abstract: This paper proposes a blockchain-based node authentication model for the Internet of sensor things (IoST). The nodes in the network are authenticated based on their credentials to make the network free from malicious nodes. In IoST, sensor nodes gather the information from the environment and send it to the cluster heads (CHs) for additional processing. CHs aggregate the sensed information. Therefore, their energy rapidly depletes due to extra workload. To solve this issue, we proposed distance, degree, and residual energy-based low-energy adaptive clustering hierarchy (DDR-LEACH) protocol. DDR-LEACH is used to replace CHs with the ordinary nodes based on maximum residual energy, degree, and minimum distance from BS. Furthermore, storing a huge amount of data in the blockchain is very costly. To tackle this issue, an external data storage, named as interplanetary file system (IPFS), is used. Furthermore, for ensuring data security in IPFS, AES 128-bit is used, which performs better than the existing encryption schemes. Moreover, a huge computational cost is required using a proof of work consensus mechanism to validate transactions. To solve this issue, proof of authority (PoA) consensus mechanism is used in the proposed model. The simulation results are carried out, which show the efficiency and effectiveness of the proposed system model. The DDR-LEACH is compared with LEACH and the simulation results show that DDR-LEACH outperforms LEACH in terms of energy consumption, throughput, and improvement in network lifetime with CH selection mechanism. Moreover, transaction cost is computed, which is reduced by PoA during data storage on IPFS and service provisioning. Furthermore, the time is calculated in the comparison of AES 128-bit scheme with existing scheme. The formal security analysis is performed to check the effectiveness of smart contract against attacks. Additionally, two different attacks, MITM and Sybil, are induced in our system to show our system model's resilience against cyber attacks.

Keywords: blockchain; clustering; authentication; malicious node detection; LEACH protocol; service provisioning; interplanetary file system; security

Citation: Amjad, S.; Abbas, S.; Abubaker, Z.; Alsharif, M.H.; Jahid, A.; Javaid, N. Blockchain Based Authentication and Cluster Head Selection Using DDR-LEACH in Internet of Sensor Things. *Sensors* **2022**, *22*, 1972. https://doi.org/10.3390/s22051972

Academic Editor: Alexandru Lavric

Received: 25 December 2021
Accepted: 22 February 2022
Published: 2 March 2022

1. Introduction

The wireless sensors networks (WSNs) play an important part in the Internet of sensors things (IoST) [1]. IoST is useful in sensing data from the environment and is used in the field of energy trading, surveillance, smart grids, etc., [2,3]. It connects with the Internet and automates the monitoring system without any involvement from a third party. The IoST network consists of sensor nodes that perform environmental monitoring [4]. However, the sensor nodes in the WSNs face the issue of non-repudiation, limited resources, presence of malicious nodes, etc., [5–7]. Many studies are proposed to solve these aforementioned issues [8–10]. However, these studies have issues of single point of failure (SPOF) and performance bottlenecks due to their centralized architecture.

To overcome these aforementioned issues, many researchers provide different mechanisms to remove third parties by introducing blockchain in the WSNs. Blockchain is a secure and decentralized protocol that solves many issues such as SPOF, a third party involvement, etc., [11]. Moreover, the distributed and tamper-proof ledger in blockchain solves trust issues between unknown entities. The transactions that are performed by entities in the network are confirmed by the miners. These transactions are validated by the miners using various consensus mechanisms, such as proof of work (PoW) [12], proof of authority (PoA), proof of stake, etc., [13]. In the PoW mechanism, all the nodes participate in solving the mathematical puzzle. The node that solves it first validates the transactions. The blockchain is created by validating and storing the transactions. Moreover, a smart contract is used in blockchain in which all the terms and conditions are finalized. Additionally, it eliminates the third party. Moreover, blockchain provides security in the network by malicious nodes' detection through Merkle tree [14,15] and also through different techniques, such as trust evaluation of nodes, etc., [16–18].

Many blockchain-based schemes are proposed to solve the issues of single point of failure, huge monetary cost, and performance bottlenecks [19–22]. However, the data of all these networks are stored on blockchain, which is very costly. When 1MB of data are stored on blockchain, it costs USD 14151.68 [23]. Moreover, PoW consensus algorithm is used in [21,22], which is not suitable for resource constrained environment.

In IoST networks, routing is an important aspect in which nodes communicate and transmit data from source to the destination. The transmitted data are controlled by different nodes in IoST that are sensor nodes, CHs and base stations (BSs). In [19], the data are processed by CHs and are forwarded to the BSs. However, authentication of network nodes is not performed. Therefore, any node can enter the network and behave maliciously. Moreover, in [19], no cost-effective data storage mechanism is proposed, which leads to expensive data storage in blockchain. As data are permanently stored on blockchain then the issues of limited storage arises. Additionally, in an IoST network, CHs fail due to high energy depletion, which affects the whole network's performance. In [21], no mechanism is proposed for the selection of new CHs. Furthermore, in PoW, the miners solve the puzzle for validating the transactions and adding the blocks into the blockchain [17,22]. This mining process takes considerable time due to puzzle's complexity, which ultimately increases network's computational cost. This paper is the extension of [24] and the contributions in the proposed work are as follows:

- The identity authentication of nodes is performed to remove the external unauthenticated nodes;
- CHs are selected from ordinary nodes using the proposed minimum distance, highest degree, and highest residual energy (DDR) based LEACH protocol;
- IPFS is used to provide distributed storage for IoST;
- A payment method is proposed to motivate IPFS for long term data storage;
- A blockchain based secure service provisioning mechanism is proposed;
- An advanced symmetric encryption algorithm (AES) 128-bit is used for the integrity of data;
- Comparison of DDR-LEACH is performed with the LEACH protocol;
- Formal security analysis is performed for the smart contract to check its effectiveness;
- Man in the middle (MITM) and Sybil attacks are induced in the network, which show that our proposed system is resilient against these attacks.

The rest of the paper is organized as follows. The related work is discussed in Section 2 and Table 1. The proposed system model is presented in Section 3. The simulation results and the validation of system model are discussed in Section 4. The formal security analysis is discussed in Section 5. Section 6 contains the conclusion of the proposed work and future work.

Table 1. Related work.

Limitations Already Addressed	Contributions Already Provided	Validations Already Done	Limitations to be Addressed
Data security and data privacy, huge energy consumption of resources, low computation power of resources, nodes' authentication, trust issue [13]	Decentralized blockchain, public key infrastructure for resolving trust issue, nodes' authentication	Reputation level	Authors will evaluate all parts of authentication process
Malicious nodes' detection, malicious nodes' traceability [17]	Trust mechanism, consortium blockchain	Sensor nodes' data input and output parameters, credit of sensors	PoW uses more computational power, no reward for sink nodes
Mobile nodes' management, data protection [19]	Uncertainty principle, Voronoi cell architecture, Blockchain	Network lifetime, energy consumption, average end-to-end delay, packet delivery ratio	No storage mechanism, no registration and authentication
No encryption and certificate scheme, nodes' authentication [20]	Blockchain, SHA 64-bit algorithm, crypto based authentication	Security analysis	Node battery issue, storage issue
Node authentication, security issue, centralized system [21]	Hybrid structure is performed, Keccak hash function, consortium blockchain	Security analysis	PoA should be used for each validation and private blockchain
Data latency, limited data bandwidth, data security [22]	Blockchain based SDN, PoA, Argan2	Transactions per second, average time per block, latency	PoW consumes more computational power
Trust issue, central authority, gray hole and black hole attacks in an untrusted network [25]	Blockchain based routing protocol for route establishment, reward to minimize selfish behavior	Route overhead, packet delivery ratio, gray hole attack, black hole attack	Proposed solution must be used for ad-hoc network
Data privacy, untrusted nodes [26]	Decentralized blockchain based authentication scheme	Energy consumption	N/A
PoW takes more computational power [27]	Blockchain incentive mechanism, SHA-256	Pairing is performed by the hyper elliptic curve for the finite field	Proof of retrievability is used for recovering data in less time
Computationally extensive PoW-based mining [28]	Computation offloading mechanism	Net revenue of computing, average delay	Try different consensus mechanisms
Single point of failure, data storage [29]	Block offloading filter, blockchain	Comparison of PoW and synergistic multiple proof	N/A
Data storage, slow information validation in blockchain [30]	Blockchain distributed ledger, Tangle based technology to minimize computational time	Age of information vs sampling interval, processing power vs sampling interval	N/A
Data transparency [31]	Decentralized blockchain	Probability of attack detection by system, falsification attack, authentication delay and probabilistic scenario	No routing path is defined in order to reach the manager
No data privacy protection [32]	Blockchain-based privacy protection mechanism, double SHA-256	Data about noise	Scaled experimental data will be collected for better and complete judgment, algorithm will be improved for better result
Data privacy and data security [33]	Information centric network, public key cryptographic scheme, two-tier structure, SHA-1	Processing time, response time	Scheme should be used as practical implementation
Localization, network security [34]	Decentralized blockchain-based trust management model	Energy consumption, localization error, average error ratio	Dynamic behavior of nodes
Nonrepudiation [35]	Nonrepudiation mechanism, homomorphic hash function	Transaction latency, throughput, gas consumption	No user authentication, double spending
Malicious nodes' detection, data security [36]	Trust aware routing algorithm	Time complexity, throughput	No authentication mechanism

2. Related Work

The studies related to the blockchain integrated with WSNs are discussed in this section. The studies are categorized based on the limitations they have addressed.

2.1. Nodes' Authentication

The sensor nodes perform an important role in performing many tasks in IoT networks and nodes' identities authentication is one of them. The nodes in the IoT network work together and provide the services to the buyers. However, the nodes' identities authentication is not performed in [13,21], which leads to malicious nodes becoming the part of the network and affecting its performance.

Authentication is required to restrict unauthenticated nodes from entering the network. The unauthenticated nodes behave maliciously by tampering data during routing, as well as refusing to forward data packets toward the destination. Therefore, in [20], the authors propose a lightweight authentication mechanism for WSNs. The sensor nodes use unique sequence numbers during data transmission based on the concept of a Merkle tree. Secure hashing algorithm 1 (SHA-1) is used to authenticate the nodes. Although in [25], the network nodes are authenticated using routing protocol. However, due to centralized authority, a trust issue is created. In [26], the wireless body area network is comprised of sensor nodes, which collect the information of human body parts and publicly forward it to the local node. Health is a very critical and sensitive matter, and the malicious nodes can enter in the network and misuse the data. However, these nodes are not authenticated in the network.

2.2. Lack of Data Storage

The authors in [27] propose an incentive based data storage mechanism. Each node stores data in the blockchain; however, computational cost increases due to the usage of PoW consensus mechanism. Although, in [29], mining is performed using PoW. The PoW increases the computational cost. Therefore, a lightweight blockchain network is proposed to reduce blockchain storage and computational requirements in IoST environment. The blockchain is merged with IoTs in [30] to aggregate the blocks' header information and transmit it to the IoT nodes. However, keeping a copy of data in a resource constrained network is not appropriate.

2.3. Lack of Data Privacy

In [19], no technique is proposed to prevent the network data from being stolen by malicious nodes. Additionally, in [31], the products are controlled and monitored by workers in the industry. However, the issue of data transparency is created. The important information of the products may be stolen by the workers. In addition, the misuse of important products' record is also possible. Additionally, in the WSNs, the security of data and its privacy is compromised [13]. In [32], the authors state that collecting the information from crowdsensing is essential for the network nodes. However, privacy protection of the data is not considered.

Whereas, in [33], the data-driven network is converged with WSN and the reserving information is copied for sharing it in the network. However, the security of data is being compromised by the malicious nodes. Moreover, in [22], the growth of IoT in the smart city creates data latency, scalability, and huge bandwidth issues. Therefore, hybrid blockchain network is proposed and SDN controllers are used as an interface between the IoT. Additionally, digital signatures are used for the data security in the network. The sensor nodes transmit the data to IoT nodes in [37]; however, no mechanism for data security is proposed, which causes data security issue.

2.4. Lack of Resources

In [27], the chances of malicious nodes' existence are high in the network, which do not allow the legitimate nodes to participate in the network. Additionally, blockchain technology is used for different purposes, such as content caching in [28]. Moreover, in [27,28], the authors use PoW consensus mechanism, in which high computational power is required. The blockchain is integrated with IoT for secure routing in [30]. However, the nodes in the network have very low storage capability and these resource constrained nodes cannot keep the copied records in them. Whereas, in [38], PoW is replaced with Tangle based technology to provide fast and secure information. However, the frequency of transaction is very low. Moreover, the IoT sensor nodes' energy depletes very fast due to the high computational overhead. Low computational power of IoT nodes hinders the validation of transactions.

2.5. Malicious Nodes' Existence

Obtaining the exact location of sensor nodes in the network is an emerging domain nowadays. However, malicious behavior of sensor nodes lead to the broadcast of wrong location information. Due to this, the security of the network is compromised [34]. Additionally, in [17], no mechanism is proposed to detect the malicious nodes in the network. Different fields, such as manufacturing products and healthcare use IoT [35]. However, still some challenges are faced by provisioning process, such as provision of malicious services. Furthermore, the client can behave maliciously by repudiating on behalf of services. In [36], the sensor nodes find the shortest path for communication. However, there is no mechanism to secure the data and to find the malicious nodes.

2.6. Single Point of Failure Issue

The network performance is affected when the identity authentication of nodes is compromised. In [21], the nodes' identity depends on central authority servers that become the reason for SPOF. Whereas, in [39], the authors integrate the software-defined networking and blockchain to detect attacks without any involvement of a third party. The data are sent directly to the centralized cloud. However, bandwidth and latency issues arise. The smart contracts in the blockchain system share the trained classifiers with the cloud layer for fusion. However, SPOF issue arises due to central authority. In [25], a centralized authority is used to authenticate the routing nodes. Due to the central authority used in the system, SPOF issue arises. In [40], the data are stored in the network by a centralized system, which leads to SPOF issue.

Table 2 presents problems identified in existing literature, their proposed solutions and their validations. An authentication scheme is proposed to prevent from the unauthenticated nodes, so that only authenticated nodes are allowed to perform an action. In DDR-LEACH protocol, motivated from [41], the highest degree node is selected, which solves the node battery issue. Moreover, IPFS is used to solve the costly data storage issue of the blockchain. IPFS stores the data cost-effectively and distributively. Additionally, the issue of high computational cost is resolved using PoA consensus mechanism.

Table 2. Mapping between limitations, solutions, and validations.

Limitations Identified	Solutions Proposed	Validations Done
L1. Nodes are not authenticated [19]. L2. No mechanism for malicious nodes' detection [19]	S1. Authentication mechanism	V1. Message size, as shown in Figure 3 V2. Transaction cost, as depicted in Figure 4
L3. Inefficient energy consumption [21]	S2. CHs' selection considering nodes' residual energy, minimum distance from BS and degree	V3. Energy consumption, as depicted in Figure 5 V4. Throughput, as shown in Figure 6 V5. Network lifetime, as shown in Figure 7
L4. High computational cost [17,22]	S3. PoA	V6. Average transaction cost, as shown in Figure 8
L5. Costly data storage [19]	S4. IPFS	V7. Average transaction cost, as shown in Figure 9 V8. Encryption time, as depicted in Figure 10

3. Proposed System Model

In this section, the assumptions, components and work flow of the proposed system model are discussed.

3.1. Assumptions

The system model is based on following assumptions:

- BSs are considered legitimate. As they are peers of blockchain; therefore, they provide secure services to buyers;
- Symmetric keys are exchanged securely in the network.

3.2. System Components

In this section, the components of the proposed system model are discussed as depicted in Figure 1. The components include IoST, buyers, IPFS, and blockchain.

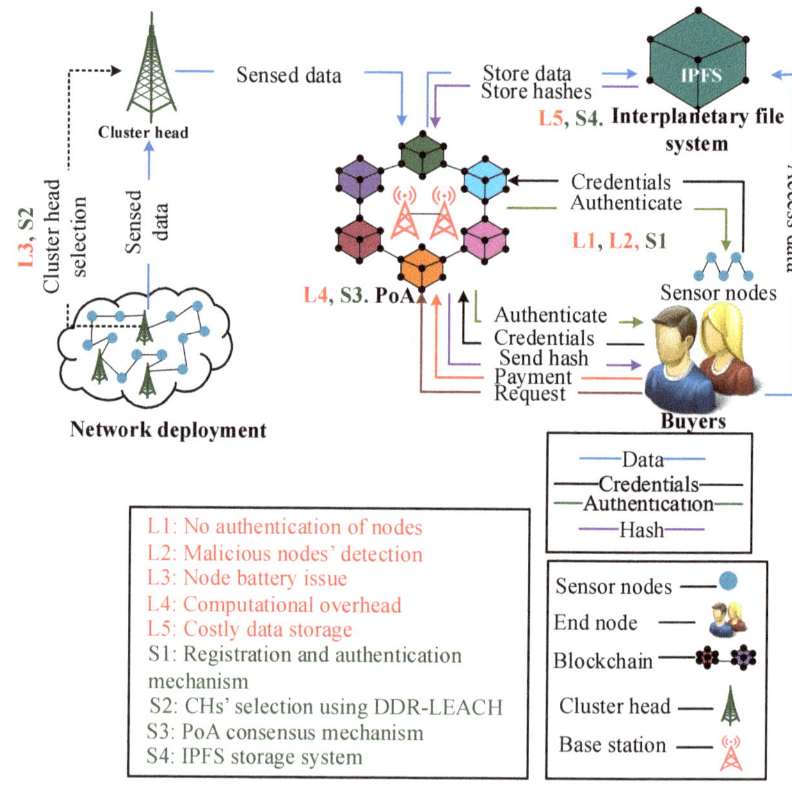

Figure 1. Blockchain based nodes' authentication and CHs' selection in IoST.

Internet of sensor things: The IoST is an emerging technology, which consists of sensor nodes deployed for collecting the environmental data [32]. The sensor nodes sense the surrounding information such as the data of humidity, pressure, and temperature, etc., [42]. In the proposed system model, the IoST consists of sensor nodes, CHs, and BSs. Their working is described in Section 3.3.

Buyers: To prevent the network from malicious activities, the buyes are registered and authenticated in the blockchain network. For that purpose, registration and authentication schemes are used, motivated from [21].

Interplanetary file system: It is a distributed platform where data are stored in the form of chunks. Whenever the data are stored on IPFS, a hash is generated. IPFS generates the 32-bit hash in result of data storage, which is stored on the blockchain as a record.

3.3. *Workflow of the System Model*

The system model is discussed in the steps given below.

Step 1. Initialization: The blockchain technology introduces a smart contract, which is a digital agreement that works without the involvement of any third party. It is deployed on BSs that handle the network transactions. The blockchain is used for registering sensor nodes in the network by storing their credentials for authentication. Credentials are sent in the form of a message shown in following equation.

$$(ID_{Node}, MACAddr_{Node}, Reputation_{Node})_{Packet} = Message \tag{1}$$

In the registration and authentication process, MAC address, ID, and Reputation of nodes are used as credentials. $Reputation_{Node}$ is the reputation value given to a specific node on the basis of its previous history of interaction with the network. If the node provides accurate data to the network, its reputation increases; otherwise, it decreases. The credentials are stored in the blockchain using a asymmetric scheme. The blockchain also keeps their addresses to prevent the network free from malicious activities. Therefore, during authentication process, the credentials are matched with already stored data. If the credentials are not matched with already stored data, then the node is considered as a malicious node.

Step 2. CH selection: The IoST network consists of sensor nodes, CHs, and BSs. The sensor nodes collect the data from the environment and send it to CHs for performing computation. CHs receive data from sensor nodes, process the data, and send it to BSs. BS requests IPFS for storing data; IPFS calculates hash value of the data and sends it to BS. In processing and storing the data, the energy of CHs depletes rapidly. To solve this issue, our proposed model provides a mechanism that selects CHs from ordinary nodes using DDR-LEACH. The CHs are selected based on three parameters: residual energy, minimum distance from BSs, and maximum degree of a node. If a node satisfies the criteria mentioned earlier, then it is selected as CH. CH aggregates the data packet and transmits it to the destination, as given in Algorithm 1. If more than one nodes meet the criteria, then CH is randomly selected.

Algorithm 1: CH selection.

```
Inputs: Deployment of N nodes, BSs and CHs
Outputs: CHselection
/*Er is the residual energy of a node
   Select maximum energy node from set N
   Select the node having minimum distance from BS
   Select maximum degree node from set N
for i=1:1:N do
  | Select (Max (S(i).Er), Max (S(i).Degree) and Min (S(i).Distance))
  | if New selected CH = Max(Degree, Er) and Min(Distance) then
  |   | Selected node is CH
  | end
  | Check next node
end
```

Step 3. Nodes′ authentication: In this step, sensor nodes are authenticated in the blockchain. The sensor nodes' authentication is important to revoke malicious nodes from the network and provide the secure services. The messages are exchanged between the WSN nodes and the distributed ledger, i.e., blockchain, after being encrypted using asymmetric encryption. The encryption is performed over the registration data for providing more security. The node encrypts the data with the public key of the BS (known to everyone in the network) and then BS decrypts the encrypted data with its private key (known to BS only) and stores data in the blockchain. Moreover, the solution for providing the keys for encryption and decryption remains the same because the keys can neither be invalidated nor can be changed. Additionally, whenever any node performs any activity in the network, its credentials are verified by the BS. Therefore, this solution is sufficient to provide the security to the overall network. There are many papers that are based on nodes' authentication, i.e., [43,44]. As in [43], the nodes' identity authentication is performed using an encryption scheme. However, due to centralized authority, the SPOF and performance bottleneck issues arise. Although, in [44], a lightweight authentication scheme is used. However, due to centralized authority, the SPOF issue occurs. Few authentication-based papers are mentioned in the related work, i.e., [20,21]. As in [20], authentication of nodes is performed by their acknowledgment to the sink node. The nodes acknowledge the sink node based on their provided sequence numbers. However, during nodes' authentication, the PoW consensus mechanism is used, which incurs high computational cost. Although, in [21], hybrid blockchain-based nodes' authentication is performed. However, PoA is used, which incurs high computational cost. Moreover, data storage in blockchain is very costly. In our proposed system model, we have used nodes' identity authentication using a PoA consensus mechanism, which reduces computational cost. We do not claim that this authentication model is the novel work. We have embedded this authentication model with an efficient CH selection and secure storage of sensors' data. This integration of authentication model is a novel combination, as authentication mechanism is used in different scenarios [20,21,44]. Moreover, a distributed IPFS data storage mechanism is used to reduce overall monetary cost of the network. The authentication is performed by matching the credentials of nodes that are already stored in the blockchain.

When a node is taken over by a malicious node, then malicious node can transmit these data in the network to perform malicious activities. As we know, all legitimate nodes are registered with BSs and their credentials are already stored in the blockchain. When any malicious node transmits the data, it can be easily detected in authentication process because its credentials are not stored in the blockchain. In this way, the malicious node cannot perform malicious activities in our network by taking over any node. Moreover, when a malicious node is removed from the network, it is revoked from transmitting any kind of data in the network. The malicious node is removed and revoked by BSs because blockchain is deployed on them and they are responsible for validating the transactions. In the proposed model, PoA is used as the consensus mechanism for mining the transactions. It incurs less computational cost as compared to the traditional PoW consensus mechanism. Additionally, the miners are pre-selected nodes in PoA. The registered nodes are authenticated after the authentication request. The request contains ID_{Node}, $MACAddr_{Node}$, and $Reputation_{Node}$, which are already stored on the blockchain. Blockchain checks whether the credentials provided by nodes are matched with the credentials already stored or not in the blockchain. If the credentials match the provided information, the nodes become authentic and are broadcasted as legitimate nodes. Otherwise, they are broadcasted as malicious or unauthentic nodes. The step-wise process is according to Algorithm 2.

Step 4. Data storage: Storing the large amount of data on the blockchain is not suitable because the storage cost is high on the blockchain. The cost of storing 1 MB data on the blockchain is approximately USD 14151.68 [23]. Therefore, BSs send data to IPFS and keep its hash values and the credential of registered nodes in the blockchain. In the blockchain, the transaction of each entity is stored in a ledger. This ledger is distributed to all entities of the network. All entities act as a foundation of the network. When any transaction is performed, the ledgers of all entities are updated simultaneously. When malicious node tries to manipulate the transaction record in a ledger of any entity, then it can easily be detected because this ledger is not matched with already stored ledger. The data are stored on IPFS in the form of chunks. IPFS does not store the data for long time. Therefore, we propose a payment method for long time data storage of IPFS. The IPFS is incentivized in order to motivate the peer nodes for storing the data. The hashes are stored on the blockchain. Only authenticated nodes obtain data using the provided hash. However, the data storage on IPFS is temporary. Therefore, we propose an incentive method for IPFS to store a huge amount of data for a long time.

Algorithm 2: Nodes' authentication process.

Inputs: $ID_{Node}, MACAddr_{Node}, Reputation_{Node}$
Outputs: *Nodes authentication message*
  Registration← Input ($ID_{Node}, MACAddr_{Node}, Reputation_{Node}$)
  Authentication
if ID_{Node} *and* $MACAddr_{Node}, Reputation_{Node}$ *are stored in blockchain* **then**
  The Node is authenticated;
  if ID_{Node} *and* $MACAddr_{Node}, Reputation_{Node}$ *are not stored in blockchain* **then**
    The node is unauthenticated
  end
  Recommend for registration
end

Step 5. Service provisioning: We use a private blockchain, deployed on BSs, to register and authenticate the nodes. In the beginning, the blockchain receives the registration request from the buyers. Then, the nodes are checked by the blockchain whether they are already registered or not. If a buyer is already registered, it discards the request. Otherwise, it allows the node for registration. Algorithm 2 shows how the authentication process works.

The already stored credentials in the blockchain are used to authenticate the buyers. A buyer must first provide its credentials for verification in the blockchain. The blockchain checks the credentials to confirm whether the node's credentials exist in the blockchain or not. If the provided credentials match with the stored one, it is considered as an authentic user; otherwise, it is considered a malicious node, which is immediately removed from the network. Afterward, if the buyer is authenticated, then the ethers are checked according to the threshold. If ethers are enough to buy the data, the hash of the requested data is sent to the buyer; otherwise, request will be rejected. After the authentication process, the buyer receives services from the network. Whenever a buyer requests the service, BS encrypts the service with buyer's secret key and sends the cypher text to the buyer.

The buyer receives the encrypted data and then decrypts it with the private Figure 2.

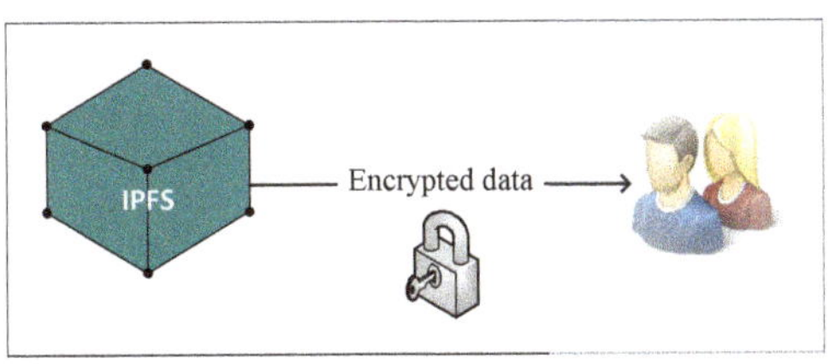

Figure 2. Interaction of buyers with IPFS.

We use AES 128-bit encryption for ensuring data security in the network. Moreover, SHA-256 is used with AES encryption to ensure data integrity. Initially, the BS calculates the hash by the SHA-256 hashing algorithm and then uploads this hash on the blockchain. After this, the BS encrypts the data with a secret key and sends these encrypted data to a client. The buyer receives the data and decrypts it with the secret key provided by the sender. After decryption of data, the buyer calculates the hash by itself using SHA-256 algorithm and compares this hash with the hash already stored in blockchain. In this way, the SHA-256 hashing technique together with the AES encryption technique ensures data security and data integrity. Moreover, we have used the AES 128-bit encryption technique in our scenario because our primary goal is to provide real time data to buyers and AES 128-bit encryption consumes a very small amount of time in encryption and decryption of data. Furthermore, the efficiency of the AES 128-bit encryption scheme in terms of time is shown in Figure 10 in the simulation section.

In the proposed system model, 5% of the buying amount is given to IPFS as an incentive. The service provisioning mechanism is shown in Algorithm 3.

Algorithm 3: Service provisioning.

```
Inputs: senseddata, dataindex, buyeraddr
Outputs: Buyandsellservices
senseddata temperature, pressure, humidity;
  Function datainIPFS Input(dataindex, senseddata)
  IPFS generates hash
  Store blockchain←hash
function Getservices Input(dataindex, buyerAddr)
if buyingamount is less than threshold then
  | "low Amount"
else
  if dataindex = index.temperature then
    PushRequestvalue in temperatureArray
    buydata=temperature[temperatureindex]x5%
    sendpaymenttoIPFS
    buyingamounttemperatureindex − payIPFS
  else
    if dataindex = index.pressure then
      PushRequestvalue in pressureArray
      buydata=pressure [pressureindex]x5%
      sendpaymenttoIPFS
      buyingamountpressureindex − payIPFS
    end
    if dataindex = index.humidity then
      PushRequestvalue in humidityArray
      buydata= humidity[humidityindex]x5%
      sendpaymenttoIPFS
      buyingamounthumidityindex − payIPFS
    end
  end
end
```

In Table 2, the mapping between limitations, proposed solutions and their validations is provided. The limitations L1 and L2 represent the unauthenticated and the malicious activities of nodes in the network, which are solved by S1. The parameter that is used to validate the nodes' authentication is message size, which is evaluated by transaction cost. The message size indicates that how many bytes it takes to authenticate a node while the transaction cost means the cost required for nodes' authentication. The third limitation L3 is the inefficient energy consumption of CHs. In S2, CHs are selected based on nodes' residual energy, minimum distance from BS and degree. If these conditions are satisfied for a node, then it becomes a CH. The network lifetime, packet delivery ratio and energy consumption are used as validation parameters. L4 indicates the high computational cost when using PoW consensus mechanism. In S3, PoA consensus mechanism is used to solve this issue, which consumes less computational resources as compared to PoW. The cost incurred when using PoA is depicted by the average transaction cost. The limitation L5 shows that the data storage on the blockchain is very costly. Therefore, a large amount of data are stored in the IPFS in S4. The average transaction cost and encryption time are taken as validation parameters. The average transaction cost shows the cost used to store data on

IPFS. At the same time, encryption time shows the time, which is used in encryption and decryption of data during service provisioning.

4. Simulation Results and Discussion

In this section, the performance evaluation of the proposed model is described. The specifications for the simulation setup include an Intel(R) Core (TM) i5-2520M CPU @ 2.50 GHz processor, 64-bit operating system, x64-based processor, and 8 GB RAM. Remix IDE is used to develop a smart contract, whereas, Ganache is used to manage the transactions. Although MetaMask is used for providing virtual currency to perform transactions. Moreover, for simulating our WSN, MATLAB is used. The use of a cryptographic scheme is simulated in Visual Studio using Python.

In the network, three types of nodes are considered for the simulations: 100 sensor nodes, 4 CHs, and 2 BSs in the network area of 100×100 m^2. Moreover, there are three input data in our system model: sensed data, data index, and buyer's address. The data are sensed by sensor nodes in the network. They send the sensed data to CHs for processing. The CHs after processing the data send it to BSs for further processing. The BS stores the data with data index. The sensed data with data index are provided by BSs. Furthermore, the buyer's address is the Ethereum address that is provided to each buyer when it joins the network. Initially, the nodes are registered and authenticated to provide malicious nodes free network. The simulation parameters are mentioned in Table 3.

Table 3. Simulation parameters.

Parameters	Value of Parameters
Sensing area	100×100 m^2
Deployment	Random
Total nodes	100
CHs	4
BSs	2
Network interface	Wireless

Figure 3 illustrates the message size of network nodes. During the registration phase, unknown nodes request to become a part of the network. For the registration process, a new node first registered in the blockchain network before its participation. The nodes send their credentials to the blockchain. BS stores its credentials, and perform authentication process. The message size is large during the registration phase as compared to authentication phase because in registration phases, more resources are required to store the data on the blockchain. Whereas, in authentication phase, BS only matches the nodes' credentials with the already stored credentials, which incurs less resources.

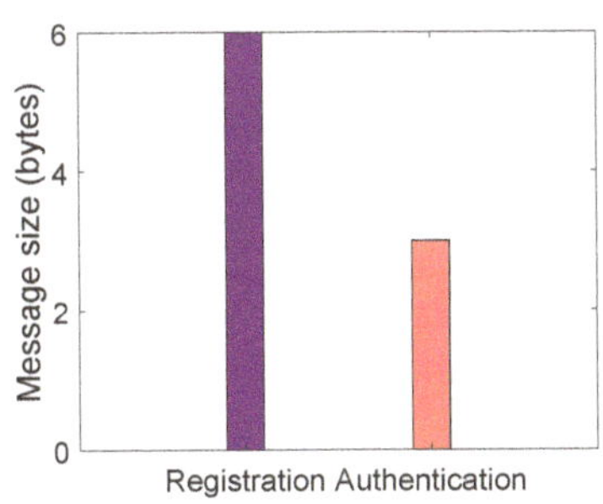

Figure 3. Message size.

Figure 4 shows the transaction cost incurred during registration and authentication phases. When smart contracts are deployed, then transaction cost is incurred. The results show that the transaction cost increases with the increase in the number of packets. Similar behavior is observed for authentication and registration phases in Figure 4 as that of Figure 3. The transaction cost increases with the increased number of nodes. The authentication process takes less transaction cost because already stored registration information has to be verified.

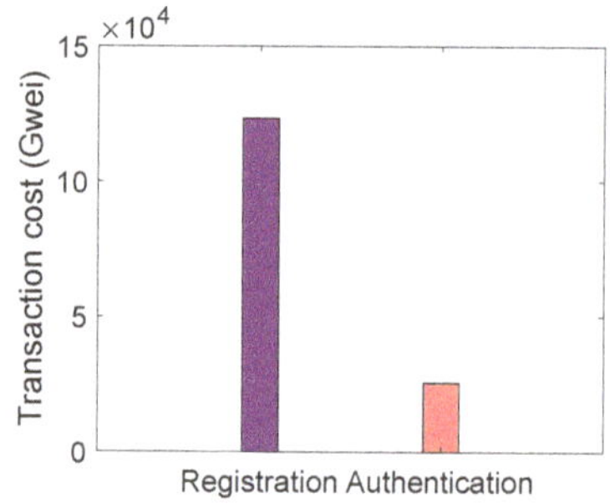

Figure 4. Transaction cost during registration and authentication of nodes.

Figure 5 shows the energy consumption against the number of rounds. The comparison of LEACH protocol with DDR-LEACH is performed in terms of energy consumption. In the DDR-LEACH, energy is consumed by the nodes till 1400 rounds. Whereas, in the LEACH, the energy is consumed by the nodes till 1000 rounds. The energy consumption is high in LEACH because random CHs' selection is performed. The sensor nodes near to the BS may not participate in the network and may not be selected as CHs. Therefore, maximum energy is utilized by any of the nodes to send the data packets from source to destination. As compared to LEACH, the maximum energy is consumed by the nodes till 1400 rounds. The DDR-LEACH considers the three parameters that are maximum degree, minimum distance, and minimum energy consumption to select the CHs. The energy usage is efficient in the starting rounds because CHs are alive and working. CHs aggregate data and send it to BS. Therefore, energy consumption is decreased, as shown in Figure 5.

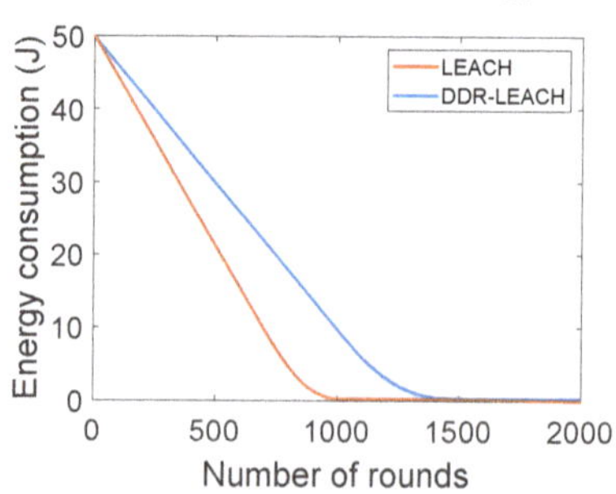

Figure 5. Energy consumption.

Figure 6 depicts that the network throughput of both schemes is zero at the start. The reason is that no data packet is sent at initials rounds. It continues to increasing with the number of rounds because large amount of packets are sent at these rounds. In DDR-LEACH, the amount of data sent from ordinary nodes to BSs is increased gradually because all the nodes are participating in the network. It gradually decreases at 1500th round, trend becomes constant. The throughput is maximum because CHs are selected using DDR-LEACH and maximum nodes are alive to send the data packets. The data packets are increased with the decrease in the number of rounds. Whereas, in LEACH, the nodes are randomly selected and if the CHs are selected that are away from BSs, then nodes have to utilize more energy. Additionally, in DDR-LEACH, the selection of CHs is based on three parameters. Whereas, in LEACH, the randomly selected CHs die because these parameters are not considered in the CHs' selection process. Therefore, nodes inefficiently perform operations in the network and minimum data packets are sent from source to destination.

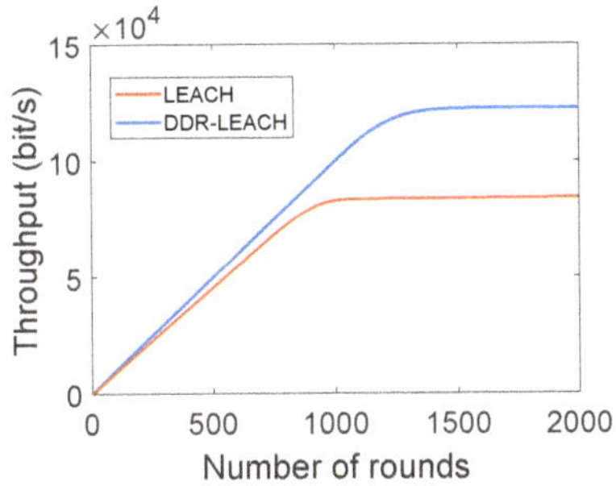

Figure 6. Network throughput.

Figure 7 depicts the network lifetime of nodes. The DDR-LEACH is compared with LEACH in terms of network lifetime. In LEACH, the nodes die at an early stage like the first node dies at 600th round. Although, the 10th node dies at 750th round. All the nodes in LEACH die at 1000th round. The nodes die early in the LEACH because these nodes consume more energy and there is no mechanism for efficient CH selection. Therefore, random selection of CHs makes the network inefficient and it affects the networks' performance. In the comparison of LEACH, the total number of rounds is 2000 in which the network nodes operate. The energy of the first node depletes at an early stage. Whereas, the 10th node dies at 1150th round. Although, all nodes die at the 1500th round. It shows that the network has a good lifetime as it operates for a large number of rounds.

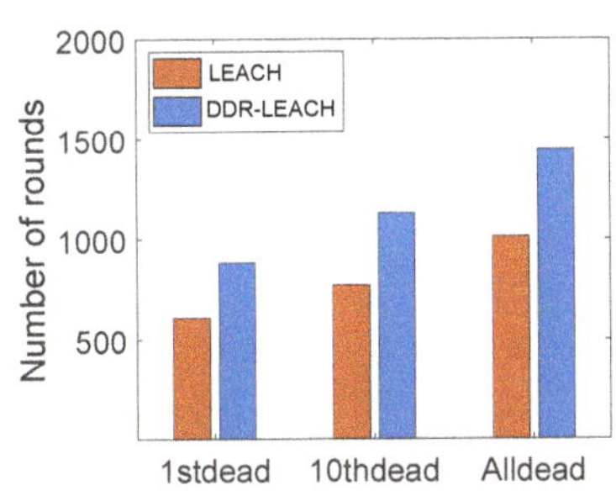

Figure 7. Network lifetime.

Figures 8 and 9 illustrate the comparison of PoW and PoA consensus mechanism in service provisioning and data storage, respectively. In the proposed system model, PoA consensus mechanism is used that incurs less computational cost because in the proposed system model, private blockchain is used. Whereas, the PoW works efficiently in the public blockchain. Both consensus mechanisms are compared and their computational cost is evaluated in terms of Gwei. In the comparison of both consensus mechanisms, PoA performs better than PoW. It is because, in PoA, no mathematical puzzle is being solved by the mining nodes. In PoA, the miners are pre-selected and are responsible for validating the transactions. It is the reason that PoW incurs a large computational cost as compared to PoA. In Figure 8, when the buyers request blockchain for services, a smart contract is deployed in which PoA consensus mechanism is used. PoA, in terms of service provisioning, is compared with PoW and it is observed that PoA consumes less computational cost than

PoW. On the other side, in Figure 9, when data are stored in IPFS, its average transaction cost is calculated. As discussed above, the average transaction cost of PoW is more than the average transaction cost of PoA.

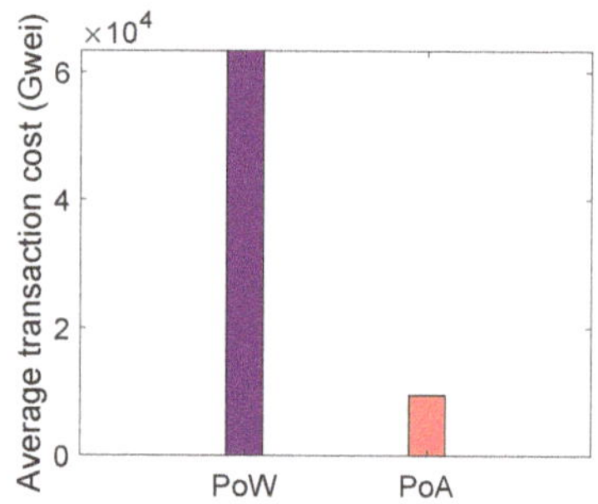

Figure 8. Average transaction cost for service provisioning.

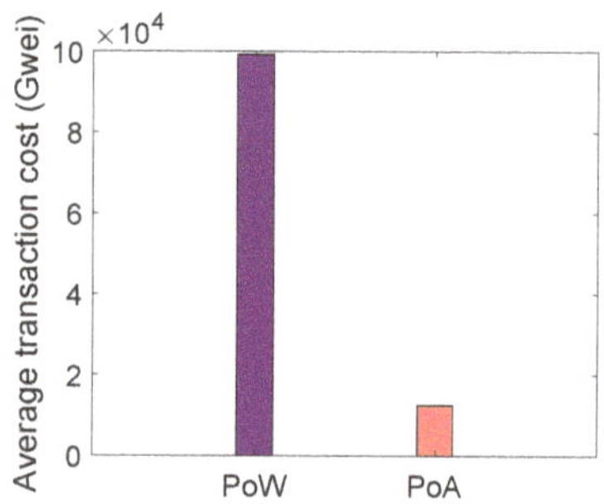

Figure 9. Average transaction cost for data storage in IPFS.

In the proposed system model, AES 128-bit symmetric encryption and decryption scheme is used for service provisioning. Figure 10 depicts the encryption and decryption time comparison using AES 128-bit and rivest shamir adleman (RSA) schemes. AES 128-bit scheme is compared with RSA to show that AES 128-bit works efficiently in terms of time complexity. Decryption time is less than encryption times of AES, which indicates that AES performs efficiently during encryption and decryption processes. In both 256-bit and 192-bit schemes, the last rounds are asymmetric due to the absence of a mix column layer. An AES 128-bit scheme takes input data and uses a 128-bit length key K_{Pr} to encrypt the plain text. The key size varies according to different key lengths in AES. Other AES schemes have different key lengths, such as 256-bit, 192-bit, etc. This paper uses AES 128-bit encryption because it takes a short time for encryption and decryption as compared to AES 192-bit and AES 256-bit [45]. In AES 256-bit scheme, there are 14 rounds while there are 12 rounds in AES 192-bit scheme to meet the encryption and decryption processes [46,47]. Moreover, the symmetric technique takes less time to encrypt the data as compared to asymmetric encryption. Additionally, for normal security purposes, AES 128-bit is enough while AES 192-bit and AES 256-bit are made to resist against the quantum computing based brute force attack, which is usually used in military based critical matters. When buyers request for the services, they are provided with the services in an encrypted form. The reason to encrypt the services is to prevent the data from unauthorized access.

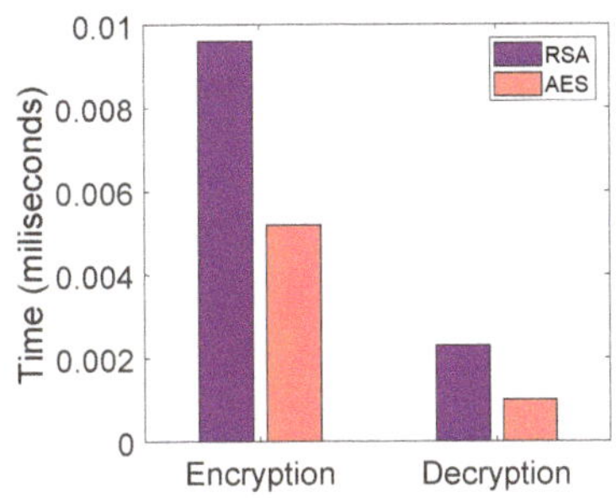

Figure 10. Comparison of execution time for AES and RSA.

5. Formal Security Analysis

The formal security analysis is performed to detect the malicious nodes in the network. The authentication mechanism is used specifically for the malicious nodes' detection. The Sybil and MITM attacks are induced in the network to check the robustness of the network. Moreover, the identity information of the network nodes is stored on the blockchain and is analyzed using Oyente tool. The attacks' analyses are given below.

Sybil attack: The authentication mechanism is used to make the network free from malicious nodes. In the authentication mechanism, the nodes are registered and authenticated in the network before any task is performed by them. The nodes' information is stored in the blockchain network. A unique identity is provided to every node and its credentials are also stored in the blockchain. Therefore, this attack is not possible in this system model because nodes are mutually authenticated before performing any task.

MITM attack: The MITM attack is induced in the DDR-LEACH protocol, which interrupts the existing conversion in the network. To make the network free from these types of attacks, nodes' authentication mechanism is used. Only those nodes take part in the network whose credentials are stored in the blockchain network. Before communication, these nodes have to be authenticated first. When the attacker intercepts during the registration process, the node is validated with the provided information. In the registration process, the node is provided with a unique key that is used during the authentication process. The provided key is already stored in the blockchain network. Whenever, an attacker wants to become a part of the network, it first needs to provide the exact information as the legitimate node has, which is not possible in this scenario. Therefore, attacker node is detected during the registration phase. Additionally, in the authentication phase, the attacker node needs the unique identity that is stored in the blockchain. Therefore, attacker must be verify its information using the blockchain provided key.

Figure 11 shows that the attacker intercepts during the communication and tamper the data packets. In the MITM attack, the attacker performs the malicious activity by sending the wrong information or tampering the data packet. When the data are sent from sensor nodes to CHs and CHs to BSs, due to malicious activities, the original packet is not received at the destination. The attackers send the malicious packets again and again towards the destination. Therefore, the nodes are provided with unique key identities. Every node in the network is authenticated before it communicates with any other legitimate node. Additionally, in the Sybil attack, the attacker node makes multiple identities and manipulates the whole network. When the attacks are induced in the network, its throughput decreases because only malicious packets are sent to the destination point. After the detection of attackers through mutual registration and authentication, the network performance is improved. The throughput increases when the network is free from the attackers.

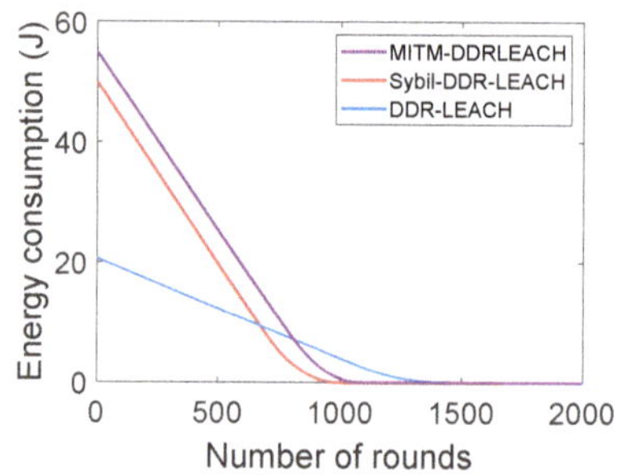

Figure 11. Security analysis of attacks with the proposed solution in terms of energy consumption.

In Figure 12, when the attackers are present in the network, the energy consumption is maximum. The attackers send the malicious packets to the destination point, due to which much energy is consumed. Whereas, after the nodes' registration and authentication, only legitimate nodes become the of the network. Therefore, their energy consumption is minimum as compared to the energy consumed in the presence of attackers.

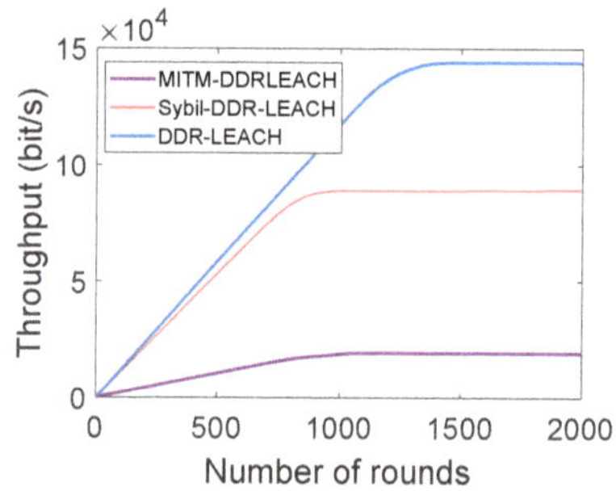

Figure 12. Security analysis of attacks with the proposed solution in terms of throughput.

Figure 13 illustrates the network lifetime in the presence of attackers and without their presence. When the attackers are present in the network, they only send the malicious data packets in the network. Additionally, they consume high energy. Therefore, the data packets sent by the legitimate nodes do not reach the destination. When the nodes' authentication is performed, the malicious nodes are removed from the network. Therefore, the energy consumed by the legitimate nodes is less and the nodes do not die early. Whereas, in the attacker model, when the malicious nodes are present in the network, the nodes die early because their energy depletes in sending the wrong data packets to disturb the traffic.

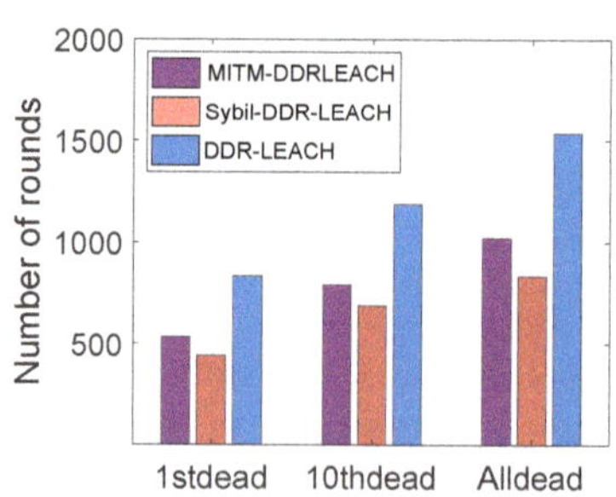

Figure 13. Security analysis of attacks with the proposed solution in terms of network lifetime.

Smart Contract Analysis

To register and authenticate the network nodes, a smart contract is written in the solidity language. Additionally, service provisioning mechanism is used for providing the services to the buyers. The security analysis is performed on the smart contract to check its vulnerability. The tool that is used for the security analysis is Oyente. It is an open source tool that checks the possible vulnerabilities in the smart contract. The reason to perform the security analysis is the bad programming practices. Therefore, attacks are also performed such as DAO [48]. Additionally, all the business rules of our network are stored on smart contract. All sellers and buyers communicate with each other by following these business rules. Moreover, all the conditions for detecting and revoking malicious nodes in the network are also stored in the smart contract. The smart contracts provide basic infrastructure for our blockchain-based authentication and CH selection scheme. Therefore, we have provided the formal security analysis of our smart contract because it is responsible for managing every single transaction of our network. Furthermore, in blockchain based schemes, the security analysis of entire solution is performed by formal analysis of smart contact, as performed in [49–51]. The attacks that are possible on the smart contract are discussed below.

Figure 14 explains the security analysis of nodes' registration and authentication smart contract using the Oyente tool. It shows that the smart contract is resilient against the attacks that are shown in the figure. The result shows that the mentioned attacks in the figure are not possible on this smart contract. However, some attacks that are very close to vulnerabilities for the smart contract are discussed below.

Re-entrancy attack: When a function runs in the smart contract, the malicious node in the network calls the external function and stops the working of actual function. In our smart contract, this attack is not possible because all nodes in the network are authenticated before performing any task. The IDs of legitimate nodes are stored on the blockchain. Therefore, when nodes are not authenticated due to wrong information, then they are removed from the network.

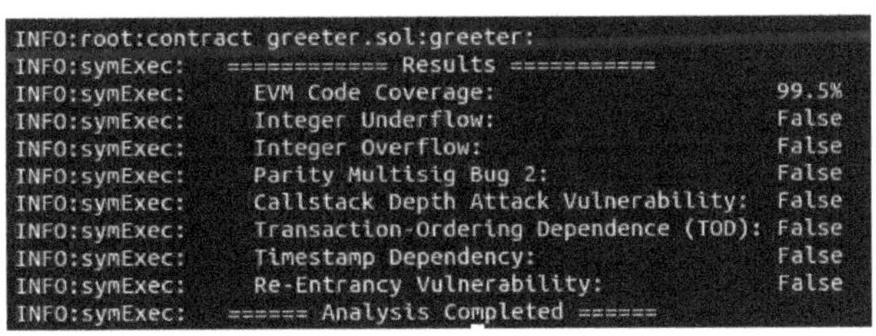

Figure 14. Security analysis of smart contract during registration and authentication of nodes.

No double spending: In our system model, the buyers are authenticated using the secret key that is provided to them. Therefore, no malicious node can obtain the data and perform any malicious activity in the network.

Denial of service attack: The denial of service attack is not possible in our system model because the buyers are authenticated by providing the secret keys for communication and they have to exchange the keys before receiving data.

Single point of failure: In our system model, IPFS is used for storage, which is a distributed network. Whereas, in the centralized storage system, the system is not able to give responses frequently. In our proposed system, due to distributed use IPFS, this attack is not possible. The system quickly responds to data storage and data provisioning.

6. Conclusions and Future Work

This paper presents a blockchain and IoST based network to minimize malicious activities and incur less computational cost. Nodes' authorization is ensured using authentication scheme. Only authorized nodes are allowed to take part in the network and send data to CHs for further operations. CHs aggregate the data due to which their energy depletes rapidly and they die. In that case, we propose a DDR-LEACH protocol in which CHs are selected from the ordinary nodes based on their maximum degree, minimum distance from BS and maximum residual energy. CHs with low energy are replaced with nodes that satisfy the above criteria. Moreover, the aggregated data are stored in the IPFS. For long-term storage, blockchain gives incentives to IPFS. The services are provided in an encrypted form using AES 128-bit encryption scheme. The simulation results show that less computational cost is incurred during data storage and service provisioning. Moreover, low transaction cost is incurred during the registration and authentication phases. The average transaction costs incurred by PoW and PoA during data storage and service provisioning are compared and it is observed that PoA outperforms PoW. Furthermore, efficient energy consumption, network lifetime and throughput are shown in the results that are conducted by comparing LEACH with DDR-LEACH. The results show that DDR-LEACH DDR-LEACH outperforms LEACH. The encryption scheme AES 128-bit used in the proposed work shows better performance than RSA in terms of execution time. The formal security analysis is performed to check the effectiveness of smart contract against attacks. Two attacks, MITM and Sybil, are also induced in the proposed network to show the its resilience against cyber attacks. In the future work, efficient machine learning technique will be used for the malicious nodes' detection in the network.

Author Contributions: Conceptualization, methodology and validation is done by S.A. (Sana Amjad). S.A. (Shahid Abbas) contributed in formal analysis and investigation. Z.A. has proof read and edited the manuscript. M.H.A. performed the visualization. Project administration was under A.J. Overall work was supervised by N.J. All authors have read and agreed to the published version of the manuscript.

Funding: This study did not receive any funding.

Data Availability Statement: This study did not report any data.

Conflicts of Interest: There is no conflict of interest associated with this article.

Abbreviations

Notation	Description
AES	Advance encryption standard
BSs	Base stations
CHs	Cluster heads
DDR	Distance degree and residual energy
IoT	Internet of things
IoST	Internet of sensor things
IPFS	Interplanetary file system
LEACH	Low energy adaptive clustering hierarchy
MITM	Man in the middle
PoA	Proof of authority
PoW	Proof of work
RSA	Rivest shamir adleman
SHA	Secure hashing algorithm
SPOF	Single point of failure
WSNs	Wireless sensors networks
K_{Pr}	Private key

References

1. Javaid, N. Integration of context awareness in Internet of Agricultural Things. *ICT Express* 2021, *in press*. [CrossRef]
2. Fu, M.H. Integrated technologies of blockchain and biometrics based on wireless sensor network for library management. *Inf. Technol. Libr.* 2020 *39*, 3. [CrossRef]
3. Kumari, S.; Om, H. Authentication protocol for wireless sensor networks applications like safety monitoring in coal mines. *Comput. Netw.* **2016**, *104*, 137–154. [CrossRef]

4. Prabu, P.; Ahmed, A.N.; Venkatachalam, K.; Nalini, S.; Manikandan, R. Energy efficient data collection in sparse sensor networks using multiple mobile data patrons. *Comput. Electr. Eng.* **2020**, *87*, 106778. [CrossRef]
5. Jiang, Q.; Zeadally, S.; Ma, J.; He, D. Lightweight three-factor authentication and key agreement protocol for internet-integrated wireless sensor networks. *IEEE Access* **2017**, *5*, 3376–3392. [CrossRef]
6. Abbas, S.; Javaid, N.; Almogren, A.; Gulfam, S.M.; Ahmed, A.; Radwan, A. Securing genetic algorithm enabled SDN routing for blockchain based Internet of Things. *IEEE Access* **2021**, *9*, 139739–139754. [CrossRef]
7. Javaid, N. A Secure and Efficient Trust Model for Wireless Sensor IoTs using Blockchain. *IEEE Access* **2022**, *10*, 4568–4579. ACCESS.2022.3140401. [CrossRef]
8. Kumar, M.; Mukherjee, P.; Verma, K.; Verma, S.; Rawat, D.B. Improved deep convolutional neural network based malicious node detection and energy-efficient data transmission in wireless sensor networks. *IEEE Trans. Netw. Sci. Eng.* 2021, *in press*. [CrossRef]
9. Kar, J.; Liu, X.; Li, F. CL-ASS: An efficient and low-cost certificateless aggregate signature scheme for wireless sensor networks. *J. Inf. Secur. Appl.* **2021**, *61*, 102905. [CrossRef]
10. Verma, N.; Kaushik, A.; Nayak, P. A lightweight secure authentication protocol for wireless sensor networks. In *International Conference on Innovative Computing and Communications*; Springer: Singapore, 2021; pp. 291–299.
11. Padmavathi, U.; Rajagopalan, N. Concept of Blockchain Technology and Its Emergence. *In Blockchain Applications in IoT Security*; IGI Global: Raipur, India, 2021; pp. 1–20. [CrossRef]
12. Abubaker, Z.; Khan, A.U.; Almogren, A.; Abbas, S.; Javaid, A.; Radwan, A.; Javaid, N. Trustful data trading through monetizing IoT data using BlockChain based review system. *Concurr. Comput. Pract. Exp.* **2021**, *34*, e6739. [CrossRef]
13. Moinet, A.; Darties, B.; Baril, J.L. Blockchain based trust and authentication for decentralized sensor networks. *arXiv* **2017**, arXiv:1706.01730.
14. Goyat, R.; Kumar, G.; Saha, R.; Conti, M.; Rai, M.K.; Thomas, R.; Hoon-Kim, T. Blockchain-based Data Storage with Privacy and Authentication in Internet-of-Things. *IEEE Internet Things J.* 2020, *in press*. [CrossRef]
15. Noshad, Z.; Khan, A.U.; Abbas, S.; Abubaker, Z.; Javaid, N.; Shafiq, M.; Choi, J.G. An Incentive and Reputation Mechanism Based on Blockchain for Crowd Sensing Network. *J. Sens.* **2021**, *2021*, 1798256. [CrossRef]
16. Christidis, K.; Devetsikiotis, M. Blockchains and smart contracts for the Internet of Things. *J. Fintech Blockchain Smart Contract.* **2018**, *1*, 7–12. [CrossRef]
17. She, W.; Liu, Q.; Tian, Z.; Chen, J.S.; Wang, B.; Liu, W. Blockchain trust model for malicious node detection in wireless sensor networks. *IEEE Access* **2019**, *7*, 38947–38956. [CrossRef]
18. Magazzeni, D.; McBurney, P.; Nash, W. Validation and verification of smart contracts: A research agenda. *Computer* **2017**, *50*, 50–57. [CrossRef]
19. Haseeb, K.; Islam, N.; Almogren, A.; Din, I.U. Intrusion prevention framework for secure routing in WSN-based mobile Internet of Things. *IEEE Access* **2019**, *7*, 185496–185505. [CrossRef]
20. Hong, S. P2P networking based internet of things (IoT) sensor node authentication by Blockchain. *Peer-to-Peer Netw. Appl.* **2020**, *13*, 579–589. [CrossRef]
21. Cui, Z.; Fei, X.U.E.; Zhang, S.; Cai, X.; Cao, Y.; Zhang, W.; Chen, J. A hybrid BlockChain- based identity authentication scheme for multi-WSN. *IEEE Trans. Serv. Comput.* **2020**, *13*, 241–251. [CrossRef]
22. Sharma, P.K.; Park, J.H. Blockchain based hybrid network architecture for the smart city. *Future Gener. Comput. Syst.* **2018**, *86*, 650–655. [CrossRef]
23. Stack Exchange. Available online: https://ethereum.stackexchange.com/questions/872/what-is-the-cost-to-store-1kb-10kb-100kb-worth-of-data-into-the-ethereum-block (accessed on 6 March 2021).
24. Amjad, S.; Aziz, U.; Gurmani, M.U.; Awan, S.; Sajid, M.B.E.; Javaid, N. Blockchain based Authentication for end-nodes and efficient Cluster Head selection in Wireless Sensor Networks. In *Conference on Complex, Intelligent, and Software Intensive Systems*; Springer: Cham, Switzerland, 2021; pp. 195–205.
25. Ramezan, G.; Leung, C. A blockchain-based contractual routing protocol for the internet of things using smart contracts. *Wirel. Commun. Mob. Comput.* **2018**, *2018*, 4029591. [CrossRef]
26. Xu, J.; Meng, X.; Liang, W.; Zhou, H.; Li, K.C. A secure mutual authentication scheme of blockchain-based in WBANs. *China Commun.* **2020**, *17*, 34–49. [CrossRef]
27. Ren, Y.; Liu, Y.; Ji, S.; Sangaiah, A.K.; Wang, J. Incentive mechanism of data storage based on blockchain for wireless sensor networks. *Mob. Inf. Syst.* **2018**, *2018*, 6874158. [CrossRef]
28. Liu, M.; Yu, F.R.; Teng, Y.; Leung, V.C.; Song, M. Computation offloading and content caching in wireless blockchain networks with mobile edge computing. *IEEE Trans. Veh. Technol.* **2018**, *67*, 11008–11021. [CrossRef]
29. Liu, Y.; Wang, K.; Lin, Y.; Xu, W. LightChain: A Lightweight Blockchain System for Industrial Internet of Things. *IEEE Trans. Ind. Inform.* **2019**, *15*, 3571–3581. [CrossRef]
30. Danzi, P.; Kalør, A.E.; Stefanović, Č; Popovski, P. Delay and communication tradeoffs for blockchain systems with lightweight IoT clients. *IEEE Internet Things J.* **2019**, *6*, 2354–2365. [CrossRef]
31. Rathee, G.; Balasaraswathi, M.; Chandran, K.P.; Gupta, S.D.; Boopathi, C.S. A secure IoT sensors communication in industry 4.0 using blockchain technology. *J. Ambient. Intell. Humaniz. Comput.* **2020**, *12*, 533–545. [CrossRef]
32. Jia, B.; Zhou, T.; Li, W.; Liu, Z.; Zhang, J. A blockchain-based location privacy protection incentive mechanism in crowd sensing networks. *Sensors* **2018**, *18*, 3894. [CrossRef]

33. Mori, S. Secure caching scheme by using blockchain for information-centric network-based wireless sensor networks. *J. Signal Process.* **2018**, *22*, 97–108. [CrossRef]
34. Kim, T.H.; Goyat, R.; Rai, M.K.; Kumar, G.; Buchanan, W.J.; Saha, R.; Thomas, R. A novel trust evaluation process for secure localization using a decentralized blockchain in wireless sensor networks. *IEEE Access* **2019**, *7*, 184133–184144. [CrossRef]
35. Xu, Y.; Ren, J.; Wang, G.; Zhang, C.; Yang, J.; Zhang, Y. A blockchain-based nonrepudiation network computing service scheme for industrial IoT. *IEEE Trans. Ind. Inform.* **2019**, *15*, 3632–3641. [CrossRef]
36. Kumar, M.H.; Mohanraj, V.; Suresh, Y.; Senthilkumar, J.; Nagalalli, G. Trust aware localized routing and class based dynamic block chain encryption scheme for improved security in WSN. *J. Ambient. Intell. Humaniz. Comput.* **2020**, *12*, 1–9. [CrossRef]
37. Guerrero-Sanchez, A.E.; Rivas-Araiza, E.A.; Gonzalez-Cordoba, J.L.; Toledano-Ayala, M.; Takacs, A. Blockchain mechanism and symmetric encryption in a wireless sensor network. *Sensors* **2020**, *20*, 2798. [CrossRef]
38. Rovira-Sugranes, A.; Razi, A. Optimizing the age of information for blockchain technology with applications to IoT sensors. *IEEE Commun. Lett.* **2019**, *24*, 183–187. [CrossRef]
39. Rathore, S.; Kwon, B.W.; Park, J.H. BlockSecIoTNet: Blockchain-based decentralized security architecture for IoT network. *J. Netw. Comput. Appl.* **2019**, *143*, 167–177. [CrossRef]
40. Lee, Y.; Rathore, S.; Park, J.H.; Park, J.H. A blockchain-based smart home gateway architecture for preventing data forgery. *Hum.-Centric Comput. Inf. Sci.* **2020**, *10*, 1–14. [CrossRef]
41. Heinzelman, W.R.; Chrakasan, A.; Balakrishnan, H. Energy-efficient communication protocol for wireless microsensor networks. In Proceedings of the IEEE 33rd annual Hawaii International Conference on System Sciences, Maui, HI, USA, 4–7 January 2000. [CrossRef]
42. Tian, Y.; Wang, Z.; Xiong, J.; Ma, J. A blockchain-based secure key management scheme with trustworthiness in DWSNs. *IEEE Trans. Ind. Inform.* **2021**, *16*, 6193–6202. [CrossRef]
43. Cao, C.; Tang, Y.; Huang, D.; Gan, W.; Zhang, C. IIBE: An improved identity-based encryption algorithm for WSN security. *Secur. Commun. Netw.* **2021**, *2021*, 8527068. [CrossRef]
44. Abdi Nasib Far, H.; Bayat, M.; Kumar Das, A.; Fotouhi, M.; Pournaghi, S.M.; Doostari, M.A. LAPTAS: Lightweight anonymous privacy-preserving three-factor authentication scheme for WSN-based IIoT. *Wirel. Netw.* **2021**, *27*, 1389–1412. [CrossRef]
45. Al-Hazaimeh, O.M.A. A new approach for complex encrypting and decrypting data. *Int. J. Comput. Netw. Commun.* **2013**, *5*, 95. [CrossRef]
46. Pancholi, V.R.; Patel, B.P. Enhancement of cloud computing security with secure data storage using AES. *Int. J. Innov. Res. Sci. Technol.* **2016**, *2*, 18–21.
47. Ali, S.; Javaid, N.; Javeed, D.; Ahmad, I.; Ali, A.; Badamasi, U.M. A blockchain-based secure data storage and trading model for wireless sensor networks. In *International Conference on Advanced Information Networking and Applications*; Springer: Cham, Switzerland, 2020; pp. 499–511.
48. Ghaleb, B.; Al-Dubai, A.; Ekonomou, E.; Qasem, M.; Romdhani, I.; Mackenzie, L. Addressing the DAO insider attack in RPL's Internet of Things networks. *IEEE Commun. Lett.* **2018**, *23*, 68–71. [CrossRef]
49. Lu, N.; Wang, B.; Zhang, Y.; Shi, W.; Esposito, C. NeuCheck: A more practical Ethereum smart contract security analysis tool. *Softw. Pract. Exp.* **2021**, *51*, 2065–2084. [CrossRef]
50. Brent, L.; Jurisevic, A.; Kong, M.; Liu, E.; Gauthier, F.; Gramoli, V.; Scholz, B. Vandal: A scalable security analysis framework for smart contracts. *arXiv* **2018**, arXiv:1809.03981.
51. Praitheeshan, P.; Pan, L.; Yu, J.; Liu, J.; Doss, R. Security analysis methods on ethereum smart contract vulnerabilities: A survey. *arXiv* **2019**, arXiv:1908.08605.

sensors

MDPI

Article

Novel Scoring for Energy-Efficient Routing in Multi-Sensored Networks

Wooseong Kim [1], Muhammad Muneer Umar [2], Shafiullah Khan [1,2,*] and Muhammad Altaf Khan [2]

1 Computer Engineering Department, Gachon University, Seongnam 13120, Korea; wooseong@gachon.ac.kr
2 Institute of Computing, Kohat University of Science & Technology, Kohat 26000, Pakistan; muneer.umar@kust.edu.pk (M.M.U.); dr.altaf@kust.edu.pk (M.A.K.)
* Correspondence: skhan@kust.edu.pk

Abstract: The seamless operation of inter-connected smart devices in Internet of Things (IoT) wireless sensor networks (WSNs) requires consistently available end-to-end routes. However, the sensor nodes that rely on a very limited power source tend to cause disconnection in multi-hop routes due to power shortages in the WSNs, which eventually results in the inefficiency of the overall IoT network. In addition, the density of the available sensor nodes affects the existence of feasible routes and the level of path multiplicity in the WSNs. Therefore, an efficient routing mechanism is expected to extend the lifetime of the WSNs by adaptively selecting the best routes for the data transfer between interconnected IoT devices. In this work, we propose a novel routing mechanism to balance the energy consumption among all the nodes and elongate the WSN lifetime, which introduces a score value assigned to each node along a path as the combination of evaluation metrics. Specifically, the scoring scheme considers the information of the node density at a certain area and the node energy levels in order to represent the importance of individual nodes in the routes. Furthermore, our routing mechanism allows for incorporating non-cooperative nodes. The simulation results show that the proposed work gives comparatively better results than some other experimented protocols.

Keywords: IoT; WSN; routing; load balancing; energy efficiency

Citation: Kim, W.; Umar, M.M.; Khan, S.; Khan, M.A. Novel Scoring for Energy-Efficient Routing in Multi-Sensored Networks. *Sensors* **2022**, *22*, 1673. https://doi.org/10.3390/s22041673

Academic Editors: Alexandru Lavric, Liliana Anchidin and Adrian I. Petrariu

Received: 22 December 2021
Accepted: 18 February 2022
Published: 21 February 2022

1. Introduction

The Internet of Things (IoT) [1] was recently made practical with the adoption of some state-of-the-art technologies, such as wireless sensor networks [2] and intelligent sensing [3]. The applications of IoT include health care, inventory tracking, smart grid networks, security systems, and maintainable transportation. The interconnected smart objects with embedded sensors in the IoT network cooperate and coordinate with one another to send the collected data to a gateway sink. For IoT-based applications, such as industrial control, environmental sensing, smart homes, and logistics management, the wireless sensor network (WSN) is an essential part of the infrastructure [1]. The WSN can be represented as a graph of multiple interconnected sensor nodes, where each node senses some data from the environment and transfers them to an ultimate station. The infrastructure of IoT-based WSNs can be autonomously organized without any complicated time-consuming installation and configuration compared to typical wired networks for a variety of purposes [2].

In WSNs, nodes operate with limited powered batteries and cannot be recharged or replaced in a short period since the sensor nodes are typically unattended. The various applications of WSNs in IoT environments suffer from this limitation. Accordingly, most of the previous research works have focused on the extension of the lifetime of the nodes while achieving peak throughput [4]. In WSNs, data transmission is done through the nodes cooperating with one another since most of the nodes may not have a direct connection to a sink node; the nodes use other nodes as relays for transferring their sensed data, which is known as multi-hop communication. For multi-hop communication in a WSN, a node probably has multiple options to select a path towards a destination. Many researchers have proposed various routing schemes considering routing parameters such as the nodes' energy levels, transmission rate, security, and so forth [5].

In IoT-based WSNs, the energy consumption of sensors is a major concern. Therefore, the effects on energy consumption have been investigated in most of the legacy routing protocols. Moreover, many routing schemes are designed with particular focus on the energy preservation and elongation of the network lifetime [6]. The goal of energy management is to ensure that the sensors perform for longer periods of time and all the sensors consume their energies equally [7]. Different techniques have been developed to balance the load and energy consumption among the nodes [8]. However, it is unavoidable that some nodes in the network do not cooperate for the sake of saving their energies. Such non-cooperative nodes behave selfishly either temporarily or forever. These selfish nodes severely degrade the overall network performance. In most of the legacy mechanisms, therefore, the selfish nodes are either isolated or blocked [9].

For energy-efficient routing, various mechanisms have been proposed [10]. Sleep scheduling approaches are of recent special interest to the scientific community, as they allow for some nodes to be idle for a particular period of time [11]. In most sleep scheduling mechanisms, the density of the nodes at various locations in a network is considered. In [12], the authors proposed the usage of some approaches combined in a genetic algorithm to formulate a discrete particle swarm optimization algorithm. The main objective of such mechanisms is to preserve the idle nodes for future operations that are redundantly deployed in the network. Moreover, it was shown that the sleeping nodes cause no negative impact on the overall performance of the network. Thus, the sleep scheduling mechanisms are very efficient for energy optimization in WSNs. However, these mechanisms purely rely on the density of nodes and become ineffective when there are no redundant nodes in the network. Moreover, some nodes that die over time also reduce the redundancy and degrade the impact of the sleeping scheduling.

In some proposed mechanisms [13], it is assumed that the nodes cooperate with each other while conducting a common routing protocol. However, in ad hoc and IoT networks, the smartness of nodes is very common. Therefore, this aspect must be properly addressed in such types of networks while designing an energy-efficient scheme. Many schemes focus on the individual contribution of each node towards energy efficiency by adapting a routing protocol [13], sleep behavior [11], coordination mechanism, data aggregation procedure [7], hop division [9], and cluster divisions [14], and so forth. The nodes may intelligently coordinate with each other considering each node's status. The nature of the deployment of the nodes also has a significant impact on the performance and lifetime of the networks. The energy efficiency techniques should adequately utilize the density or redundancy of the nodes in such types of networks [15]. There should be sufficient space in the mechanism to consider as many parameters as possible for designing an energy-efficient routing in a WSN-based IoT network. The parameters can be the selfishness of the node, neighborhood, connectivity through hop levels, density of the nodes, redundancy of nodes, energies, distances, and surrogate values such as points, score, or credit values for nodes, and so forth.

In this work, we propose the node status and score-based route optimization protocol (NSSROP), where each node keeps some additional data to balance the routing load among all the nodes. In an IoT setup, a sensor node may have the shortest available route towards a sink. However, it should wisely choose a route that balances the load and elongates the life of the entire network. Some nodes may be placed at a location where they may get a higher rate of relaying requests compared to other nodes. This situation can highly degrade the network performance by unbalancing the load among the nodes. For this purpose, each node calculates some values for itself that are referred to as scores. Unlike other typical routing protocols, the proposed mechanism addresses various parameters associated with the routing and energy optimization in the network. These parameters are used for the calculation of the scores. During the selection of a route by the source node towards the central control, each forwarding node builds the route by considering these scores of the relay nodes. Modified route request (RREQ) and route reply (RREP) packets are used to exchange the variations in these scores.

The remainder of this paper is ordered as follows: In Section 2, related works are described. In Section 3, the preliminary formations are described and the whole mechanism is explained. In Section 4, the simulation results are discussed. Lastly, Section 5 includes the conclusion and future work.

2. Related Work

The routing in IoT sensor-based networks is one of the most remarkable research areas in communication networks. There are lots of research articles related to this field. There are various parameters in the network that can be used to optimize the routing, for example, optimization with load balancing (traffic load distribution), as discussed in [16].

A proactive tree-based routing protocol, the routing protocol for low-power and lossy networks (RPL), is defined by RoLL [17]. RPL is a standard protocol that operates on an IPv6-based IoT network. It brought an opportunity to develop WSNs on a very large scale. Routing and message control are the RPL's most important mechanisms for establishing and maintaining an effective and stable network. Despite its standing as the standard routing protocol for IoT networks, RPL has had various flaws since its inception, and other approaches have emerged to address them [18]. Among these, routing loops are critical.

Most of the recent studies that have aimed at energy efficiency and load balancing in WSNs and WSN-based IoT networks preferred the cluster-based approach. In [14], the authors proposed the integration of the bat algorithm and low-energy adaptive clustering hierarchy (LEACH) for the efficient cluster head selection to reduce energy consumption and balance load among the network nodes. The nodes are also bound to follow a schedule for the transmission of their data packets. This mechanism primarily focuses on the cluster head selection by considering the nodes' energy levels. Each cluster head is bound to have a particular number of connected cluster members. However, unlike this mechanism, the distribution of nodes in a network can be random, which makes it difficult to specify the number of nodes for each cluster.

Turgut and Altan [19] introduced a fully distributed energy-aware multi-level (FDEAM) routing and clustering mechanism for WSN-based IoT networks. The two-level and multi-level inter-cluster transmission methods are represented in this work. In the second level, the communication and transmission strength are determined by considering the distance between the nodes and the base station (BS). The clusters are statically distributed over the entire network. However, the option for re-clustering the network is also defined. Self-arranged nodes elect the limits for clustering, and cluster heads are selected by executing the FDEAM method. However, this method is inappropriate for non-uniform node distribution and has the shortcoming of being reliant on a dominant source.

The authors of [20] presented an energy-efficient architecture of a self-sustaining WSN based on an energy-collecting BS and a mobile charger considering the cost of deployment. They conducted extensive simulations and demonstrated the efficacy of their proposed strategy by showing that it maximizes the expected network lifetime while minimizing deployment costs. The main idea is focused on the usage of mobile chargers and the energy-harvesting BS. However, the work did not primarily deal with energy-efficient routing.

The position responsive routing protocol (PRRP) was proposed in [21]. The main objective of this proposed work was to minimize energy consumption by incorporating the global positioning system (GPS) into the nodes. The network is divided into equally sized grids with a static or dynamically distributed number of nodes. The nodes can adjust their transmission power by using GPSs while communicating with each other.

To balance energy consumption within each cluster, Wang et al. [22] suggested uneven cluster generation and distributed cluster head rotation based on residual energy and relative location. The authors also designed a routing path updating system to prevent node energy depletion. The selection of the cluster head is based on the level of residual energy of the nodes. The routing paths are dynamic and also associated with the nodes' energy levels.

An energy-efficient regional source routing protocol was proposed in [23], which balances the network's energy usage by dynamically picking cluster heads with the most remaining energy among the WSN nodes. Furthermore, the ant colony algorithm based on distance is employed to determine the global ideal transmission path for each node, which reduces data transmission distance and energy consumption. The experiment results show that the proposed approach outperformed the compared approaches in terms of network lifetime and throughput.

The authors of [24] proposed open vehicle routing (OVR) based on fundamental WSNs parameters, in which a data collection protocol called EAL improves the energy efficiency by balancing the lifetime of the network nodes while considering latency.

Han et al. [25] proposed a cross-layer routing protocol for optimizing the routing in geographic node disjoint multi-paths. The routing layer performs according to the underlying energy demand of the network nodes while the physical layer adjusts the transmission power according to the energy levels. The authors also applied sleep and awake states for energy saving. In [26], the higher level of traffic generated by several source nodes in an IoT environment was considered. Three factors are used to determine optimal routes by taking the next hop nodes. These factors include (1) the signal to interference and noise ratio, and (2) the survivability factor and congestion level of the preferred forwarding node.

The Path Operator Calculus Centrality (POCC) routing protocol was proposed in [27]. POCC is used to determine the nodes' centrality scores, which are further used for path determination. The approximation of the centrality score uses the operator calculus method based on the topology of the network. The authors argue that this technique provides optimal paths towards the BS. The article [28] proposed a directional transmission-based energy-aware routing protocol (PDORP) to find energy-efficient routes. The DSR protocol is used as a base protocol in this mechanism. Moreover, a hybrid of bacterial foraging optimization and a genetic algorithm is used to efficiently collect node information. The authors presented comparatively better results for energy consumption, bit error rate, delays, and throughput from their experiments. The objective of this work was to attain a better quality of service and extend the network life. The predicted remaining delivery (PRD) protocol, based on the path weighting technique, was proposed in [29]. PRD considers the fundamental parameters, such as route quality, residual energy, end-to-end delays, and inter-node distance for designing the weightage system.

A well-known approach for selfish node management was introduced in the watchdog and pathrater method [30]. In this work, the watchdog detects the non-cooperative behavior of the nodes and the pathrater blocks the selfish nodes from being part of the routes. The presence of non-malicious selfishness is potentially higher in unlicensed entities in an IoT infrastructure. Therefore, it is critical to block the unwanted nodes in such a network.

Many research works have described mechanisms for determining and utilizing nodes' individual importance in a network. Sun et al. [31] proposed an important assessment mechanism for a particular node with respect to the energy field. They determined key nodes based on the average length and density of nodes for the stability of a network. For this, the authors used graph theory for the properties and correlation of the nodes with the energy field. In another work [15], an evaluation index was introduced based on the topology of the network, which eventually determines the nodes' locations within a network. Additionally, supernodes are designated to manage multiple key nodes within the network.

In a mechanism proposed in [32], the selection of relay nodes is made by a concept of "equivalent nodes" based on a proposed energy consumption model. The network life can be lengthened by applying a probabilistic dissemination algorithm among those relay nodes.

Some fuzzy logic-related articles have also been proposed to improve the energy efficiency in WSNs. Sheriba et al. [33] proposed a fuzzy logic and black widow optimization clustering protocol. However, the black widow optimization's ideal performance is modest. Later, the authors proposed a strategy for designing the optimal interval type 2 fuzzy logic by involving the evolutionary algorithms [34]. This solution technique is suitable for WSNs with limited energy since it helps to extend the network's lifespan. In reference [35], a trust-aware energy-saving stable clustering algorithm based on the fuzzy type-2 algorithm was devised to solve the constraint of the shortening lives of the cluster heads in clustering algorithms.

Various studies have proposed game-theoretic approaches for the establishment of a tradeoff between the desired signal-to-noise ratio (SNR) and energy consumption [36,37]. These approaches focus on optimal route selection while considering communication quality. The game-theoretic approach is effective in the sense of getting a payoff for individual nodes. However, the entire network's performance cannot be optimized by these approaches. Moreover, the nodes are self-focused in such approaches, and these do not give any length to the network life. The node selection mechanisms such as those in [27,38] were also proposed for choosing the best nodes among others for energy optimal efficiency in the network.

Various nodes and network scoring mechanisms have been proposed by many articles mainly focused on energy efficiency, node behavior, and security. The GoNe scheme, proposed in [39], was designed for enforcing data security and privacy in WSNs. Nodes are given some scores based on their reputation in the network. These reputation scores are managed by CHs, which are later used to manipulate the behavior of nodes. In another score-based load management scheme [40], the authors proposed a mechanism to compress the data through CHs to reduce the load on the nodes with low scores. The best nodes are chosen based on their remaining energy and distance from the BS. The CHs use compressive sensing to compress data and then forward information towards the sink through the best nodes. The authors claim that in this way, the load is balanced among all the nodes. The SBRR protocol [41] considers many factors to score paths for nodes. The parameters are the hop count, the remaining energy of nodes, link quality, and the buffer sizes on the nodes. All the parameters are integrated to form the path score. The main focus of the work was to reduce the pack loss in the transmission. Still, there is space for load balancing and energy efficiency in the work.

3. The Proposed Mechanism

Unlike RPL and other existing routing protocols, the proposed mechanism addresses various parameters associated with routing and energy optimization in the network. A node scoring mechanism is introduced based on the nodes' existence in the network. The neighborhood of a node is further classified as closed or identical neighbors. Some of the procedures in this study were influenced by our previous work on reward-based mechanisms (RwBMs) [9]. The RwBM is a game-based approach that uses the Rubinstein bargaining game for the management of virtual currencies which are referred to as scores. Our previous work deals with the management of selfish nodes in WSNs using the RwBM mechanism. In this study, we proposed a novel scoring scheme to select the best relay nodes while choosing a path. Herein, the key algorithm for the score manipulation and calculation, which involves entirely different procedures, is distinguishable from the RwBM.

This section is divided into two major subsections. In Section 3.1, the preliminaries are discussed, and in Section 3.2, the entire mechanism is explained in detail.

3.1. Preliminary Formation for the Proposed Work

This section presents the basic details of the main mechanism of the NSSROP. Each node in the network contains the following information:

3.1.1. Hop Level

The networks are divided into a hierarchical format of interconnected nodes. The intermediate nodes with a direct connection to a sink are denoted as having a hop level of one, while the nodes at the end boundaries have the maximum level values. Each node keeps its own hop level to determine its distance from the sink. The hop level can be used to define the number of nodes in a route towards the sink. Equation (1) shows the hop level for a node within a network of ***n*** nodes, as follows:

$$1 \leq HL_i \leq n \tag{1}$$

3.1.2. Neighbors

In multi-hop communication, the presence of neighboring nodes plays a vital role. A higher number of neighbors leads to a higher availability of routes towards the sink. Each node in the network keeps a list of all possible nodes that are in the transmission range. The neighbors can be classified as upward, downward, or sibling nodes. Upward nodes are neighbor nodes one hop level up. Downward nodes are the neighbors with a lower hop level, while the siblings are the nodes with the same hop level. A node with HL equal to HL^{max} indicates that this node is at the very bottom of the network. Such nodes do not have downward nodes so they only transmit data through forwarding or sibling nodes.

3.1.3. Closed Neighbors

Among the neighboring nodes, some nodes are relatively placed closer than other neighbors. Such nodes can be considered as closed neighbors. However, it is more appropriate to consider this distance based on the received signal strength indicator (RSSI) mechanism than the physical distance. An RSSI-based distance value is calculated by Equation (2) and is used to determine the set of neighbor nodes [42].

$$DST_{i,j} = P_{i,j}^{-1}(d), \quad P_{i,j}(d) = \frac{p_i\, G_i\, G_j\, \Lambda^2}{(4\pi)^2\, d^2} \tag{2}$$

where p_i denotes the transmission power, and G_i and G_j denote the antenna gains of nodes I and j, respectively. Nodes i and j are the transmitter and receiver, respectively. Λ indicates the wavelength (meter) of the transmission signal. $p_{i,j}$ is the receiving power at the node j when the inter-node distance is d . In a constrained situation where the sensor nodes are deployed in a controlled environment, the Pythagoras two-dimensional distance formula can also be used for creating the set of CNs. Moreover, if nodes are deployed in irregular, unaligned, or not plane areas, then the same can be converted into the three-dimension distance formula.

Each node keeps a separate set of closed neighbors (CNs) with their estimated location and energy information. If a node has a frequent number of CNs, then it means that the node has less opportunity to become a forwarder of other nodes. Nodes with a fewer number of CNs are vital and can perform more than others. The set of CNs can be calculated as follows:

$$CN_i = \{j : DST_{i,j} \leq DST^{tr}\} \tag{3}$$

CN_i are all the nodes that are located within a distance value DST^{tr} with a specified RSS threshold from node i.

In Algorithm 1, the distance of the inputted node is compared with all the nodes from 1 to n. In each iteration, the computed RSSI-based distance is checked for whether it is less than a predefined threshold distance for the CNs. Stack memory is used to store all the nodes that are at a concerning distance with the inputted node, that is, equal or less than the threshold distance. In case no CN of a node exists, this function returns 0.

Algorithm 1. Calculation of CNs for a given node (node_ID).

CALCULATE_CN(node ID)
for $i = 1$ to n
  StackCNs.Top $= 0$
  if (ID $\neq i$) //case of a sin gle node
    $DST_{i,ID} = P_{i,ID}(d) = \frac{p_i\ G_i\ G_{ID}\ \Lambda^2}{(4\pi)^2\ d^2}$
    if $0 < DST_{i,ID} < DST^{tr}$
      *Push*Ii to StackCNs.Top
      StackCNs.Top = StackCNs.Top + 1
    end if
  end if
end for
Return StackCNs

DST_{tr} is a threshold value used to limit the succeeding nodes to being *CNs* with a specific node. This value can be wisely defined by the consideration of the total number of nodes, nodes' placements, and the transmission range of the nodes. If the number of nodes is higher in a particular field, then we can assume that the nodes are more densely located. Similarly, the nodes with lower transmission power will make most of the nodes directly unreachable to each other. The value of DST^{tr} should be less than the maximum RSSI value in the network; otherwise, the parameters of the *CNs* will have no or an erroneous impact on the mechanism. Moreover, if we use a very small DST_{tr} value, then it will return none or a smaller number of *CNs*. The proposed work uses the *CNs* for the network optimization; therefore, the reasonable number of *CNs* has a greater impact on the overall performance of our proposed work. In our work, for the experiments, we kept this value as half of the maximum possible RSSIs.

3.1.4. Identical CNs

It is probable that at some particular locations in the sensor field, some nodes reside at a very near distance to some other nodes. Such closely located sensor nodes generally sense data in parallel and get RREQs from the same source nodes. Two or more closely deployed nodes with the same connections to other nodes can be denoted as identical to each other. Such nodes will have relatively similar sensed data and similar RREQ from other nodes. It is feasible to utilize such nodes by assigning them more load compared to other nodes. For closely related nodes, an appropriate distance must be configured. A higher distance leads to a larger number of identical nodes, which may cause a negative impact. Contrarily, a lower value reduces the sets of identical nodes, making the proposed work ineffective. In our experiments, we assumed that the nodes with a distance of a quarter of the maximum transmission range of the nodes were identical CNs.

In Figure 1, nodes a and b are CNs of each other and also have similar connections to other nodes, while node c is alone and has a similar hop level. In case one of the identical nodes exhausts its energy completely, the effect is limited compared to a stand-alone node. Therefore, it is wise to utilize nodes a and b more frequently than node c.

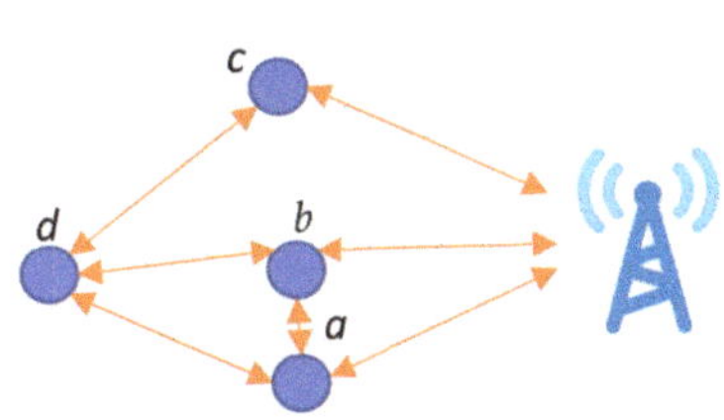

Figure 1. Identical CNs (a and b).

3.1.5. Energies of Neighbors

The main concern is energy optimization in the IoT sensors. Therefore, each node keeps its own as well as its neighbors' energy information. The energy of node i can be denoted as E_i. The values of E_i can be determined as follows:

$$0 \leq E_i \leq E^{max} \tag{4}$$

The nodes with energy levels equal to 0 are considered dead nodes. Such nodes are automatically omitted from the network. The features of the dead nodes cannot be used in the formation of scores for routing. Each node keeps track of its neighbors and deletes the dead nodes from the neighbors' lists.

3.1.6. Scores

In the network, each node governs a routing table that is used for sending its data. For each source to its destination, a sequence of nodes is kept in this table. The routing table keeps the sequence of nodes in each possible route for a destination. A set, R, can be defined to show all the possible routes for node i via the presence of intermediate nodes. Each route can be denoted by r with a sequence of nodes.

$$R_i = \{r_1, r_2, r_3, \ldots, r_j\} \tag{5}$$

The nodes present in the routes can be distinguished by their importance using a scoring mechanism. The routes can be determined by evaluating the scores of the nodes.

The efficiency of a route can be calculated by considering the nodes' density, that is, the number of CNs and their energies at each hop level towards the sink. Each transmission of the data packets from a source consumes the energies of all involved nodes. The deduction of energies induces a variation in the scores of nodes. The score of each node i can be calculated by its energy level, the sum of the energies of its CNs, and their size m at time t. The score of node i at time t can be derived by Equation (6) as follows:

$$\lambda_i^t = \frac{E_i^t}{E^{max} + \sum_{j=1}^{m} E_{CNj}^t} \times (m+1) \tag{6}$$

The high level of energy of node i and/or a higher number of CNs leads to a higher value of λ. Additionally, CNs' energies also influence this value. The higher-level CN energies reduce the value of λ. For an exceptional case in which a node does not have any connected or dead CNs, λ is calculated by dividing the energy of node i by the maximum level of energy, as follows:

$$\lambda_i^t = \frac{E_i^t}{E^{max}} \tag{7}$$

Since each route may have multiple nodes, the λ of relay nodes is considered at each level. The main aim of considering the node density is to reduce the load on single or scarce nodes. Once the system starts putting the load on the densely located nodes, it is obvious that at some following stages, some of the nodes will exhaust their energies and will ultimately no longer operate. The load is finally diverted to other nodes.

Algorithm 2 is used to calculate the λ values for the nodes. This algorithm uses a subroutine, CALCULATE_CN(ID), to get the list of all the CNs for a specific node. The number of all CNs is retrieved and then according to Equation (6), these CNs are processed using a simple loop.

Algorithm 2. Calculation of λ for each node in the network.

```
LAMBDA(node ID)
1. ID.CNs = CALCULATE_CN(ID)
2. λ_D = MAXEnergy
3. TotalCNs = Size (ID.CNs) //0 is assigned if no CN exits
4. if (TotalCNs ≠ 0)
5.    for i = 1 to TotalCNs
6.       TempCN = ID.CNs(i)
7.       λ_D = λ_D + TemptCN.Energy
8. λ_N = (TotalCNs + 1)*ID.Energy
9. λ = λ_N/λ_D
10. Return λ
```

3.1.7. The Reputation of Hop Level Neighbors

Some nodes may not cooperate due to their selfish behavior. Such nodes are enlisted during the data transmission by a source node. If a node does not reply to a route request, then it is considered a selfish node. The selfish nodes are not served by other relay nodes for data transfer requests. In detail, if a node continues non-cooperative behavior for a long period of time T, then it is black-listed. In addition, the blacklisted nodes are not requested for relaying services. A counter Cr is used when the source node declares a node as a blacklisted one after being selfish. A list of selfish and blacklisted nodes is broadcasted among the neighbors so that they might be contacted adaptively by all neighbor nodes. The states of selfish and blacklisted nodes can revert to normal after a specified period of time.

Due to the existence of selfish nodes in the WSN, unfavorable situations may exist. For instance, it is possible that a node may not be able to communicate or transfer its data to the sink due to the presence of one or more selfish nodes along the route. In the worst case, a node can meet relay nodes that are all selfish. In such cases, these nodes cannot communicate with their destinations [9]. Such nodes typically attempt to retransmit data repeatedly during a particular period of time. In this study, we assumed that multiple alternate paths for transmissions were available in order to cope with such network partition by selfish nodes. Due to the fact that node intelligence encourages selfish behavior, a node must be able to select only the reliable ones among multiple relays to prevent excessive packet drops.

3.1.8. Routing Information Formats

The format of the modified DSR route table is given in Figure 2. This table demonstrates the format of a single entry for a destination node *i*. In the routing cache, an additional 2-byte Lambda (λ) value is concatenated with each address. Nodes can easily update these according to their own knowledge. Each node transmits its own λ field through the routing and topology control messages.

Dest	Source Route Record				...
Dest[i]	Addr(1)|Lam(1)	Addr(2)|Lam(2)	...	Addr of Dest(i)	...

Figure 2. Modified routing table format.

Figures 3 and 4 show the RREQ and RREP formats in this work. An additional 1-byte field for the *Lambda Option* is added in the header. The presence of this field specifies the addition of λ values with each address. Usually, the *Lambda Option* is kept as null in the RREQ to avoid any additional bandwidth and energy. In the RREP, we used the modified addresses fields. Accordingly, each relay node adds its own address with its λ score for the RREP.

Option Type	Opt Data Length	Identification	Lambda Op
Target Address			
Addr(1) | Lam(1)			
Addr(2) | Lam(2)			
. . .			
Addr(n) | Lam(n)			

Figure 3. Modified DSR RREQ format.

Option Type	Opt Data Length	L | Reserved
Target Address		
Addr(1) | Lam(1)		
Addr(2) | Lam(2)		
. . .		
Addr(n) | Lam(n)		

Figure 4. Modified DSR RREP format.

Most of the time, the nodes estimate their neighbors' locations and energies by analyzing the sequence of the involved nodes in the flow of data transmission from a source towards a sink. However, sometimes nodes may require an update for these values after a specified period of time. OLSR topology control messages are used to get the neighbors' locations and their energies. For this, we modified the OLSR topology control message as shown in Figure 5, where each address is combined with the nodes' energies for the sake of information sharing.

ANSN	RESERVED
Advertised neighbor address and Remaining Energy	
Advertised neighbor address and Remaining Energy	
. . .	

Figure 5. Information exchange through modified OLSR topology control message.

An unsigned integer 2 bytes in size is used to represent the value of ʎ. According to Equation (6), the value of Lambda must be a decimal ranging from 0 to the maximum number of nodes. The possible value range for 2 bytes is from 0 to 6.5535. To adjust the decimal values into an integer of 2 bytes, the value of ʎis rounded to 4 decimal points and then multiplied by 1000. After applying this procedure, the mechanism can use the maximum value of ʎup to 6.5535. However, there is the possibility that this value will be near the total number of nodes in the network. When we used an appropriate value for calculating the set of CNs, there was a much lower chance of this value being greater than 2 in most of the experimental cases. For example, a sample of calculated values can be seen in the next section of Simulation Results. A value greater than normal will have the same impact no matter how much greater it is. So, during the process, we assumed that any value of ʎgreater than 6.5535 must be considered 6.5535.

3.2. Proposed Mechanism

The flowchart for the entire process of our algorithm is depicted in Figure 6. Our proposed mechanism is divided into sub-parts of the initial configuration, the selection of relay nodes, handling the selfish nodes, and the information exchange.

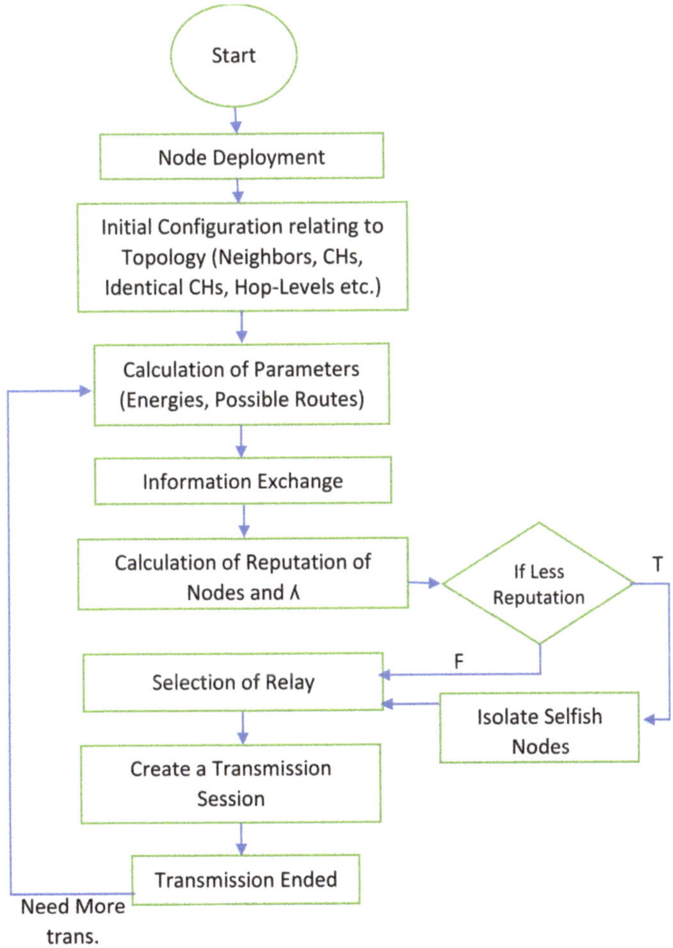

Figure 6. Flowchart of the proposed mechanism.

3.2.1. Initial Configuration

After the deployment of the network, initially, each node broadcasts control packets to determine its location, neighbors, and CNs and their energy levels. After learning the values of other connected nodes, each node builds its routing table with routes towards a sink through the relay nodes. Since the main concern is to select the optimal route towards the sink, each node i calculates its λ at time t. This value is then shared with the neighbors. The initial configuration can be seen in Figure 7, where two sample nodes are shown with their recorded parameter values.

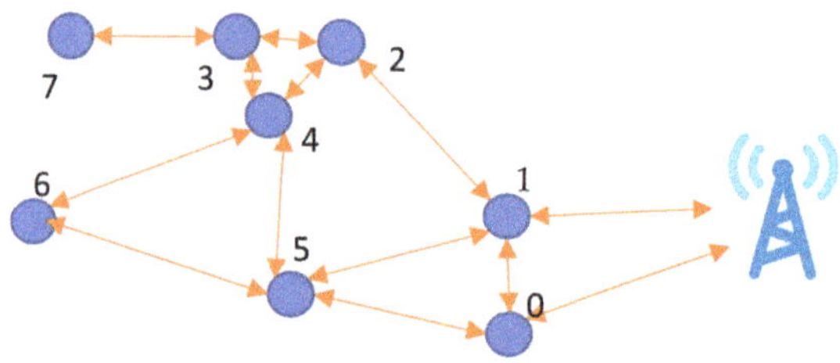

Sample Configuration values		
ID	Node 4	Node 5
Hop level	2	2
Energy	95	80
CNs	2,5,3	-
$\sum_{j=1}{}^{n E}_{CN j}$ $\sum_{j=1}^{n} E(CN_j)$	90 + 60 + 73	0
λ	1.1764	0.80

Figure 7. Initial self-configuration of nodes.

3.2.2. Selection of Relay Nodes

While selecting a route, the source node sends RREQs to an upper-hop node that has the highest λ value. Upon receiving the RREQ, the relay node then further requests its ascendant node, which has a higher λ value. This process is repeated by all forwarders until the intermediate nodes with a hop level of one receive the RREQ. The last-mile node in the route, as the nearest node to the sink, then replies the RREPs to the requested nodes. Subsequently, the RREP is acknowledged to the source node by the forwarders along the reverse route. In Figure 8, the possible connections are shown, among which a route has been selected based on the calculated λ scores.

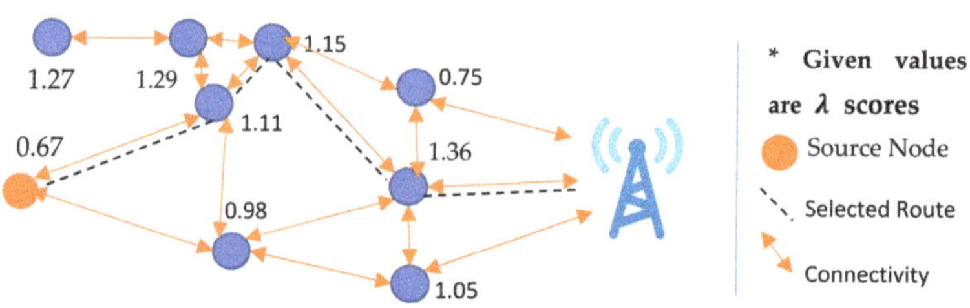

Figure 8. Selection of relay nodes.

3.2.3. Selfish Node Management

It is possible that a node with the highest λ does not respond to an RREQ during n number of attempts. The source node then marks such a node as a selfish node and recalculates the value of λ by considering the remaining nodes. Moreover, the source node piggybacks the address of the selfish node with its RREQ to let the other nodes know about it. The nodes stop entertaining the selfish nodes once they get their information. However, selfish nodes can still be requested for the route. If the selfish node entertains an RREQ, then the source node announces it as a normal node again. In case a node does not respond to an s number of RREQs, it is then broadcasted as a blacklisted node. The network nodes omit the blacklisted nodes from their connected nodes' lists for a specified period of time t. The features and capabilities of nodes in terms of their cooperation levels are indicated with blue, yellow, and red in Figure 9.

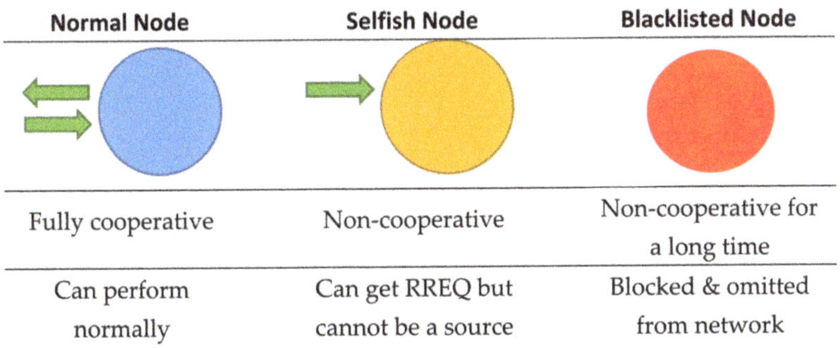

Figure 9. Stages for managing non-cooperative nodes.

3.2.4. Information Exchange

Upon each data transmission, the involved nodes consume their energies. These nodes update their λ values according to their knowledge. Using the DSR protocol [43] as a base protocol, each involved node can get the list of all the relay nodes. They update the values of the involved nodes in their routing tables. However, to avoid further energy consumption, these nodes do not broadcast their updated values. The nodes update their routing tables with approximate values by predicting the energy consumption and the number of transmitted packets.

If a node has not received any updated information about other nodes in a possible route, it selectively sends a route confirmation to a preferred forwarder according to its knowledge. The source node does not proactively update the values, the λ of the relay node, which may be changed due to previous data transmissions. Therefore, the relay node selectively forwards the RREQ of the source node to its sibling neighbor node with the highest λ as shown in Figure 10.

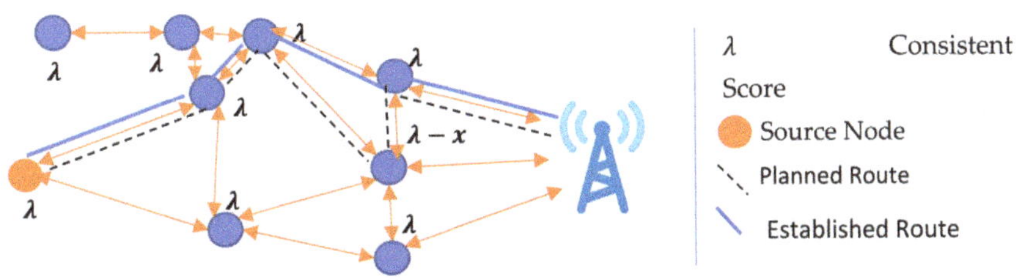

Figure 10. Updating statistics through RREP.

The node also sends additional information with the RREQ that reflects the values of the nodes involved in previous data transmissions. The nodes in such a case search out the route by contacting upper-hop nodes with an RREQ along with the additional information. In this mechanism, the fundamental DSR-based technique of the RREQ broadcast is not used. The source and relay nodes selectively send the RREQ to the preferred nodes in the upper or similar hops.

Sometimes the source node does not receive acknowledgment of a selfish node after several attempts. In such a case, the source node either broadcasts the RREQ to its neighbors or selects another node according to its λ score. Each entry in the routing table is associated with a time stamp. This decision is made according to the time stamp attached to the next node's stored λ score in the routing table. Each entry in the routing table is updated along with the current time.

4. Simulation Results

The proposed work was simulated using MATLAB 2018a. The list of simulation parameters is given in Table 1. The associations of the λ values in the first experiment are shown with the targeted parameters.

Table 1. List of simulation parameters.

Parameter	Value
Number of nodes	50 to 250
Area	500 × 500, 750 × 750, and 1000 × 1000 m^2
Max propagation	100 meters
Max RSSI	100
CN threshold	50
Distance for identical nodes	25
Location of sink	250, 250
Energy max	100 J
Base protocols	DSR, OLSR
Node distribution	Random
Rx power	0.6 W
Tx power	0.6 W
Movement trace	Off
Comparisons	LEACH, DSR, PDORP, PRRP, NSSROP

The placements of 100 nodes can be seen in Figure 11. All the nodes were distributed evenly, while the location of the sink, labeled as BS, was kept at the center of the simulation space. We observed that the nodes were densely deployed in some places, while some nodes had low neighbor density according to their location. The nodes' placements highly affected the network throughput, especially the availability and lifetime of the routes.

In the experiment scenario of Figure 11, the λ values for a sample of 12 nodes were derived, as shown in Table 2. The node that had an ID = 10 with a higher number of CNs had a comparatively higher λ value. This is because nodes with multiple CNs will get more route requests than others. Since their elimination from the network will not affect much due to the presence of multiple CNs, such nodes will be frequently utilized. Moreover, a node that had multiple CNs but less energy compared to its CNs had a lower value. For such cases, the node ID = 6 can be compared with the node ID = 7. Both had an equal number of CNs but different levels of energy. Therefore, different λ values were assigned to ID = 6 and ID = 7. Figure 12 shows a clear relationship among the nodes' energies, their CNs' energies, and the computed values of λ.

This work was further compared with some other protocols such as the PDORP, PRRP, DSR, and LEACH [44]. Experiments were performed to check the energy consumption, network life, throughput, and delays.

Figure 13 shows the comparative results for energy consumption in all the experimented protocols. LEACH is not very sophisticated compared to modern protocols, but it is very famous for creating a baseline for other protocols. Many studies adopt LEACH as a base protocol for designing and comparing their work. In our results, its performance decreased with the increased number of nodes. The PDORP and PRRP obtained consistent results in terms of energy efficiency. The authors who developed the PDORP claimed to obtain encouraging results by using the genetic algorithm with a modified DSR. The PRRP was better than the DSR and LEACH but could not compete with the others. The key reason for this is the incorporation of a typical GPS in the nodes. The proposed mechanism did not perform well with a lower number of nodes, such as 50, because the NSSROP operates on the scores that are based on the nodes' neighborhoods and densities. In the case of a smaller number of nodes, there were fewer or no CNs and identical CNs. Therefore, the proposed mechanism failed to obtain distinctive features from its key parameters with a lower number of nodes. However, a WSN-based IoT network mostly consists of a large number of devices. In such a dense network, therefore, the NSSROP worked much better than the other protocols; the NSSROP outperformed the other protocols with a number of nodes greater than 100.

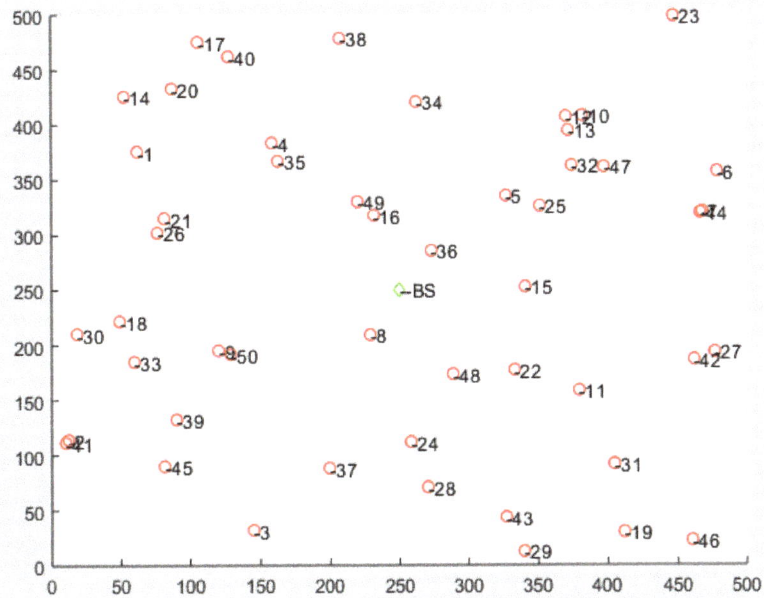

Total Area: 500 m × 500 m. *Total Node:* 50 *CN_Threshold:* 50
Nodes' placement: Random *BS Location:* 250, 250 *Transmission Range:* 200

Figure 11. Placement of 100 nodes in an area of 500 m × 500 m.

Table 2. λ scores influenced by other parameters.

ID	Energy	CNs	CNs Energy	CN Energy Sum	λ
1	62.7347	0	0	0	0.6273
2	02.1650	41	94.5579	94.5579	0.0223
3	91.0570	0	0	0	0.9106
4	80.0559	35	10.6942	10.6942	1.4464
5	74.5847	25	68.3839	68.3839	0.8859
6	81.3113	7	38.3306	115.0137	1.1345
		44	76.6831		
7	38.3306	6	81.3113	157.9944	0.4457
		44	76.6831		
8	61.7279	0	0	0	0.6173
9	57.5495	50	52.7847	52.7847	0.7533
10	53.0052	12	24.8629	116.2235	1.2257
		13	45.1639		
		32	21.7802		
		47	24.4165		

Table 2. *Cont.*

11	27.5070	0 10	0 53.0052	0	0.2751
12	24.8629	13 32	45.1639 21.7802	119.9493	0.4522

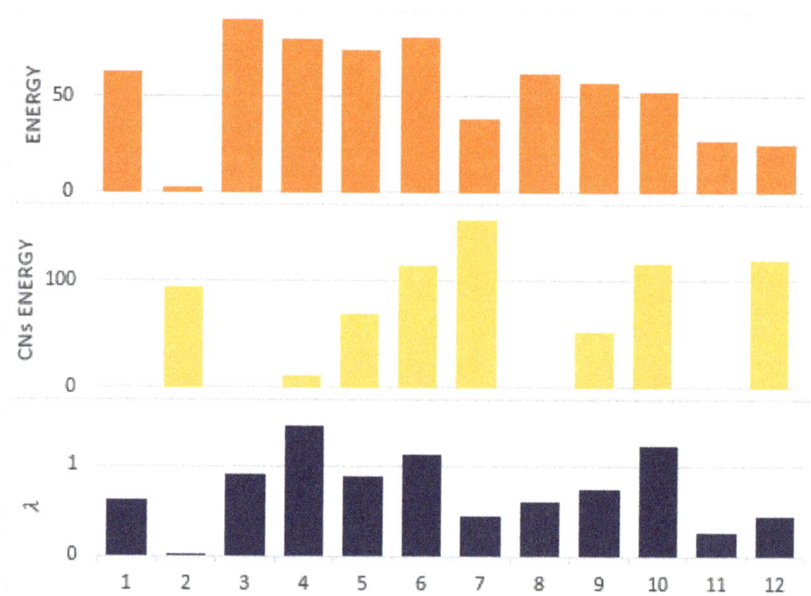

Figure 12. λ values according to nodes' energies and CNs' energies.

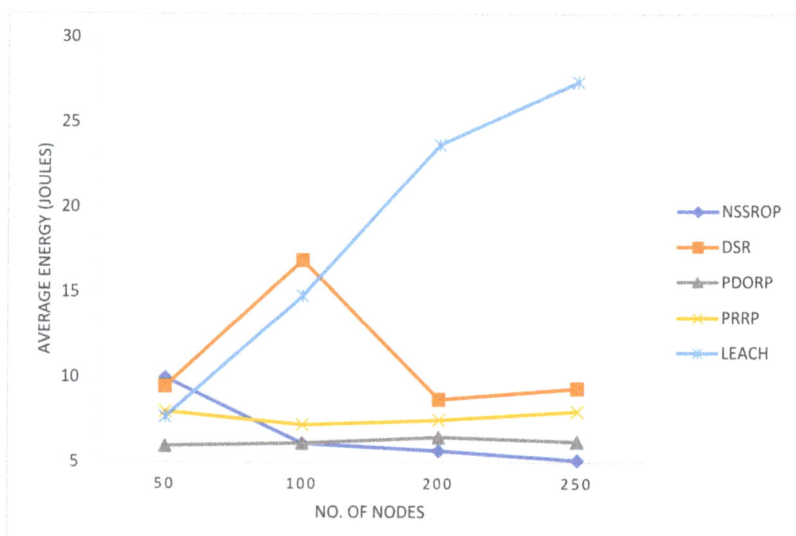

Figure 13. Average energy consumption by 5 protocols with a varying number of nodes.

The results for the ratio of dead nodes against five pause times can be seen in Figure 14. These results were reflected by the previous experiment on energy consumption. The NSSROP also outperformed in this test. Due to the equal load balancing, our protocol allowed all the nodes to equally participate in the network. Moreover, the blacklisting mechanism was also effective by not letting other nodes waste their energies on contacting them.

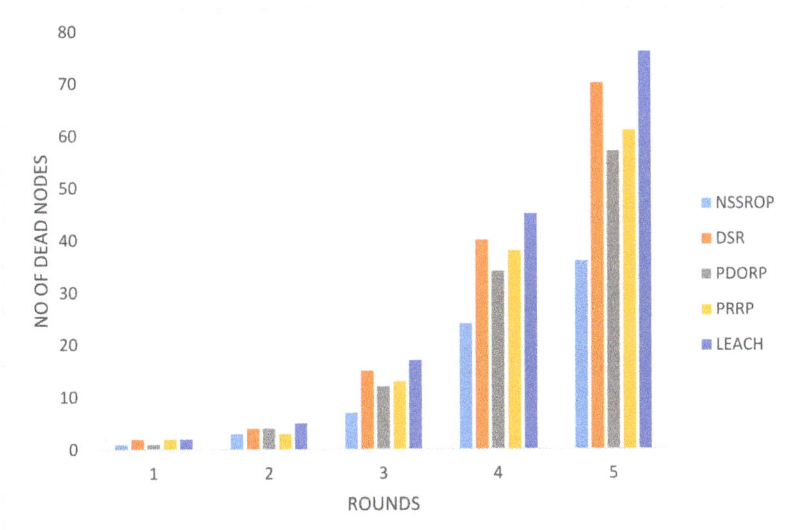

Figure 14. Number of dead nodes with different pause times.

Figure 15 shows the comparative results for the throughput of the experimented protocols. The PDROP and PRRP somehow obtained similar results. The DSR and LEACH had very poor throughput in the experiment. The PRRP has a mechanism that operates on fixed-sized grids and does not rely on the time duration; therefore, it had a relatively consistent level of throughput. The PDORP initially took time to implement its hybrid mechanism of a genetic algorithm and bacterial foraging optimization. As shown in the figure, the NSSRP achieved a comparatively higher throughput than the others. The main reason for this is the implementation of modified control packets, that is, RREQ, RREP, and OLSR-based topology control messages. With these modifications and incorporation of scoring, the packet drops decreased, and the exchange of data increased, causing a higher throughput.

An experiment to check the impact of the density of nodes was carried out by varying the area up to 1000 m^2 with a set of 250 nodes. The results in Figure 16 show that the performance of all the protocols degraded with an increased area size. This is because the scattered nodes may not have been able to gain multiple routes due to longer inter-node distances. The NSSROP obtained almost similar results to the PDORP for 1000 m^2. However, due to the utilization of the density aspect, the performance of the NSSROP was much better for 500 m^2 against the PDORP. Such a higher density increased the number of CNs and subsequently caused equal load distribution among the nodes and better selection among multiple routes.

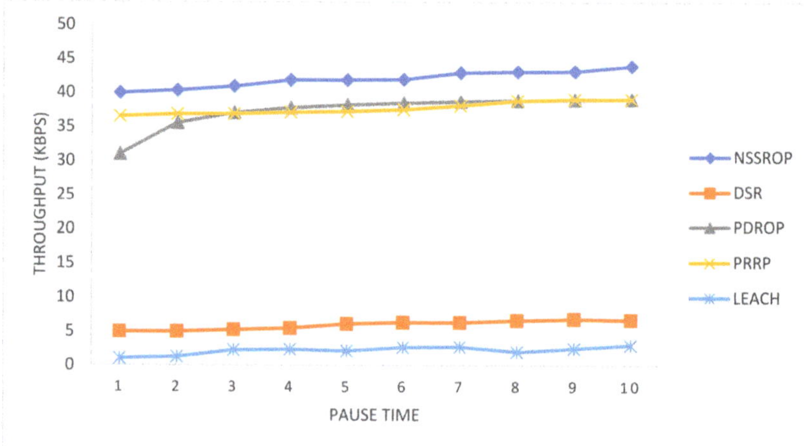

Figure 15. Throughput (packets per second) at time intervals.

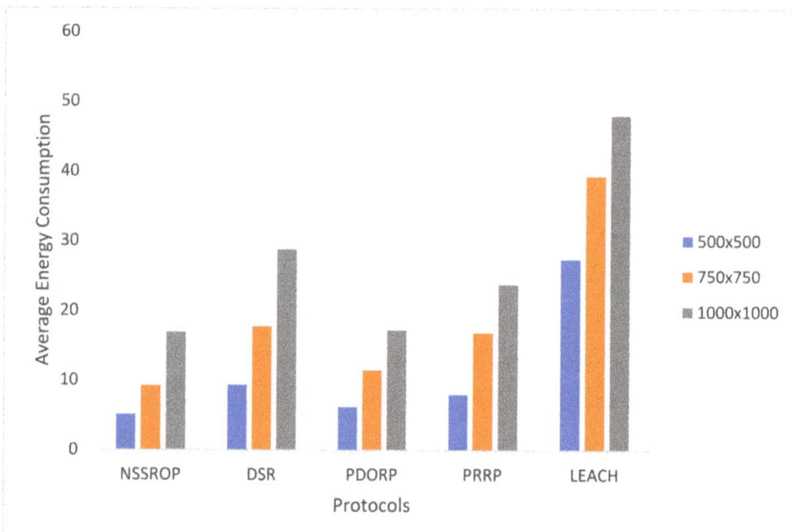

Figure 16. Average energy consumption with varying area size.

In Figure 17, the end-to-end delays are shown. The DSR and LEACH had the worst results due to their outdated routing procedures. DSR uses the typical route discovery mechanism that has a drawback of higher delays. The PRRP had moderate values for this experiment. The PDORP and NSSROP had almost similar results for end-to-end delays. The main reason for this is the incorporation of modified routing tables and the exchange of periodic topology control messages along with on-demand route discoveries. Moreover, the selection of appropriate routes also had a significant impact on the end-to-end delays.

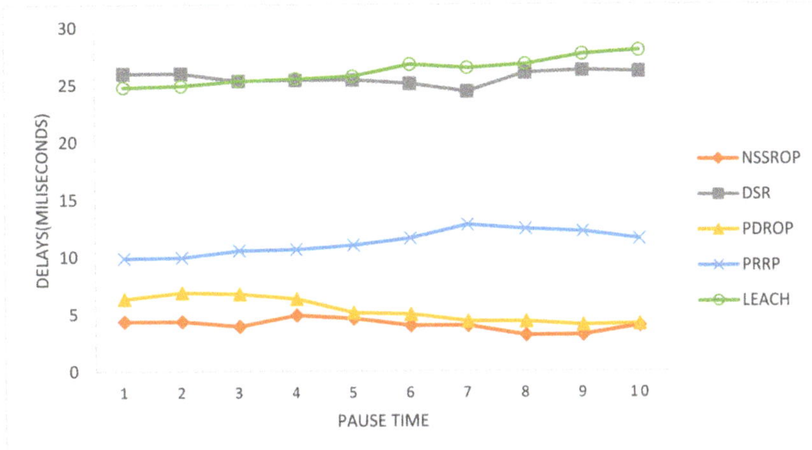

Figure 17. Average end-to-end delays with time pauses.

5. Conclusions and Future Work

In this work, we proposed a new routing protocol, the NSSROP, which balances the load efficiently among the nodes in a WSN-based IoT environment. We implemented the NSSROP on top of two base protocols, the DSR and OLSR, with the novel scoring mechanism for path selection. Each node is scored considering its energy and CNs to indicate the nodes' densities. In addition, the blacklisting mechanism is defined to deal with non-cooperating nodes in the WSN. In the experiment, the NSSROP showed outstanding results in terms of average energy consumption, throughput, and end-to-end delay.

This work can be further expanded by incorporating game theory and using clusters or groups in the network. A Stackelberg or evolutionary game can be incorporated into the mechanism for cluster formation and cluster head selection processes. Moreover, the same mechanism can be modified to design a scheme for the development of trusted routes while considering selfish nodes in the network. Many trust management systems have been proposed. The existing trust management systems, focusing on trust development from node to node, can be extended to the trust development for entire routes in the network.

Author Contributions: W.K. introduced the main idea and finalized the methodology. W.K. supervised the entire work and also did the funding acquisition. M.M.U. conducted investigation, simulations and validations. S.K. conducted the formal analysis, implementation of methods, administrations. M.A.K. is expert in writing, carried out original draft preparation, review, editing and visualization. All authors have read and agreed to the published version of the manuscript.

Funding: This work was supported by the Technology development Program (No. G21S286191902) funded by the Ministry of SMEs and Startups (MSS, Korea).

Conflicts of Interest: The authors declare no conflict of interest.

References

1. Atzori, L.; Lera, A.; Morabito, G. A survey on Internet of Things architectures. *J. King Saud Univ.-Comput. Inf. Sci.* **2018**, *30*, 291–319.
2. Kandris, D.; Nakas, C.; Vomvas, D.; Koulouras, G. Applications of Wireless Sensor Networks: An Up-to-Date Survey. *Appl. Syst. Innov.* **2020**, *3*, 14. [CrossRef]
3. Deng, R.; Chen, J.; Cao, X.; Zhang, Y.; Maharjan, S.; Gjessing, S. Sensing-Performance Tradeoff in Cognitive Radio enabled Smart Grid. *IEEE Trans. Smart Grid* **2013**, *4*, 302–310. [CrossRef]
4. Guerroumi, M.; Pathan, A.-S.K.; Badache, N.; Moussaoui, S. On the Medium Access Control Protocols Suitable for Wireless Sensor Networks—A Survey. *Int. J. Commun. Netw. Inf. Secur. (IJCNIS)* **2014**, *6*, 89–103.
5. Melodia, T.; Akyildiz, I.F. Research Challenges for Wireless Multimedia Sensor Networks. *Distrib. Video Sens. Netw.* **2011**, 233–246. [CrossRef]

6. Pasricha, S.; Ayoub, R.; Kishinevsky, M.; Mandal, S.K.; Ogras, U.Y. A Survey on Energy Management for Mobile and IoT Devices. *IEEE Des. Test* **2020**, *37*, 7–24. [CrossRef]
7. Dehkordi, S.A.; Farajzadeh, K.; Rezazadeh, J.; Farahbakhsh, R.; Sandrasegaran, K.; Dehkordi, M.A. A survey on data aggregation techniques in IoT sensor networks. *Wirel. Netw.* **2020**, *26*, 1243–1263. [CrossRef]
8. Lloret, J.; Diaz, J.; Jimenez, J.; Boronat, F. An architecture to connect disjoint multimedia networks based on node's capacity. In Proceedings of the Pacific-Rim Conference on Multimedia, Hangzhou, China, 2–4 November 2006; pp. 890–899.
9. Umar, M.M.; Khan, S.U.; Ahmad, R.; Singh, D. Game Theoretic Reward Based Adaptive Data Communication in Wireless Sensor Networks. *IEEE Access* **2018**, *6*, 28073–28084. [CrossRef]
10. Kundaliya, B.L.; Hadia, S.K. Routing Algorithms for Wireless Sensor Networks: Analysed and Compared. *Wirel. Pers. Commun.* **2020**, *110*, 85–107. [CrossRef]
11. Mostafaeia, H.; Montierib, A.; Persicoc, V.; Pescapé, A. A sleep scheduling approach based on learning automata for WSN partial coverage. *J. Netw. Comput. Appl.* **2017**, *80*, 76–78.
12. Dong, L.; Tao, H.; Doherty, W.; Young, M. A Sleep Scheduling Mechanism with PSO Collaborative Evolution for Wireless Sensor Networks. *Int. J. Distrib. Sens. Netw.* **2015**, *11*, 517250. [CrossRef]
13. Umar, M.M.; Alrajeh, N.; Mehmood, A. SALMA: An Efficient State-Based Hybrid Routing Protocol for Mobile Nodes in Wireless Sensor Networks. *Int. J. Distrib. Sens. Netw.* **2016**, *12*, 2909618. [CrossRef]
14. Khadim, R.; Maaden, A.; Ennaciri, A.; Erritali, M. An Energy-Efficient Clustering Algorithm for WSN Based on Cluster Head Selection Optimization to Prolong Network Lifetime. *Int. J. Future Comput. Commun.* **2018**, *7*, 51–57. [CrossRef]
15. Zheng, G.; Dong, X. Node Importance Evaluation in Wireless Sensor Networks based on Topology Contribution and Energy Efficiency. In Proceedings of the 7th International Conference on Communication and Network Security, Toyko, Japan, 24–26 November 2017.
16. Popa, L.; Rostamizadeh, A.; Karp, R.; Papadimitriou, C.; Stoica, I. Balancing traffic load in wireless networks with curveball routing. In Proceedings of the 8th ACM International Symposium on Mobile Ad Hoc Networking and Computing—MobiHoc, Montreal, QC, Canada, 9–14 September 2007; Volume 7.
17. Brandt, A.; Hui, J.; Kelsey, R.; Levis, P.; Pister, K.; Struik, R.; Alexander, R. *RPL: IPv6 Routing Protocol for Low-Power and Lossy Networks*; IETF RFC 6550; Internet Engineering Task Force (IETF): Fremont, CA, USA, 2012.
18. Sobral, J.V.V.; Rodrigues, J.J.P.C.; Rabêlo, R.A.L.; Al-Muhtadi, J.; Korotaev, V. Routing Protocols for Low Power and Lossy Networks in Internet of Things Applications. *Sensors* **2019**, *19*, 2144. [CrossRef]
19. Abasıkeleş-Turgut, İ.; Altan, G. A fully distributed energy-aware multi-level clustering and routing for WSN-based IoT. *Trans. Emerg. Telecommun. Technol.* **2021**, *32*, e4355. [CrossRef]
20. Abid, K.; Jaber, G.; Lakhlef, H.; Lounis, A.; Bouabdallah, A. An Energy Efficient Architecture of self-sustainable WSN based on Energy Harvesting and Wireless Charging with Consideration of Deployment Cost. In Proceedings of the 16th ACM Symposium on QoS and Security for Wireless and Mobile Networks, Alicante, Spain, 16–20 November 2020.
21. Zaman, N.; Abdullah, A.B.; Jung, L.T. Optimization of energy usage in Wireless Sensor Network using Position Responsive Routing Protocol (PRRP). In Proceedings of the IEEE Symposium on Computers & Informatics, Kuala Lumpur, Malaysia, 20–23 March 2011.
22. Wang, Z.; Qin, X.; Liu, B. An Energy-Efficient Clustering Routing Algorithm for WSN-Assisted IoT. In Proceedings of the IEEE Wireless Communications and Networking Conference (WCNC), Barcelona, Spain, 15–18 April 2018.
23. Xu, C.; Xiong, Z.; Zhao, G.; Yu, S. An energy-efficient region source routing protocol for lifetime maximization in WSN. *IEEE Access* **2019**, *7*, 135277–135289. [CrossRef]
24. Yao, Y.; Cao, Q.; Vasilakos, A.V. EDAL: An energy-efficient, delay-aware, and lifetime-balancing data collection protocol for heterogeneous wireless sensor networks. *IEEE/ACM Trans. Netw. (TON)* **2015**, *23*, 810–823. [CrossRef]
25. Han, G.; Dong, Y.; Guo, H.; Shu, L.; Wu, D. Cross-layer optimized routing in wireless sensor networks with duty cycle and energy harvesting. *Wirel. Commun. Mob. Comput.* **2014**, *15*, 1957–1981. [CrossRef]
26. Elappila, M.; Chinara, S.; Parhi, D.R. Survivable Path Routing in WSN for IoT applications. *Pervasive Mob. Comput.* **2018**, *43*, 49–63. [CrossRef]
27. Syarif, A.; Abouaissa, A.; Lore, P. Operator Calculus Approach for Route Optimizing and Enhancing Wireless Sensor Network. *J. Netw. Comput. Appl.* **2017**, *97*, 1–10. [CrossRef]
28. Brar, G.S.; Rani, S.; Chopra, V.; Malhotra, R.; Song, H.; Ahmed, S.H. Energy efficient direction-based PDORP routing protocol for WSN. *IEEE Access* **2016**, *4*, 3182–3194. [CrossRef]
29. Lai, X.; Ji, X.; Zhou, X.; Chen, L. Energy Efficient Link-Delay Aware Routing in Wireless Sensor Networks. *EEE Sens. J.* **2018**, *18*, 837–848. [CrossRef]
30. Marti, S.; Giuli, T.J.; Lai, K.; Baker, M. Mitigating routing misbehavior in mobile ad hoc networks. In Proceedings of the 6th Annual International Conference on Mobile Computing and Networking, Boston, MA, USA, 6–11 August 2000.
31. Sun, Q.; Qiao, Y.; Wang, J.; Shen, S. Node importance evaluation method in wireless sensor network based on energy field model. *EURASIP J. Wirel. Commun. Netw.* **2016**, *199*, 2016. [CrossRef]
32. Luo, J.; Wu, D.; Pan, C.; Zha, J. Optimal Energy Strategy for Node Selection and Data Relay in WSN-based IoT. *Mob. Netw. Appl.* **2015**, *20*, 169–180. [CrossRef]

33. Sheriba, S.; Rajesh, D.H. Energy-efficient clustering protocol for WSN based on improved black widow optimization and fuzzy logic. *Telecommun. Syst.* **2021**, *77*, 213–320. [CrossRef]
34. Sheriba, S.; Rajesh, D. Improved hybrid cuckoo black widow optimization with interval type 2 fuzzy logic system for energy-efficient clustering protocol. *Int. J. Commun. Syst.* **2021**, *34*, e4730. [CrossRef]
35. Mittal, N.; Singh, S.; Singh, U.; Salgotra, R. Trust-aware energy-efficient stable clustering approach using fuzzy type-2 Cuckoo search optimization algorithm for wireless sensor networks. *Wirel. Netw.* **2021**, *27*, 151–174. [CrossRef]
36. Long, C.N.; Zhang, Q.; Li, B.; Yang, H.; Guan, X. Non-Cooperative Power Control for Wireless Ad hoc Networks with Repeated Games. *IEEE J. Sel. Areas Commun.* **2007**, *25*, 1101–1112. [CrossRef]
37. Tsiropoulou, E.E.; Katsinis, G.K.; Papavassiliou, S. Distributed Uplink Power Control in Multi-Service Wireless Networks via a Game Theoretic Approach with Convex Pricing. *IEEE Trans. Parallel Distrib. Syst.* **2012**, *23*, 61–68. [CrossRef]
38. Cheng, H.; Su, Z.; Zhang, D.; Lloret, J.; Yu, Z. Energy-efficient node selection algorithms with correlation optimization in wireless sensor networks. *Int. J. Distrib. Sens. Netw.* **2014**, *10*, 576573. [CrossRef]
39. Sicari, S.; Rizzardi, A.; Grieco, L.A.; Coen-Porisini, A. GoNe: Dealing with node behavior. In Proceedings of the 2015 IEEE 5th International Conference on Consumer Electronics, Berlin (ICCE-Berlin), Berlin, Germany, 6–9 September 2015.
40. Gattani, V.S.; Jafri, S.M.H. Data collection using score based load balancing algorithm in wireless sensor networks. In Proceedings of the 2016 International Conference on Computing Technologies and Intelligent Data Engineering (ICCTIDE'16), Kovilpatti, India, 7–9 January 2016.
41. Yousefi, H.; Dabirmoghaddam, A.; Mizanian, K.; Jahangir, A.H. Score based reliable routing in wireless sensor networks. In Proceedings of the 2009 International Conference on Information Networking, Chiang Mai, Thailand, 21–24 January 2009.
42. Xu, J.; Liu, W.; Lang, F.; Zhang, Y.; Wang, C. Distance Measurement Model Based on RSSI in WSN. *Wirel. Sens. Netw.* **2010**, *2*, 606–611. [CrossRef]
43. Johnson, D.B.; Maltz, D.A.; Broch, J. DSR: The dynamic source routing protocol for multi-hop wireless ad hoc networks. *Ad Hoc Netw.* **2001**, *5*, 139–172.
44. Tandel, R.I.; Reshma. Leach Protocol in Wireless Sensor Network: A Survey. *Int. J. Comput. Sci. Inf. Technol.* **2016**, *7*, 1894–1896.

sensors

MDPI

Article

Internet of Things Concept in the Context of the COVID-19 Pandemic: A Multi-Sensor Application Design

Alexandru Lavric [1,*], Adrian I. Petrariu [1,2], Partemie-Marian Mutescu [1], Eugen Coca [1] and Valentin Popa [1,2]

1 Computers, Electronics and Automation Department, Stefan Cel Mare University of Suceava, 720229 Suceava, Romania; apetrariu@usm.ro (A.I.P.); marian.mutescu@usm.ro (P.-M.M.); eugen.coca@usv.ro (E.C.); valentin.popa@usm.ro (V.P.)

2 MANSiD Research Center, Stefan Cel Mare University of Suceava, 720229 Suceava, Romania

* Correspondence: lavric@usm.ro

Abstract: In this paper, we present the design, development and implementation of an integrated system for the management of COVID-19 patient, using the LoRaWAN communication infrastructure. Our system offers certain advantages when compared to other similar solutions, allowing remote symptom and health monitoring that can be applied to isolated or quarantined people, without any external interaction with the patient. The IoT wearable device can monitor parameters of health condition like pulse, blood oxygen saturation, and body temperature, as well as the current location. To test the performance of the proposed system, two persons under quarantine were monitored, for a complete 14-day standard quarantine time interval. Based on the data transmitted to the monitoring center, the medical staff decided, after several days of monitoring, when the measured values were outside of the normal parameters, to do an RT-PCR test for one of the two persons, confirming the SARS-CoV2 virus infection. We have to emphasize the high degree of scalability of the proposed solution that can oversee a large number of patients at the same time, thanks to the LoRaWAN communication protocol used. This solution can be successfully implemented by local authorities to increase monitoring capabilities, also saving lives.

Keywords: Internet of Things; COVID-19 sensors; remote healthcare; telemedicine; quarantine monitoring; IoT wearable device

Citation: Lavric, A.; Petrariu, A.I.; Mutescu, P.-M.; Coca, E.; Popa, V. Internet of Things Concept in the Context of the COVID-19 Pandemic: A Multi-Sensor Application Design. *Sensors* **2022**, *22*, 503. https://doi.org/10.3390/s22020503

Academic Editor: Marco Picone

Received: 17 December 2021
Accepted: 5 January 2022
Published: 10 January 2022

Publisher's Note: MDPI stays neutral with regard to jurisdictional claims in published maps and institutional affiliations.

1. Introduction

In recent years, we have witnessed continuous discussions about the IoT (Internet of Things) concept, which involves the connection of various objects that surround us in our everyday life to the Internet. The main purpose of the IoT concept is closely related to the smart city topic, enabling an increase in quality of life by contributing to the efficient use of resources and environment protection.

IoT technologies are sufficiently enhanced to enable the development of integrated solutions for the challenges humankind is facing today. All this knowledge and technological progress does not seem to have prepared us enough for the current context of the pandemic we are experiencing. Considering the pandemic context, all of these IoT devices that surround us in our daily life are powerless in the face of the spread of the SARS-CoV2 virus infection. Until the start of the pandemic, there was not a strong connection between the IoT and the healthcare industry. This aspect must change and allow for the integration of healthcare services in current existing IoT infrastructure; this change is also forced by the current pandemic situation. Thus, it is our responsibility, as a research community, to develop new systems and find solutions to current problems.

The COVID-19 disease, which is a severe acute respiratory syndrome generated by the SARS-CoV2 virus, is an ongoing pandemic [1]. This has led to an immense public health concern in the international community, as the World Health Organization (WHO)

has stated, that the outbreak was a public health emergency. International concern led to declaring the COVID-19 outbreak a global pandemic on the 11th of March 2020 [2]. Since then, things have progressed quite rapidly, with the number of infections growing exponentially. Globally, as of 8th December 2021, there have been approximately 300 million confirmed cases of COVID-19, including more than 5 million deaths, reported to WHO [3].

In this paper, we present the design, development, and implementation of an integrated system for COVID-19 illness management that is associated with the SARS-CoV2 virus infection. The urgent need for this system is related to the fact that the architecture can enable local authorities to reduce the pressure on the medical system by integrating telemedicine facilities allowing for remote monitoring of patients. Thus, the health status of patients is closely monitored by medical staff and depending on the current load on the medical system, suspect patients, quarantine, isolated patients and even mild cases can be remotely monitored while at home.

The proposed system integrates an IoT multi-sensor approach that offers the possibility of monitoring different vital signs of the patient. The patient-monitoring node is integrated in a wearable device that is easy to wear and fully reconfigurable, allowing further development by integrating other sensor types. Thus, the proposed system can be seen as a powerful IoT multi-sensor platform. The implemented and tested novel architecture allows for the integration of a high number of wearable devices, in an effective and cost-efficient manner. The system also monitors the persons that are quarantined due to infection or suspicion of infection with the SARS-CoV2 virus.

The paper is organized as follows: first, a brief introduction related to the state-of-the-art technology, followed by Section 2, where the main IoT challenges regarding the SARS-CoV2 pandemic are presented. The IoT communication protocols, and challenges for pandemic control are discussed in Section 3. In Section 4, the IoT wearable multi-sensor architecture for remote patient monitoring is presented and discussed in detail. The final section of the paper is represented by the conclusions and the overall performance discussion of the system. From the analysis conducted in this paper and the obtained results, we can see that the proposed system ensures a high-level of performance and can be easily developed and implemented by the local authorities to scale up the remote patient monitoring capabilities, ultimately saving lives.

The main contributions and originality of this paper are the following:

(1) The design of an in-depth analysis of IoT challenges regarding the SARS-CoV2 pandemic;
(2) The performance evaluation of the communications protocol that can be integrated in the patient monitoring architecture distributed over a large geographical area;
(3) The design, development, and implementation of a novel integrated system for COVID-19 illness management;
(4) The patient monitoring IoT multi-sensor platform integration in a wearable device that is easy to wear and fully reconfigurable;
(5) The multi-sensor IoT wearable device can monitor the health condition related parameters like pulse, blood oxygen saturation and body temperature, as well as the patient's current location increasing the monitoring capabilities of the local authorities and saving lives.

2. SARS-CoV2 and IoT Integration

The SARS-CoV2 virus infection can be the cause of a potentially fatal disease that currently threatens humanity as a global public health issue. The rapid person-to-person transmission of COVID-19 infection is extremely concerning. It is precisely this aspect that led to many people being isolated or quarantined. Thus, extensive measures to reduce person-to-person transmission have been implemented to control the current outbreak. Special attention and significant efforts to protect or reduce transmission must be applied in sensitive populations, including children, healthcare providers, older people, and people with serious health issues.

The new B.1.1.529 or Omicron variations of SARS-CoV2 may be more transmissible than other variants and are partially resistant to existing vaccines. According to Markets and Markets, the Internet of Medical Things (IoMT) market is currently worth USD 26.5 billion and is expected to reach USD 94.2 billion by 2026 with a CAGR (Compound Annual Growth Rate) of 28.9% [4]. The major challenges in using the IoT concept are most often related to ensuring a high autonomy (obtaining the lowest possible energy consumption), the possibility of integrating as many wireless sensors as possible, scalability, and ensuring the highest possible communication range.

Another aspect that should not be neglected is that of communication system deployment costs, that should be kept to a minimum. Thus, it would be ideal for the sensors to be able to transmit data within an unlicensed frequency spectrum, avoiding additional costs required by cellular data subscriptions. Nowadays, IoT technologies are mature enough to be integrated into healthcare diagnostic systems that can contribute effectively to the fight against this unseen enemy named COVID-19. All these aspects have generated a series of new problems that we have never faced before. Thus, it is our duty to identify new solutions that contribute to increasing the patient's medical care by providing support to local government authorities.

Worldwide, there are several applications that use machine learning (ML) technologies into real-time healthcare decisions, contributing a great deal to saving lives. Improving the efficiency and quality of hospital care services has proved to be an important and critical challenge during the pandemic. Machine learning is also included in the informational technology family along with artificial intelligence (AI) [5–8], as well as AI prediction [9] and it can be integrated within the diagnostic process, contributing to the rapid detection of the SARS-CoV2 infection [10,11]. These mechanisms can also help reduce misdiagnosis, reduce the diagnosis time and ultimately save patients' lives. All these techniques along with the above-mentioned mechanisms must help in the context of the pandemic and contribute to slow down the spread of infection and make the diagnosis process more efficient. Figure 1 presents the main identified advantages of using IoT devices to fight against the current COVID-19 pandemic. The advantages that can contribute to saving lives are related to the increase in treatment quality [12], remote monitoring of the patients [13], quarantine control systems [14], telemedicine [15], new enhanced diagnosis techniques and improving the virus detection techniques [16].

Figure 1. Advantages of using IoT for COVID-19 pandemic monitoring.

During the pandemic, local authorities also faced situations where people violate the imposed quarantine or isolation conditions. Thus, the designed system proposed in this paper allows for the active monitoring of the patient's position while also monitoring the health condition. The main advantages are to limit the spread of the disease and avoid serious complications by administering the optimal treatment on time.

3. Communication Protocols, and IoT Challenges for Pandemic Control

The current situation demands a deep analysis for a better understanding of humankind's preparation level in facing the SARS-CoV2 virus. This analysis was performed taking into account the informational technologies that can be used and integrated into applications and systems which will allow the flattening of the pandemic evolution curve and can even help eradicate it. Humankind has invested enormously in a variety of technologies that surround us in our daily life, contributing to the increase in its quality, but it seems that in the face of such an enemy we are powerless.

The introduction of communication capabilities to simple objects involves the development and implementation of large-scale, high-density complex network topologies. Usually, the wireless sensor devices are distributed over a large geographical area so specialized communication protocols are needed to integrate a large number of devices in the same network. All these protocols must also ensure a high level of performance and be cost efficient.

The first challenge when a large-scale patient monitoring system is developed is related to the selection of the communication protocol. Thus, to select the best candidate for the communication protocol, functional requirements of the system must be correlated with the current available technologies. The second step is related to the design and development of a wearable IoT multi-sensor platform that can integrate a multi-sensor approach in a flexible manner, while also considering the power efficiency of the design.

Currently, there are all sorts of standards, protocols, and communication mechanisms that promise to solve the main problems raised by the IoT concept like SigFox [17], NB-IoT [18], Symphony Link [19], IEEE 802.15.4 [20], Z-Wave [21], IEEE 802.15.1 [22] and cellular technologies [23]. Table 1 contains a comparison of the main IoT communication protocols that can be implemented in the IoT patient monitoring system.

Table 1. IoT communication protocols that can be implemented in patient monitoring system.

Application Requirements	SigFox	NB-IoT	Symphony Link	IEEE 802.15.4	Z-Wave	IEEE 802.15.1	Cellular Technologies	LoRaWAN
Communication range	~4 km	~1 km	~5 km	~150 m	~100 m	~100 m	~1 km	~5–10 km
Large-scale scalability	✓	✓	✓	-	-	-	✓	✓
Popularity	✓	✓	-	-	✓	✓	✓	✓
Frequency band	Unlicensed	Licensed	Unlicensed	Unlicensed	Unlicensed	Unlicensed	Licensed	Unlicensed
Carrier independent	-	-	✓	✓	✓	✓	-	✓
Power consumption Efficiency	Low	Medium	Low	Low	Low	Low	High	Low
Fast deployment	-	-	-	✓	✓	✓	-	✓
Cost-effective solution	-	-	-	-	-	✓	-	✓
Resistance to interferences	✓	-	-	-	-	✓	✓	✓

The particularity of the patient monitoring system involves the integration of many sensors that can communicate and are distributed over a large geographical area. The challenges are great considering the small communication distance due to the limited access resources of sensors like processing capabilities, available data storage or limited power sources. At the same time, to ensure the highest possible level of performance, as well as the integration of multiple sensors, it is mandatory that we improve and try to enhance the communication mechanism through intense research while constantly evaluating the possibilities it offers in stopping the current pandemic context.

To summarize, the communication protocol that can be integrated in the patient monitoring architecture has to meet the following characteristics:

- Ensure a long communication distance and a high-level of performance within a city-specific architecture;
- Have the possibility to integrate wireless sensors distributed over a large geographical area;
- Be suitable for non-LoS (line of sight) urban specific conditions;
- Have resistance to radio interference;
- Confirm the technology is already available and implemented by many municipalities;
- Facilitate massive deployment within a low complexity, communication architecture, with a short implementation time;
- Contribute to the active monitoring of the patient's health condition;
- Have low-power consumption to allow the integration in a wearable multi-sensor smart device;
- Provide a high level of performance as a cost-effective solution.

From the previously presented information we can conclude that the LoRaWAN communication protocol ensures a high level of performance and can be integrated in development of the patient monitoring system.

LoRaWAN is a communication protocol for media access control (MAC) designed for IoT applications and wide area networks that use LoRa modulation. LoRa modulation uses orthogonal spreading factors (SF) for individual wireless nodes to increase the communication range by reducing the data rate. LoRa modulation uses communication channels with fixed bandwidths of 125 kHz or 250 kHz for uplink channels and 500 kHz for downlink radio channels [24]. The SF can be varied from 7 to 12. The higher the SF, the higher the value of the packet time on air (ToA) will be, lowering the data rate and increasing the communication range by lowering the sensitivity level of the radio transceiver, from -123 dBm for using SF7 to -137 dBm when SF12 is used, respectively [25]. Thus, LoRa wireless nodes that are located in the proximity of the gateway will use SF7 and nodes located at longer distances will use SF12. The SF allocation is defined by the LoRaWAN specifications [26] in an adaptive data rate (ADR) algorithm where the SF is increased if the communication link budget is high meanwhile the SF is reduced if the communication link budget is low. The ADR and SF particularity of the LoRaWAN communication protocol offers a real advantage in urban environmental conditions, achieving long communication distances and is ideal for the patient monitoring system.

The increase in COVID-19 infected people brought along with it a general overload of the medical system. In many regions the hospitals are overwhelmed with severe cases, leaving no room for the mild cases. Taking this into consideration, recently we have seen great interest in remote patient monitoring devices and in telemedicine, as is discussed in the previous section. In the scientific literature, multiple approaches are presented for remote patient monitoring, each one with unique features, advantages or disadvantages that must be overcome in future iterations.

Zhang et al. [27] present a device for COVID-19 prevention that monitors and records the daily activity of a patient. The main monitored parameters are the movement of the body recorded with a 3-axis accelerometer and the body temperature recorded with a CMOS analog temperature sensor. The device is attached to the subject's wrist with two elastic bands to ensure constant contact of the temperature sensor. The data are transmitted to a PC using a BLE (Bluetooth low energy) module. The system also integrates an AI approach as to detect patterns of human activities. One disadvantage of the system is the short range of the communication transceiver, covering only approximately 10 m. The developed architecture is not scalable due to the BLE integration thus, long-distance communication was not considered.

Ullah et al. [28] propose a patient quarantine monitoring system based on multiple sensors approach. The sensors used in the application (temperature, respiratory, accelerometer, pulse, SpO_2 and GPS—global positioning system) are distributed on the patient's body. The communication with the main microcontroller unit (MCU) is achieved using a BLE communication link, same as the above related solution. The MCU forwards the measurement information to a local server through an Internet connection. The system provides a solution for the indoor GPS signal limitation by measuring the RSSI (receive signal strength indicator) of the Wi-Fi signal between the sensor device and the locally deployed server. Alerts and notifications are sent to the local authorities if the quarantine conditions are violated. This system is presented as being able to monitor multiple patients, but the capability is reduced due to the low coverage area of the Wi-Fi communication. Another disadvantage is the high complexity of the system on the sensor level, as the sensors are distributed on different areas of the body being mostly a proof-of-concept system, unavailable as an integrated wearable device.

Mukhtar et al. [29] present an IoT enabled solution for patient monitoring, using a rule-based approach for determining the current state of the patient health. As with the previous presented system [28], the sensors are distributed on the patient's body to collect data, regarding the pulse, SpO_2, temperature and cough rate. The measurements are sent to the cloud by an 802.11n communication protocol (Wi-Fi), where they are processed and analyzed. The authors define four patient classes, each patient being attributed a class through a rule-based classification process. The classes are summarized in Table 2.

Table 2. COVID-19 symptom classes.

Patient Class	SpO_2	Cough Rate	Heartbeat	Temperature
Non-symptomatic	≥95%	No cough	≤90 bpm	≤37.2 °C
Mild symptoms	≥95%	≤5/min	≤100 bpm	36 °C ≤ T ≤ 38 °C
Moderate clinical symptoms	93% ≤ SpO_2 ≤ 94%	5/min ≤ Cough Rate < 30/min	>100 bpm	≥38 °C
Serious clinical symptoms	≤92%	≥30/min	>120 bpm	>38 °C

A similar patient monitoring system is presented in [30]. The system uses an accelerometer sensor and two temperature sensors (a contact one and an IR—infrared sensor) for movement and body temperature measurement, respectively. Additionally, the system includes an ambient temperature and a humidity sensor to detect and warn the patient if the room parameters are off the limits. The measurements are sent to the web-based application via MQTT (message queue telemetry transport) using a BLE connection. The whole system is implemented into a M5stickC device, being an off-the-shelf embedded solution for a multi-application fast prototyping system.

Some published works propose COVID-19 tracing applications using Bluetooth low energy in order to track and monitor the spread of the virus [31,32]. These applications enhanced with ML capabilities can determine persons that are at risk of being infected because they were in the proximity of a confirmed infected person.

During our evaluation regarding the literature survey, three major disadvantages have been identified, one being related to the communication protocol used for patient monitoring systems. In this category, the lack of scalability due to the low communication range of the technology used to send the sensor data to the monitoring center, is a serious problem, which is not available when LoRaWAN protocol is used. Another disadvantage is related to the lack of energy efficiency of the proposed systems because they use a constant communication link, that drains the battery of the monitoring device very fast. The last major disadvantage is related to the hardware architecture modularity. Few monitoring systems use sensors in a wearable device that are comfortable and easy to wear. Thus, most of the solutions are related as a proof-of-concept design, being impossible to add new sensors to extend their functionality.

4. IoT Multi-Sensor System for Remote Patient Monitoring

This section presents the design of an IoT multi-sensor patient monitoring system that uses and integrates information technologies related to the IoT domain. The functionalities of the system are also described in detail. Figure 2 presents the main COVID-19 symptoms [33,34]. To ensure a high level of performance, these main symptoms must be monitored remotely so that the lives of quarantined patients are not endangered. The sooner a patient receives treatment, the lower the risk of developing severe life-threatening complications.

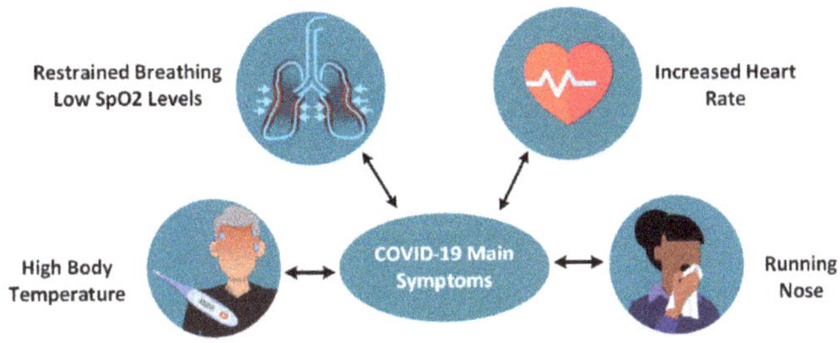

Figure 2. COVID-19 disease symptoms.

According to the CDC (Centers for Disease Control and Prevention), the main symptoms may appear on average 2–14 days after the exposure (based on the incubation period of MERS-CoV viruses), the symptoms being fever, cough [33], and shortness of breath due to impaired lung capacity [34]. Figure 3 shows some of the symptoms caused by the SARS-CoV2 virus. The latest studies show that the incubation period is about 5.2 days, so quarantine monitoring is crucial [35,36].

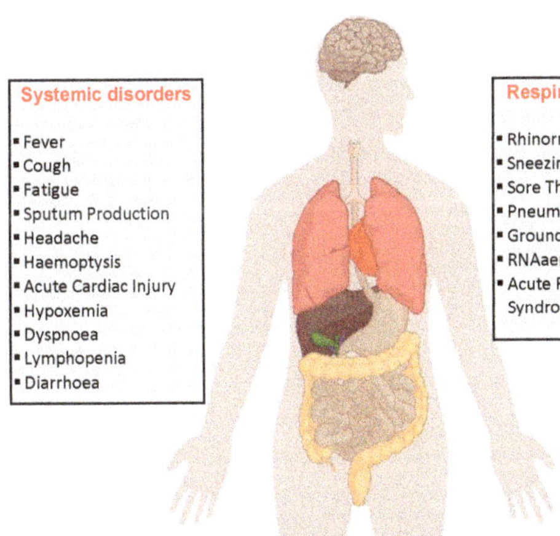

Figure 3. COVID-19 effects on the human body.

4.1. Patient Monitoring System

Considering the previously identified communication protocol requirements, the LoRaWAN protocol [37,38] meets performance criteria such as high coverage radius, low costs, the possibility of integrating a large number of nodes, and developing a scalable system that can comply with the requirements of high-density wireless sensor network scenario suitable for urban non-LoS conditions.

Figure 4 presents the proposed system architecture, which includes, the LoRaWAN gateway (GW), the sensors installed in the wearable device that monitors the patient's vital signs and the network server (NS) that acts as a relay between the LoRaWAN network and the local monitoring center. All the collected data are stored at the monitoring center where medical staff oversee the health status of the patients. The information is conveyed through the LoRa modulation technology for long communication distances between the

GW and the patient's wearable device using IP-based technologies for the data transmission between multiple GWs and the NS.

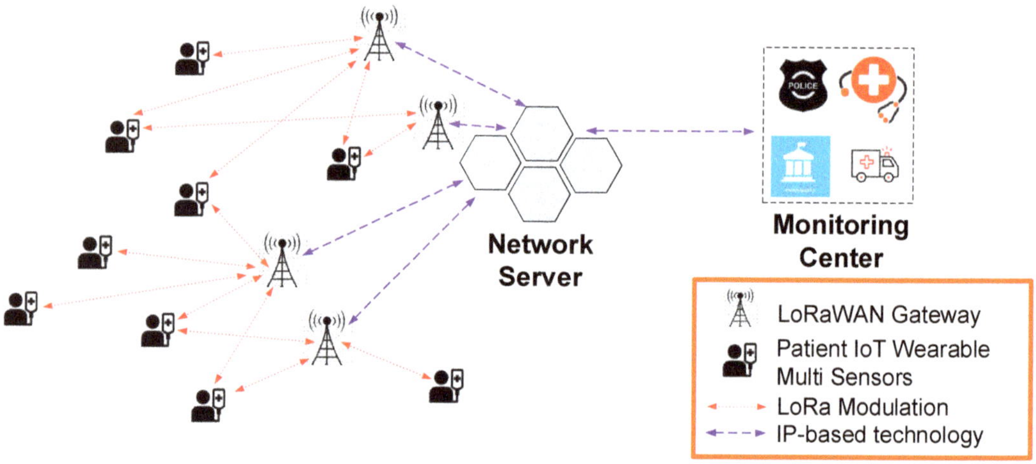

Figure 4. COVID-19 IoT multi-sensor patient monitoring architecture.

The LoRaWAN architecture consists of LoRa transceivers installed in the IoT multi-sensor wearable devices and GW modules that are communicating directly with the NS. The integrated IoT multi-sensors send management and configuration commands to the LoRa transceiver whose main purpose is to transmit the message to the connected GW module. Since the communication protocol is an aloha-type wireless sensor network, end-devices are allowed to transmit arbitrarily [39]. The transceiver integrated in the IoT multi-sensor device is of class A type, meaning the communication is initiated only from the multi-sensor node side. This ensures that the energy consumption of the multi-sensor node is the lowest possible from all LoRaWAN communication classes available, allowing for long-term operation.

The main compromise of the LoRaWAN communication protocol is the limited time intervals in which the node can receive messages from the application server, but this mechanism is not an issue. Thus, after a message is sent to the NS, the node listens for any server messages only for a limited time interval. This disadvantage does not reduce the level of performance for the designed architecture.

Many local municipalities worldwide have already installed or have access to the city developed infrastructures so the wearable devices can be easily enrolled and added as an extension to the existing configuration with no supplementary costs. The medical staff can easily monitor many patients, and if a patient's condition worsens, rapid intervention is possible. If the LoRaWAN communication protocol is used, then the system has no additional licensing costs or monthly fees like other communication protocols presented in Table 1, so the maintenance cost for the proposed system is low.

The system can be used both in institutionalized quarantine areas and in the case of people who are in solitary confinement at home. By setting up automatic alerts, a very large number of people can be monitored by a small number of medical staff in a centralized manner. Thus, we provide smart usage of the limited resources available in the pandemic context, contributing to saving as many lives as possible.

4.2. Wearable IoT Multi-Sensor Device for Patient Monitoring

Figure 5 presents the block diagram of the wearable device that includes an OLED display, the GPS sensor of Neo-6M type, the MAX30102 vital signs measurement sensor and the SX1276 LoRaWAN transceiver that allows for the collected data to be transmitted at the monitoring center. The wearable IoT device can monitor health conditions related parameters like pulse, blood oxygen saturation and body temperature as well as the patient's current location. At the center of our wearable IoT multi-sensor device we have the ATmega328p microcontroller unit that processes the information collected from different sensors linked with I2C, SPI or UART busses.

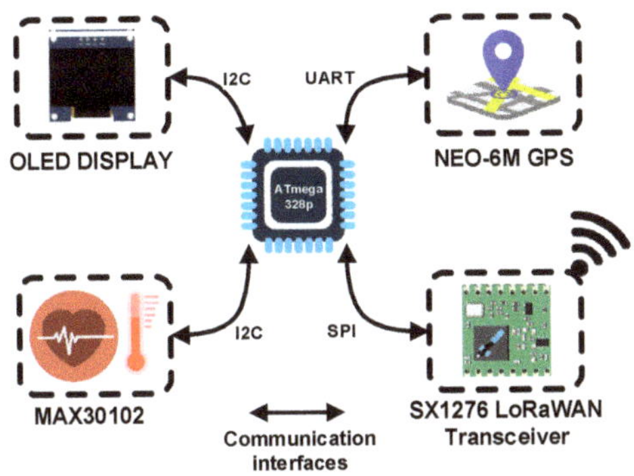

Figure 5. Wearable device block diagram of the multi-sensor approach.

For the health-related measurements, the MAX30102 [40] sensor was used as it offers an integrated solution for all three body parameters: pulse, blood oxygen saturation and body temperature, respectively. The sensor is based on the principle of spectrophotometry, which means that it detects the pulse and the blood oxygen saturation by measuring the amount of red and infrared light absorbed by the deoxygenated and oxygenated blood using photodiodes. The body temperature is measured using the temperature sensor integrated into the MAX30102. The signal provided by the MAX30102 is sent to the MCU via the I2C (Inter-Integrated Circuit) communication bus to be processed. The acquisition rate may be adjusted from the monitoring center based on the patient's health status. Different patient classes can be implemented as presented in Table 2.

Monitoring the patient's GPS position is also mandatory since maintaining a quarantine period involves isolation. Thus, a GPS module, named NEO-6M, is integrated into the sensor's architecture, being connected to the MCU via the UART (universal asynchronous receiver-transmitter) communication interface. The location data are sampled at random time intervals as to obtain energy efficiency. By monitoring the patient's current position, the authorities can be alerted in the eventuality that the patient does not comply with the quarantine conditions. If the current GPS position changes, the authorities are notified by a standard alert message transmitted to the monitoring center.

The wireless transceiver used for long-range communication and LoRa modulation is SX1276 from Semtech [41]. The SF is adjustable using the ADR mechanism provided by LoRaWAN specifications, thus, a high-level of performance and power consumption optimization is obtained. Another feature of the wearable device is an OLED display used only for local parameter checks or a debug option in the testing scenarios. The monitoring platform is implemented in a compact wearable form factor case that can be attached to the patient's wrist and is easily worn. Using a Li-Po power cell of 400 mA, the designed wearable device can operate for up to 20 days without recharging the battery, so the patient can wear the device during the 14-day quarantine period without power loss issues. This is possible by using a power management algorithm integrated into the MCU. The following rule can be used; for COVID-19 confirmed patients the time between measurements can be lower and offer more in-depth information about the evolution of the case. The physical implementation of the IoT multi-sensor wearable is presented in Figure 6.

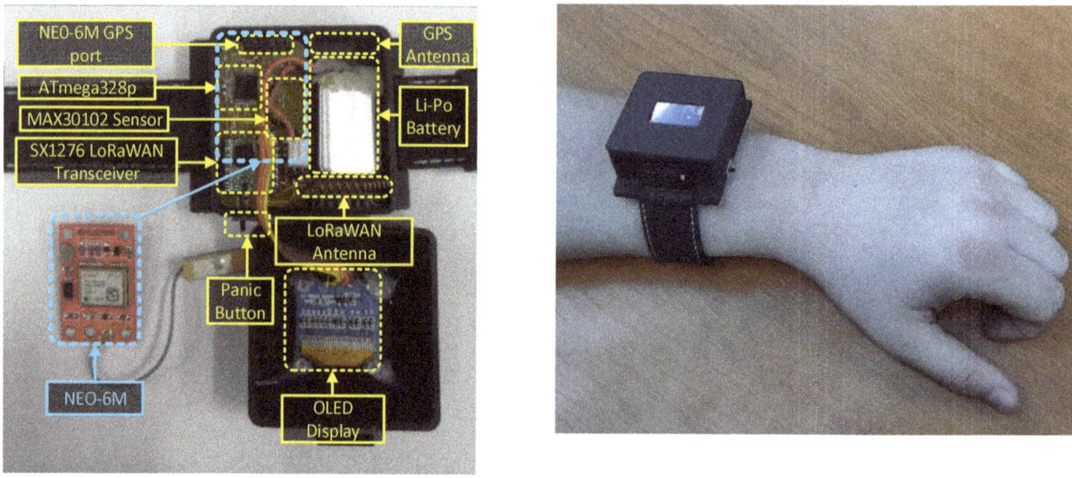

(a) (b)

Figure 6. Physical implementation of the wearable device. (**a**) IoT multi sensor configuration. (**b**) IoT wearable device installed on patient.

Depending on the monitored case, by using the application from the monitoring center the local authorities can set a minimum predefined period per day for the wearable device to be worn. Thus, the proposed wearable multi-sensor device can be removed by the patient for short intervals. The developed system triggers alerts when the health of the person in quarantine deteriorates. Thus, the onset of symptoms should be closely monitored so that treatment is administered as soon as possible. The medical staff can adjust, from the monitoring application, different threshold values that can be applied individually per each monitored vital sign. These threshold values of monitored parameters allow the system to generate only relevant alarms and notifications, excluding false triggers. Additionally, the monitored person can make an emergency call for urgent medical services in case of rapid deterioration of their health status. This is achieved by simply pressing the panic button on the device worn by the patient. After the IoT wearable device joins the network, it starts monitoring the parameters: GPS location, heart rate, oxygen saturation level, and temperature of the patient.

4.3. LoRaWAN Gateway Placement and Field Tests

The LoRaWAN gateway module (GW) allows for the connection of a very large number of wireless sensors. This is one of the main advantages that determined the selection of LoRa technology as being at the center of the proposed system and used as the communication protocol. The optimal placement of the GW module is very important and has an impact on the global performance of the patient monitoring system. In order to determine the optimal position of the GW, RadioPlanner [42] software was used with the propagation models presented in [43]. The developed model included the urban geographical area that we wanted to monitor, the terrain configuration, building positions, heights, and other interferences including vegetation path losses.

From the obtained data depicted in Figure 7, the optimal placement of the GW module for the field measurement tests is on the roof top of Suceava County Hospital (GW04 from the Figure 4), at a height of 360 m above sea level. In this configuration, a single GW will cover about 5 km^2 in the urban environment, so with a small number of GWs the surface of an entire city may be covered. RadioPlanner can be used by LoRaWAN integrators to determine the optimal placement of the GW modules, thus will reduce the implementation costs and the performance level of the wireless sensor networks.

Figure 7. Field coverage measurements—single GW configuration.

The next step was to perform field measurements as to validate the simulated coverage measurements. In this test setup, a RakWireless 7249 gateway [44] was used with an omnidirectional antenna having gain of 12 dBi. Figure 8 presents the obtained results. One can see that the recorded values are very close to the values obtained from our radio propagation model.

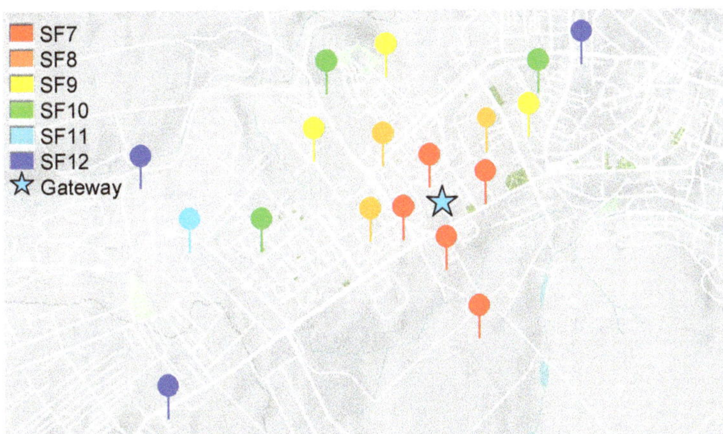

Figure 8. Field coverage measurements of the LoRaWAN gateways installed on the roof top of Suceava Hospital.

If the IoT multi-sensor system is to be extended to cover Suceava town entirely, we need to place multiple gateways. Thus, we evaluated the scenario to find out the number of LoRaWAN gateways needed to cover the entire area. As seen in Figure 9, the highest level of performance is obtained when using three gateways: GW01, GW02 and GW03. The distance between the gateways is approximately 5 km and can ensure a coverage area of approximately 20 km^2, including the neighboring villages.

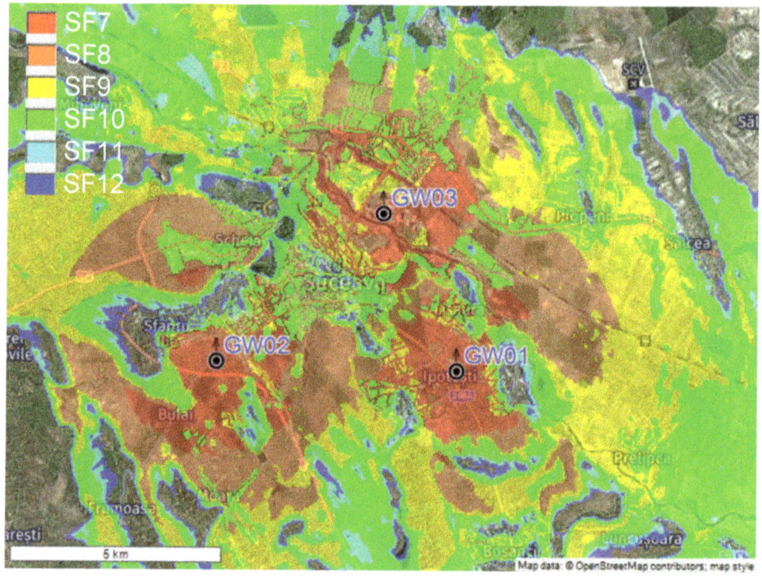

Figure 9. Field coverage measurements—multiple GW configuration.

The GW used in the field tests of the system runs ChirpStack Gateway OS [45], which has the capabilities to decode, store and forward the LoRaWAN packets to external applications under JSON application format. Figure 10 presents the monitoring and control application user interface of the application dashboard. The application is developed as an interactive map that can be easily displayed to the medical staff. The information shown includes the GPS location of each individual patient, his/her vital signals in a build-in graph window, the battery level of the device, the isolation period for each patient and specific alarms based on the threshold setup by the medical staff in the configuration settings of the integrated system.

Figure 10. IoT multi-sensor patient monitoring application dashboard.

Figure 11 presents the information acquired by the wearable IoT device for two different patients under quarantine. The measurements were acquired for the whole 14-day standard quarantine time. For the first patient, Patient A (Figure 11a,b), the measurements were within normal defined boundaries, while the second patient, Patient B (Figure 11c,d), showed an increase in body temperature and a slight drop in SpO_2 level, which is associated with a mild form of COVID-19 illness. Further tests and examinations performed on Patient B showed that he was positive for SARS-CoV2 virus infection.

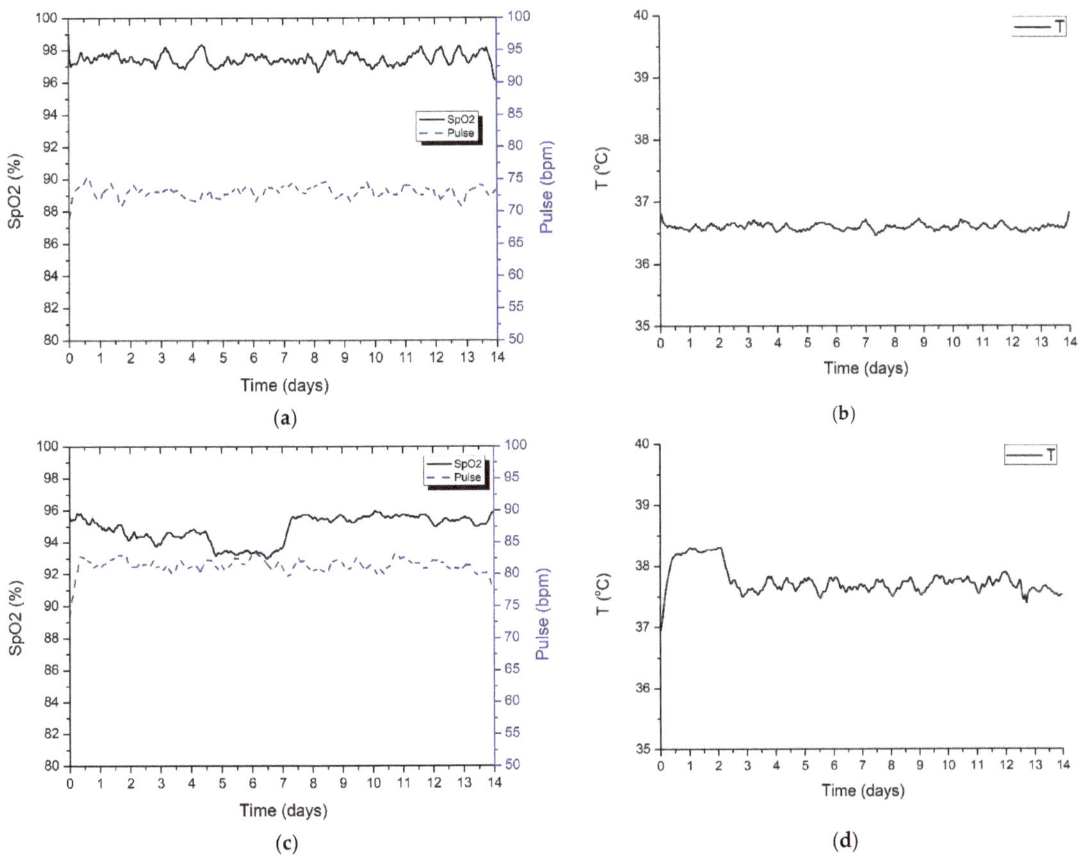

Figure 11. Patients' vital signs measurement obtained from the IoT wearable device. (**a**) SpO_2 and pulse (Patient A). (**b**) Body temperature (Patient A). (**c**) SpO_2 and pulse (Patient B). (**d**) Body temperature (Patient B).

5. Conclusions

COVID-19 is caused by the infection of the SARS-CoV2 virus and can be a potentially fatal disease that currently threatens humanity as a global public health issue. The rapid person-to-person transmission of the infection is extremely concerning; thus, a collective effort is needed to find new innovative solutions that can save lives and stop the spread of the disease. Nowadays, the IoT ecosystem has become an extension of the Internet, its impact on how humanity approaches the current problems will be huge in the years to come. This paper presents the design, development, and implementation of an integrated system for COVID-19 illness management. The increase in the number of COVID-19 infected people brought along with it a general overload of the medical system. In many world regions, the hospitals are overwhelmed with severe infection cases, leaving no room for the mild cases or other common diseases.

The paper also analyzes the main IoT challenges regarding the SARS-CoV2 pandemic and presents an in-depth performance evaluation of the IoT communication protocols that can be used for pandemic control. The designed IoT wearable device can monitor the health condition-related parameters like pulse, blood oxygen saturation and body temperature, as well as the patient's current location. The system can offer great advantages, allowing for remote monitoring of COVID-19 symptoms.

For testing the proposed system performance, two persons under quarantine were monitored, for a complete 14-day standard quarantine time interval. Based on the data transmitted to the monitoring center, the medical staff decided, after several days of monitoring, when the measured values were out of the normal range, to do an RT-PCR test for one of the two persons, confirming the SARS-CoV2 virus infection. The field tests evaluated scenarios also took into account the optimum placement of the LoRaWAN gateway to obtain the highest level of performance.

The proposed solution that uses the LoRaWAN communication protocol is scalable and can oversee a large number of patients. Table 3 contains a summary comparison of our proposed IoT multi-sensor approach and other systems presented in the scientific literature [27–30]. The wearable device uses an IoT multi-sensor approach that is easily reconfigurable, offering a flexible and cost-efficient solution for local authorities in the fight against COVID-19 pandemic. The data collected using the developed IoT wearable device can help us to fully understand the spread of the SARS-CoV2 virus.

Table 3. Performance evaluation of the developed system compared with other solutions.

Application Parameters	Zhang et al. [27]	Ullah et al. [28]	Mukhtar et al. [29]	Hoang et al. [30]	Our Proposed Approach
SpO_2	-	✓	✓	-	✓
Pulse	-	✓	✓	-	✓
Temperature	✓	✓	✓	✓	✓
Location	-	✓	-	-	✓
Scalability of the system	-	-	-	-	✓
High coverage area	-	-	-	-	✓
Modular architecture	-	-	-	-	✓
Wearable device integration	-	-	-	✓	✓
IoT communication technology	BLE	BLE/802.11n	802.11	BLE/802.11n	LoRaWAN

Author Contributions: Conceptualization, A.L., A.I.P.; methodology, A.L., E.C., P.-M.M., V.P.; software, A.I.P., P.-M.M.; validation, A.I.P., E.C. and A.L.; formal analysis, A.L.; investigation, A.L., A.I.P. and P.-M.M.; resources, A.I.P. and E.C.; data curation, A.I.P., A.L., P.-M.M.; writing—original draft preparation, A.L., E.C., A.I.P., and P.-M.M.; writing—review and editing, A.L., E.C., A.I.P., P.-M.M.; visualization, A.I.P., E.C. and A.L.; supervision, E.C. and V.P.; project administration, A.L. and V.P.; funding acquisition, A.L. and V.P. All authors have read and agreed to the published version of the manuscript.

Funding: This paper was made using the infrastructure from the project "Integrated Center for Research, Development and Innovation in Advanced Materials, Nanotechnologies, and Distributed Systems for Fabrication and Control (MANSiD)", Contract No. 671/09.04.2015, Sectoral Operational Program for Increase of the Economic Competitiveness co-funded from the European Regional Development Fund, and was supported by a grant of the Romanian Ministry of Research and Innovation, CCCDI-UEFISCDI, project number PN-III-P2-2.1-PTE-2019-0642, within PNCDI III.

Institutional Review Board Statement: Not applicable.

Informed Consent Statement: Not applicable.

Conflicts of Interest: The authors declare no conflict of interest. The funders had no role in the design of the study; in the collection, analyses, or interpretation of data; in the writing of the manuscript, or in the decision to publish the results.

References

1. Tsang, H.F.; Chan, L.W.C.; Cho, W.C.S.; Yu, A.C.S.; Yim, A.K.Y.; Chan, A.K.C.; Ng, L.P.W.; Wong, Y.K.E.; Pei, X.M.; Li, M.J.W.; et al. An Update on COVID-19 Pandemic: The Epidemiology, Pathogenesis, Prevention and Treatment Strategies. *Expert Rev. Anti-Infect. Ther.* **2021**, *19*, 877–888. [CrossRef] [PubMed]
2. Cucinotta, D.; Vanelli, M. WHO Declares COVID-19 a Pandemic. *Acta Biomed.* **2020**, *91*, 157. [PubMed]
3. Dashboard, W.C. (COVID-19) No Title. Available online: https://covid19.who.int (accessed on 4 January 2022).
4. Inc. M. and Markets IoT Medical Devices Market by Product (Blood Pressure Monitor, Glucometer, Cardiac Monitor, Pulse Oximeter, Infusion Pump), Type (Wearable, Implantable, Stationary), Connectivity Technology (Bluetooth, Wifi), End User (Hospital)—Global Forecast to 2026. Available online: https://www.marketsandmarkets.com/ (accessed on 4 January 2022).
5. López-Cabrera, J.D.; Orozco-Morales, R.; Portal-Díaz, J.A.; Lovelle-Enríquez, O.; Pérez-Díaz, M. Current Limitations to Identify COVID-19 Using Artificial Intelligence with Chest x-Ray Imaging (Part Ii). The Shortcut Learning Problem. *Health Technol.* **2021**, *11*, 411–424. [CrossRef]
6. Nguyen, D.C.; Ding, M.; Pathirana, P.N.; Seneviratne, A. Blockchain and AI-Based Solutions to Combat Coronavirus (COVID-19)-Like Epidemics: A Survey. *IEEE Access* **2021**, *9*, 95730–95753. [CrossRef] [PubMed]
7. Hussain, A.A.; Bouachir, O.; Al-Turjman, F.; Aloqaily, M. AI Techniques for COVID-19. *IEEE Access* **2020**, *8*, 128776–128795. [CrossRef] [PubMed]
8. Naudé, W. Artificial Intelligence vs. COVID-19: Limitations, Constraints and Pitfalls. *AI Soc.* **2020**, *35*, 761–765. [CrossRef] [PubMed]
9. Ardabili, S.F.; Mosavi, A.; Ghamisi, P.; Ferdinand, F.; Varkonyi-Koczy, A.R.; Reuter, U.; Rabczuk, T.; Atkinson, P.M. COVID-19 Outbreak Prediction with Machine Learning. *Algorithms* **2020**, *13*, 249. [CrossRef]
10. Fani, M.; Zandi, M.; Soltani, S.; Abbasi, S. Future Developments in Biosensors for Field-Ready SARS-CoV-2 Virus Diagnostics. *Biotechnol. Appl. Biochem.* **2021**, *68*, 695–699. [CrossRef]
11. Griffin, J.H.; Downard, K.M. Mass Spectrometry Analytical Responses to the SARS-CoV2 Coronavirus in Review. *TrAC—Trends Anal. Chem.* **2021**, *142*, 116328. [CrossRef]
12. Maese, J.R.; Seminara, D.; Shah, Z.; Szerszen, A. Perspective: What a Difference a Disaster Makes: The Telehealth Revolution in the Age of COVID-19 Pandemic. *Am. J. Med. Qual.* **2020**, *35*, 429–431. [CrossRef]
13. Watson, A.R.; Wah, R.; Thamman, R. The Value of Remote Monitoring for the COVID-19 Pandemic. *Telemed. E-Health* **2020**, *26*, 1110–1112. [CrossRef]
14. Patel, A.; Patel, S.; Fulzele, P.; Mohod, S.; Chhabra, K. Quarantine an Effective Mode for Control of the Spread of COVID19? A Review. *J. Fam. Med. Prim. Care* **2020**, *9*, 3867. [CrossRef]
15. Moazzami, B.; Razavi-Khorasani, N.; Dooghaie Moghadam, A.; Farokhi, E.; Rezaei, N. COVID-19 and Telemedicine: Immediate Action Required for Maintaining Healthcare Providers Well-Being. *J. Clin. Virol.* **2020**, *126*, 104345. [CrossRef]
16. Gilanie, G.; Bajwa, U.I.; Waraich, M.M.; Asghar, M.; Kousar, R.; Kashif, A.; Aslam, R.S.; Qasim, M.M.; Rafique, H. Coronavirus (COVID-19) Detection from Chest Radiology Images Using Convolutional Neural Networks. *Biomed. Signal Process. Control* **2021**, *66*, 102490. [CrossRef]
17. Lavric, A.; Petrariu, A.I.; Popa, V. SigFox Communication Protocol: The New Era of IoT? In Proceedings of the 2019 International Conference on Sensing and Instrumentation in IoT Era, ISSI 2019, Lisbon, Portugal, 29–30 August 2019.
18. Harwahyu, R.; Cheng, R.G.; Liu, D.H.; Sari, R.F. Fair Configuration Scheme for Random Access in NB-IoT with Multiple Coverage Enhancement Levels. *IEEE Trans. Mob. Comput.* **2021**, *20*, 1408–1419. [CrossRef]
19. Peña Queralta, J.; Gia, T.N.; Zou, Z.; Tenhunen, H.; Westerlund, T. Comparative Study of LPWAN Technologies on Unlicensed Bands for M2M Communication in the IoT: Beyond Lora and Lorawan. *Proc. Procedia Comput. Sci.* **2019**, *155*, 343–350. [CrossRef]
20. Lavric, A.; Popa, V.; Males, C.; Finis, I. A Performance Study of ZigBee Wireless Sensors Network Topologies for Street Lighting Control Systems. In Proceedings of the 2012 International Conference on Selected Topics in Mobile and Wireless Networking, ICOST 2012, Avignon, France, 2–4 July 2012.
21. Danbatta, S.J.; Varol, A. Comparison of Zigbee, Z-Wave, Wi-Fi, and Bluetooth Wireless Technologies Used in Home Automation. In Proceedings of the 7th International Symposium on Digital Forensics and Security, ISDFS 2019, Barcelos, Portugal, 10–12 June 2019.
22. Lavric, A.; Popa, V.; Finis, I.; Gaitan, A.M.; Petrariu, A.I. Packet Error Rate Analysis of IEEE 802.15.4 under 802.11 g and Bluetooth Interferences. In Proceedings of the 2012 9th International Conference on Communications, COMM 2012—Conference Proceedings, Washington, DC, USA, 7–8 January 2012.
23. Varahachalam, S.P.; Lahooti, B.; Chamaneh, M.; Bagchi, S.; Chhibber, T.; Morris, K.; Bolanos, J.F.; Kim, N.Y.; Kaushik, A. Nanomedicine for the SARS-CoV-2: State-of-Theart and Future Prospects. *Int. J. Nanomed.* **2021**, *16*, 539. [CrossRef] [PubMed]
24. Lavric, A.; Petrariu, A.I.; Coca, E.; Popa, V. LoRaWAN Analysis from a High-Density Internet of Things Perspective. In Proceedings of the 2020 15th International Conference on Development and Application Systems, DAS 2020 Proceedings, Suceava, Romania, 21–23 May 2020.

25. Semtech Understanding LoRa Adaptive Data Rate. Available online: https://lora-developers.semtech.com/uploads/documents/files/Understanding_LoRa_Adaptive_Data_Rate_Downloadable.pdf (accessed on 20 December 2021).
26. LoRa Alliance Technical Commitee LoRaWAN 1.1 Specification. *LoRaWAN 1.1 Specif.* 2017. Available online: https://lora-alliance.org/wp-content/uploads/2020/11/lorawantm_specification_-v1.1.pdf (accessed on 15 December 2021).
27. Zhang, L.; Zhu, Y.; Jiang, M.; Wu, Y.; Deng, K.; Ni, Q. Body Temperature Monitoring for Regular COVID-19 Prevention Based on Human Daily Activity Recognition. *Sensors* **2021**, *21*, 7540. [CrossRef]
28. Ullah, F.; Haq, H.U.; Khan, J.; Safeer, A.A.; Asif, U.; Lee, S. Wearable Iots and Geo-Fencing Based Framework for COVID-19 Remote Patient Health Monitoring and Quarantine Management to Control the Pandemic. *Electronics* **2021**, *10*, 2035. [CrossRef]
29. Mukhtar, H.; Rubaiee, S.; Krichen, M.; Alroobaea, R. An Iot Framework for Screening of COVID-19 Using Real-Time Data from Wearable Sensors. *Int. J. Environ. Res. Public Health* **2021**, *18*, 4022. [CrossRef]
30. Hoang, M.L.; Carratù, M.; Paciello, V.; Pietrosanto, A. Body Temperature—Indoor Condition Monitor and Activity Recognition by Mems Accelerometer Based on IoT-Alert System for People in Quarantine Due to COVID-19. *Sensors* **2021**, *21*, 2313. [CrossRef] [PubMed]
31. Bahle, G.; Rey, V.F.; Bian, S.; Bello, H.; Lukowicz, P. Using Privacy Respecting Sound Analysis to Improve Bluetooth Based Proximity Detection for COVID-19 Exposure Tracing and Social Distancing. *Sensors* **2021**, *21*, 5604. [CrossRef] [PubMed]
32. Aljohani, A.J.; Shuja, J.; Alasmary, W.; Alashaikh, A. Evaluating the Dynamics of Bluetooth Low Energy Based COVID-19 Risk Estimation for Educational Institutes. *Sensors* **2021**, *21*, 6667. [CrossRef]
33. Fernández-De-las-peñas, C.; Palacios-Ceña, D.; Gómez-Mayordomo, V.; Cuadrado, M.L.; Florencio, L.L. Defining Post-COVID Symptoms (Post-Acute COVID, Long COVID, Persistent Post-COVID): An Integrative Classification. *Int. J. Environ. Res. Public Health* **2021**, *18*, 2621. [CrossRef]
34. Zaki, N.; Mohamed, E.A. The Estimations of the COVID-19 Incubation Period: A Scoping Reviews of the Literature. *J. Infect. Public Health* **2021**, *14*, 638–646. [CrossRef]
35. Jiang, X.; Niu, Y.; Li, X.; Li, L.; Cai, W.; Chen, Y.; Liao, B.; Wang, E. Is a 14-Day Quarantine Period Optimal for Effectively Controlling Coronavirus Disease 2019 (COVID-19)? *medRxiv* **2020**. [CrossRef]
36. Lauer, S.A.; Grantz, K.H.; Bi, Q.; Jones, F.K.; Zheng, Q.; Meredith, H.R.; Azman, A.S.; Reich, N.G.; Lessler, J. The Incubation Period of Coronavirus Disease 2019 (COVID-19) from Publicly Reported Confirmed Cases: Estimation and Application. *Ann. Intern. Med.* **2020**, *172*, 577–582. [CrossRef]
37. Lavric, A. LoRa (Long-Range) High-Density Sensors for Internet of Things. *J. Sens.* **2019**, *2019*, 3502987. [CrossRef]
38. Lavric, A.; Popa, V. Performance Evaluation of LoRaWAN Communication Scalability in Large-Scale Wireless Sensor Networks. *Wirel. Commun. Mob. Comput.* **2018**, *2018*, 6730719. [CrossRef]
39. Beltramelli, L.; Mahmood, A.; Osterberg, P.; Gidlund, M. LoRa beyond ALOHA: An Investigation of Alternative Random Access Protocols. *IEEE Trans. Ind. Inform.* **2021**, *17*, 3544–3554. [CrossRef]
40. Ahmed, M.F.; Hasan, M.K.; Shahjalal, M.; Alam, M.M.; Jang, Y.M. Design and Implementation of an OCC-Based Real-Time Heart Rate and Pulse-Oxygen Saturation Monitoring System. *IEEE Access* **2020**, *8*, 198740–198747. [CrossRef]
41. Abouzaid, L.; Sabir, E.; Errami, A.; Elbiaze, H. Semtech Corporation Datasheet SX1276/77/78/79 LoRa Transciever | Semtech. In Proceedings of the IEEE 5th World Forum on Internet of Things, WF-IoT 2019—Conference Proceedings, Limerick, Ireland, 15–18 April 2019.
42. Radio Planner. Available online: https://www.wireless-planning.com/radioplanner (accessed on 4 January 2022).
43. Petrariu, A.-I.; Mutescu, P.-M.; Coca, E.; Lavric, A. A Study on LoRa Signal Propagation Models in Urban Environments for Large-Scale Networks Deployment. *Adv. Electr. Comput. Eng.* **2021**, *21*, 61–68. [CrossRef]
44. RakWireless 7249 Gateway. Available online: https://www.rakwireless.com/en-us/products/lpwan-gateways-and-concentrators/rak7249 (accessed on 4 January 2022).
45. Chirp Stack Gateway OS. Available online: https://www.chirpstack.io/gateway-os/ (accessed on 4 January 2022).

sensors

MDPI

Article

Large-Scale Internet of Things Multi-Sensor Measurement Node for Smart Grid Enhancement

Adrian I. Petrariu [1,2,*], Eugen Coca [1] and Alexandru Lavric [1]

[1] Computers, Electronics and Automation Department, Stefan cel Mare University of Suceava, 720229 Suceava, Romania; eugen.coca@usv.ro (E.C.); lavric@usm.ro (A.L.)
[2] MANSiD Research Center, Stefan cel Mare University of Suceava, 720229 Suceava, Romania
* Correspondence: apetrariu@usm.ro

Abstract: Electric power infrastructure has revolutionized our world and our way of living has completely changed. The necessary amount of energy is increasing faster than we realize. In these conditions, the grid is forced to run against its limitations, resulting in more frequent blackouts. Thus, urgent solutions need to be found to meet this greater and greater energy demand. By using the internet of things infrastructure, we can remotely manage distribution points, receiving data that can predict any future failure points on the grid. In this work, we present the design of a fully reconfigurable wireless sensor node that can sense the smart grid environment. The proposed prototype uses a modular developed hardware platform that can be easily integrated into the smart grid concept in a scalable manner and collects data using the LoRaWAN communication protocol. The designed architecture was tested for a period of 6 months, revealing the feasibility and scalability of the system, and opening new directions in the remote failure prediction of low voltage/medium voltage switchgears on the electric grid.

Keywords: smart grid; LoRaWAN; partial discharge; predictive maintenance; scalability

Citation: Petrariu, A.I.; Coca, E.; Lavric, A. Large-Scale Internet of Things Multi-Sensor Measurement Node for Smart Grid Enhancement. *Sensors* **2021**, *21*, 8093. https://doi.org/10.3390/s21238093

Academic Editor: Danda B. Rawat

Received: 16 November 2021
Accepted: 1 December 2021
Published: 3 December 2021

1. Introduction

Electric power infrastructure has revolutionized our world. With the advent of electricity, our way of living completely changed. But, as our energy needs increase faster than we realize, our energy sources depleted at a similar rate. Our grid is running against its limitations and blackout conditions are more frequently met. Due to the increase of greenhouse gas emission, our carbon footprint is also increasing, which is leading to climate change and numerous associated problems [1]. Therefore, a solution is needed to tackle all these problems. A solution is required for using electricity in a sustainable manner. We can use information technology to overhaul the electric grid and to monitor it [2]. With the ever-increasing demand of electrical vehicles, the real time monitoring and measurement of the power grid is mandatory [3]. Thus, urgent solutions need to be found as to meet this higher and higher energy demand.

The conventional grid consists of electromechanical components which cannot be controlled in real time. However, the smart grid comes into the picture when ICT (information and communication technology), electrical, and power systems are used collectively. In the existing grid, communication is one way; information goes from power generating units and utilities to the consumer, but almost never from the consumer to the utilities. The smart grid concept allows two-way communications with all stakeholders, so that information can be shared in a productive way. Utility companies may understand the consumption patterns of consumers and decide the price per electricity unit depending on peak loads and usage time.

The conventional grid does not integrate many sensors in its network, due to which it is considered blind, being unable to self-monitor and self-heal. That is why, much of the time, utility companies are not aware of failures or blackouts and cannot predict them so as to prepare for consequence to the end users. Testing and restoration tasks are also usually not automatic, involving human presence in multiple remote locations. On the other hand, in a smart grid framework, a few sensors are attached throughout the network, which enable it to self-monitor. By extending the concept, utility companies will be able to pick-up information from the network remotely and the network will have the capability to even self-heal in an automatic manner [4,5].

Analyzing the demands from the abovementioned, we can say that the main contribution of this work is the design of a fully reconfigurable wireless sensor node that can sense the smart grid environment. The proposed prototype can be easily integrated into the smart grid concept in a scalable manner, using a modular hardware platform developed by the authors.

In the specialized literature, there are a series of scientific papers which describe various smart grid architectures. This paper comes to fill in the gap by implementing and designing a smart grid architecture based a multi-sensor wireless node that uses state of the art technologies for sensing the environment. The designed concept allows the integration of a high-density of sensors distributed over a large-scale geographical area. This is possible by using LPWAN (Low-Power Wide-Area Network) technologies. Some of the most used technologies are SigFox, LoRaWAN, or NB-IoT. Of these, LoRaWAN is the most suitable for our scenario, due to its numerous advantages compared with the others, such as unlicensed spectrum transmitting, easy deployment by any hardware/software developer for both components, nodes, and gateways. SigFox is a technology that is only available in selected countries, being a mobile carrier supported solution, and requires subscription fees to register new nodes in the network. Furthermore, several message transmission restrictions are in place. Thus, only 140 messages of 12 bytes are possible for the uplink, and only four messages of eight bytes are possible for the downlink. These aspects limit the applicability of the SigFox communication protocol in different geographical regions. Meanwhile, NB-IoT operates in the license spectrum and is mostly driven by leading telecommunication companies around the world, so flexibility for the end-user is limited. Furthermore, to transmit messages, each node must be registered on the network, generating additional costs for each installed node.

The paper is organized as follows: first, a brief introduction related to the state-of-the-art, followed by Section 2, where the main challenges of the smart grid concept are presented. The LoRa modulation and the LoRaWAN communication protocol are discussed in Section 3. In Section 4, the smart grid architecture design is presented and analyzed with emphasis on the developed multi-sensor measurement node. The experimental results and discussion of them are presented in Section 5. The final section of the paper represents the conclusions and the overall performance evaluation of the proposed multi-sensor monitoring node.

2. Smart Grid Challenges and Solutions

The concepts of cyber-physical systems (CPS) [6] or the internet of things (IoT), which have been around for more than a decade now, are currently creating a great deal of buzz in the marketplace and media, with promises to enhance the way we live, travel and work. There are three major areas of IoT applications: in the consumer, industrial, and public sectors. Recent interest has mainly focused on the consumer side, including consumer appliances, home area networks, and other small office applications. Industrial applications promise to improve business outcomes for many sectors, including manufacturing, asset management, and healthcare. In the case of public sector applications, the internet of things is a major enabling concept to accelerate the development and deployment of smart city solutions, including electric vehicles charging stations [7–9], utilities monitoring, or public transport fleet management [10].

In the smart city concept, the most important element is represented by the electric grid, which basically transports electricity to each consumer. The electric grid parameters can be monitored using wired or wireless communication systems.

For wired communication, PLCs (power line communications) are the best option due to the working principle and use of the existing grid infrastructure. They are efficient if they are used in a small architecture configuration. In cases where two communication points are located between a transformer, PLC systems are useless due to the galvanic isolation between primary and secondary windings. Thus, to optimize communication, some hybrid solutions exist, where low-range wireless communications are attached to each transformer side PLC system to ensure data transmission. Other solutions are based on PLC and cellular technologies, extending the communication range between isolated power nodes, or using it to increase coverage of the non-signal or poor-signal areas encountered in on-site implementation [11–13]. These solutions have a monthly fee and are dependent on cellular coverage.

Below, some wireless communication solutions are briefly listed from the scientific literature, with applications for smart city and smart grid concepts.

In [14], the authors propose a method to detect the partial discharge that may occur into a switchgear by using a synchronization mechanism between multiple switchgears in a row with LoRa (long-range) technology and TEV (transient earth voltage) sensors. This method uses the attenuation principle in the process of electromagnetic wave propagation, so by comparing measured amplitude values received at each switchgear from the same discharge source, it can detect the faulty switchgear from a row. Thus, LoRa technology is used only to detect which switchgear is faulty or not, the information being transmitted and processed locally.

Gao et al. [15] use LPWAN technologies to transmit any grid issue that may occur in the medium voltage (MV) distribution power network. In the paper presented, an IoT-based HFCT (high frequency current transformer) sensor with LoRa and NB-IoT capabilities transmits any partial discharge issue from the MV cables.

Another hybrid approach is made in [16], where LoRa and SigFox technologies are used to communicate in a smart grid infrastructure. The authors propose a general architecture for a local implemented project called MAIGE. Several sensors are attached to some key points (high voltage/medium voltage switchgears or transformation centers) that transmit information regarding electrical discharge, grounding problems, or battery electrolyte levels to a SCADA (supervisory control and data acquisition) monitoring center.

The hybrid communication approach is tackled in other papers [17–21], combining the well-known and used LPWAN, ZigBee, or 5G communication systems to transmit smart grid issues to a control and command center.

Figure 1 centralizes the main communication protocols that can be integrated in the smart grid: PLC, 5G, LoRa, ZigBee, LPWAN, NB-IoT, and LoRaWAN (Long-Range Wide Area Network) in a cost effective and scalable manner.

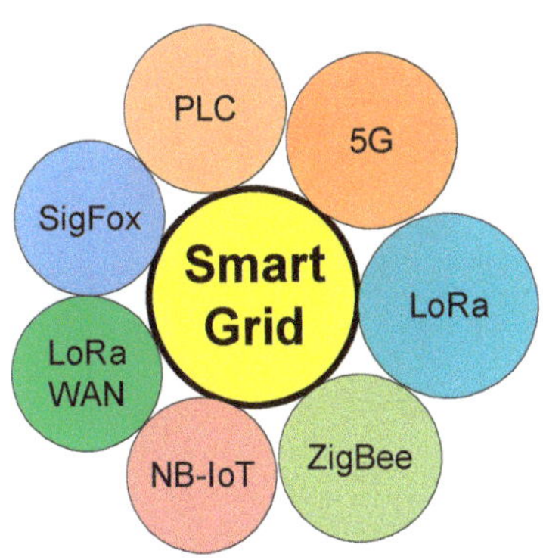

Figure 1. Technologies identified in the smart grid concept.

In this paper we present the design implementations and testing of a standalone smart WSN (wireless sensor node), called the multi-sensor measuring node (MSMN), that can be used to monitor different parameters from an electric grid, where remote monitoring of failures or blackouts in low voltage/medium voltage (LV/MV) switchgears is needed for grid efficiency management, This monitoring multisensory node can be integrated into a smart grid architecture using LoRa technology and can oversee power lines, detecting the eventual failure of the LV/MV switchgears in crowded cities. The proposed solution is called "multi-sensor" due to the use of different sensors for data acquisition, such as temperature, humidity, air pressure or ozone concentration, that are integrated into a modular hardware platform developed by the authors. Those sensors cannot be integrated into a single sensor type, so for our goal, we use more than one sensor, connected through different communication busses in the hardware platform. The obtained node has the main advantage of being flexible and easy to use, allowing the end user to add new sensors depending on the monitored environment. This reconfigurability and modularity of the proposed MSMN node entails cost efficiency and must be considered when evaluating its performance level. The high level of performance of the developed node is possible only by using the integrated acquisition algorithm, presented in detail in Section 4, where each individual sensor is controlled separately.

3. LoRa Technology Overview

LoRa is a wireless transmitting technology that used chirp spread spectrum (CSS) modulation, ensuring long-range communication links between modules. This type of modulation offers robustness against interferences and a very low signal-to-noise ratio (SNR) for the receiver to be able to demodulate the received signal. For wirelessly transmitting the information, LoRa uses the unlicensed radio spectrum with the ISM (Industrial, Scientific, Medical) or SRD (Short-Range Devices) band, using 433 MHz, 868 MHz, or 915 MHz as the main frequencies.

LoRa is a long-range technology [22] that has a communication range up to 10 km and 50 km in urban or rural areas, respectively, depending upon the spreading factor (SF) used. It is a low-power consumer (can use a few mA to hundreds of mA when transmitting) [23,24], is scalable (can be reconfigured remotely, depending on the infrastructure) [25,26], and does not generate costs during its lifetime. LoRa is thus a suitable solution for applications that require a very long battery lifetime and reduced cost.

Due to its unique modulation technique that implies CSS, LoRa technology allows users to negotiate between data rate and the maximum range by varying the spreading factor of the transmitter. The standard bandwidth is 125 kHz for the European region and 250 kHz for the US region.

Compared to other existing LPWAN technologies, LoRa uses the same phase between two chirp symbols. Thus, the synchronization mechanism between the node and the gateway is improved, resulting in a cheaper hardware for the gateway. Some related LoRa specifications are in Table 1, considering the 125 kHz bandwidth.

Table 1. Chirp length for each SF used in the LoRa modulation.

Spreading Factor	Chirp Length (bits)	Throughput (bps)
7	128	5469
8	256	3125
9	512	1758
10	1024	997
11	2048	537
12	4096	293

LoRa is the modulation technique used in the LoRaWAN systems. LoRaWAN uses long-range star architecture in which gateways are used to relay messages between the end nodes and a central core network, being a protocol for medium access control (MAC) designed for internet of things (IoT) applications. In a LoRaWAN architecture, the transmitters (nodes) are not associated with a specific gateway, their transmissions being received by all the gateways available in the communication range. Furthermore, LoRaWAN uses some specific communication algorithms that imply adaptive data rate (ADR) in coding rate (CR) and spreading factor estimations. Although it uses a free radio frequency band, some limitations are required in transmitting duty cycle, being imposed by the current international laws governed by the ETSI (European Telecommunications Standards Institute-Sophia-Antipolis, France) and FCC (Federal Communications Commission—Washington, DC, USA). Thus, a duty cycle less or equal than 1% is allowed for all European sub-channels.

4. Smart Grid Monitoring System

4.1. Monitoring Architecture

The proposed monitoring architecture from this paper uses the LoRaWAN specification. LoRaWAN is the communication protocol defined by the LoRa Alliance. Figure 2 presents the LoRaWAN communication stack. The physical layer of the protocol uses the LoRa modulation, discussed in the previous section, combined with a DSSS (direct sequence spread spectrum) techniques inspired from radar communication systems. The frequency band used belongs to ISM or SRD—thus, no license fees are needed—and it entails a cost reduction, which is a major advantage that LoRaWAN technology offers. The MAC (medium access control) layer of the LoRaWAN protocol defines three classes of sensors named class A, B, and C that determine different communication constraints regarding transmission time and transmission mode. Most of the sensor architectures for the smart grid concept use class C of the LoRaWAN sensor due to the existing electricity in the monitored center, which means continuous reception and a bidirectional link between the monitored point and the LoRaWAN gateway. In our proposed architecture, the multi-sensor node will be powered from a battery with a renewable energy source backup and the selected communication class will be A, to obtain energy efficiency for the whole system.

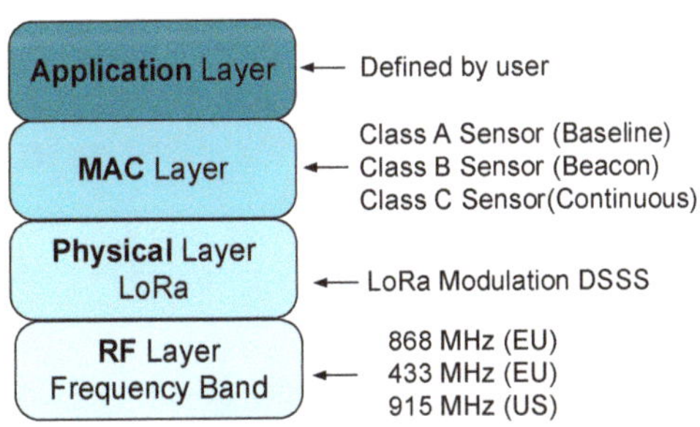

Figure 2. The LoRaWAN communication stack.

The LoRaWAN gateway acts as a relay between the smart grid and the internet network on which the information is transmitted. The information is stored and processed on a web server with the ability to display different reports and statistics. The main advantage of the proposed monitoring architecture is the decentralized data collection process with the ability of integrating an exceptionally large number of monitoring points.

Figure 3 presents the proposed smart grid architecture. The architecture is distributed over a large geographical area specific to urban non-LoS (line-of-sight) communication conditions. We can observe that the monitoring nodes that are located at the edge of the network are communicating using an SF set to 12. To increase the number of integrated measuring nodes, the star network topology is used, in which the sink node is represented by the LoRaWAN gateway.

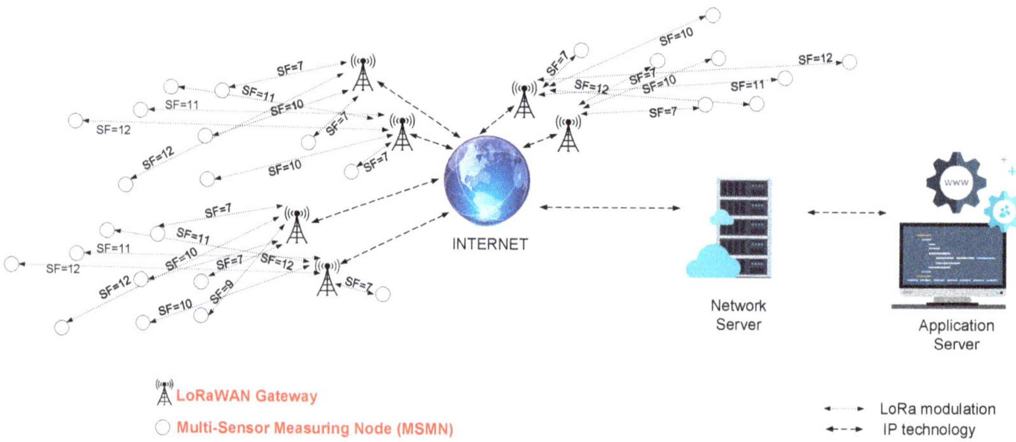

Figure 3. The proposed smart grid architecture based on the LoRaWAN communication.

The optimal placement of the sink node will be addressed in the next section of the paper using some measurements and with some simulation scenarios based on the RadioPlanner simulator.

Each multi sensor measuring node (MSMN) can sense the monitoring environment. Advantages of the MSMN node:

- Modular platform with multi sensors integration capabilities: any type of analogue, digital, or I2C interfaced sensor;
- High reconfigurability of the developed platform, which means that new sensors can be easily added to the developed platform;
- The MSMN can be placed without any previous configuration in an area with LoRaWAN coverage and will automatically connect to the network;
- The MSMN node can function as standalone wireless module for a long period of time without any maintenance due to renewable energy integration;
- The MSMN node can be customized with other communication protocols based on the usability scenario.

When the MSMN node sends measurement data, the information is transmitted to the closest LoRaWAN gateway and can be processed by a monitoring center.

4.2. Multi-Sensor Monitoring Node

The monitoring architecture uses the LoRaWAN specification. The general architecture of the proposed LoRaWAN node is listed in Figure 4.

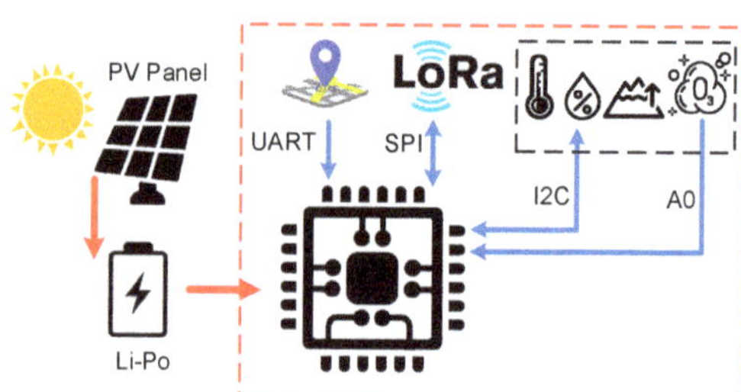

Figure 4. Proposed MSMN architecture.

The MSMN uses the platform proposed in [27] as the main board. This platform has a LoRa communication module to provide a long-distance communication and to cover the areas where other communication types like GSM/GPRS have a poor coverage. The MSMS also has a NEO-6G GPS module for location estimation. This module has a particular feature: a very low-power consumption of only 72 mW at a 1.8 V voltage rating. microcontroller, a ATmega88A is used, and the clock is set to 1 MHz; it works with a power supply from 1.8 V to a maximum of 3 V. All the external features of the platform are programmed to work in sleep mode. Thus, all the system architecture is low-power consumption, only demanding approximately 63 mA in active mode when SF7 is used or 172 mA when SF12 is used. In sleep mode, the entire platform drains only 10 μA.

The solution proposed in [27] is a modular one. Thus, miscellaneous sensors for data acquisition can be integrated on the main board. Environmental data acquisition sensors like air temperature, air humidity, air pressure, air quality (BME680 connected using I2C bus), and ozone (MQ-131 connected using A0 analogic input) are available in the architecture from the MSMN node. These sensors are required for the LV/MV (low voltage/medium voltage) switchgear monitoring, to avoid problems caused by malfunction of these electric energy distribution cells. According to some scientific papers [28–30], the main cause of electrical grid malfunction is the interruption of the power supply, due to damage that may occur in the LV/MV switchgear cells (which supply household users). These problems occur due to improper administration of the MV switchgear cells because in most of the cases they are not supervised, being mounted in the open field with different atmospheric conditions. Variable temperature and humidity can cause changes in the insulation of the conductors over time, which can lead to the corrosion of metal elements or even the appearance of the corona effect. If certain compartments in a switchgear are monitored, faults can be predicted, thus avoiding the problems mentioned above. Corona effect, which can be found as corona discharges, can be easily monitored because it is an electrical discharge caused by the ionization of a fluid such as air surrounding a conductor carrying a high voltage. In many high voltage applications, like MV switchgears where high voltages pass through electric wires, corona is an unwanted side effect. In the air, coronas generate gases such as ozone (O_3) and nitric oxide (NO), and in turn, nitrogen dioxide (NO_2) and thus nitric acid (HNO_3) if water vapor is present. These gases are corrosive and can degrade and embrittle nearby materials such as the switchgear high voltage conductor's isolation or even the metal case.

Water vapor can occur due to sudden changes in temperature and high humidity conditions. If a sensor is used to monitor the mentioned parameters, the value of the dew point can be determined exactly and thus the appearance of water vapor in a switchgear, which can cause corona discharge for instance, can be anticipated. In this case, the BME680 sensor is used, which through an I2C connection can transmit information related to temperature, humidity, air quality, and atmospheric pressure.

Thus, the sensors mentioned in the MSMN node can transmit data from the field, and decisions can be made at a monitoring center, history files can be created, or alarms can be set for the prevention of problems and the optimal management of LV/MV cells.

The power supply of the node is a rechargeable Li-Po battery with a capacity of 1000 mAh. This battery capacity is necessary mainly due to the ozone measuring sensor, MQ-131. According to the datasheet, the sensor is a metal oxide semiconductor (MOS) type gas sensor also known as a chemiresistor, because detection is based on the change of resistance of the sensing material when the gas encounters the material. Using a simple voltage divider network, the gas concentration can be easily detected. Thus, for good operation, the sensor heater resistance needs approximately 900 mW, consuming the most battery capacity.

To optimize the energy consumption of the node, the algorithm proposed in [27] is used. Firstly, the MCU, the BME680, and the MQ-131 sensors are woken up. To obtain the correct values, the MQ-131 heater resistance needs to preheat at least 2 min before obtaining the measurements. During this time, the BME680 sensor values are read and stored in the MCU internal EEPROM memory. The next step is to put the BM680 sensor in sleep mode and verify the NEO-6G GPS coordinates. For this to happen, an internal counter is checked to verify if a 24-h interval has elapsed from the last GPS fix measurement. If this condition is fulfilled, the GPS module is powered-up, otherwise this step is passed over. This process may take up to 26 s (according to the GPS datasheet) only once a day. After the GPS location coordinates are collected and saved, the GPS module is deactivated. The next step is acquiring the ozone values from the MQ-131 sensor and storing them in the same EEPROM memory. After that, all the integrated sensors are deactivated and at the same time the LoRa communication module is woken-up to transmit the saved information to the monitoring center through the LoRaWAN gateway.

During the day, a charging circuit can be used from a photovoltaic panel mounted in the system, supplementing the energy requirements of the system for a longer lifetime. The whole system is designed for autonomous functionality, being able to be moved or reused for other monitoring areas without the need for interruptions from the power supply in case the power supply would be used directly from the switchgear. Using the GPS module, we can easily determine the position of the node and possibly assign new limits for the monitored parameters, changes that can be made from the control center.

The hardware platform from [27] with the new features added will need approximately 243 mA in active mode with SF7 and 352 mA with SF12. These consumption values are given for the worst-case scenario, when all the components from the platform are active at the same time.

For programming the platform, an open-source IDE programming environment is used. This is an easy way of integrating already existing functions and using existing libraries to perform communication within LoRaWAN protocol.

The hardware development of the node can be seen in Figure 5.

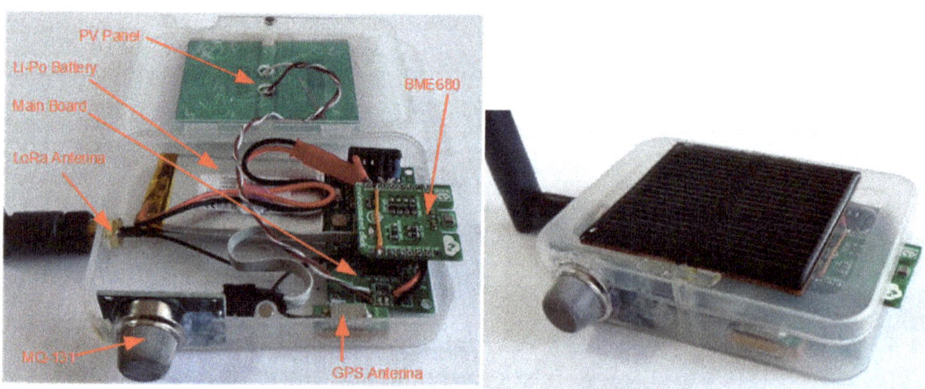

Figure 5. Hardware implementation of the MSMN node.

4.3. *Communication Coverage of the Proposed Architecture*

Given that this paper proposes a monitoring architecture for the smart grid concept, the proposed node must cover a large area for LoRaWAN communication. For proper communication, the gateways must cover the areas where switchgears for monitoring purpose are mounted. Thus, some tests are performed to ensure gateway communication and optimize the functionality parameters. The communication architecture is listed in Figure 6.

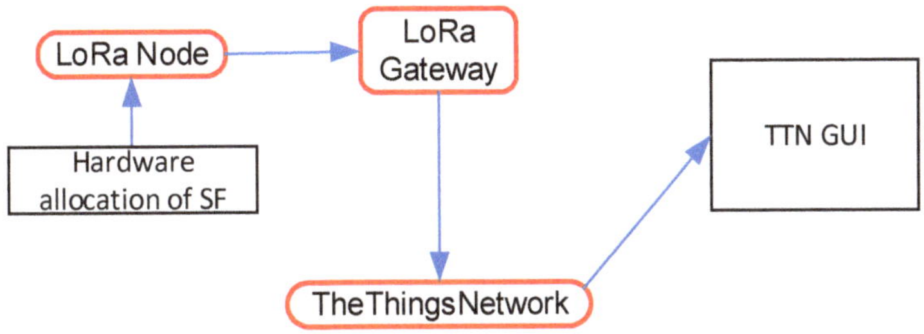

Figure 6. Communication architecture for the coverage estimation.

For this test, the node was programmed with a test code, with the communication spreading factor manually chosen with the use of some external hardware jumpers. Each time the packet was received, the message appeared in the TheThingsNetwork cloud IoT platform interface (TTN GUI). The used gateway was the one discussed in detail in [31], with an omnidirectional antenna from Taoglas with a 12 dBi gain and was placed in the Stefan cel Mare University Campus.

From the obtained results depicted in Figure 7, we can see that the communication range in the tested zone, which is an urban one, is strongly affected by the spreading factor, covering only approximately 2 km^2 when all variations of the spreading factor are linked, from 7 to 12. Thus, to deploy a large-scale LoRaWAN network for the smart grid concept, we need to choose the gateways' locations on the required infrastructure. The easiest method for this step is to use a wireless communication emulator. This network emulator must ensure the option for attenuation losses due to the geographical terrain variation, because LoRaWAN can be deployed for more than 10 km in the urban areas or more than 50 km in rural areas, where terrain variation with buildings and vegetation is available. Such a simulator is RadioPlanner [32] and it was used in this paper for the LoRaWAN coverage simulation. This software framework can deploy large-scale architectures using long-range communications like LoRa, GSM, LTE, UHF, or VHF.

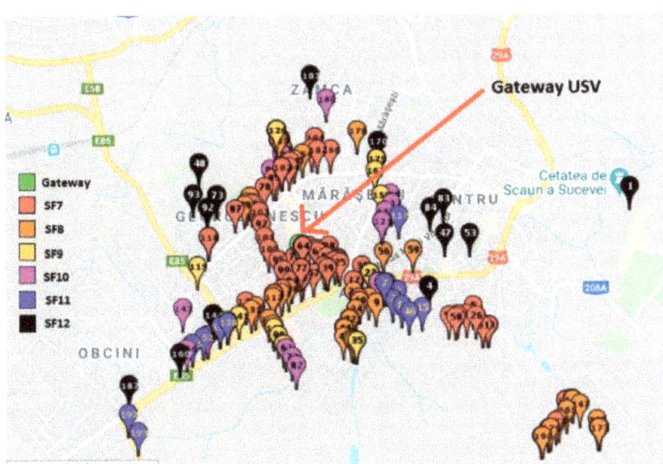

Figure 7. LoRaWAN coverage map measurement.

To ensure the simulation validation and calibration process, we chose the values measured with the test setup mentioned earlier. All the parameters were changed for the transmitter (the LoRaWAN node) and for the receiver (the LoRaWAN gateway), choosing the antenna pattern, gain, cable loss, beam tilt and the height where is mounted from the ground point. Also, for accurate results, some attenuation loss parameters were considered with the specifications from the ITU-R P.1812-4 report [33], which represents a path-specific propagation prediction method for point-to-area terrestrial services in the VHF and UHF bands. The simulated results are depicted in Figure 8.

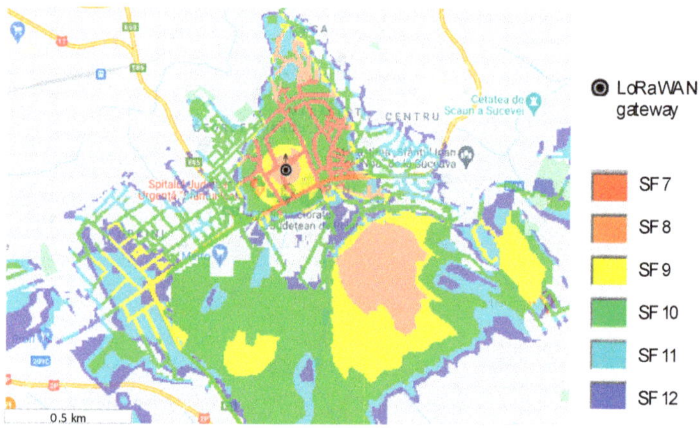

Figure 8. LoRaWAN coverage map simulation using RadioPlanner.

5. Results and Discussion

To ensure the feasibility of the proposed sensor node, we chose some LV/MV switchgears from our regional electric distribution network, located in a neighborhood in the Suceava town. The monitored area is around 5 km^2, it has 18 monitoring points, and it is an urban area where there are buildings and other steel/concrete structures, forest trees, and different elevations.

The first step was to locate the best position of the LoRaWAN gateway. Taking the optimized parameters for the simulator, we made some gateway location estimates. The first scenario was placing the gateway in the middle of the monitored area (point IT242), with the results obtained in Figure 9.

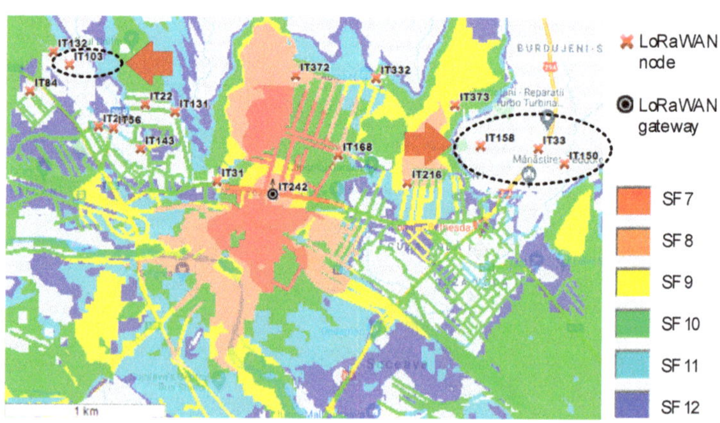

Figure 9. LoRaWAN Coverage Scenario 1.

From the first simulation, we can see that not all the LoRaWAN nodes were communicating with the gateway, even if the gateway was in the middle of the monitored area. This is due to the geographical terrain variation, which causes diffraction, free space, or clutter losses. A result from IT242 (the gateway location) and IT33 (one uncovered node location) is depicted in Figure 10.

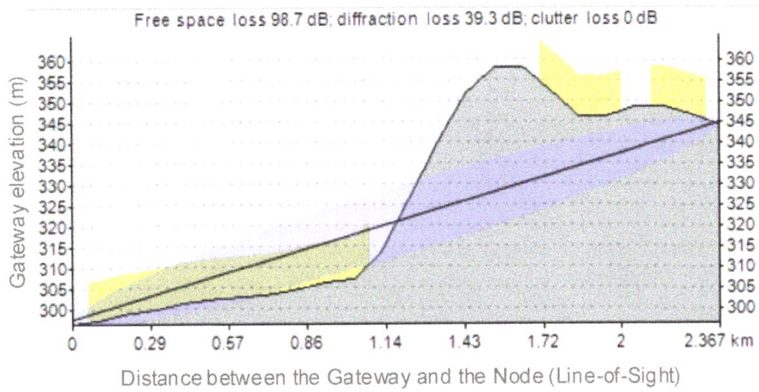

Figure 10. Losses due to the geographical terrain variation for Scenario 1.

The next scenario was made using other locations from the monitored area, but not linked directly to a LV/MV switchgear point. If we take the elevation parameter when making the simulation, we can find the best point for the gateway to be located (Figure 11).

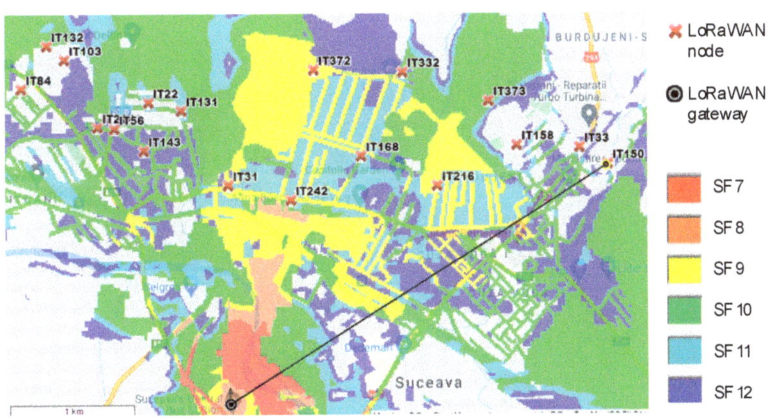

Figure 11. LoRaWAN Coverage Scenario 2.

Here we can see that all the monitored points are covered by the gateway, even if the distance between each monitored point and the gateway is greater than the previous scenario, covering a much larger area than the Scenario 1. Figure 12 reveals the elevation between the gateway location and the IT150 (the node location), with a LoS (line-of-sight) distance of 3.5 km. Even if the free space losses are greater than the previous scenario, the node can transmit data to the gateway using SF12 due to the elevation difference between the measuring points.

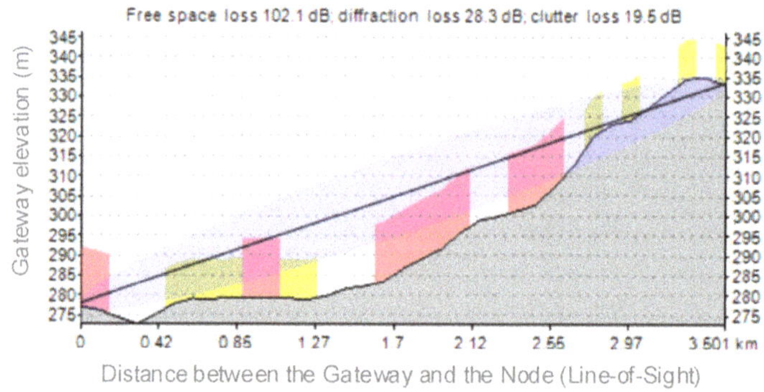

Figure 12. Losses due to the geographical terrain variation for Scenario 2.

Once the gateway location is set, we can place the sensor nodes on the LV/MV switchgears for monitoring the environmental parameters.

For testing purpose, we chose a faulty LV switchgear that needed maintenance on the cable side so that the transmitted values were accurate to the switchgear issues (appearance of the partial discharge inside), marked IT131 on the coverage map. The measurements were performed during a 6-month period, from May to October (Figure 13).

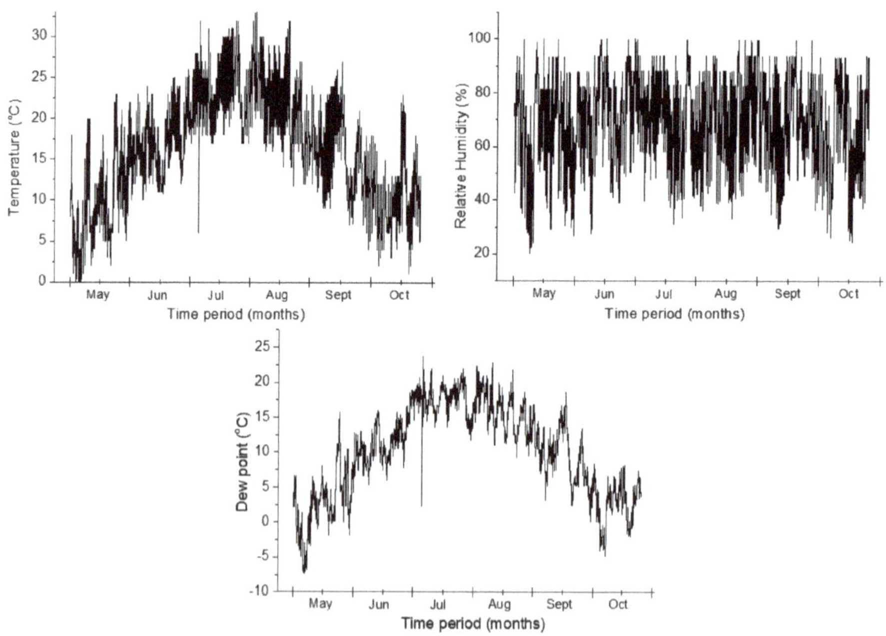

Figure 13. Environmental parameters from the MSMN (temperature, relative humidity, and dew point).

All the parameters mentioned in Section 4 were transmitted using the LoRaWAN communication protocol. The GPS coordinates were used to place the exact position of the nodes in the Google Maps view, so that each node can be easily identified. The temperature and the humidity were transmitted to obtain the dew point value inside the switchgear, to make an estimation regarding an eventual rising of its value that can cause partial discharge and degrade the conductor isolation. In technical terms, the dew point is the temperature at which the water vapor in a sample of air at constant barometric pressure condenses into liquid water at the same rate at which it evaporates. At temperatures below the dew point, the rate of condensation will be greater than that of evaporation, forming more liquid water. To obtain the exact value, some calculations are made using the equations from [34] as reference:

$$t_d \approx t - \left(\frac{100 - RH}{5}\right), \quad (1)$$

where t_d is the dew point temperature measured in Celsius, t is the air temperature measured in Celsius, and RH is the relative air humidity in percentage. Listed below are the transmitted values and the dew point calculation according to Equation (1).

Some issues regarding the switchgear can be noticed from the proposed MSMN node and this is the corona discharge effect, discussed in Section 4.2 of the paper. This is possible with the help of ozone values received from the node. In some periods, because of the increased relative humidity value, the dew point exceeds the air temperature, so water vapor condenses on the cable isolators. Thus, locally partial discharges occur for short periods, causing the increase of ozone levels. The obtained results received from the monitored switchgear can be seen in Figure 14.

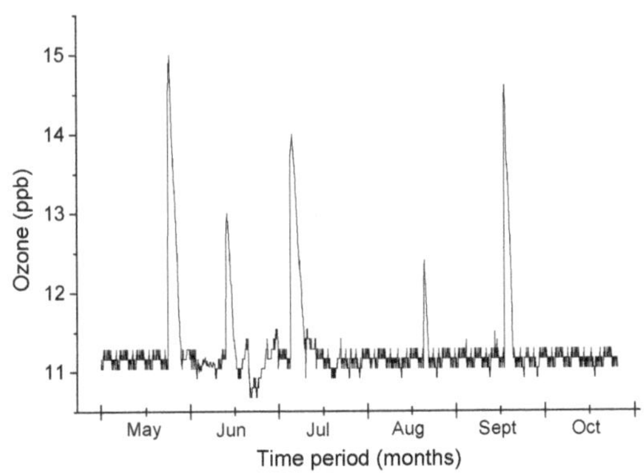

Figure 14. Ozone level received from the MSMN node.

The node was programmed to transmit the data three times a day (in the morning, in the afternoon, and in the evening); the rest of the time the node was in sleep mode, waking up only to sample the measurement parameters. This decision for transmitting the data was taken for power optimization, resulting an autonomous functionality node that does not need a continuous power supply, being a standalone LoRaWAN node. In the daytime, the solar panel will charge the node battery to increase power efficiency.

6. Conclusions

The maintenance of switchgear parameters from an electric grid is essential to avoid failures or blackouts. At present, the population is demanding of electricity, so it is necessary to find monitoring solutions for the distribution points.

The ever-increasing demand for electrical energy is also sustained by aggressive electrical vehicle technology integration to meet green energy certifications and carbon free emissions, according to EU regulations.

By using the internet of things infrastructure, we can remotely manage the electricity distribution points, receiving data that can predict any future failure points of the grid. By extending the concept, utility companies will be able to pick-up information from the network remotely and the network will even have the capability to self-heal in a cost-effective automatic manner.

In this paper, we present a smart grid monitoring architecture with a multi-sensor monitoring node from which we can collect data with the LoRaWAN communication protocol. The designed architecture was tested for a period of 6 months, revealing the feasibility and scalability of the system, and opening new directions in the remote prediction of failures and blackouts of the LV/MV switchgears on the electric grid. Furthermore, the monitoring solution is still transmitting data and we will continue to analyze its functionality in the cold wintertime. The measurements have been made during a period including the rainiest months in our region, May, and June. During these intervals, solar radiation was at an average compared to that of the sunniest days, which are in August. As per our laboratory tests with variable exposure to light, the sensor will operate with a minimum of 5 days per month of solar radiation.

This novel architecture can be integrated successfully into the smart city concept with smart grid applicability. Monitoring some normal environment parameters like air humidity and air temperature, we can obtain important information related to the condition of the network, allowing the evaluation of the performance level of the smart grid.

The novel designed multi-sensor monitoring node uses a modular platform with a high degree of reconfigurability that is also able to function in a standalone mode using state of the art communication protocols like LoRaWAN suitable for crowded city topologies.

Author Contributions: Conceptualization, A.I.P., E.C. and A.L.; methodology, A.I.P. and E.C.; software, A.I.P.; validation, A.I.P., E.C. and A.L.; formal analysis, A.I.P. and A.L.; investigation, A.I.P.; resources, A.I.P. and E.C.; data curation, A.I.P. and A.L.; writing—original draft preparation, A.I.P., E.C. and A.L.; writing—review and editing, A.I.P., E.C. and A.L.; visualization, A.I.P., E.C. and A.L.; supervision, E.C.; project administration, A.I.P. and E.C.; funding acquisition, A.I.P. and E.C. All authors have read and agreed to the published version of the manuscript.

Funding: This paper was made using the infrastructure from the project "Integrated Center for research, development and innovation in Advanced Materials, Nanotechnologies, and Distributed Systems for fabrication and control (MANSiD)", Contract No. 671/09.04.2015, Sectoral Operational Program for Increase of the Economic Competitiveness co-funded from the European Regional Development Fund, and was supported by a grant of the Romanian Ministry of Research and Innovation, CCCDI-UEFISCDI, project number PN-III-P2-2.1-PTE-2019-0642, within PNCDI III.

Institutional Review Board Statement: Not applicable.

Informed Consent Statement: Not applicable.

Conflicts of Interest: The authors declare no conflict of interest. The funders had no role in the design of the study; in the collection, analyses, or interpretation of data; in the writing of the manuscript, or in the decision to publish the results.

Abbreviations

ICT	Information and communication technology
LPWAN	Low-power wide area network
LoRaWAN	Long-range wide area network
NB-IoT	Narrowband-internet of things
CPS	Cyber-physical system
IoT	Internet of things
PLC	Power line communication
TEV	Transient earth voltage
LV/MV	Low voltage/medium voltage
HFCT	High frequency current transformer
SCADA	Supervisory control and data acquisition
WSN	Wireless sensor network
CSS	Chirp spread spectrum
SNR	Signal to noise ratio
ISM	Industrial, scientific, medical

SRD	Short-range devices
SF	Spreading factor
MAC	Medium access control
ADR	Adaptive data rate
CR	Coding rate
ETSI	European telecommunications standards institute
FCC	Federal communications commission
DSSS	Direct sequence spread spectrum
LoS	Line-of-sight
MSMN	Multi-sensor measuring node
GSM/GPRS	Global system for mobile communications/general packet radio service
GPS	Global positioning system
MOS	Metal oxide semiconductor
MCU	Micro controller unit
LTE	Long-term evolution
UHF	Ultra-high frequency
VHF	Very-high frequency
MHz	Megahertz
kHz	Kilohertz
bps	Bits per second
mW	Milliwatt
mA	Milliampere
μA	Microampere
I2C	Inter-integrated circuit
mAh	Milliampere hour
t_d	Dew point temperature
t	Air temperature
RH	Air humidity

References

1. Fu, G.; Wilkinson, S.; Dawson, R.J.; Fowler, H.J.; Kilsby, C.; Panteli, M.; Mancarella, P. Integrated Approach to Assess the Resilience of Future Electricity Infrastructure Networks to Climate Hazards. *IEEE Syst. J.* **2018**, *12*, 3169–3180. [CrossRef]
2. Morello, R.; Capua, C.; Fulco, G.; Mukhopadhyay, S.C. A Smart Power Meter to Monitor Energy Flow in Smart Grids: The Role of Advanced Sensing and IoT in the Electric Grid of the Future. *IEEE Sens. J.* **2017**, *17*, 7828–7837. [CrossRef]
3. Raboaca, M.S.; Meheden, M.; Musat, A.; Viziteu, A.; Creanga, A.; Vlad, V.; Filote, C.; Rata, M.; Lavric, A. An overview and performance evaluation of open charge point protocol from an electromobility concept perspective. *Int. J. Energy Res.* **2021**, 1–21. [CrossRef]
4. Bansal, P.; Singh, A. Smart metering in Smart Grid Framework: A Review. In Proceedings of the Fourth International Conference on Parallel, Distributed and Grid Computing (PDGC), Waknaghat, India, 22–24 December 2016.
5. Ma, J.; Zheng, Y.; Ning, H.; Yang, L.T.; Huang, R.; Liu, H.; Mu, Q.; Yau, S.S. Top Challenges for Smart Worlds: A Report on the Top10Cs Forum. *IEEE Access* **2015**, *3*, 2475–2480. [CrossRef]
6. Sun, C.; Cembrano, G.; Puig, V.; Meseguer, J. Cyber-Physical Systems for Real-Time Management in the Urban Water Cycle. In Proceedings of the International Workshop on Cyber-physical Systems for Smart Water Networks (CySWater), Porto, Portugal, 10–13 April 2018.
7. Mierlo, V.J.; Berecibar, M.; El Baghdadi, M.; De Cauwer, C.; Messagie, M.; Coosemans, T.; Jacobs, V.A.; Hegazy, O. Beyond the State of the Art of Electric Vehicles: A Fact-Based Paper of the Current and Prospective Electric Vehicle Technologies. *World Electr. Veh. J.* **2021**, *12*, 20. [CrossRef]
8. Shen, Y.; Fang, W.; Ye, F.; Kadoch, M. EV Charging Behavior Analysis Using Hybrid Intelligence for 5G Smart Grid. *Electronics* **2020**, *9*, 80. [CrossRef]
9. Qiang, T.; Mingzhong, X.; Kun, Y.; Yuansheng, L.; Dongdai, Z.; Yun, S. A Decision Function Based Smart Charging and Discharging Strategy for Electric Vehicle in Smart Grid. *Mob. Netw. Appl.* **2019**, *24*, 1722–1731. [CrossRef]
10. Sokowoo, R. Catalyzing the Internet of Things and Smart Cities: Global City Teams Challenge. In Proceedings of the 1st International Workshop on Science of Smart City Operations and Platforms Engineering (SCOPE) in Partnership with Global City Teams Challenge (GCTC) (SCOPE-GCTC), Vienna, Austria, 11 April 2016.
11. Bavarian, S.; Lampe, L. Communications and Access Technologies for Smart Grid. In *Smart Grid Communications and Networking*; Hossain, E., Han, Z., Poor, H.V., Eds.; Cambridge University Press: Cambridge, UK, 2012; pp. 111–146.
12. Wang, Z.; Ge, W.; Wang, C.; Zeng, C. The applications of networking of consumption data acquisition system by combining broadband powerline communication and wireless communication. In Proceedings of the IEEE PES Innovative Smart Grid Technologies, Tianjin, China, 21–24 May 2012; pp. 1–4.

13. Lasagani, K.; Iqbal, T.; Mann, G. Data Logging and Control of a Remote Inverter Using LoRa and Power Line Communication. *Energy Power Eng.* **2018**, *10*, 351–365. [CrossRef]
14. Zhang, H.; Yan, Y.; Xu, P.; Xu, X.; Lu, Y.; Yang, S. A Wireless Synchronization Locating Method for Partial Discharge in Switchgear. In Proceedings of the 6th Global Electromagnetic Compatibility Conference (GEMCCON), Xi'an, China, 20–23 October 2020; pp. 1–4.
15. Xu-Ze, G.; Tianxin, Z.; Ming, R.; Bo, S.; Wenguang, H.; Ming, D. IoT-based On-line Monitoring System for Partial Discharge Diagnosis of Cable. In Proceedings of the IEEE Electrical Insulation Conference (EIC), Calgary, AB, Canada, 16–19 June 2019; pp. 54–57.
16. De Campo, G.; Gomez, I.; Cañada, G.; Santamaria, A. Hybrid LPWAN Communication Architecture for Real-Time Monitoring in Power Distribution Grids. In Proceedings of the IEEE 5th World Forum on Internet of Things (WF-IoT), Limerick, Ireland, 15–18 April 2019; pp. 920–924. [CrossRef]
17. Wu, J.; Zhang, J.; Ma, J.; Zhao, M.; Wang, X.; Gao, X.; Zhang, H. Energy Efficient 5G LoRa Ad-Hoc Network for Smart Grid Communication. In Proceedings of the IEEE 11th International Conference on Electronics Information and Emergency Communication (ICEIEC), Beijing, China, 18–20 June 2021; pp. 1–4. [CrossRef]
18. Tang, J.; Li, J.; Zhong, A.; Xiong, B.; Bian, X.; Li, Y. Application of LoRa and NB-IoT in Ubiquitous Power Internet of Things: A Case Study of Fault Indicator in Electricity Distribution Network. In Proceedings of the 4th International Conference on Intelligent Green Building and Smart Grid (IGBSG), Hubei, China, 6–9 September 2019; pp. 380–383. [CrossRef]
19. Gallardo, J.L.; Ahmed, M.A.; Jara, N. LoRa IoT-Based Architecture for Advanced Metering Infrastructure in Residential Smart Grid. *IEEE Access* **2021**, *9*, 124295–124312. [CrossRef]
20. Mazur, K.; Wydra, M.; Ksiezopolski, B. Secure and Time-Aware Communication of Wireless Sensors Monitoring Overhead Transmission Lines. *Sensors* **2017**, *17*, 1610. [CrossRef] [PubMed]
21. Zidan, A.; Khairalla, M.; Abdrabou, A.M.; Khalifa, T.; Shaban, K.; Abdrabou, A.; El Shatshat, R.; Gaouda, A.M. Fault Detection, Isolation, and Service Restoration in Distribution Systems: State-of-the-Art and Future Trends. *IEEE Trans. Smart Grid* **2017**, *8*, 2170–2185. [CrossRef]
22. Lavric, A.; Petrariu, A.I.; Coca, E.; Popa, V. LoRa Traffic Generator Based on Software Defined Radio Technology for LoRa Modulation Orthogonality Analysis: Empirical and Experimental Evaluation. *Sensors* **2020**, *20*, 4123. [CrossRef] [PubMed]
23. Lauridsen, M.; Nguyen, H.; Vejlgaard, B.; Kovacs, I.Z.; Mogensen, P.; Sorensen, M. Coverage Comparison of GPRS, NB-IoT, LoRa, and SigFox in a 7800 km^2 Area. In Proceedings of the IEEE 85th Vehicular Technology Conference (VTC Spring), Sydney, NSW, Australia, 4–7 June 2017; pp. 1–5. [CrossRef]
24. Lavric, A.; Popa, V. Internet of Things and LoRa™ Low-Power Wide-Area Networks: A survey. In Proceedings of the International Symposium on Signals, Circuits and Systems (ISSCS), Iasi, Romania, 13–14 July 2017. [CrossRef]
25. Shnayder, V.; Hempstead, M.; Chen, B.-R.; Allen, G.W.; Welsh, M. Simulating the Power Consumption of Large-Scale Sensor Network Applications. In Proceedings of the 2nd International Conference on Embedded Networked Sensor Systems, SenSys'04, Baltimore, MD, USA, 3–5 November 2004; pp. 188–200. [CrossRef]
26. Lavric, A.; Petrariu, A.I.; Popa, V. Long Range SigFox Communication Protocol Scalability Analysis Under Large-Scale, High-Density Conditions. *IEEE Access* **2019**, *7*, 35816–36825. [CrossRef]
27. Petrariu, A.I.; Lavric, A.; Coca, E.; Popa, V. Hybrid Power Management System for LoRa Communication using Renewable Energy. *IEEE Internet Things J.* **2021**, *8*, 8423–8436. [CrossRef]
28. Zhang, C.; Dong, M.; Ren, M.; Huang, W.; Zhou, J.; Gao, X.; Albarracín, R. Partial Discharge Monitoring on Metal-Enclosed Switchgear with Distributed Non-Contact Sensors. *Sensors* **2018**, *18*, 551. [CrossRef] [PubMed]
29. Hussain, G.A.; Kumpulainen, L.; Klüss, J.V.; Lehtonen, M.; Kay, J.A. The Smart Solution for the Prediction of Slowly Developing Electrical Faults in MV Switchgear Using Partial Discharge Measurements. *IEEE Trans. Power Deliv.* **2013**, *28*, 2309–2316. [CrossRef]
30. Perdon, K.; Scarpellini, M.; Magoni, S.; Cavalli, L. Modular online monitoring system to allow condition-based maintenance for medium voltage switchgear. *CIRED—Open Access Proc. J.* **2017**, *1*, 346–349. [CrossRef]
31. Petrariu, A.I.; Lavric, A.; Coca, E. LoRaWAN Gateway: Design, Implementation and Testing in Real Environment. In Proceedings of the IEEE 25th International Symposium for Design and Technology in Electronic Packaging (SIITME), Cluj-Napoca, Romania, 23–26 October 2019; pp. 49–53. [CrossRef]
32. Radioplanner Software. Available online: https://www.wireless-planning.com/radioplanner (accessed on 1 November 2021).
33. International Telecommunication Unit. *A Path-Specific Propagation Prediction Method for Point-to-Area Terrestrial Services in the VHF and UHF Bands*; ITU-R P.1812-4 Report; International Telecommunication Unit: Geneva, Switzerland, 2015. Available online: https://www.itu.int/dms_pubrec/itu-r/rec/p/R-REC-P.1812-4-201507-I!!PDF-E.pdf (accessed on 30 November 2021).
34. Lawrence, M.G. The Relationship between Relative Humidity and the Dewpoint Temperature in Moist Air: A Simple Conversion and Applications. *Bull. Am. Meteorol. Soc.* **2005**, *86*, 225–233. [CrossRef]

sensors

MDPI

Communication

Antenna Impedance Matching Using Deep Learning

Jae Hee Kim [1] and Jinkyu Bang [2,*]

1 School of Electrical, Electronics and Communication Engineering, Korea University of Technology and Education, Cheonan 31253, Korea; jaehee@koreatech.ac.kr
2 Department of Electrical and Electronic Engineering, Youngsan University, Yangsan 50510, Korea
* Correspondence: jinkyu.bang@ysu.ac.kr

Abstract: We propose a deep neural network (DNN) to determine the matching circuit parameters for antenna impedance matching. The DNN determines the element values of the matching circuit without requiring a mathematical description of matching methods, and it approximates feasible solutions even for unimplementable inputs. For matching, the magnitude and phase of impedance should be known in general. In contrast, the element values of the matching circuit can be determined only using the impedance magnitude using the proposed DNN. A gamma-matching circuit consisting of a series capacitor and a parallel capacitor was applied to a conventional inverted-F antenna for impedance matching. For learning, the magnitude of input impedance S_{11} of the antenna was extracted according to the element values of the matching circuit. A total of 377 training samples and 66 validation samples were obtained. The DNN was then constructed considering the magnitude of impedance S_{11} as the input and the element values of the matching circuit as the output. During training, the loss converged as the number of epochs increased. In addition, the desired matching values for unlearned square and triangular waves were obtained during testing.

Keywords: antenna impedance matching; artificial neural network; deep learning; input impedance (S_{11})

Citation: Kim, J.H.; Bang, J. Antenna Impedance Matching Using Deep Learning. *Sensors* **2021**, *21*, 6766. https://doi.org/10.3390/s21206766

Academic Editors: Liliana Anchidin, Adrian I. Petrariu and Alexandru Lavric

Received: 23 September 2021
Accepted: 9 October 2021
Published: 12 October 2021

1. Introduction

The most recent electronic devices support wireless communication, for which an antenna operating at a specific frequency band must be used. As the resonant frequency of an antenna is affected by its shape and surrounding materials in a device, the antenna must be modified whenever the device design is changed. To avoid antenna redesign, commercial communication devices can be used. However, such devices often include bulky external dipole antennas. In addition, the required operation frequency may not be available because commercial devices are intended for predefined frequencies, such as the ISM (industrial, scientific, and medical) band. If an antenna with a fixed shape could automatically operate at the desired frequency, the development time of wireless devices could be notably reduced along with the development cost of antennas.

Antennas used to tune the resonant frequency can be divided into reconfigurable and tunable antennas. Reconfigurable antennas [1–5] adjust the resonant frequency by changing their shape through a switch. Thus, small antennas or multiband antennas should often be reconfigurable given the difficulty to obtain a wide bandwidth. For instance, the tuning of the resonant frequency while reducing the size of a slot loop antenna has been achieved by using varactor diodes [3]. In addition, selection of the LTE (Long Term Evolution) band of 1.8 or 2.6 GHz has been achieved by inserting a PIN diode at the end of a loop antenna [4]. A reconfigurable antenna can change its radiation pattern by modifying its structure, providing high radiation efficiency depending on its shape. However, reconfigurable antennas are generally difficult to design given their complex structure.

On the other hand, tunable antennas operate at various frequency bands by changing the element values in the matching circuit [6–12]. As their structure is fixed and only the element values in the matching circuit change to obtain a resonant frequency, tunable antennas are simple to design. By applying a tunable matching circuit to an ultrawideband antenna, resonant frequencies from 1.8 to 2.8 GHz have been set at applied voltages from 0 to 23 V [9]. In addition, a tunable matching circuit has been applied to provide the required service frequency bandwidths for small antennas [12]. However, tunable antennas deliver suboptimal efficiency due to losses in the matching elements at any operating frequency [10]. To improve performance and properly tune the resonant frequency, the magnitude and phase of the antenna impedance should be accurately determined. In addition, the factor causing the change in the resonant frequency over the range of element values should be identified. Despite their simple structure, tunable antennas must be carefully designed by considering the antenna characteristics for the matching circuit.

Recently, machine learning has been applied to optimize antenna performance [13–19] and implement impedance matching [20–22]. A machine learning method can determine the element values without requiring a mathematical description of the matching circuit. In wireless power transmission, a neural network has recently been used to achieve the maximum efficiency [20,21]. Specifically, the matching element value according to the impedance of a wireless power transfer (WPT) coil was learned, and matching was performed automatically based on the measured impedance. The efficiency can be maximized by automatically compensating the matching value according to the distance between WPT coils. To date, however, no machine learning method has been devised for antenna impedance matching.

We propose a deep learning method that determines the element values of the matching circuit for a given magnitude of input impedance S_{11}. The input is only the impedance magnitude, and the output is the corresponding element values of the matching circuit. Unlike the conventional approach, the proposed method determines the appropriate matching element values, and it can solve even unimplementable input impedances.

The remainder of this paper is organized as follows. Section 2 presents the antenna structure and matching circuit used in this study. Section 3 describes the method for acquiring input impedance S_{11} according to the capacitor values of the matching circuit. In Section 4, we introduce the proposed deep neural network (DNN) for antenna impedance matching. Section 5 reports the deep learning results and presents the corresponding discussion. Finally, we draw conclusions in Section 6.

2. Antenna and Matching Circuit

Figure 1 shows the antenna structure to simulate the matching circuit effect. The basic structure is an inverted-F antenna, which is the most common type for mobile devices. The resonant frequency of the inverted-F antenna is determined by the length of the antenna, and the matching of the antenna is determined by the distance between the feeding point and the shorting stub. In general, the inverted F antenna is designed in the form of a meander line to include the length of the antenna in a narrow space in order to lower the resonant frequency. However, if there is not enough space for metal patterning, a matching circuit should be used to adjust the resonance frequency. The antenna is patterned on an FR4 substrate with a dielectric constant of 4.3 and a thickness of 1 mm. The obtained inverted-F antenna has a length of 28 mm and a height of 10 mm from the ground. The line width is 1 mm, and the shorting stub at the left end is connected to the ground. The feeding point is 2 mm from the shorting stub. The matching circuit is directly connected to the feeding point. The dimensions of the antenna are arbitrary. If there is no matching circuit, the antenna resonates at 1.9 GHz. The resonant frequency of the antenna can be tuned from 0.9 to 1.4 GHz using a matching circuit.

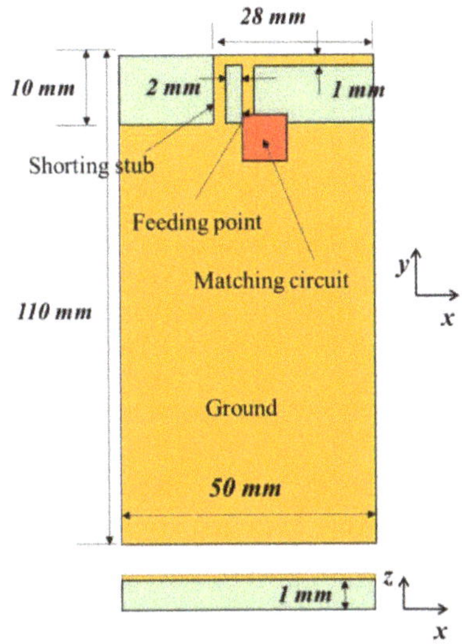

Figure 1. Antenna structure for impedance simulation.

We performed simulations using Ansys HFSS (3D high-frequency simulation software). Figure 2 shows input impedance S_{11} of the antenna without a matching circuit on the Smith chart and the real part and imaginary part of the impedance for the frequency range of 0.8–1.5 GHz. As can be seen from Figure 2b, the imaginary part has values higher than 50 ohm as positive values. This means that in order to make resonance, the imaginary part should be compensationed through capacitors. The antenna impedances at the lowest and highest frequencies are located in the upper-right corner of the Smith chart. Therefore, the impedance can be matched by combining a series capacitor and a parallel capacitor, as shown in the gamma-matching circuit of Figure 3. Impedance matching is possible at the designed frequencies according to the capacitor values. Specifically, we used a series capacitor C_S of 0.9–3.3 pF and parallel capacitor C_P of 1–15 pF.

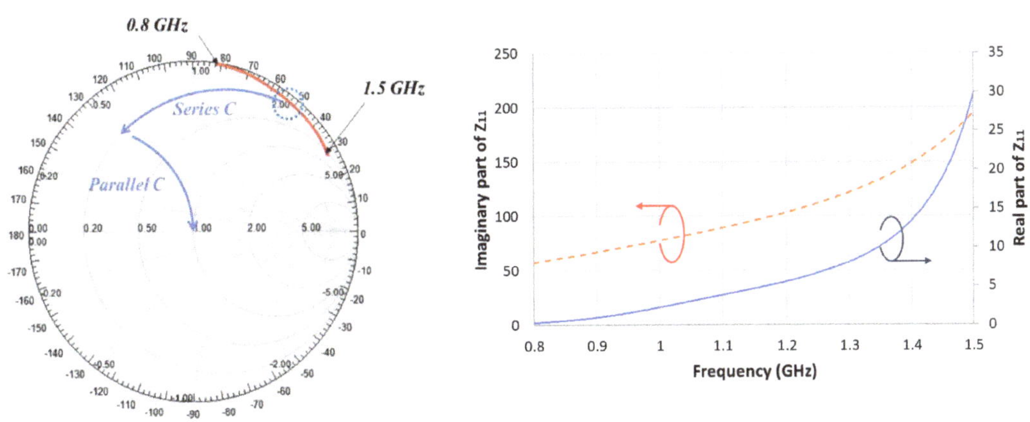

(**a**) (**b**)

Figure 2. Simulated antenna impedance without matching circuit (**a**) Smith chart, (**b**) real and imaginary parts.

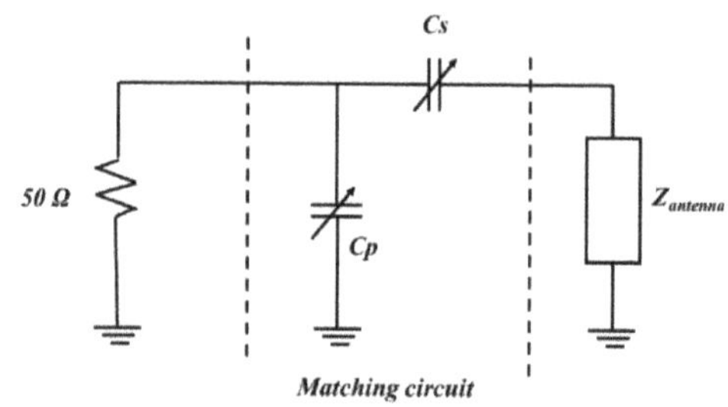

Figure 3. Gamma-matching circuit.

Matching should be applied for the capacitor values to match at the lowest and highest resonant frequencies of 0.9 and 1.4 GHz, respectively. For resonance at 1.4 GHz, the values of the series (C_S) and parallel (C_P) capacitors should be 0.9 and 3 pF, respectively. For resonance at 0.9 GHz, the respective values should be 3 and 20 pF. These values for the matching circuit were determined by mathematical calculations based on accurate information about the real and imaginary parts of the antenna impedance.

Although it is possible to measure S_{11} including its real and imaginary parts by using a network analyzer, expensive equipment is required. Instead, we propose a method for determining the matching element values using the S_{11} magnitude in a DNN. As the magnitude does not include phase information, an accurate matching value cannot be determined mathematically. However, through learning, the proposed method determines the matching element values solely from the input impedance magnitude.

3. Data Acquisition

The effectiveness of machine learning depends on the availability of large amounts of data. However, manually obtaining input impedance S_{11} according to the matching element values is time-consuming. Therefore, automated data acquisition should be performed. To this end, we linked MathWorks MATLAB and Ansys HFSS. In MATLAB, series capacitor C_S and parallel capacitor C_P were set as variables, and these values were linked with HFSS. According to the matching element values, the S_{11} magnitude was extracted as a text file. The matching element values for training are listed in Table 1, and those for validation are listed in Table 2. The magnitude of input impedance S_{11} is a scalar value ranging from 0 to 1 over 401 datapoints, corresponding to a frequency range from 0.8 to 1.5 GHz. For the training data, as 13 series capacitors and 29 parallel capacitors were used, $13 \times 29 = 377$ samples were obtained. In addition, the validation samples were $11 \times 10 = 110$. The postprocessing time to obtain S_{11} per setting of matching element values was 12 s, taking approximately 90 min to obtain all the training and validation samples. Figure 4 shows the S_{11} magnitude for all the training (Figure 4a) and validation (Figure 4b) samples. It is important to match the antenna impedance at the designed resonant frequency. The reason for graphing all samples in Figure 4 is to indicate that the resonant frequency of validation samples is different from the resonant frequency of the training samples. This is to investigate how well the DNN learns for these different resonant frequencies.

Table 1. Capacitor values in matching circuit for training.

Element	Values (pF)	No. Cases
Series capacitor C_S	0.9, 1.1, 1.3, 1.5, 1.7, 1.9, 2.1, 2.3, 2.5, 2.7, 2.9, 3.1, 3.3	13
Parallel capacitor C_P	1, 1.5, 2, 2.5, 3, 3.5, 4, 4.5, 5, 5.5, 6, 6.5, 7, 7.5, 8, 8.5, 9, 9.5, 10, 10.5, 11, 11.5, 12, 12.5, 13, 13.5, 14, 14.5, 15	29

Table 2. Capacitor values in matching circuit for validation.

Element	Values (pF)	No. Cases
Series capacitor C_S	1, 1.2, 1.4, 1.6, 1.8, 2, 2.2, 2.4, 2.6, 2.8, 3	11
Parallel capacitor C_P	1.1, 1.7, 2.3, 3.7, 5.5, 7.5, 9.5, 11.5, 13.5, 16	10

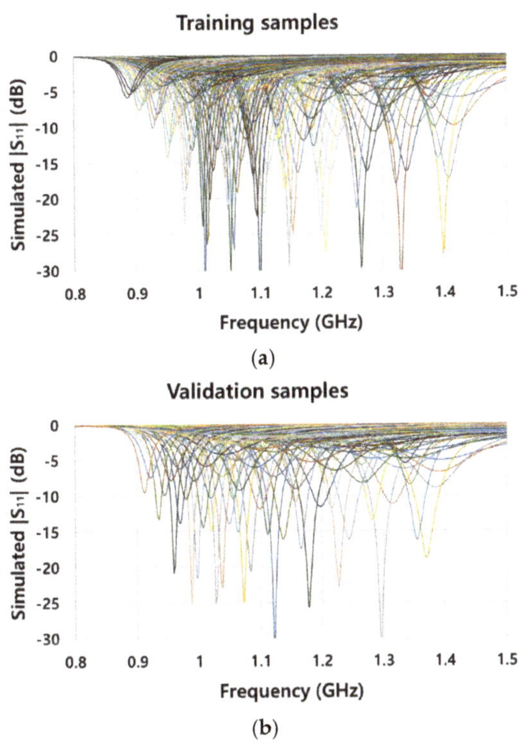

Figure 4. Magnitude of input impedance S_{11} for (**a**) all training and (**b**) all validation samples according to the combination of series capacitor C_S and parallel capacitor C_P.

4. DNN Modeling and Training

Deep learning allows us to obtain the correct output for both learned data and previously unseen data. We used high-level Keras API in TensorFlow 2.0 to construct a DNN using Python. Figure 5 shows the structure of the proposed DNN. The input for deep learning is S_{11}, whose magnitude is generally expressed in decibels. For implementation, the S_{11} magnitude was converted into a scalar value to normalize the input. In this study, the number of input samples was 401, with values ranging from 0 to 1 corresponding to frequencies from 0.8 to 1.5 GHz. The DNN output is given by the values of the series and parallel capacitors. As these values influence each other in the matching circuit, we considered two branches followed by addition (ADD layer) to reflect the influence, as shown in Figure 5. Each output value of the DNN for the corresponding capacitor value was obtained from one layer. As the DNN output should also be normalized, each capacitor value should be weighted. Input impedance S_{11} is highly sensitive to small values of the series capacitor. Therefore, we use the reciprocal of the series capacitor value as output. On the other hand, the value of the parallel capacitor was weighted by 0.1, as a larger value has a greater influence on the impedance. As a result, the weighted values of the two capacitors for training ranged from 0 to 1.5. The activation function of the output layer was linear, and the remaining layers used rectified linear unit (ReLU) activation to prevent the vanishing gradient problem. Each stage in the DNN implements a dense layer that fully connects the input and output neurons. As processing through the layers proceeded, the number of output neurons decreased. The number of neurons is expressed as a number in parentheses under the layer in Figure 5. The ADD layer functions to add two input values. RMSProp was used as the optimizer for learning DNN. The RMSProp does not simply accumulate gradients, but uses an exponentially weighted moving average to reflect the latest gradients larger. The loss function for DNN training was based on the mean squared error to perform optimization via root mean square propagation. The learning rate was set to 0.00005. For the DNN, 377 samples (Table 1) were used for training, and 110 samples (Table 2) were used for validation. Training proceeded for 2000 epochs with a batch size of 10.

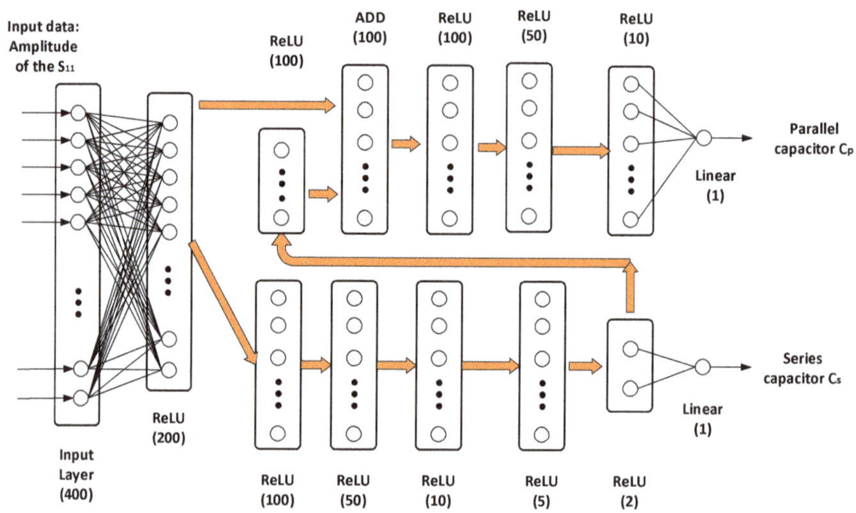

Figure 5. Architecture of the proposed DNN for antenna impedance matching.

The loss throughout training is shown in Figure 6. As training proceeds, the loss values converged at 0.0010 for training and 0.013 for validation. In this study, it took approximately 10.5 min to train the DNN in a computer equipped with an Intel(R) Xeon(R) processor at 2.30 GHz and 16 GB memory.

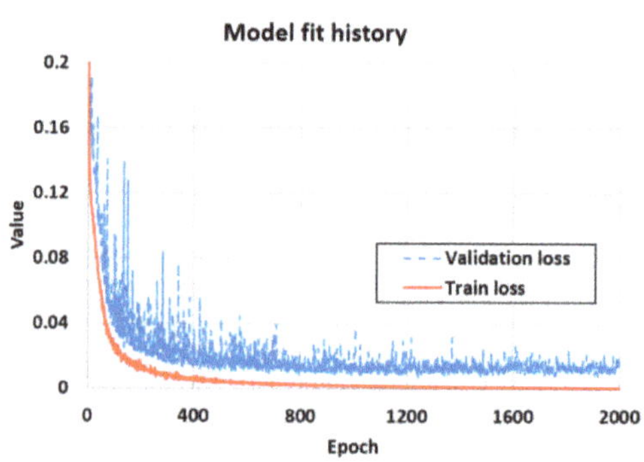

Figure 6. Training and validation losses according to epochs.

5. Simulation Results and Discussion

To validate the proposed DNN, two test sets with ground truths (i.e., calculated values) were considered. The selected capacitor values in the matching circuit are listed in Table 3. Samples with S_{11} magnitude up to 0.3 were selected, as antenna design requires S_{11} to be small. The S_{11} magnitudes from the test sets were used as input for the proposed DNN to obtain the corresponding capacitor values as outputs, as listed in Table 3. The capacitor values obtained from the DNN have some errors with respect to the calculated values. To analyze the effect of this error on antenna impedance matching, we conducted a simulation using the output capacitor values in HFSS. Figure 7 shows the comparison of S_{11} between the ground truths and DNN predictions, which are very similar.

Table 3. Capacitor values in matching circuit for validation.

Ground Truth	DNN Output	Sample
C_S = 1.2 pF C_P = 5.5 pF	C_S = 1.2191441 pF C_P = 5.201703 pF	14
C_S = 1.8 pF C_P = 7.5 pF	C_S = 1.8085649 pF C_P = 7.569376 pF	45

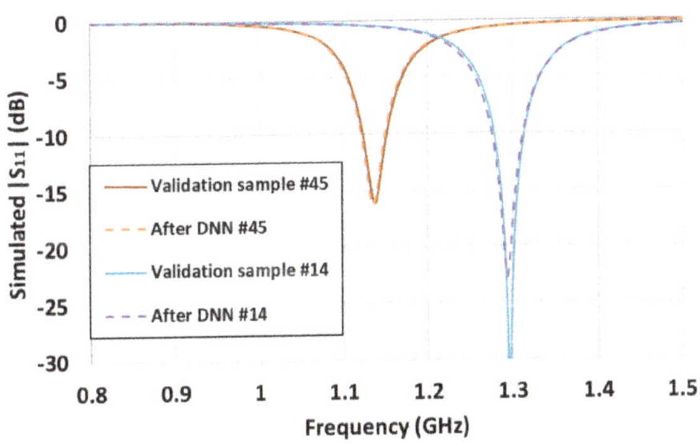

Figure 7. Comparison of results from ground truth and DNN output.

Using the trained DNN, the output was derived using ideal S_{11} patterns that cannot be implemented in practice as inputs. Table 4 lists the DNN outputs for three ideal patterns. The first ideal pattern is a square wave with S_{11} having magnitudes of 0.3 in 0.9–1.0 GHz and 0.97 in the other frequencies. The second ideal pattern is a triangular wave with the S_{11} magnitude decreasing linearly from 1.1 GHz until a minimum value of 0.3 at 1.15 GHz and then increasing linearly up to 1.2 GHz with a magnitude of 0.99. The third ideal pattern is also a square wave, but in a frequency range of 1.3–1.4 GHz. Figure 8 shows the comparison between the ideal S_{11} patterns and the simulated S_{11} patterns that use the output matching values obtained from the DNN. The DNN provides appropriate matching values for an ideal input. For the first, second, and third ideal patterns, the resonant frequencies were 0.95, 1.16, and 1.34 GHz, respectively. However, a completely consistent solution is infeasible because ideal patterns cannot be implemented in practice. Nevertheless, the DNN manages to determine the matching element values that approximate the desired S_{11} waveform.

Table 4. Capacitor values obtained from DNN for ideal inputs.

Ideal Waveform	DNN Output	Sample Number
Square (0.9–1.0 GHz)	C_S = 2.74568 pF C_P = 13.05057 pF	1
Triangular (1.1–1.2 GHz)	C_S = 1.5721719 pF C_P = 13.184719 pF	2
Square (1.3–1.4 GHz)	C_S = 1.0054374 pF C_P = 7.1495605 pF	3

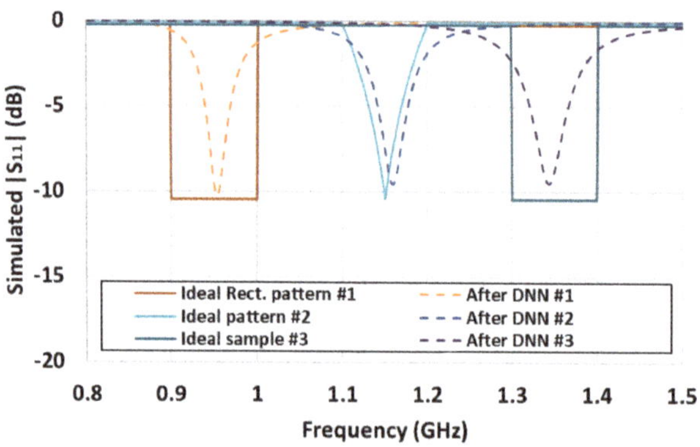

Figure 8. Comparison of ideal S_{11} waveform with that obtained from DNN results.

Machine learning has been applied for impedance matching using neural networks, as listed in Table 5. However, those applications consider frequencies in the order of megahertz, which is relatively lower than the gigahertz band required for antenna impedance matching. Moreover, those applications are limited to implementable impedance patterns. On the other hand, the proposed DNN can perform antenna impedance matching in the gigahertz frequency band. Unlike conventional methods, it uses only the magnitude instead of the complex impedance value to learn antenna matching values. Moreover, reasonable capacitor values for antenna impedance matching can be obtained even for S_{11} magnitudes that cannot be implemented in practice. In reference papers [20,21], matching values for ideal inputs were not presented.

Table 5. Comparison of machine learning methods for impedance matching.

Study	Method	Array Geometry	Neural Network Size	Application	Network Type
[20]	Back-propagation neural network	1D	5 (3 hidden layers)	Wireless power transfer	Gamma matching
[21]	Feedforward neural network	1D	12 (10 hidden layers)	Wireless power transfer	Three cascading L-type stages
This study	DNN	1D	12	Antenna	Gamma matching

This study applied deep learning to antenna matching through simulation. For experimental verification, it is necessary to implement the tunable matching network with a control circuit including a DNN, and the magnitude of the S_{11} should be measured at the rear end of the matching circuit using a device that can measure the reflection coefficient. Using the switching value of tunable matching circuit and magnitude of the measured impedance, the applicability of the proposed DNN can be verified experimentally. It is also necessary to research whether the matching value is properly found when there is noise in the impedance data. The practical performance of the deep learning method in selecting the value of the matching circuit element is an interesting future work.

6. Conclusions

We proposed a DNN to determine the capacitor values in the circuit for antenna impedance matching. The matching circuit consists of a series capacitor and a parallel capacitor and is intended for an inverted-F antenna, which is often used in small wireless devices. S_{11} data were acquired by simulating the antenna structure for various capacitor values. Then, the DNN was constructed using the S_{11} magnitude as input and the capacitor values of the matching circuit as outputs. After training on 377 training samples and 64 validation samples, the DNN achieved a loss of 0.001. The trained DNN was then applied to S_{11} magnitudes of ideal square and triangular waves. The simulated S_{11} obtained from DNN outputs shows the desired resonant frequency even for physically impossible patterns, suggesting that deep learning can be used for robust antenna impedance matching.

Author Contributions: Conceptualization, J.H.K.; methodology, J.H.K.; validation, J.H.K.; data curation, J.B.; supervision, J.B. All authors have read and agreed to the published version of the manuscript.

Funding: This work was supported by Institute of Information & communications Technology Planning & Evaluation (IITP) grant funded by the Korea government(MSIT) (No.2021-0-00763, Innovative Fusion Technologies of Intelligent Antenna Material/Structure/Network for THz 6G) and This work was supported by the Education and Research Promotion Program of KOREATECH in 2021.

Institutional Review Board Statement: Not applicable.

Informed Consent Statement: Not applicable.

Data Availability Statement: Not applicable.

Conflicts of Interest: The authors declare no conflict of interest.

References

1. Sun, C.; Zheng, H.; Zhang, L.; Liu, Y. A Compact Frequency-Reconfigurable Patch Antenna for Beidou (COMPASS) Navigation System. *IEEE Antennas Wirel. Propag. Lett.* **2014**, *13*, 967–970.
2. Nie, Z.; Zhai, H.; Liu, L.; Li, J.; Hu, D.; Shi, J. A Dual-Polarized Frequency-Reconfigurable Low-Profile Antenna With Harmonic Suppression for 5G Application. *IEEE Antennas Wirel. Propag. Lett.* **2019**, *18*, 1228–1232. [CrossRef]
3. Chi, P.-L.; Waterhouse, R.; Itoh, T. Compact and Tunable Slot-Loop Antenna. *IEEE Trans. Antennas Propag.* **2011**, *59*, 1394–1397. [CrossRef]
4. Kulkarni, A.N.; Sharma, S.K. Frequency Reconfigurable Microstrip Loop Antenna Covering LTE Bands with MIMO Implementation and Wideband Microstrip Slot Antenna all for Portable Wireless DTV Media Player. *IEEE Trans. Antennas Propag.* **2013**, *61*, 964–968. [CrossRef]
5. Li, T.; Zhai, H.; Wang, X.; Li, L.; Liang, C. Frequency-Reconfigurable Bow-Tie Antenna for Bluetooth, WiMAX, and WLAN Applications. *IEEE Antennas Wirel. Propag. Lett.* **2015**, *14*, 171–174. [CrossRef]
6. Chen, Y.; Manteuffel, D. A Tunable Decoupling and Matching Concept for Compact Mobile Terminal Antennas. *IEEE Trans. Antennas Propag.* **2017**, *65*, 1570–1578. [CrossRef]
7. Ko, J.B.; Kim, D. A Wideband Frequency-Tunable Dipole Antenna Based on Antiresonance Characteristics. *IEEE Antennas Wirel. Propag. Lett.* **2017**, *16*, 3067–3070. [CrossRef]
8. Firrao, E.L.; Annema, A.J.; van Vliet, F.E.; Nauta, B. Hardware Implementation Overhead of Switchable Matching Networks. *IEEE Trans. Circuits Syst. I* **2017**, *64*, 1152–1163. [CrossRef]
9. Lee, C.; Yang, C. Matching Network Using One Control Element for Widely Tunable Antennas. *Prog. Electromagn. Res. C* **2012**, *26*, 29–42. [CrossRef]
10. Rahola, J. Optimization of frequency tunable matching circuits. In Proceedings of the 2015 9th European Conference on Antennas and Propagation (EuCAP), Lisbon, Portugal, 13–17 April 2015; pp. 1–4.
11. Wang, H.; Wu, Z.; Wang, Y.; Sim, C.; Yang, G. Small-Size Folded Monopole Antenna with Switchable Matching Circuit for Ultra-Thin Mobile Applications. *Prog. Electromagn. Res. C* **2016**, *65*, 131–138. [CrossRef]
12. Melde, K.; Park, H.-J.; Yeh, H.-H.; Fankem, B.; Zhou, Z.; Eisenstadt, W.R. Software Defined Match Control Circuit Integrated With a Planar Inverted F Antenna. *IEEE Trans. Antennas Propag.* **2010**, *58*, 3884–3890. [CrossRef]

13. Kim, J.H.; Choi, S.W. A Deep Learning-Based Approach for Radiation Pattern Synthesis of an Array Antenna. *IEEE Access* **2020**, *8*, 226059–226063. [CrossRef]
14. El Misilmani, H.M.; Naous, T.; Al Khatib, S.K. A review on the design and optimization of antennas using machine learning algorithms and techniques. *Int. J. RF Microw. Comput. Eng.* **2020**, *30*, e22356. [CrossRef]
15. Bang, J.; Kim, J.H. Predicting Power Density of Array Antenna in mmWave Applications with Deep Learning. *IEEE Access* **2021**, *9*, 111030–111038. [CrossRef]
16. Zheng, B.; Zhang, H. Deep Learning Based Multi-layer Metallic Metasurface Design. In Proceedings of the IEEE International Symposium on Antennas and Propagation, Montreal, QC, Canada, 5–10 July 2020; pp. 2049–2050.
17. Misilmani, H.M.E.; Naous, T. Machine Learning in Antenna Design: An Overview on Machine Learning Concept and Algorithms. In Proceedings of the 2019 International Conference on High Performance Computing & Simulation (HPCS), Dublin, Ireland, 15–19 July 2019; pp. 600–607.
18. Erricolo, D.; Chen, P.-Y.; Rozhkova, A.; Torabi, E.; Bagci, H.; Shamim, A.; Zhang, X. Machine Learning in Electromagnetics: A Review and Some Perspectives for Future Research. In Proceedings of the 2019 International Conference on Electromagnetics in Advanced Applications (ICEAA), Granada, Spain, 9–13 September 2019; pp. 1377–1380.
19. Yao, H.M.; Li, M.; Jiang, L. Applying Deep Learning Approach to the Far-Field Subwavelength Imaging Based on Near-Field Resonant Metalens at Microwave Frequencies. *IEEE Access* **2019**, *7*, 63801–63808. [CrossRef]
20. Li, Y.; Dong, W.; Yang, Q.; Zhao, J.; Liu, L.; Feng, S. An Automatic Impedance Matching Method Based on the Feedforward-Backpropagation Neural Network for a WPT System. *IEEE Trans. Ind. Electron.* **2018**, *66*, 3963–3972. [CrossRef]
21. Jeong, S.; Lin, T.-H.; Tentzeris, M.M. A Real-Time Range-Adaptive Impedance Matching Utilizing a Machine Learning Strategy Based on Neural Networks for Wireless Power Transfer Systems. *IEEE Trans. Microw. Theory Tech.* **2019**, *67*, 5340–5347. [CrossRef]
22. Hemminger, T.L. Understanding Transmission Line Impedance Matching Using Neural Networks and PowerPoint. In Proceedings of the Frontiers in Education 35th Annual Conference, Indianopolis, IN, USA, 19–22 October 2005; p. T4E.

MDPI
St. Alban-Anlage 66
4052 Basel
Switzerland
Tel. +41 61 683 77 34
Fax +41 61 302 89 18
www.mdpi.com

Sensors Editorial Office
E-mail: sensors@mdpi.com
www.mdpi.com/journal/sensors

www.ingramcontent.com/pod-product-compliance
Lightning Source LLC
Chambersburg PA
CBHW061547220926
51850CB00005B/271
9783036577517